The Fukushima Daiichi Nuclear Accident

Atomic Energy Society of Japan
Editor

The Fukushima Daiichi Nuclear Accident

Final Report of the AESJ Investigation Committee

Springer

Editor
Atomic Energy Society of Japan
Tokyo, Japan

Original Japanese edition
The Fukushima Daiichi Nuclear Accident - Final Report of the AESJ Investigation Committee By Atomic Energy Society of Japan, Copyright © 2014 Published by Maruzen Publishing Co., Ltd.
2-17 Kanda Jimbo-cho, Chiyoda-ku, Tokyo, Japan

ISBN 978-4-431-55159-1 ISBN 978-4-431-55160-7 (eBook)
DOI 10.1007/978-4-431-55160-7
Springer Tokyo Heidelberg New York Dordrecht London

Library of Congress Control Number: 2014948905

Printed on acid-free paper

Springer is part of Springer Science+Business Media (www.springer.com)

Preface

On March 11, 2011, the accident at the Fukushima Daiichi Nuclear Power Stations of Tokyo Electric Power Company occurred by the Great East Japan Earthquake induced the worst case scenario of releasing a massive amount of radioactive materials, causing the devastating effects of nuclear disasters to make known to the world. Even now, after three and a half years have passed since the accident, many people are forced to live outside the evacuation area and this tells how deep the ravages of the accident have left behind. And still, the decommissioning of nuclear power reactors has many tough challenges, including such as the treatment of contaminated water and unloading of molten fuels.

The Atomic Energy Society of Japan (AESJ) feels the strong responsibility for the nuclear disaster of this kind as a nuclear science/engineering specialist group, and since the accident, we have been actively involved in a convergence of the accident and a restoration of the surrounding environment.

As a part of this effort, the "AESJ Investigation Committee on the Accident at the Fukushima Daiichi Nuclear Power Station of Tokyo Electric Power Company" (AESJ Investigation Committee) was organized at the AESJ General Assembly held on June 22, 2012. When the criticality accident at the Tokai Plant of JCO Co., Ltd occurred in 1999, the AESJ organized the "AESJ Investigation Committee on the JCO Accident" and conducted a thorough investigation. Following the practice of this precedent, the AESJ has gathered representatives from all technical divisions, relevant Liaison Meetings/Standing Committees, and the Board of Directors of the AESJ, and has actively carried out the investigation.

Since the first Investigation Committee held on August 21, 2012, the Committee had convened 17 meetings by the end of 2013, based on the investigation outcomes provided by the Subcommittees, and these efforts have now compiled into this final report. The Committee members hope that this report will be used by the experts as well as by the public broadly and contribute to nuclear safety enhancement. It would be our great pleasure if this report in English could provide the insight of the accident to the international society.

Tokyo, Japan
August 2014

Satoru Tanaka

Contents

Chapter 1
Introduction

Abstract This report was completed by the "AESJ Investigation Committee on the Accident at the Fukushima Daiichi Nuclear Power Station of Tokyo Electric Power Company" (AESJ Investigation Committee). In this introduction the overview and the overall structure of the report are explained.

Keywords Activities of Investigation Committee • Committee configuration • Establishment of Investigation Committee • History of activities • Perspective on investigation • Structure of report

1.1 Background to the Establishment of the AESJ Investigation Committee

(1) **Responses of the AESJ immediately after the accident**
The Atomic Energy Society of Japan (hereinafter called the "AESJ") acknowledged the significance of the accident of Fukushima Daiichi Nuclear Power Station of Tokyo Electric Power Company immediately after the accident, established the Team 110 comprising specialists, organized a system to respond to inquiries, etc. as well as the "Nuclear Safety" Investigation Committee (chair: Takashi Sawada) and started investigating and examining the development process and causes of the accident. Subsequently, based on the actual severity of the nuclear disaster, such as the radioactive environmental pollution caused by the accident, the AESJ established the Technology Analysis Subcommittee, the Clean-up Subcommittee, and the Radiation Effects Subcommittee, and studied countermeasures for various challenges, etc.

(2) **Establishment of the AESJ investigation committee**
Public and private sector organizations have investigated the causes of the accident and summarized reports. In particular, as national formal bodies, independent investigative committees comprising experts were established in the government and diet respectively to pursue the investigation of the accident. The reports of two investigative committees were valuable, with abundant human and financial resources allocated to both committees, and an enormous amount of collated data analyzed. Still, the involvement of nuclear power specialists remained limited.

The Fukushima Daiichi Nuclear Accident: Final Report of the AESJ Investigation Committee, DOI 10.1007/978-4-431-55160-7_1

Under the circumstances, the AESJ decided to establish its own accident investigative committee during the AESJ General Assembly and the Meeting of the board of directors on June 22, 2012, based on the following considerations:

- With deep remorse that the academic group of nuclear specialists could not prevent the accident, the AESJ has responsibilities to investigate, analyze and find measures to ensure such nuclear disaster would never recur from specialists' perspectives, apart from the above-mentioned various accident investigations, facing seriously all the problems in the nuclear community.
- In particular, AESJ witnessed the severity of the environmental pollution caused by the radioactive release after the accident and the collapse of local communities and their local infrastructures of inhabitants and wanted not only to analyze the accident but also study activities including current environmental restoration efforts to improve the situation as quickly as possible.

(3) **Examples of previous AESJ accident investigations**

When the JCO Criticality Accident occurred on September 30, 1999, AESJ established an accident investigative committee. This experience was useful reference in the case of Fukushima Daiichi accident.

As for the JCO accident, the "Nuclear Safety" Investigation Committee (chair: Hiroshi Sekimoto) investigated for about 1 year after the accident, summarized and released the report from the committee: the "Investigation Report on the Criticality Accident at the JCO Uranium Processing Factory" in September 2000 based on expert seminars and questionnaires to the AESJ members.

To adopt a more comprehensive investigation system as well as inherit the activities, the "AESJ Investigation Committee on the JCO Accident" (Chairman: Hideki Nariai; Secretary: Yutaka Kukita; Secretariat: Satoshi Mori), was formed from members selected from all of Technical Divisions in specialized fields in the AESJ. It was established in December 2000 and exerted all the AESJ's efforts to promote investigation activities. A report entitled the "JCO Criticality Accident in Tokai-mura: Facts, Causes and Responses" (published by Tokai University Press) was compiled and published after devoted deliberations through more than 30 Committee meetings for more than 4 years.

1.2 Activities of the AESJ Investigation Committee

(1) **Purpose**

The AESJ Investigation Committee clearly set out the following purposes for the establishment, decides to compile, review, and discuss the results of investigations and examinations by subcommittees and others to create the final report.

- As an academic institution comprising specialists in nuclear science/engineering, the AESJ established the AESJ Investigation Committee, which aims to assume responsibility for analyzing the accident at Fukushima

Daiichi Nuclear Power Station of Tokyo Electric Power Company and the actual conditions of the nuclear disaster associated with the accident from scientific and specialist perspectives and revealing its background and root causes as well as proposing a philosophy and measures for safety achievements, which is a strategy and basis for ensuring nuclear safety and achieving continuous safety improvements.

- At the same time, another important aim for the AESJ is to propose necessary reforms by confronting its own organizational and social issues and clarifying factors that could not prevent the nuclear disaster.
- The AESJ has to strive to reflect suggestions from the AESJ Investigation Committee on various activities; including organizational and operational reforms in the nuclear community and nuclear safety research.

(2) **The committee configuration**

The AESJ has set out AESJ Investigation Committee as a special committee directly connected to the Board of Directors to exert utmost effort to pursue investigations, based on the experience of the JCO accident, and determined that the committee should comprise members representing all of the Technical Divisions, and relevant Liaison Meetings, Standing Committees, etc. (Refer to Appendix 1 for the committee member list.)

(3) **Investigative method**

The Committee has proceeded with basic investigations within Technical Divisions, Committees, and Liaison Meetings established according to specialized fields in the AESJ to cover a wide scope of challenges associated with the accident at Fukushima Daiichi Nuclear Power Station and analyze issues thoroughly from a specialist's perspective based on the above-mentioned committee configuration. The AESJ Investigation Committee has reviewed each investigation result and executed overall adjustments, further analysis and examination if necessary to compile a report. It also actively exchanges opinions with other AESJ members at its conventions and annual meetings to reflect results on deliberations in the committee. In addition, it tried to take in the views and knowledge, etc. of global specialists such as non-Japanese nuclear academia.

Moreover, as for the data forming the basis of the investigation, the Committee has made the maximum use of the information released by the government and TEPCO as well as the information revealed by various accident investigative committees.

(4) **Perspective on investigation and examination**

The purpose of safety assurance in nuclear facilities was to "protect people and the environment." To achieve this, the Committee investigated and examined causes leading to the release of radioactive materials from nuclear facilities as well as problems of emergency measures with a view to protecting residents against radiation exposure from radioactive materials. In addition, to determine root causes, it tried to conduct more thorough investigations.

As nuclear power technology deals with complex giant artifacts, a bird's eye view response is required to form a whole picture. This view includes responses to external/internal events, multilayer protection measures, namely, DiD (defence in depth) and many perspectives from humans, software, technology, etc. In analyzing the accident and studying the safety assurance measures, the Committee acknowledged the characteristics of nuclear science/engineering as a general technology encompassing technologies in many different fields and tried to add a bird's-eye view as well as interdisciplinary coordination.

In addition to these technical and specialist perspectives, another important aim for the AESJ is to propose the necessary reforms while confronting its own organizational and social issues and clarifying factors that made the experts fail to prevent the nuclear disaster. Accordingly, the Committee analyzed the results of a questionnaire presented to AESJ senior members who once served as board members, chairpersons of divisions/committees, etc., and heard opinions broadly from members to consider the organizational reform of the society.

Safety assurance in nuclear facilities is a common goal shared by operators, regulatory authorities, manufacturers, academics, etc. and should be achieved through their efforts and cooperation. The AESJ is an academy featuring the participation of specialists belonging to various stakeholders in their private capacity. It should be added that this report was compiled under cooperation from a wide range of individuals concerned.

(5) **History of activities**

Since the AESJ Investigation Committee held the 1st meeting on August 21, 2012, 17 further meetings had been convened and engaged in discussion by the end of 2013 based on the results of deliberations of Technical Divisions, etc. Additionally, during this time, the core group of the Committee held up to 40 preparatory meetings (see Appendix 2).

The investigative progress was reported at AESJ conventions and Annual/Fall Meetings for the Committee to reflect opinions from members and others. The interim report was published during a public session at the Annual Meeting of the AESJ in March 2013 (at the Kinki University), followed by a draft final report in its Fall Meeting on September 4 (at the Hachinohe Institute of Technology) to exchange opinions. For a draft of the final report, a symposium was held in Tokyo on September 2, 2013, at which not only AESJ members but also the general public could exchange opinions. We also tried to exchange opinions with non-Japanese. For example, the draft of the final report was translated into English and sent to non-Japanese atomic energy academies.

1.3 Structure of the Report

The first half of the report, Chaps. 2–5, organizes factual records related to the accident.

Chapter 2 summarizes the facilities and equipment deployed at the Fukushima Daiichi Nuclear Power Station such as safety roles including differences by unit. In addition, measures for severe accidents and seismic and tsunami-resistant designs are explained to facilitate understanding of the accident progress shown in and after Chap. 3. Moreover, an overview of other power stations' facilities damaged by the tsunami is also described.

Chapter 3 explains an overview of the accident at the Fukushima Daiichi Nuclear Power Station, starting from the influence of the earthquake and tsunami to the progress of respective accidents of Nuclear Reactor Units 1–4 based on measurement data, etc. The status of Units 5 and 6 is also described.

Chapter 4 covers power stations other than the Fukushima Daiichi Nuclear Power Station: the Fukushima Daini Nuclear Power Station, the Onagawa Nuclear Power Station, and the Tokai Daini Power Station, and explains the influence of the earthquake and tsunami and an overview of the phenomena there while focusing on factual records.

Chapter 5 describes the actual nuclear disaster mainly as accident correspondence outside the power station. Specifically, it covers the actual circumstances of overall control, including the emergency plan prepared before the accident and the emergency actions taken when accidents occur, individual challenges, including residents' evacuation and radiation/radiation exposure measurements, actions to estimate the amount of radiation of radioactive material, and communication after the accident.

Chapter 6 analyzes and evaluates problems why we could not prevent the accident based on the factual records organized in Chaps. 2–5. We tried to identify problems through two different approaches not to omit relevant issues. One approach was to pick up the issues following the chronic progress of accident events, and another was to systematically arrange disputed points from the nuclear safety scheme. We also analyzed the relation between the results of simulation analysis of the inside of the nuclear power reactor based on the progress of the accident and the release of radioactive materials, and then verified the scenario of the accident progressing. After that, the following items were analyzed and evaluated in detail:

(1) concept of nuclear safety, (2) defence in depth, (3) plant design, (4) accident management, (5) external events, (6) radiation monitoring and environment remediation activities, (7) simulation analysis, (8) emergency preparedness and response, (9) nuclear security, physical protection, and safeguards, (10) human resources and human factors, (11) relationship with international society, and (12) information dissemination. The chapter also shows remaining issues to be thoroughly investigated and examined in future concerning the progress of the accident.

Chapter 7 analyzes the aspects of organizations, namely the safety regulation system, the industry system, the R&D/safety research system, the global system, and also the role of the AESJ itself.

Based on analysis through Chap. 7, Chap. 8 shows the direct causes behind the accident and organizational underlying causes that brought about the direct causes and determined these as the root causes, based on which we make recommendations.

Chapter 9 analyzes and evaluates the current decommissioning status of reactors at the Fukushima Daiichi Nuclear Power Station, which is still underway, and various related challenges. Given the constantly changing conditions and the long-term nature of the project, a perspective, which is different from the one used through Chap. 8 to analyze direct causes of the accident, is used to compile Chap. 9.

The Abbreviation Table, which is used in this report, is listed at the end of the report (see Appendix 3).

Chapter 2
Overview of Nuclear Power Station

Abstract Chapter 2 shows the overview of the facilities in Fukushima Daiichi Nuclear Power Station for understanding the analysis and assessment on the accident of Fukushima Daiichi NPS in Chap. 6. The tsunami generated by the earthquake resulted in the occurrence of the beyond design basis event at the plant. Chapter 2 also shows the outline of countermeasures for earthquake, tsunami and beyond design basis events which were considered before the accident.

Keywords Accident management • Design • Facilities • Seismic design • Tsunami

2.1 Overview of Facilities in the Fukushima Daiichi Nuclear Power Station

2.1.1 Main Facilities Including Safety Equipment

2.1.1.1 Overview of the Power Station

The Fukushima Daiichi Nuclear Power Station (hereinafter the "Fukushima Daiichi") comprises six boiling water reactors (BWRs). Units 1–4 are located in Okuma town, Futaba district, Fukushima prefecture, while Units 5 and 6 are located in Futaba town, Futaba district. The site is half-oval in shape, facing the Pacific to the east, and covering an area of about 3.5 million m^2.

Fukushima Daiichi is the first nuclear power station for the Tokyo Electric Power Company (hereinafter "TEPCO"). Since Unit 1 started operation in March 1971, further nuclear reactors were added in succession, with Unit 6 starting operation in October in 1979. The total installed power generation capacity is 4.696 million kW.

2.1.1.2 Safety Equipment and Others at Power Stations

The Types of BWR used by TEPCO are BWR-3, -4, -5, and ABWR, in order from oldest to newest. Unit 1 is BWR-3, Units 2–5 are BWR-4, and Unit 6 is BWR-5. ABWR was developed after completing the construction of Fukushima Daiichi.

The Fukushima Daiichi Nuclear Accident: Final Report of the AESJ Investigation Committee, DOI 10.1007/978-4-431-55160-7_2

Table 2.1 Main design specifications of Units 1–6 at the Fukushima Daiichi

	Unit 1	Unit 2	Unit 3	Unit 4	Unit 5	Unit 6
Rated power (MWe)	460	784	784	784	784	1,100
Thermal power (MWt)	1,380	2,381	2,381	2,381	2,381	3,293
Commissioning	3/1971	7/1974	3/1976	10/1978	4/1978	10/1979
Nuclear power reactor type	BWR-3	BWR-4	BWR-4	BWR-4	BWR-4	BWR-5
Reactor pressure vessel Design pressure (kg/cm^2 (gage))[a]	87.9	87.9	87.9	87.9	87.9	87.9
Reactor pressure vessel Design temperature (°C)	302	302	302	302	302	302
Fuel assembly number	400	548	548	548	548	764
No. of control rods	97	137	137	137	137	185
Containment vessel	Mark I	Mark I	Mark I	Mark I	Mark I	Mark II
Containment Vessel Design pressure (kg/cm^2 (gage))[a]	4.35	3.92	3.92	3.92	3.92	2.85
Containment vessel	138 (D/W)	138 (D/W)	138 (D/W)	138 (D/W)	138 (D/W)	171 (D/W)
Design temperature (°C)	138 (S/C)	138 (S/C)	138 (S/C)	138 (S/C)	138 (S/C)	105 (S/C)
ECCS configuration	HPCI	HPCI	HPCI	HPCI	HPCI	HPCS
	CS	CS	CS	CS	CS	LPCS
	ADS	LPCI	LPCI	LPCI	LPCI	LPCI
		ADS	ADS	ADS	ADS	ADS
Reactor core isolation cooling system	IC	RCIC	RCIC	RCIC	RCIC	RCIC

[a]Shown in units shown in the establishment license application document

Containment vessels (CV) were developed according to the types of BWR-3 through ABWR. Usually, BWR-3 and -4 have Mark I with a donut-shaped suppression pool (S/P) and BWR-5 has Mark II with a non-donut shape. The Japanese electric power companies are allowed to use Mark I containment for BWR-5 at their option (see Table 2.1).

The following explains safety and other related functions used by each power station:

(1) **Unit 1**

The emergency core cooling system (ECCS), which urgently injects cooling water into a nuclear power reactor in the event of any accident involving loss of coolant, comprises two core spray systems (CS systems), one high pressure coolant injection system (HPCI system), and an automatic depressurization system (ADS). As the decay heat produced by the nuclear fuel in a loss-of-coolant accident is transferred to S/P, two containment cooling systems (CCS) are installed to eliminate heat from S/P water and cool a containment vessel.

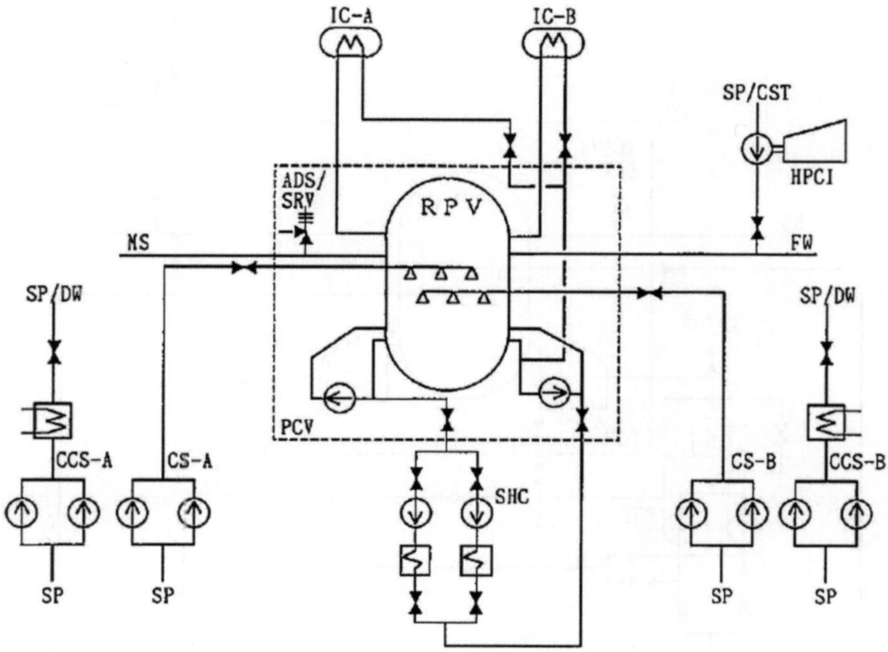

Fig. 2.1 Safety equipment and others in Unit 1

Two reactor core isolation condensers (ICs) are also installed to dissipate the heat from the nuclear power reactor when the reactor is isolated from the turbine system. Further, the reactor shutdown cooling system (SHC system) is installed to perform a cold shutdown when replacing fuel during a regular inspection (see Fig. 2.1).

(2) **Units 2–5**

The biggest change between BWR-3 and -4 systems is the installation of a residual heat removal system (RHR system). The RHR system encompasses the functions of the CCS and SHC systems of BWR-3 as well as the low pressure coolant injection system (LPCI system) as ECCS.

The ECCSs in Units 2–5 comprise two core spray systems (CS systems), one high pressure coolant injection system (HPCI system), one low pressure coolant injection system (LPCI system (LPCI mode in RHR system) (four pumps)) and an automatic depressurization system (ADS). In addition, the containment cooling mode of the RHR system is used to remove heat from containment vessels after the accident. The reactor core isolation cooling system (RCIC system) is installed as a change from IC of BWR-3 for cooling water injection if a reactor isolation event occurs, while the reactor shutdown cooling mode of the RHR system is used to dissipate heat from the nuclear power reactor during a regular inspection (see Fig. 2.2).

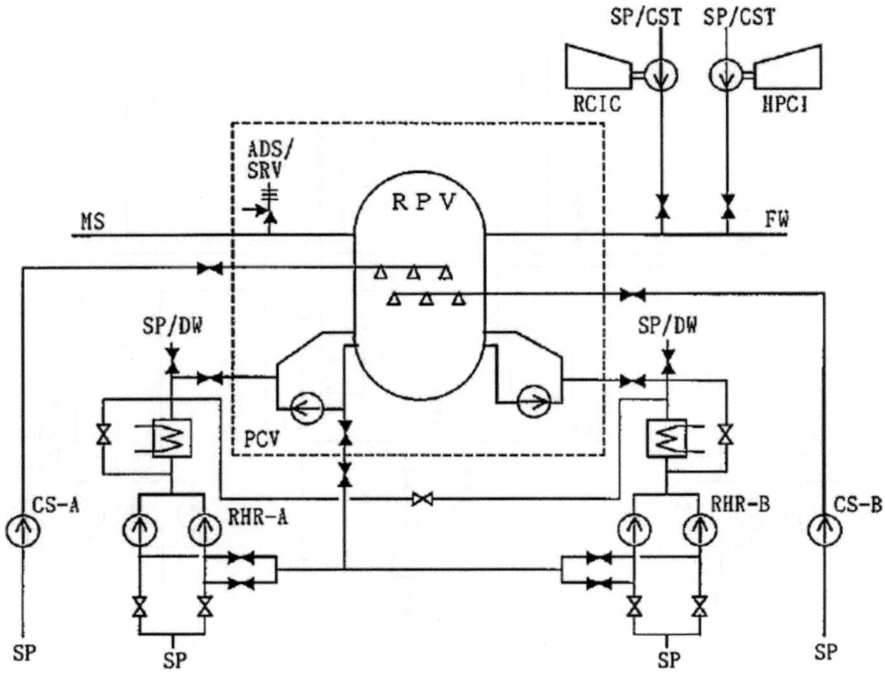

Fig. 2.2 Safety equipment and others in Units 2–5

(3) **Unit 6**

The biggest change from BWR-4 to -5 is the improved efficiency of jet pumps installed in nuclear power reactors. This can obtain larger core flows with a primary loop recirculation pump (PLR pump) with a smaller capacity.

After furthering its integration, the ECCS comprises one high pressure core spray system (HPCS system), one low pressure core spray system (LPCS system), three low pressure coolant injection systems (LPCI mode of RHR system), and an automatic depressurization system (ADS).

The reactor core isolation cooling system (RCIC system) if a reactor isolation event occurs and the reactor shutdown cooling mode of the RHR system during a regular inspection, have the same concepts as BWR-4.

2.1.1.3 Emergency Power Supply Facilities

The key facilities from safety perspectives, including ECCS, are basically designed to run on the emergency power supply, power for which is supplied from the emergency diesel generator (D/G). For the facility configuration at the power station, Units 1–5 require two D/Gs respectively while Unit 6 requires three D/Gs. In the initial design, Units 1/2, 3/4, and 5/6 each share one D/G and each D/G is of the seawater cooling type.

Subsequently, TEPCO stopped using shared D/Gs and provided each unit with its own D/G by March 1999 as one of the accident management measures. To achieve this, new three D/Gs (one each for Units 2, 4 and 6) were installed, all of which were air-cooling D/Gs.

2.1.1.4 The Cooling System in the Event of a Reactor Isolation

Described below are cooling systems in nuclear power reactors when reactors become isolated from the turbine system. The system is a usual one, not the emergency core cooling system (ECCS) but designed as an equivalent safety system. As a rule, the DC power supply should be used, and power supply and water injection capacities are specified so that the hot standby mode can be used for 8 h.

(1) **Isolation condenser (IC)**
The IC installed in Unit 1 is a system which eliminates the heat from steam produced in the reactor core, condenses it in the condenser, and returns it to the nuclear power reactor (see Fig. 2.3). Therefore, IC, which cannot supply cooling water, is designed to automatically start under reactor pressure lower than the set pressure of the main steam safety relief valve (SR valve) and avoid any decrease in reactor water level by operating the SR valve.

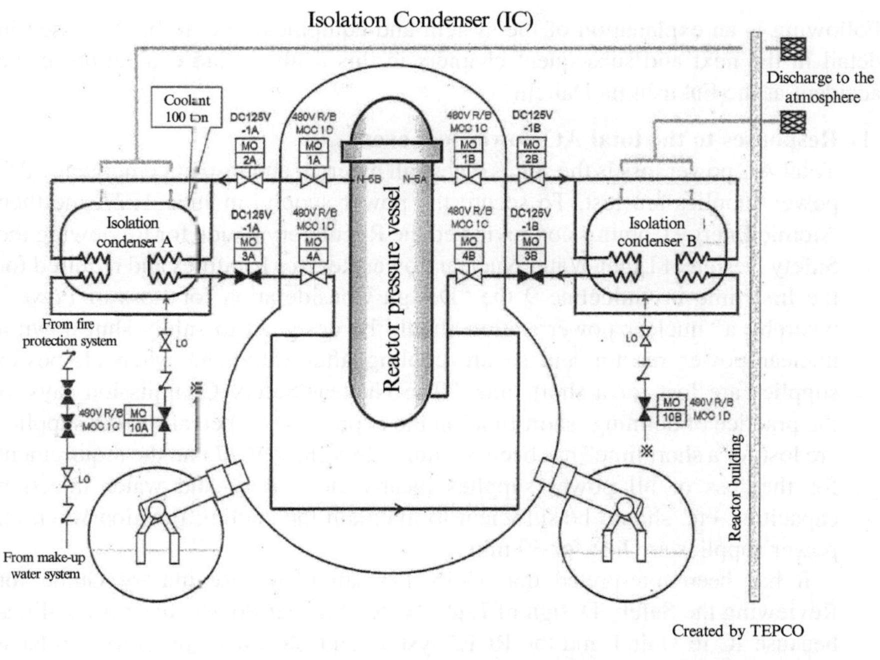

Created by TEPCO

Fig. 2.3 Isolation condenser (IC) at Unit 1. Interim Report of the Government Accident Investigation Committee (Attachments) p. 72 (26 December 2011)

IC is designed to remove the decay heat produced immediately after a reactor shutdown (3 % of the rated thermal power), and two full capacity equipment systems are installed. This enables operators to close the valves of one of the two systems after the IC becomes operational and switch to the operation of only one system because pressure in the nuclear reactor plummets if two systems automatically start up without any trouble.

Interlocks are provided in the IC steam piping and condensate return piping with elbow flow meters installed within the containment vessel to close the isolation valve according to the flow level. They detect any IC piping rupture and isolate the IC. When the supply of a control power source in the rupture detection circuit fails, an interlock generates a signal to close the isolation valve.

(2) **Reactor core isolation cooling system (RCIC system)**
The RCIC system is installed in Unit 2 and subsequent BWR plants to cool the nuclear power reactor in hot standby. It supplies mainly condensate storage tank water to the nuclear power reactor via a turbine-driven pump using nuclear power reactor steam. Cooling water can be injected under the working pressure of SR valves.

2.1.1.5 Equipment, etc. to be Discussed in the Next and Subsequent Chapters

Following is an explanation of the system and equipment, etc. to be discussed in detail in the next and subsequent chapters in this analysis and evaluation of the accident at the Fukushima Daiichi.

(1) **Responses to the total AC power loss event**
Total AC power loss is the status where all external and in-plant emergency AC power supplies are lost. To secure the power supply, in June 1977, the then Atomic Energy Commission reviewed the Regulatory Guide for Reviewing the Safety Design of Light Water Nuclear Power Reactor Facilities and required for the first time in Guideline 9 the "Design Consideration for Loss of Power" whereby a "nuclear power station should be designed to safely shut down a nuclear power reactor and secure cooling after shutdown when all power supplies are lost for a short time." The Nuclear Safety Commission says as the practice of defining 'short time' in the expression "when all power supplies are lost for a short time" has been 30 min or less since 1977 and the requirement for the loss of all power supplies means the battery and water injection capacities, etc. should be sufficient to maintain the cooling function when all power supplies are lost for 30 min.

It has been interpreted that Units 1-6 satisfy the Regulatory Guide for Reviewing the Safety Design of Light Water Nuclear Power Reactor Facilities because IC in Unit 1 and the RCIC system and SR valves in Units 2–6 have cooling ability for at least 30 min without AC power supplies.

(2) **Containment isolation valve**
The reactor containment is a boundary which prevents the release of radioactive materials outside the system. The piping system arranged by penetrating the reactor containment basically includes internal and external isolation valves. As a rule, isolation valves in the main systems are basically electric or check valves, which are automatically closed by receiving a signal triggered by a decline in the reactor water level, increase in drywell pressure, etc., closed by remote manual operation, closed as a check valve, or normally closed. In addition, some isolation valves use air-controlled valves, which close when air, etc. fails. Isolation valves in ECCS, which is used in the event of an accident, and isolation valves in systems, which can supply water to a nuclear power reactor, are designed not to close by an isolation signal.

(3) **The reactor water level indicator**
In principle, the reactor water level indicator measures based on differences between the water head pressure on the reference chamber side and that on the reactor side. In other words, the reactor water level is measured by detecting the differential pressure between the water head pressure at a defined water level formed in the steam condensation tank led to the steam portion of the reactor pressure vessel and that which varies according to the reactor water level (see Fig. 2.4). Therefore, if the density of the water changes due to a pressure change in the nuclear power reactor, the measured water level must be corrected.

(4) **Placement**
Each equipment/system with safety features is placed in an independent and distributed manner whenever possible. The emergency diesel generator (D/G) is heavy itself and usually installed underground or on the first floor.

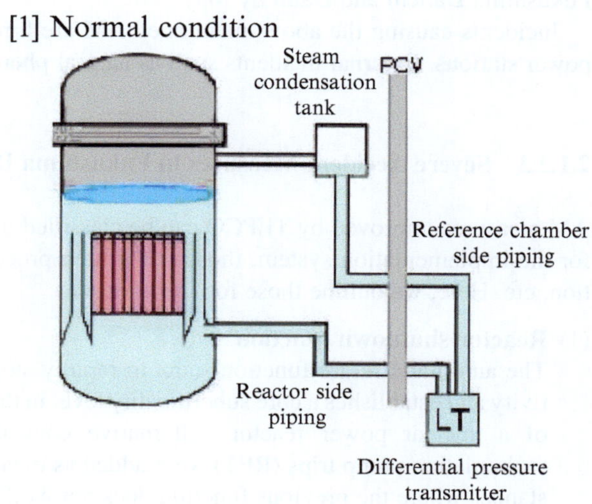

Fig. 2.4 Principle of reactor water level indicator

2.1.2 *Facilities for Accident Management Measures*

2.1.2.1 Severe Accident

Safety equipment is designed based on the safety assessment given in consideration of major large incidents called "design basis accident" that have a large impact on the nuclear power stations to cope with transients and accidents, which may occur in nuclear power stations.

If an incident considerably exceeds the design basis accidents, measures postulated in the design cannot properly control the cooling and reactivity of the reactor core, whereupon the reactor core may be significantly damaged. Such an incident is called a severe accident (SA).

In addition, measures to prevent a beyond-design-basis accident from developing into a SA using a safety margin included in the design, equipment which happens to be usable in the occurrence of the incident and equipment newly installed in case of SA, or measures for buffering against influences when the incident develops into SA are called accident management (AM).

2.1.2.2 Introduction of Severe Accident Measures to Japan

The Nuclear Safety Commission (then) proceeded to consider severe accident measures following the Three Mile Island (U.S.) accident and the Chernobyl accident in the former Soviet Union while the Ministry of International Trade and Industry (then) also issued the "Improvement of Accident Management in Nuclear Power Stations" in July 1992, whereupon AM was improved through the voluntary efforts of licensees. Licensees were expected to improve AM measures by 2000. TEPCO completed the improvement of various types of AM measures in the Fukushima Daiichi and Daini by May 2002.

Incidents causing the above-mentioned AM are limited to internal incidents in power stations. External incidents such as natural phenomena are not considered.

2.1.2.3 Severe Accident Measures in Fukushima Daiichi

AM measures improved by TEPCO can be classified into those for facilities, those for the implementation system, those for written procedures and those for education, etc. Here, we outline those for facilities.

(1) **Reactor shutdown function**
The automatic scram function starts to rapidly stop nuclear reactor core reactivity and establishes a core subcriticality level in the occurrence of abnormality of a nuclear power reactor. Alternative control rod insertion (ARI) and recirculation pump trips (RPT) were added as measures to cope under circumstances where the previous function does not work.

(2) **A function to inject water to the nuclear power reactor and the containment vessel**
Alternative water injection facilities and automatic depressurization of the nuclear reactor were added as AM measures to cope in cases where the water injection to the nuclear power reactor fails.
(3) **Function to remove heat from a containment vessel**
The containment vessel vent was added as a function to remove heat from a containment vessel. This involves setting a vent line with higher pressure resistance to prevent over-pressurization of the containment vessel.
(4) **Power supply function**
Procedure manuals for interchanging high/low voltage AC power supply from an adjacent plant and interchanging DC power supply are prepared as AM measures in the incident of loss of external power supply from a power station.

2.1.3 Seismic Resistant Design and Tsunami Resistant Design for Facilities

Following are details of how seismic and tsunami-resistant designs for the Fukushima Daiichi were initially performed:

2.1.3.1 The Regulatory Guide for Reviewing the Safety Design of Light Water Nuclear Power Reactor Facilities

Below is a summary of statements in the Regulatory Guide for Reviewing the Safety Design of Light Water Nuclear Power Reactor Facilities as regulatory requirements for earthquake and tsunamis.

- The Regulatory Guide for Reviewing the Safety Design of Light Water Nuclear Power Reactor Facilities, which was established in 1970, required consideration of the most stringent forces of nature as the consideration for natural conditions.
- In addition, nuclear power reactor facilities are required in the seismic design that it be properly classified into design groups according to importance and properly designed according to their importance, i seismic n consideration of safety impact in circumstances where the system and equipment may lose their function and get damaged.

2.1.3.2 Seismic Resistant Design

When the reactor establishment license for Fukushima Daiichi was applied for during the period 1966–1971, there was no seismic resistant design code for safety regulations, and seismic ground motion was established by TEPCO to confirm the safety function.

Subsequently, as part of safety reviews and assessments concerning applications for reactor establishment licenses, from the perspective of ensuring seismic safety to show the basis for judgment on the seismic resistant design policy, the Nuclear Safety Commission reviewed and revised the "Regulatory Guide for Reviewing Seismic Design of Nuclear Power Reactor Facilities (old guidelines)" in July 1981 based on decisions made by the then Atomic Energy Commission in September 1978. Checks were made as to whether the Fukushima Daiichi conformed to this guide.

Further, the Regulatory Guide was fully revised to reflect the accumulation of new knowledge on seismology and earthquake engineering in addition to remarkable improvements in seismic resistant design technology in September 2006.

The above-mentioned three stages of development increased the maximum acceleration of seismic ground motion, the basis for seismic resistant design, from 265 Gal at the time of constructing the Fukushima Daiichi to 370 and 600 Gal respectively. This accident occurred at the Fukushima Daiichi under circumstances where conformity to this guide is being checked.

2.1.3.3 Tsunami Resistant Design

Among the guides developed by the Nuclear Safety Commission, it is the Regulatory Guide for reviewing seismic design that provides most explicit countermeasures for tsunamis, which should be considered by nuclear power stations and the September 2006 revision says that as an incident associated with an earthquake, "facilities should be designed, fully considering the risk that even tsunamis, which could possibly occur while the facilities are in service, however unlikely, would impose a significant impact on their safety functions."

At the time of basic designs of units at the Fukushima Daiichi, the height of Onahama Peil (O.P.) +3.122 m, the highest sea level observed at Onahama Port at the time of the Chile earthquake in 1960, was considered the maximum tsunami wave height, and an establishment license was obtained. The most seaward side was prepared at a height of O.P. + 4 m and an emergency seawater pump was installed there. Subsequently, with the development of tsunami prediction technology, the predicted tsunami heights were raised: O.P. + 5.7 m and +6.1 m in 2002 and 2009. Measures were taken: for example, sea water pump motors were installed in higher location. There was another chance to review countermeasures for tsunami in 2008, but before the review, this accident occurred.

2.2 Overview of Facilities in the Power Stations Other Than the Fukushima Daiichi Nuclear Power Station

2.2.1 The Fukushima Daini Nuclear Power Station

The Fukushima Daini Nuclear Power Station of TEPCO (hereinafter the "Fukushima Daini") is located in Naraha and Tomioka towns, Futaba district, Fukushima prefecture and has four BWR plants with rated electric output of 1.1 million kW. These plants started to operate during the period 1982–1987. The reactor type of Unit 1 is BWR-5 with Mark II containment while the reactor type of Units 2–4 is BWR-5 with Mark II advanced containment. The safety system configuration is, as BWR-5, basically the same as Unit 6 at the Fukushima Daiichi (see Sect. 2.1.1).

2.2.2 The Onagawa Nuclear Power Station

The Onagawa Nuclear Power Station of the Tohoku Electric Power Company, Inc. (hereinafter the "Onagawa") is located in Onagawa town, Oshika district and Ishinomaki City, Miyagi prefecture and has three BWR plants. Unit 1 is BWR-4 with Mark I containment with the rated electric output of 524,000 kW, Units 2 and 3 are BWR-5 with Mark-I advanced containment with the rated electric output of 825,000 kW. These plants started to operate during the period 1984–2002. The safety system configuration for Unit 1 is BWR-4, basically the same as Units 2–5 at the Fukushima Daiichi. Units 2–3 have BWR-5, basically the same as Unit 6 at the Fukushima Daiichi (see Sect. 2.1.1).

2.2.3 The Tokai Daini Nuclear Power Station

The Tokai Daini Power Station (hereinafter the "Tokai Daini") of the Japan Atomic Power Company (hereinafter the "Japan Atomic Power") is located in Tokai village, Ibaraki prefecture. The plant has BWR-5 with Mark II containment and rated electric output of 1.1 million kW and started operation in 1978. The safety system configuration is BWR-5; basically the same as Unit 6 at Fukushima Daiichi (see Sect. 2.1.1).

Chapter 3
Overview of the Accident at the Fukushima Daiichi Nuclear Power Station

Abstract In the accidents of Fukushima Daiichi NPS, the core meltdown has resulted from severe accident at Unit 1–3, and the explosion of reactor building has occurred at Unit 4 which was during the periodic inspection, resulting in the substantial release of radioactive materials. Unit 5 and 6, under the periodic inspection, have cooled down. Chapter 3 describes the accident of Unit 1–4 in detail mainly.

Keywords Accident • Core damage • Explosion • Fact • Fukushima Daiichi

3.1 Damage Caused by the Earthquake and Tsunami

At 14:46 on March 11, 2011, the Tohoku District—off the Pacific Ocean Earthquake, the fourth largest in recorded world history (M9.0) occurred, with strong Level 6 seismic intensity (on the Japan Metrological Agency (JMA) scale) observed in Okuma and Futaba towns, where the Fukushima Daiichi Nuclear Power Station is located. This earthquake occurred in multiple areas: "off the coast of Miyagi prefecture", "close to the Japan trench in southern off the coast of Sanriku", "off the coast of Fukushima prefecture", "off the coast of Ibaraki prefecture", and other areas; all simultaneously. Seismometers installed on the lowest basement floors of reactor buildings of units at the Fukushima Daiichi reveal that the tremors were as large as those caused by the standard earthquake ground motion Ss or slightly larger.

Consequently, various points went out of order, including collapse of transmission steel towers, damage to circuit breakers, and loss of external power supplies. All nuclear power reactors automatically shut down. (A scram occurred.) All 12 emergency diesel generators, except that at Unit 4 under inspection, automatically started up to secure the power supply. At this stage, plant parameter values were recorded. No suspicion of any damage to the pressure boundary of reactor coolant was aroused by the reactor water level or pressure and temperature of the containment vessel. In addition, the results of walkdown inspections within the extent visually confirmable showed damage to a tiny portion of equipment with low earthquake resistance, which did not influence the safety of nuclear power reactors. It is necessary to evaluate the influence of the earthquake on the soundness of major safety equipment via detailed inspections and examinations in future.

The Fukushima Daiichi Nuclear Accident: Final Report of the AESJ Investigation Committee, DOI 10.1007/978-4-431-55160-7_3

At 15:30, just under an hour since the earthquake, a big tsunami came. The result of a TEPCO reproducing calculation revealed that the height of the tsunami in the vicinity of the tide-gage station (design height taking tsunamis into consideration) was about 13 m (TEPCO explained "a tsunami with a height of more than 15 m" in the past. However this is the inundation height.).

The Japanese Meteorological Agency (JMA) issued tsunami information immediately after the earthquake. Initially, the agency issued a major tsunami warning of 3 m but after two updates, it was 15:30 when the expected height changed to 10 m or more.

The Fukushima Daiichi station suffered widespread flooding due to the tsunami, while many seawater cooling devices were also damaged. The buildings were inundated and most of the functions of power panels at units were lost. Likewise, all AC power was cut to Units 1–5, while Units 1, 2, and 4 also lost DC power supplies.

Fukushima Daiichi devised countermeasures for the station blackout (SBO): reactor cooling through the isolation condenser (IC), which requires DC power, the reactor core isolation cooling system (RCIC system), and the high pressure coolant injection system (HPCI system) and the interchange of high-voltage power from adjacent units. Further, the interchange of low-voltage power from adjacent units was arranged in case DC power supplies were lost. However, power supplies were still lost, including in adjacent plants, and the initial requirements for the accident response were extremely severe. Moreover, great aftershocks continued and major tsunami warnings remained active. Rubble generated by the tsunami was scattered on site, which hampered the accident responses.

3.2 Unit 1

The earthquake automatically stopped the nuclear power reactor and external power supplies were lost. Both the two emergency diesel generators (D/Gs) automatically started while the main steam isolation valves (MSIV) closed. Consequently, the reactor pressure rose, the isolation condensers (IC) automatically started up, and the reactor water level was maintained.

As the IC operation lowered the reactor pressure and temperature, as assumed in the design, the operator, who feared devices might be affected, intermittently operated the IC to keep within the limits of the operational safety program and avoided a drastic drop in temperature.

Though external power supplies were lost, the necessary safety functions were secured as designed and the nuclear power reactor was kept safe until the tsunami hit the station after the earthquake.

The tsunami attacked the Fukushima Daiichi, and the seawater cooling system lost its function as well as inundation of the buildings at a height of 10 m above sea level. Almost all power panels at Unit 1 were disabled. Consequently, only emergency lighting remained on in the main control room, and most of the lighting system was unavailable, including instrumentation to detect the state of the equipment and warning lamps.

TEPCO's subsequent investigation revealed that the loss of DC power due to the tsunami triggered the IC isolation interlock and led to the closure of isolation valves. In other words, the IC cooling function was lost due to the "fail-safe design" for the confinement function.

The states of IC valves were not revealed immediately after the tsunami, but around 18:18 on the 11th, operators confirmed that two closed lamps (green light) of the external isolation valves installed at one of ICs in Unit 1 were lit and conducted an operation to open them. In operating the IC under normal condition, one of the external isolation valves should always remain open and the other should be used for the open-close operation. It was recognized in the main control room that the IC isolation interlock might have been activated, since both external isolation valves were closed. In that case, it was likely that the two internal isolation valves were also closed and it was assumed that the IC would not function with the external isolation valves open. Later, the operators closed one of the IC external isolation valves at 18:25 amid concern over the decreasing water level on the body side the condenser but re-opened it again at about 21:30. Such series of operations, however, were not communicated to the Nuclear Emergency Response Headquarters, and those at the Fukushima Daiichi and TEPCO head office believed that the IC was operating normally.

Conversely, at around 16:42, measurements via the water level indicator in the reactor pressure vessel were temporarily displayed and the reduction in water level was measured. Based on this, at around 17:15, it was expected that the water level would reach the top of active fuel (TAF) in about one hour. The Site Superintendent thus instructed staff to consider alternative water injections by diesel-driven fire pump (D/DFP) and fire engines. However, according to the Interim Report of the Gov't Accident Investigation Committee, the roles and responsibilities for conducting water injection using fire engines were unclear and until predawn on March 12, no specific preparation had been made.

As the IC lost its function due to the tsunami, the reactor pressure rose, triggering the safety valve function of the safety relief valves (SRV) and channeling steam to the suppression chamber (S/C) for condensation, whereupon the reactor water level declined. Due to the loss of the DC power source at Unit 1, plant parameters were not obtained immediately after the tsunami. It was not until 21:19 on March 11 that the water level could be measured by connecting batteries. Figures 3.1, 3.2, and 3.3 show the water level and pressure estimated by the severe accident analysis code (MAAP) referring to information on plant statuses and operator operations obtained.

It was estimated, based on analytical results, that the reactor water level fell below TAF by past 18:00 on March 11 and that the core damage started before 19:00. Conversely, the reactor pressure was confirmed at 7.0 MPa [abs (absolute pressure)] by the on-site indicator around 20:00. It was also estimated that at this stage, the pressure boundary of reactor coolant was sound and that the reactor pressure was maintained near 7.0 MPa [abs] by the SRV safety valve function.

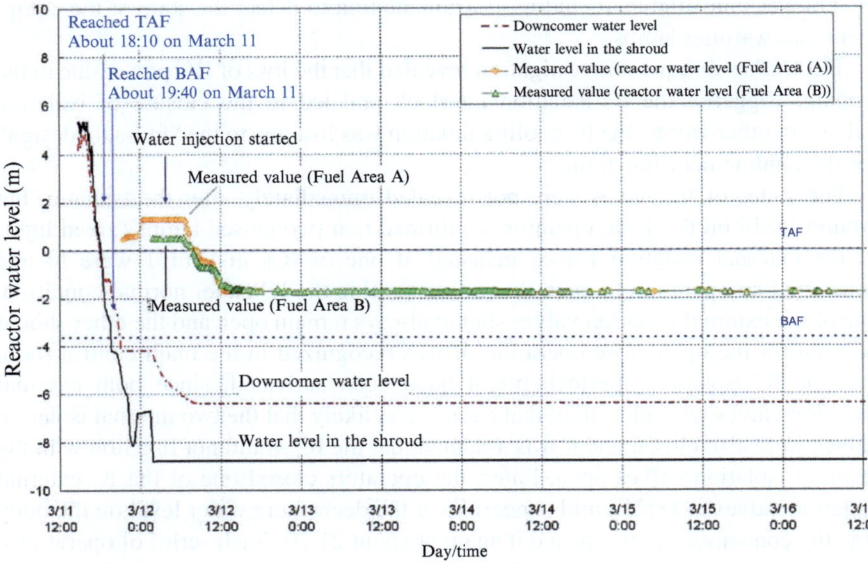

Fig. 3.1 Reactor water level (Unit 1). [TEPCO, Fukushima Nuclear Accident Analysis Report]

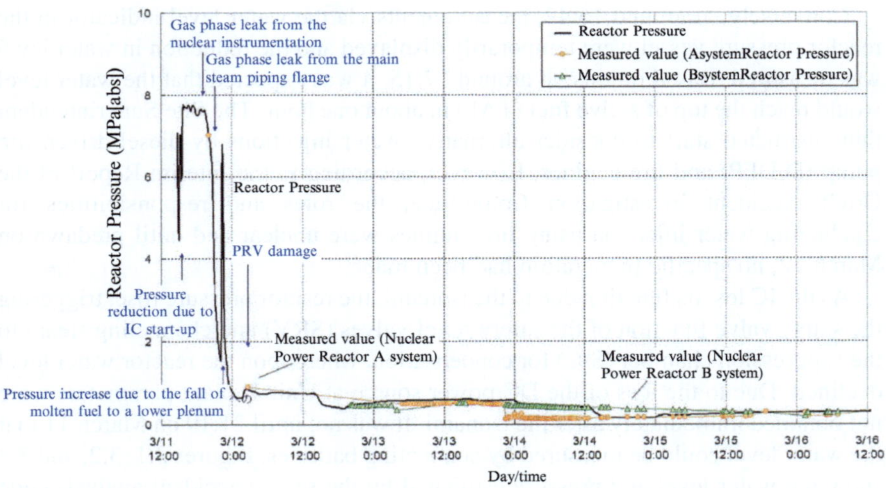

Fig. 3.2 Reactor pressure (Unit 1). [TEPCO, Fukushima Nuclear Accident Analysis Report]

Shortly after 21:00, measurements showed the reactor water level exceeding TAF, which had not been forecast at all. TEPCO's subsequent investigation revealed that after the reactor core had been exposed, the reactor water level indicator malfunctioned due to water evaporation on the reference chamber side condensation tank (see Fig. 2.4). As the actual water level in the nuclear power reactor declined, water in the water level instrumentation piping on the reactor side

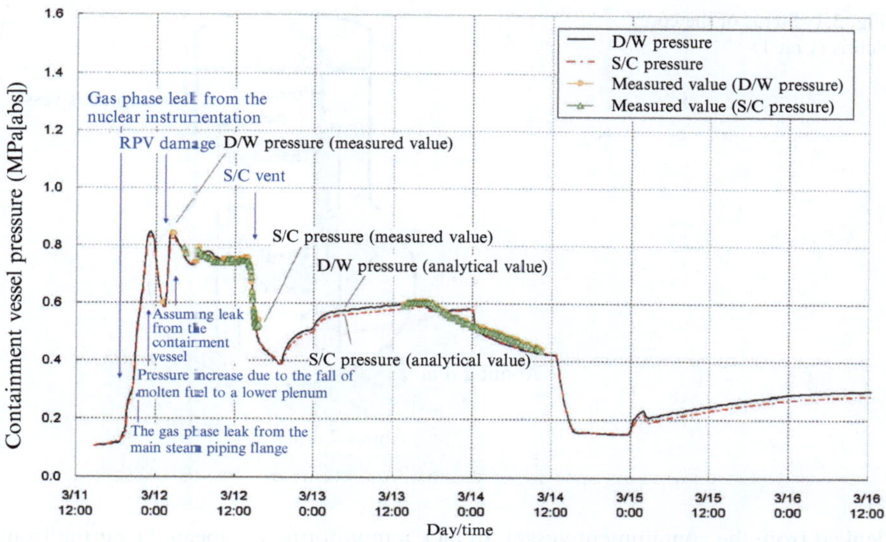

Fig. 3.3 Pressure in the containment vessel (Unit 1). [TEPCO, Fukushima Nuclear Accident Analysis Report]

started to evaporate and this state was reflected in the water level indicator readings. According to analysis, the reactor core had been exposed and started sustaining damage as of 21:00. It is assumed that this impact led to the water level reading exceeding the actual water level.

Around 0:00 on March 12, measurements in the main control room showed that drywell (D/W) pressure was 0.6 MPa [abs], exceeding the maximum operating pressure (0.531 MPa [abs]). At the time of the accident, it was assumed that the nuclear power reactor was in an abnormal condition because the maximum D/W pressure had been evaluated at 0.401 MPa [abs] in, for example, a loss of coolant accident (LOCA). Considering this, an instruction was given to prepare a PCV vent at the power station. By 2:30, the measurement value of D/W pressure rose to 0.84 MPa [abs] while the reactor pressure declined to 0.8 MPa [gage (gage pressure)]; more or less equivalent to the measured D/W pressure. Such pressure behavior in the Unit 1 reactor and the containment vessel and the fact that no operator depressurized the reactor led to the conclusion that the pressure boundary of reactor coolant had already been damaged. In addition, it was assumed that by this time, the core had been damaged and the reactor temperature was already high.

According to the analysis performed by TEPCO based on these premises, using the MAAP code after assuming damage to the in-core monitor guide tube and the SRV nozzle gasketseal, as shown in Figs. 3.2 and 3.3, the reactor pressure was equalized with containment vessel pressure. This analytical result assumed damage to the reactor pressure vessel due to the impact of molten fuels at around 2:00 on March 12.

Conversely, the measured value of the containment vessel pressure was maintained at around 0.75 MPa [abs]. It is assumed that radioactive materials

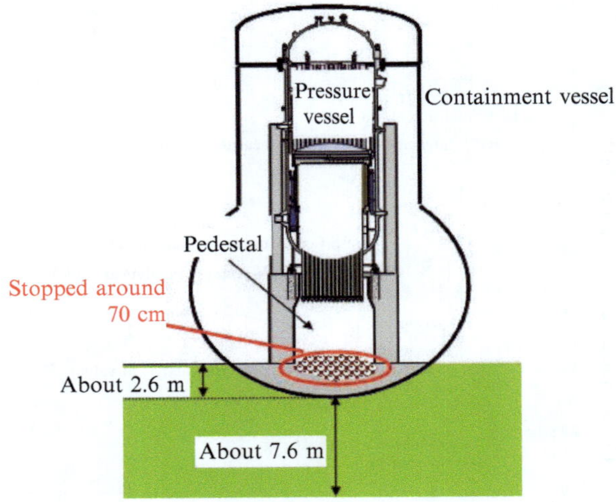

Fig. 3.4 Status of the core debris (Unit 1)

leaked from the containment vessel. In fact, a monitoring car located near the front gate detected an increase in the dose rate at dawn on March 12. Obviously, radioactive materials were released from Unit 1.

At around 4:00 on March 12, fire engines were linked to the inlet nozzle connected to the alternative water-injection line, and water injection started. Initially, fresh water in the fire protecting water tank was used as the source but this soon depleted, so seawater continued to be injected. The alternate water-injection line (fire protection system (FP) to the make-up water condensate system (MUWC) to the core spray system (CS)) was prepared as an accident management (AM) measure, assuming water supply by diesel-driven fire pumps (D/DFP) while fire engines were prepared as countermeasures after the Niigata prefecture Chuetsu-Oki earthquake. D/DFP could not be used in the Fukushima Daiichi accident, but the water-injection line and fire-engine pumps, which were prepared as AM, were combined for the core injection. However, it took considerable time for the fire engines to start water injection, including the difficulty in finding the inlet nozzle to which fire-engine hoses could be connected at the site where rubble was scattered.

Based on the MAAP code analytical results, it is assumed that this water injection was too late to prevent the core damage but reached the molten fuel transferred to the pedestal in the containment vessel. It is also assumed that consequently, the molten core concrete interaction (MCCI) was suppressed and the molten fuel remained at about 70 cm in the pedestal floor which was eroded (Fig. 3.4).

The containment vessel vent was planned concurring with the injection. As supplies of AC power and compressed air were lost, two valves had to be manually opened on site (Fig. 3.5).

The MO valve on the second floor of the reactor building could be manually opened, but to open the AO valve, access to the basement torus room was required. Shortly after 9:00, operators entered the torus room and approached the AO valve via the catwalk, but gave up the work due to the high dose, by abundant radioactive

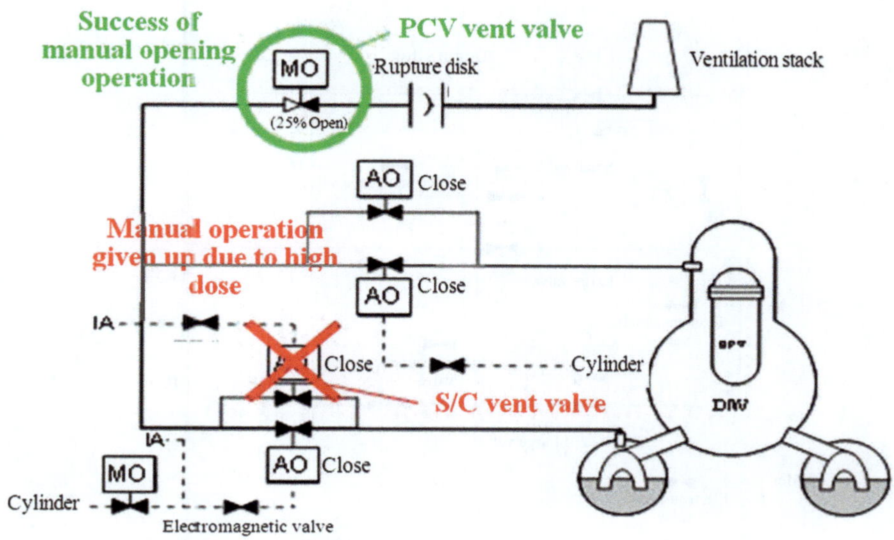

Fig. 3.5 Valves on the containment vessel vent line

materials transferred to S/C due to the core damage. Consequently, it emerged that one of the operators had been exposed to more than 100 mSv of radiation. To remotely operate the AO valve, driving compressed air and a power source to operate electromagnetic valves are required, but no compressed air was available. In expectation of the residual pressure, the operation to open the small AO valve was conducted using a small generator for temporary lighting shortly after 10:00, following which the dose rate displayed at the monitoring car in the vicinity of the front gate temporarily rose. As the containment vessel pressure was not decreased, ventilation was unsatisfactory and it was unclear whether the rupture disk on the vent line was open.

Further, it emerged that the compressed air could be supplied via the entrance of the reactor building to accommodate incoming large objects, provided the portable compressor was used. Shortly after 14:00, when the portable compressor started up, it was confirmed that the containment vessel pressure declined and steam was released from the stack, whereupon the containment vessel vent (PCV vent) was considered established. However, at this time, radiation readings shown on the monitoring car remained unchanged.

At 15:36 on March 12, the Unit 1 reactor building exploded, following which readings of the monitor car temporarily rose.

According to the past studies, reactivity accidents, steam explosions inside/outside the reactor vessel, directly containment heating, and others were known as exploding phenomena occurring in the containment vessel. However, since the positive pressure of the containment vessel was maintained to a certain degree, the occurrence of such phenomena did not occur.

Conversely, hydrogen was generated through a reaction between zirconium and water following the core damage, which was thought to be accumulated in the

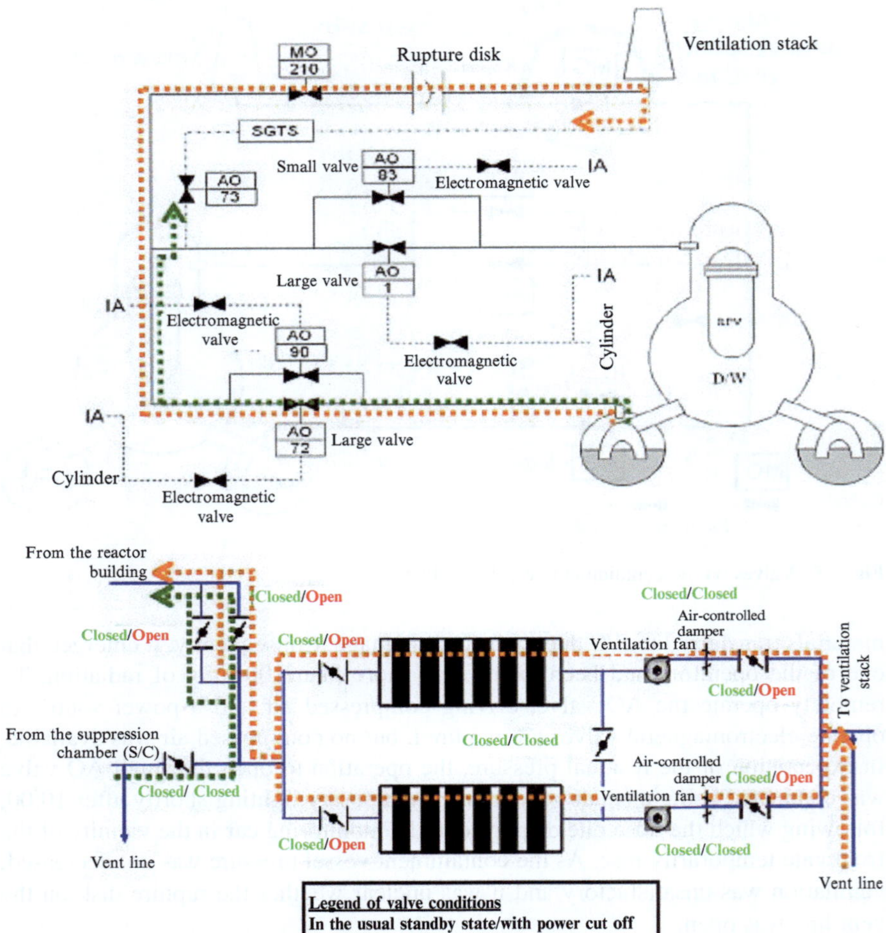

Fig. 3.6 Reverse flow of SGTS (Unit 1)

containment vessel and leaked through unknown paths to the reactor building, which caused a hydrogen explosion on the top floor. The pressure in the containment vessel once exceeded the maximum operating pressure and there were several possible gas leakage paths from the containment vessel to the reactor building, although it is currently difficult to identify precisely which was responsible.

Conversely, an explosion went off about 1 h after the containment vessel vent operation. The relation between this vent operation and the release of hydrogen gas in the reactor building should be clarified. The vent piping was connected to the standby gas treatment system (SGTS) and the SGTS inlet side was isolated by an isolation valve, but many valves, which were normally closed/open with power cut off, were mounted in the overall SGTS system (Fig. 3.6). There was also a flow-control damper on the reverse flow path from the vent piping, which was closed when the power was cut, although the damper could not perfectly stop the reverse flow. Despite not being completely airtight, it was considered able to prevent abundant flow. In addition, if the gas should be forced through the filter train of SGTS in the reverse direction, most of the specific radioactives, or fission products (FP) would have been captured, which contradicts a wide range of contamination in the reactor building. Therefore, it is difficult to completely reject the possibility of the reverse flow from SGTS, but it is assumed that SGTS was not the main path via which hydrogen flowed to the reactor building.

3.3 Unit 2

The earthquake automatically stopped the nuclear power reactor and the external power supplies were lost. Both the two emergency diesel generators (D/Gs) automatically started up while the main steam isolation valves (MSIV) closed. The reactor pressure was controlled by the safety relief valves (SRV). As for the reactor water level, the reactor core isolation cooling system (RCIC system) was expected to start automatically when the water level became low (L 2). Before the water level declined to L 2, the operator manually started the RCIC system and controlled the water level by repeating RCIC water level (L 8) trips and manual start-ups until the DC power was lost due to the tsunami.

The tsunami attacked the Fukushima Daiichi station, and the seawater cooling system as well as inundation of the flooding buildings at a site height of 10 m above sea level. Almost all power panels at Unit 2 were disabled. The main control room was blacked out and almost all instrumentation, warning lamps, and other means of determining the state of equipment were unavailable.

When the tsunami hit the station, operators had manually started up RCIC, but were subsequently unable to determine the operational state. As they could not determine the reactor water level either, they started preparing alternative water injection according to AM at around 21:00 on March 11, and finished preparing an alternative water-injection line the same day. Conversely, an operator, who headed for the RCIC room at around 1:00 on the 12th to check the RCIC operation, returned to the main control room because water was flowing from the RCIC room when he opened the door. Shortly after 2:00, an operator headed for the room again and confirmed that the RCIC remained operational because the RCIC discharge pressure was 6.0 MPa, exceeding the reactor pressure of 5.6 MPa.

At around 22:00 on the 11th, the reactor water level of Unit 2 was measured using a battery, as was done for Unit 1 and it was confirmed that the level exceeded TAF. Since then, the water level remained TAF+ 3.4 to 3.9 m. The measured values of the reactor water level as well as the analytical results according to the MAAP code are shown in Fig. 3.7. As shown in the figure, the reactor water level remained high until nearly noon on March 14. Conversely, the reactor pressure was low around 6 MPa [gage] though the SRV relief valve function did not work (Fig. 3.8). It is assumed from subsequent TEPCO's analysis that although the loss of DC

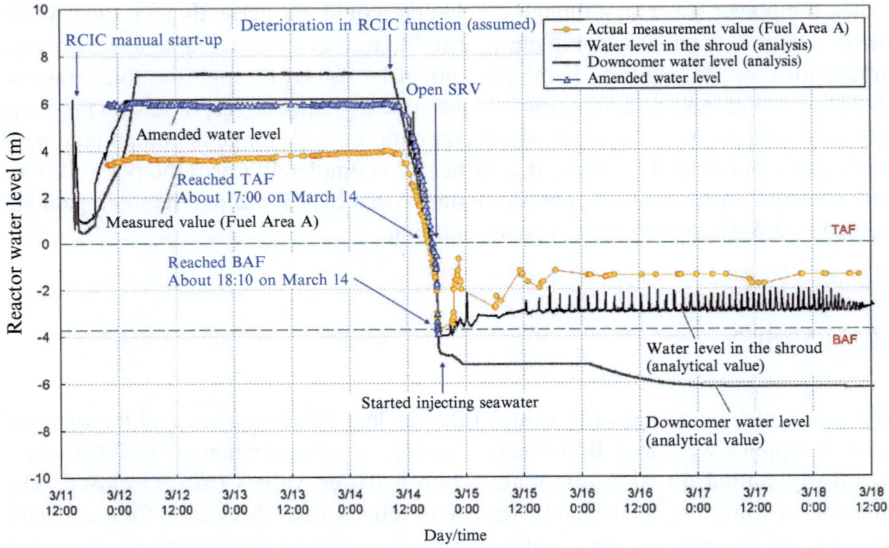

Fig. 3.7 Reactor water level (Unit 2). [TEPCO, Fukushima Nuclear Accident Analysis Report]

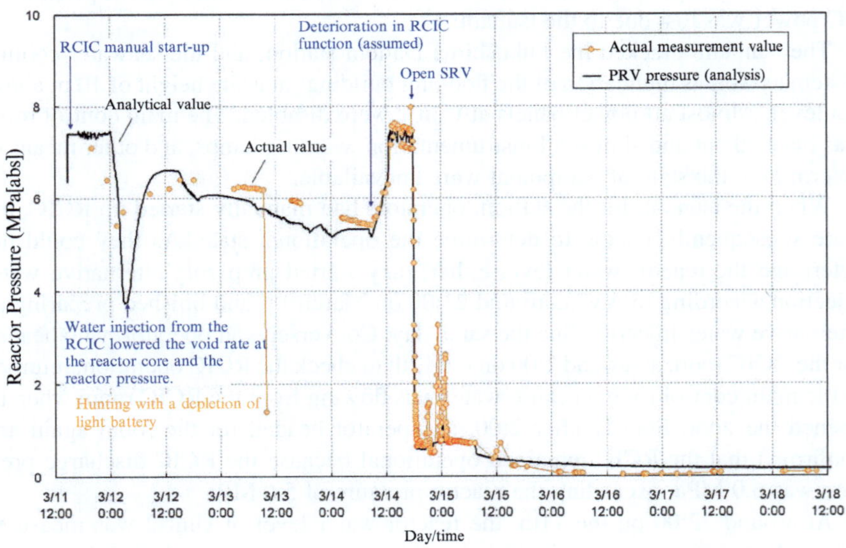

Fig. 3.8 Reactor pressure (Unit 2). [TEPCO, Fukushima Nuclear Accident Analysis Report]

power prevented the control of RCIC, plant parameters remained stable in this manner due to RCIC's specific operational state as described below.

First, the reactor water level was measured by the water level indicator in the fuel area. As this indicator was calibrated when the reactor pressure was reduced to

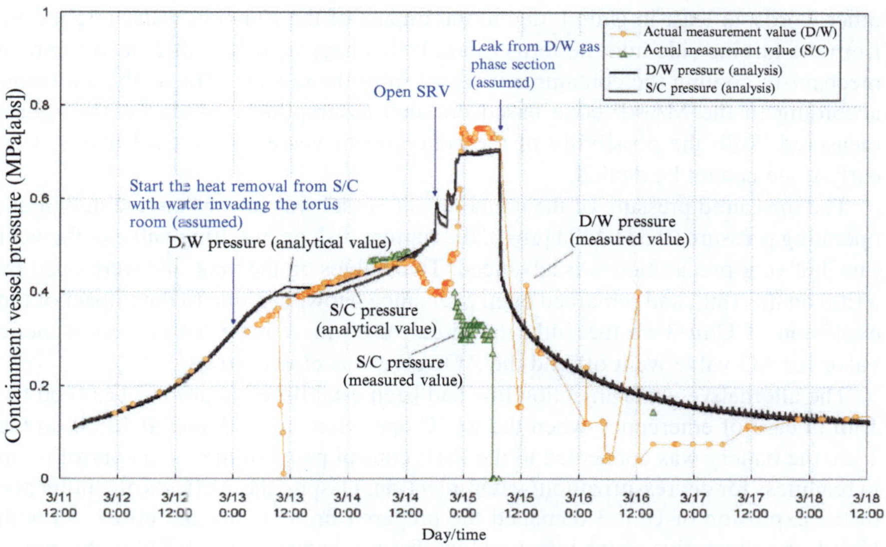

Fig. 3.9 Pressure in the containment vessel (Unit 2). [TEPCO, Fukushima Nuclear Accident Analysis Report]

atmospheric pressure, e.g. at shutdown, the instruction of the high pressure state of 6 MPa [gage] should be modified to reflect the actual water level. The modified water level corresponds to the nozzle position in the condensation tank displayed on the water level indicator. When the actual water level exceeds this nozzle, in principle, the water level indicator, which detects differential pressures, keeps the measured water level at/below the nozzle position level, while it is also assumed the actual water level during this period "exceeded" the nozzle position of the condenser tank. If the water level exceeds the nozzle position, two-phase flow came out of the main steam system and drove the RCIC turbine. It is assumed that for the RCIC, the operational condition was inefficient, whereby the water-injection flow was smaller than the rated value, and the discharge of the two-phase flow was balanced with the water injection by the RCIC. Conversely, from the nuclear power reactor, the two-phase flow, which had a larger enthalpy per unit volume, flowed out from the main steam pipe. Therefore it is assumed that the reactor pressure was balanced at around 6 MPa [gage], lower than SRV operative level.

Seeing the trend of the containment vessel pressure readings, we notice a gentler rise than the expected trend under conditions where heat removal was lost (Fig. 3.9), which suggests the presence of some heat-removal mechanism. Water was continuously injected to the nuclear power reactors at Units 1–3 subject to core damage. Although rise in water level in the containment vessels was not witnessed, polluted cooling water was seen to leak from the containment vessel into the torus rooms and then the turbine buildings. Conversely, at Unit 4, no water was injected into the nuclear reactor due to lack of necessity. Although water did not accumulate in the torus room in the same way as Units 1–3, water was still present there. In other words, at least in Unit 4, due to the impact of the tsunami, water might flow from the turbine side into the torus room. If this happened at Unit 2, it would be a mechanism cooling the containment vessel from the outside. The analytical result according to the MAAP code based on such assumption reproduced the values measured. Still, the possibility of the containment vessel at Unit 2 leaking at an early stage cannot be denied.

The measured pressure of the containment vessel was lower than the maximum operating pressure (0.531 MPa [abs]), the rupture disk open setting value of the vent line, but vent preparation was advanced. Two valves on the vent line were open by 11:00 on the 13th, and remained open ever since. However, due to the impact of the explosion of Unit 3 on the 14th, the circuit used to activate the electromagnetic valve for AO valve went off and the AO valve was closed down.

The alternative water-injection line had been established at around 12:00 on the 13th in case of emergency when the RCIC operation stopped, and at 13:00 on the 13th, the battery was connected to the SRV control panel in the main control room in readiness for depressurization/water injection. Despite these efforts, the influence of the explosion of Unit 3 damaged the prepared fire engines and hoses. As with Unit 1, the alternative water-injection line (fire protection system) (FP) to the make-up water condensate system (MUWC) and the low pressure coolant injection system (LPCI)) was prepared according to AM while fire engines were prepared as countermeasures after the Niigata prefecture Chuetsu-Oki earthquake.

Around noon on the 14th, the reactor water level started declining, and RCIC functions were deemed to be dropping, whereupon efforts were made to recompose the alternative water injection system and restore the vent line. Initially, efforts to restore S/C large vent valve were made, but since it appeared time-consuming, depressurization with the SRV and injection by fire engines was prioritized. Batteries were connected to several SRV control panels for depressurization, but the SRV was not quickly opened. After 18:00, the depressurization succeeded. Given the high on-site radiation dose, the integrity of fire engines was checked intermittently. At 19:20, it emerged that the fire engines had stopped due to fuel shortage. After the oil feed before 20:00, two fire engines started injecting water into the nuclear reactor. When the reactor pressure subsequently rose, operations to open the SRV were repeated and Unit 2 became unstable.

The analytical result by TEPCO using the MAAP code shows the reactor water level fell below the top of active fuel (TAF) at around 17:00 on the 14th when depressurization efforts were made and the core damage started around 19:20. Since then, though measured values showed that the reactor water level temporarily recovered, after the core damage, as with Unit 1, reactor water level readings were unreliable. In the analysis, the water levels measured after the core damage were unreliable and it was assumed that only part of water supplied by fire engines was injected into the nuclear reactor. As the time from the start of lowering the reactor water level to the start of water injection into the reactor is relatively shorter than Unit 1, the analysis presumed the pressure vessel was undamaged. Examination of the plant parameters and other factors show that the pressure vessel was damaged and part of the molten fuel fell to the pedestal and was cooled.

As it was assumed that the S/C vent large valve could not be opened due to some defect of the electromagnetic valves (ground fault), the small valve was fractionally opened at around 21:00 on the 14th. At this stage, the measured drywell (D/W) pressure was lower than the maximum operating pressure and the pressure was not vented. After the SRV was opened as an additional operation to depressurize the nuclear reactor at 21:20, the dose rate in the vicinity of the front gate temporarily rose, showing beyond doubt that some FP was released. It remains, however, unclear whether the rupture disk on the vent line at Unit 2 was opened or not. Subsequently, D/W pressure soared to 700 kPa [abs] or higher until 7:20 on the following 15th, which was assumed attributable to the hydrogen generation associated with the core damage. Shortly after 0:00 on the 15th, the operators attempted to open the D/W vent small valve, but it was confirmed that the valve remained closed after a few minutes and the D/W pressure remained unchanged, with no change in the dose rate displayed in the monitoring car either. Accordingly, it seems that the D/W vent did not release steam. After the D/W pressure soared at 22:00 on the 14th, the containment atmosphere monitoring system (CAMS) recovered to obtain gamma-ray dose rates and captured how the dose rate rose in line with the core damage effects.

Shortly after 6:00 on the 15th, a large impact was heard with vibration and almost simultaneously, the suppression chamber (S/C) pressure measurement value was reported as 0 kPa [abs] (The measurements in the main control room went downscale). At the time, it was thought S/C at Unit 2 might have been damaged. Staff who worked at the Seismic Isolation Building temporarily evacuated to the Fukushima Daini Nuclear Power Station except for the core necessary staff.

Later, it emerged that though the measurement value of S/C pressure in the main control room was downscale, indicating the failure of a measuring device, the value was falsely reported as 0 kPa [abs], namely a vacuum, to the Nuclear Emergency Response Headquarters at the Fukushima Daiichi. In addition, as for the impact sound, checking the proportional relationship between the arrival times of the P and S waves and the distance to the target unit from several seismometers installed in the power station, the sound was caused by the explosion of Unit 4. The Unit 2 torus room investigation video shot by a robot used by TEPCO later showed no signs of an explosion.

D/W pressure was measured as 730 kPa [abs] at 7:20 on the 15th and at 11:25, when the next instruction was obtained, had declined to 155 kPa [abs]. A picture shot at 10:00 on the 15th by a live camera set at the Fukushima Daiichi station showed white smoke emerging from the vicinity of Unit 2. In addition, as the dose rate soared in the vicinity of the front gate at the time, it was assumed that considerable radioactive materials had been released from Unit 2. Considering the high dose rate observed in the vicinity of the shield plug on the operation floor in a later TEPCO investigation and possible leakage positions according to the past test results, the main FP discharge path was assumed to be the D/W head flange seal.

The core damage occurred in Unit 2 but unlike Units 1 and 3, no hydrogen explosion occurred in the reactor building. This was considered due to the fact that the gas containing hydrogen was emitted from the blowout panel (4 × 6 m), which was opened due to the impact of the explosion in Unit 1, and hydrogen was not accumulated for an extended period in the reactor building.

3.4 Unit 3

The nuclear power reactor automatically stopped due to the earthquake, and as the external power supplies were lost, both two emergency diesel generators (D/Gs) automatically started up as well as the MSIV closing. The reactor pressure was controlled by the safety relief valves (SRV). Operators manually started RCIC before the water level declined to low water level (L2), where RCIC would automatically start up.

The tsunami attacked the power station, which meant not only was the seawater cooling system flooded, but also buildings at an elevation of 10 m above sea level. Moreover, many power panel functions were lost in Unit 3, although DC power remained available unlike Units 1 and 2, which meant DC-powered turbine-driven equipment, including the device to monitor the reactor water level, RCIC, and the high pressure coolant injection system (HPCI system) remained available. After both D/Gs tripped and the station blackout (SBO) occurred due to the tsunami, the operator controlled the water level by RCIC. At the time, the line valve was opened to return the coolant from the RCIC pump discharge line to the condensate storage tank (CST), the water source, and operations were conducted to save DC power.

At 11:36, on March 12, the RCIC automatically stopped, and the reactor water level declined. The low water level (L 2) then automatically triggered HPCI start-up. Like the RCIC, the HPCI was continuously driven by opening the test line to CST to save DC power (Fig. 3.10).

The measured values of the reactor water level as well as the analytical results according to the MAAP code are shown in Fig. 3.11. Due to the continuous operation of the HPCI, a large amount of steam generated by decay heat was continuously discharged to the HPCI turbine, and the reactor pressure was low during the HPCI operation (Fig. 3.12). During this time, the measured D/W pressure also exceeded the analytical result (Fig. 3.13). When S/C spray was performed using a diesel-driven fire pump (D/DFP) shortly after 12:00 on the 12th, the measured results were equivalent to the analytical results, but the causes of such behavior remained unknown.

Shortly after 2:00 on the 13th, the measured reactor pressure value decreased below 1 MPa, and operators were afraid the HPCI equipment might be damaged. Subsequently, a member of staff was sent to the reactor building to switch D/DFP from the spray to reactor containment to water injection to the nuclear reactor, and manually shut down the HPCI at 2:42, whereupon at 2:45, the operators tried but failed to open the SRV. The reactor pressure rose and water injection with D/DFP was impossible. The staff member returning to the main control room around 3:05

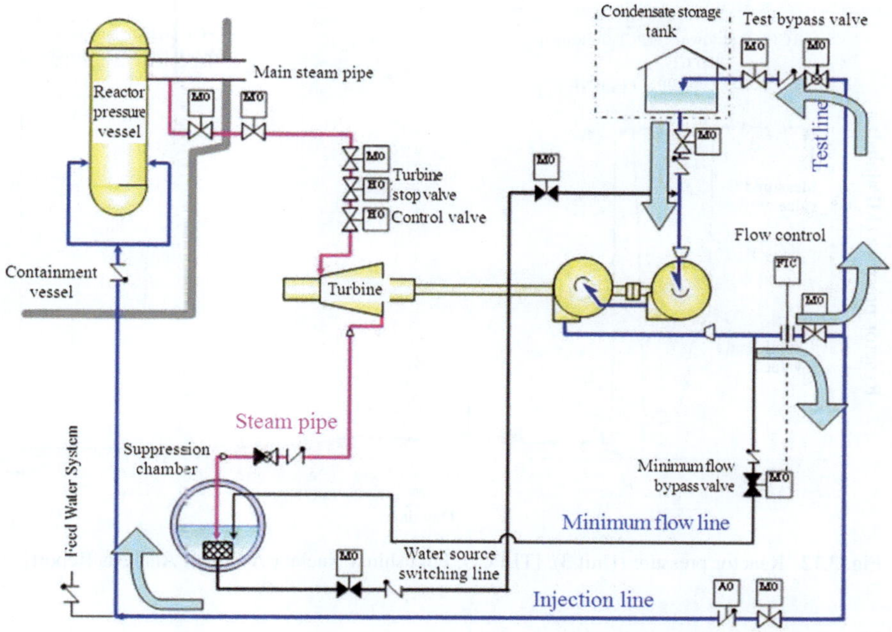

Fig. 3.10 HPCI operational state

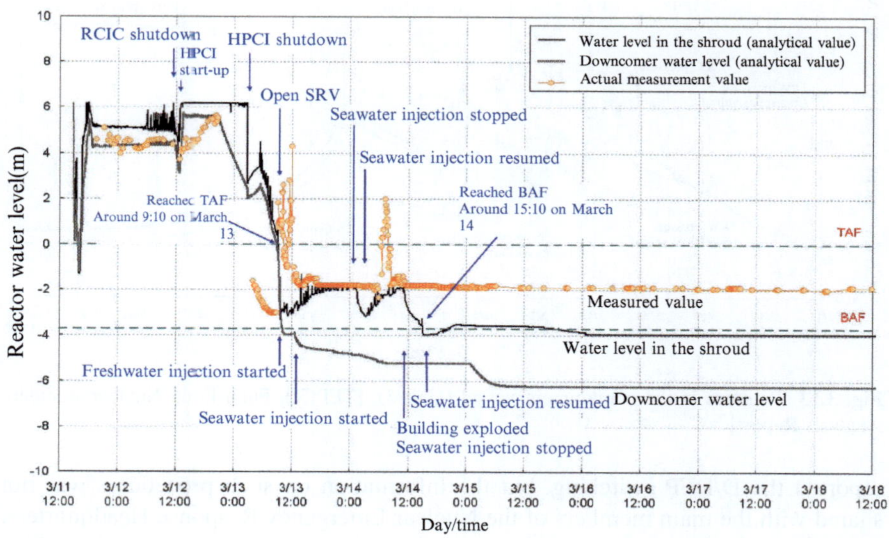

Fig. 3.11 Reactor water level (Unit 3). [TEPCO, Fukushima Nuclear Accident Analysis Report]

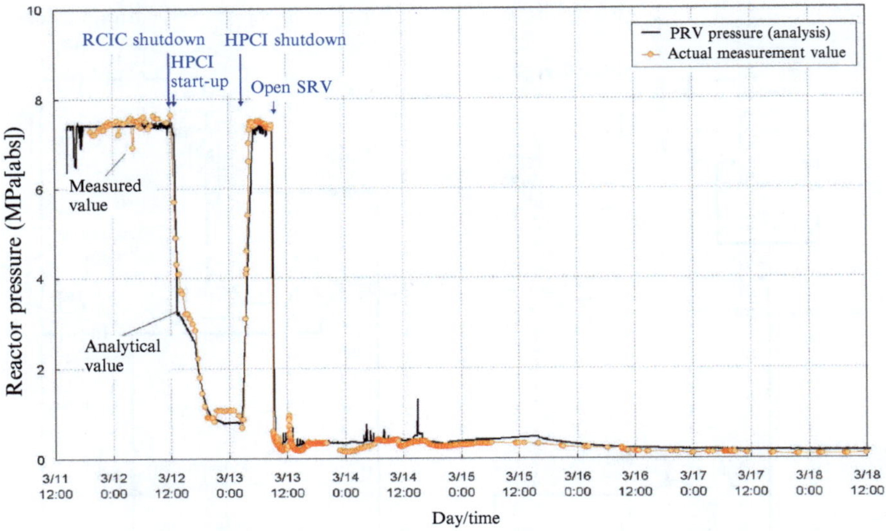

Fig. 3.12 Reactor pressure (Unit 3). [TEPCO, Fukushima Nuclear Accident Analysis Report]

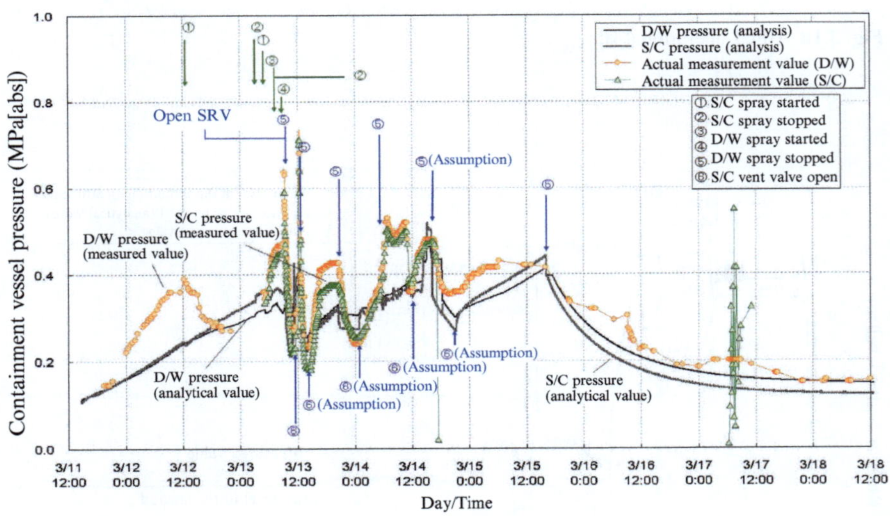

Fig. 3.13 Pressure in the containment vessel (Unit 3). [TEPCO, Fukushima Nuclear Accident Analysis Report]

reported the D/DFP switching, but the information on such procedures was not shared with the main members of the Nuclear Emergency Response Headquarters.

When the power restoration was attempted, using vehicle batteries to drive SRV electromagnetic valves, the reactor pressure abruptly dropped at around 9:00 though no SRV opening operation took place. The chart record shows a sudden

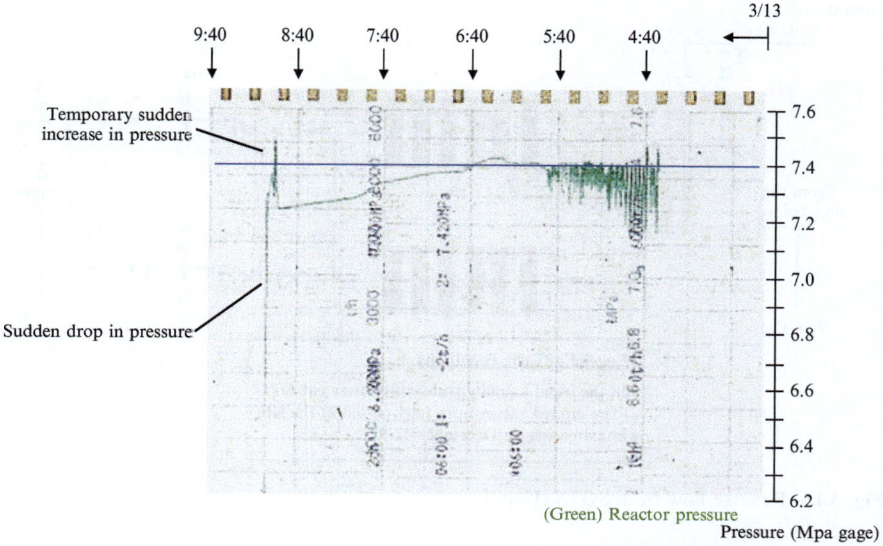

Fig. 3.14 Reactor pressure (Unit 3, magnification display from 4:40–9:40 on 3/13). [TEPCO, "Recorder Chart" (May 2011)]

drop in reactor pressure after the temporary sudden increase (Fig. 3.14). The sudden pressure drop was likely attributable to the open condition of several SRVs, due to which water injection by D/DFP and fire engines started. The analysis assumed that when the HPCI was manually stopped, the water injection stopped, which meant the reactor water level fell below TAF shortly after 9:00, and the core damage occurred at around 10:40. There might be a time when the reactor water level measured after the HPCI shutdown was lower than the MAAP analytical results, and the actual water level might fall below the top of active fuel (TAF) at an earlier stage than the analysis, whereupon the actual core damage might occur earlier. Subsequently, there was a time when the water level measurement temporarily recovered, but after the core damage, like Unit 1, the reactor water level readings seem to have been unreliable. In the analysis, since the water levels measured after the core damage were unreliable, it was assumed that only part of the flow injected by the fire engines reached the nuclear reactor.

During the process of water injection into the reactor by HPCI, the drywell (D/W) pressure measurement value was lower than the set value of the rupture disk, but with preparation of the containment vessel vent advanced, the vent line was established at 8:41 on the 13th, and the vent would start by opening the rupture disk. At around 9:20, the D/W pressure declined and the containment vessel vent (PCV vent) seemed to be effective. Immediately after this first vent, the dose rate in the vicinity of the front gate temporarily rose to around 300 μSv/h, but the later vent operations did not increase the dose rate. The specific wind directions and position of the monitoring car might have prevented the dose rate from rising, but

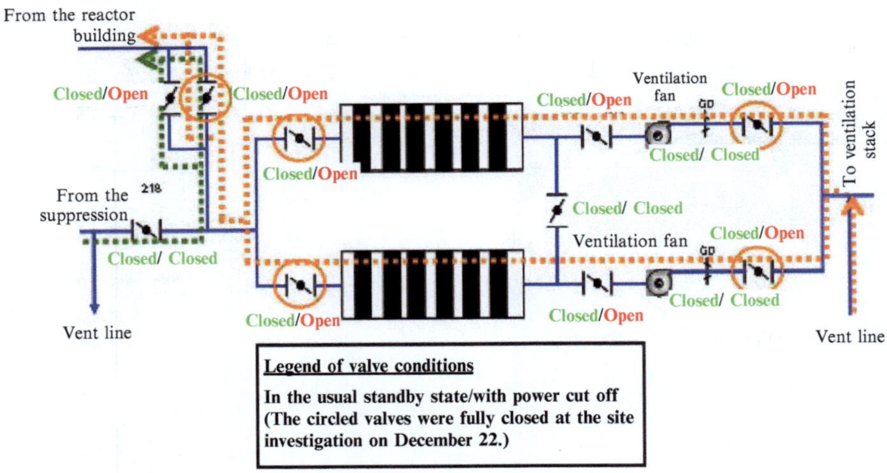

Fig. 3.15 Reverse flow from SGTS (Unit 3)

considering the fact that once abundant FP was released to the top of ventilation stack via the vent, the dose could be detected directly, regardless of wind direction, it is assumed that the FP discharge amount was limited in and after the second time. At around 8:00, immediately before the first vent, the monitoring post (MP-4) detected a rise in dose rate, but at the time, the D/W pressure in Unit 3 was below the maximum operating pressure. It seems that FP was released from Unit 1, the core of which had been damaged.

After the explosion at Unit 1, a similar explosion was expected at Unit 3. To ventilate the reactor building, measures were considered such as opening the blowout panel, forming holes in the ceiling of the reactor building, and others. Among them, a water jet, with a smaller risk of inducing explosion, was arranged, but the reactor building exploded at 11:01 on March 14 before the equipment reached the power station. The explosion at Unit 3 was more violent than Unit 1. In addition, the color of the smoke was black at Unit 3, not white, typical hydrogen explosion, as occurred in Unit 1. There was a difference between the two units. While the wall on the top floor of the reactor building at Unit 1 was steel, that at Unit 3 was reinforced concrete. It is assumed that the crushed concrete discharged black smoke. In Unit 3, more explosive hydrogen was accumulated and reinforced concrete was stronger, which was assumed to contribute to a more violent explosion. It can be assumed that like Unit 1, the hydrogen generated by the core damage moved to the reactor building via unknown paths, resulting in an explosion on the operation floor, namely the top floor. At Unit 3, four containment vessel vents were conducted during the period from the core damage to the reactor building explosion. About a day had elapsed since the first vent to the explosion, and about 4 h since the fourth vent to the explosion. As the gravity damper (GD) was installed on the standby gas treatment system (SGTS) reverse flow path from the vent piping (Fig. 3.15) and restricted the reverse flow, complete airtightness was not ensured but flow could largely be prevented. On December 22, 2011, TEPCO investigated the dose rate for the SGTS filter train, the measurement of which showed only a few mSv/h. If large quantities of reverse gas flow would take place, most of particulate FPs would have been captured. This contradicts the fact that the inside of the reactor building was widely polluted. Therefore, it is difficult to completely reject the possibility of reverse flow through SGTS, but it is assumed that SGTS was not a main hydrogen flowing path to the reactor building.

3.5 Unit 4 and Spent Fuel Pools

Unit 4 was undergoing a scheduled inspection and was already shut down. All spent fuel was in the spent fuel pool (SFP), which meant the higher heat source by the decay heat than those of other units and the shared pool (Table 3.1).

Due to the earthquake, external power supplies were lost, the cooling system for the spent fuel pool stopped, and the emergency diesel generator (B system) automatically started up (A system was undergoing an inspection).

The tsunami attacked the station, and as well as the seawater cooling system, all the DC and AC power supplies were lost, and the SFP cooling- and water supplying functions. At this stage, the situation was more acute than other pools, but the reactor core cooling at Units 1–3, which was under more serious conditions, was prioritized because it was anticipated that it would be around late March that water would have evaporated due to the decay heat, exposing the fuel at Unit 4. At around 4:00 on March 14, it was confirmed that the SFP water temperature was 84 °C, a value close to the estimated value.

At 6:12 on March 15, a large impact sound and vibration were generated and damage was confirmed in the vicinity of the fifth floor roof of the reactor building. Unit 4 was shut down, all fuel was in SFP, and the temperature was checked on the previous day. It was difficult to envisage hydrogen generated by a reaction between water and metal due to the exposure of the fuel cladding tubes. The impact of the

Table 3.1 Conditions for spent fuel storage

Unit	No. of spent fuels (No. of new fuel assemblies)	Decay heat (MW) on 3/11	Pool water quantity (m^3)
Unit 1	292 (100)	0.18	990
Unit 2	587 (28)	0.62	1,390
Unit 3	514 (52)	0.54	1,390
Unit 4	1,331 (204)	2.26	1,390
Unit 5	946 (48)	1.01	1,390
Unit 6	876 (64)	0.87	About 1,450
Shared pool	6,375 (–)	1.13	About 4,000
Cask storage facility	408 (–)	–	–

explosion on the pool and fuel was unknown, and given the lack of water level indicator for the SFP, it was unknown whether the water level was maintained or not. Under the circumstances, staff were urgently trying to determine alternative cooling measures. Conversely, it was assumed that if the SFP had no water, the sky-shine radiation from fuel would increase the radiation dosage rate in the vicinity of Unit 4, but the dose was actually low enough for staff to work. Therefore, it seemed that the fuel was not exposed.

On March 16, a helicopter checked to ensure the SFP at Unit 4 had sufficient water. Fresh water discharge by water cannon trucks started on 20th, followed by seawater discharge via concrete pump vehicles on 22nd, whereupon the seawater was changed for fresh water on the 30th, and a newly established alternative cooling system started operation on July 30.

It was assumed that the explosions were attributable to: (1) hydrogen generated by overheating of spent fuel, (2) vaporized oil, (3) introduced flammable gas, or (4) hydrogen generated by radiolysis of water. Hypothesis (1) is denied because the water level was secured, (2) is denied because it seems the temperature in the building was not hot enough to vaporize oil, (3) is denied because cylinders and others lacked sufficient flammable gas to induce a massive explosion, and (4) is denied because insufficient hydrogen was generated to induce an explosion. It is most likely that flammable gas, which caused extensive damage, was generated by the core damage in the adjacent Unit 3, and the inflow from the containment vessel vent piping shared by Units 3 and 4 was suspected (Fig. 3.16). The standby gas treatment system (SGTS) was located in the inflow path to Unit 4 building.

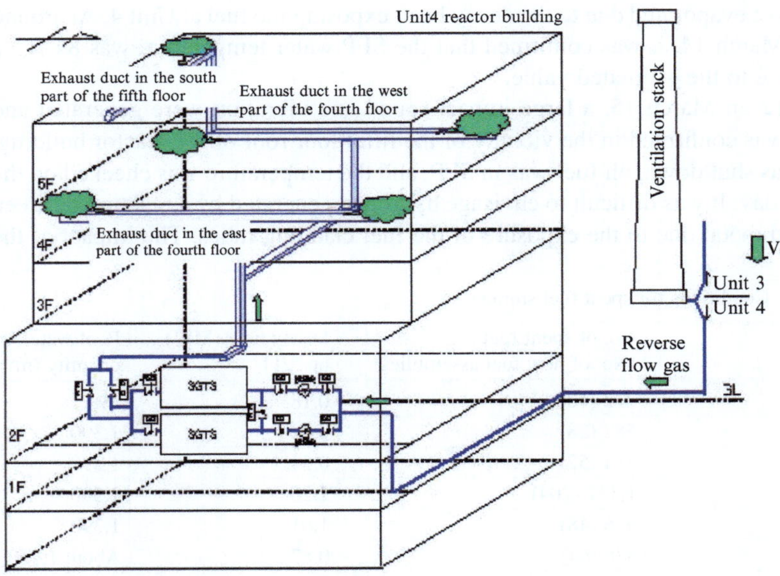

Fig. 3.16 Hydrogen flowing passage to Unit 4. [TEPCO, Fukushima Nuclear Accident Analysis Report]

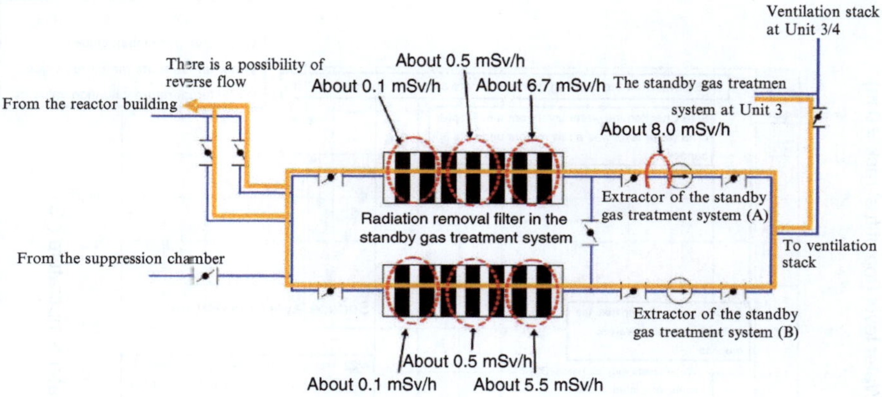

Fig. 3.17 Reverse flow from SGTS (Unit 4)

It emerged from the design information that unlike Units 1 and 3, no damper was installed to control the reverse flow. To confirm this reverse flow, the dose rate of the SGTS filter train was measured on August 25, 2011. The dose was higher on the outlet (Unit 3) side (Fig. 3.17), which supported the reverse flow from Unit 3.

The site investigation on the fourth and fifth floors of the reactor building proved that while the fifth floor surface was deformed upward, the fourth floor surface was deformed downward, and that the air-conditioning duct on the fourth floor was no longer in the original place; reduced to rubble and scattered on the floor. Under these circumstances, it seems the hydrogen explosion occurred in the air-conditioning duct on the fourth floor and propagated to the whole building via the staircase.

There was no sign of water leakage on the second floor of the lower part of the SFP, nor was any damage found in the structure to support the SFP. The structural soundness was secured. Analysis revealed that if the Fukushima Daiichi were to be exposed again to earthquake ground motion on a par with the 2011 tremor of the off the Pacific coast of Tohoku Earthquake (basic earthquake ground motion Ss) after the accident, the reactor building, including SFP, would not be destroyed. Further, with a view to enhancing the earthquake-resistance allowance, the bottom of the SFP was reinforced to achieve improvement of 20 % or more. As for the inclination of the reactor building, the distance between the pool water surface and building floor and inclination investigations on the building walls were continuously conducted by an optical device, and no significant inclination was found.

Pictures of the inside of the SFP have not shown any abnormality, and the sample taken from new fuel showed that the fuel was sound. Although in nuclide analysis of the pool water iodine and cesium were detected, their concentrations were lower than those at Units 1–3 by more than two digits. Though the concentrations were higher than the pre-accident levels, the absolute values remained small, and it was likely that no systematic mass damage occurred and that the revealed FPs were from cores at Units 1–3.

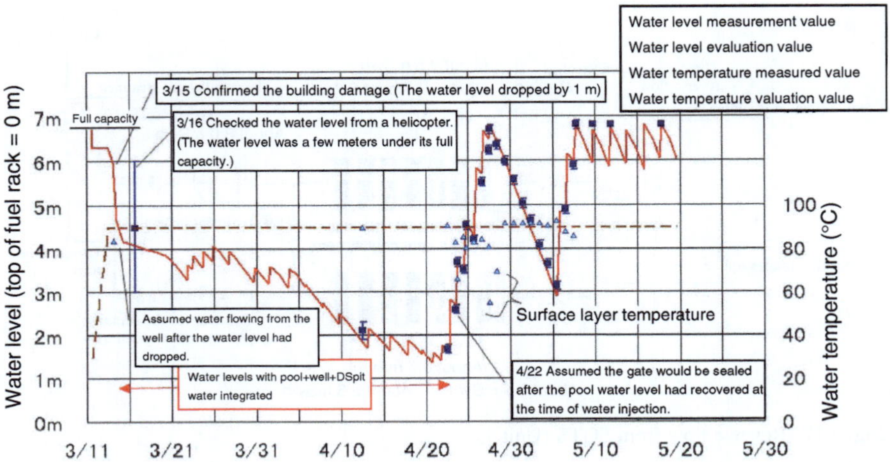

Fig. 3.18 SFP water temperature/level evaluation results (Unit 4). [TEPCO, Fukushima Nuclear Accident Analysis Report]

After the water surface had been checked by helicopter on March 16, water was discharged by water cannon trucks and concrete pump vehicles. On and after April 12, the water level was measured by using a concrete pump vehicle. Using these actual water level measurements and water discharge, and assumptions for water level decrease due to yield of water discharge (rate of water discharge injected into the SFP), sloshing by the earthquake and an explosion, the SFP water level/temperature were evaluated (Fig. 3.18). The water discharge amount was insufficient until around April 20 when the water level was the lowest (the top of fuel rack + 1.5 m). It was assumed that the pool gate was closed due to the water injection on April 22, whereupon the water level rose, and it was confirmed that on April 27, the water level of the skimmer surge tank drastically rose to reach full capacity.

SFPs at Units 1–3 were similarly evaluated. All units had smaller decay heat than that of Unit 4, and the temperatures were around 70 °C. Though the water levels decreased due to evaporation, the water levels were sufficiently secured. Reactor buildings were exploded at Units 1 and 3, but the SFP water levels were sufficiently maintained. The SFPs and shared pools at Units 5 and 6 had less decay heat than that of Unit 4. Though the water temperature temporarily rose to around 70 °C, the introduction of alternative cooling equipment maintained stable cooling. The SFP water sampling results at Units 1–3 showed that although seven months or more had elapsed since the shutdown of nuclear reactors, as short half-life nuclides were detected at the initial accident stage, and the nuclide composition resembled that of the stagnant water in the turbine building, it was assumed that SFP water was polluted by the damaged nuclear reactors.

Abundant seawater, sand, and rubble, and others flowed into the building where dry storage casks were stored, and flooding of floor surfaces, damage to louvers, doors, and others were observed. However, the radiation dose remained at the level of background, the dry storage casks were air-cooled, and the sealing performance was maintained.

3.6 Unit 5 and Unit 6

Units 5 and 6 were shut down due to a regular inspection (cold shutdown). Unit 5 underwent a pressure leak test, the reactor pressure was maintained at about 7 MPa [gage], and the RPV lid was closed at Unit 6. Like Units 1–4, the earthquake ground motion was the same level or slightly above the standard earthquake ground motion. The height of the tsunami far exceeded the latest evaluation value as well as the design standard value at the time of obtaining an establishment license. The site height of the main building installation area of Units 5 and 6, however, is O.P. + 13 m and exceeds that of Units 1–4 (O.P. + 10 m). The damage was enormous but relatively smaller.

Unit 5 lost external power supplies due to the earthquake, and two D/Gs automatically started up. Due to the loss of external power supply, the water pumps for the control rod drive stopped, and the reactor pressure temporarily fell to around 5 MPa [gage]. Under the influence of the tsunami, all high-voltage and emergency low-voltage power panels lost functions and the SBO occurred. Part of the regular low-voltage power panels as well as the DC power could be used, and the plant parameters could be checked. As the nuclear reactor had not yet started, and new fuel had been loaded into the reactor, the decay heat level was small and the rise in reactor pressure was sluggish. At night on the 11th, work to inspect and restore the power supply system started. From the 12th onward, since a pressure leakage test prevented the SRV remote operation, the nuclear reactor was depressurized by the PRV top vent. However, the PRV top vent could not depressurize the reactor sufficiently to conduct low-pressure water injection. On the 14th, staff entered the containment vessel and restored the SRV nitrogen gas feed line, so that pressure-reducing operation was intermittently given to the nuclear reactor. The power supply to the make-up water condensate system (MUWC) was restored, and the water injection to the nuclear reactor started on the 14th to maintain the reactor water level. On the 19th, a portable submerged pump started up and SFP cooling by RHR(C) started. On the 20th, RHR(C) cooled the reactor and a cold shutdown was achieved, since which time the SFP and nuclear reactor have been alternately cooled.

Unit 6 lost external power supplies due to the earthquake, whereupon three D/Gs automatically started up. Due to the tsunami, part of the high-voltage power panels could not be used, but the DC power was not flooded and remained usable, while one D/G (6B) was air-cooled and worked independently of the seawater cooling system. Unlike the reactor building, the power panel in the building where the D/G was installed was not flooded and maintained its functions, and no SBO occurred. On the 13th, the alternative water injection into the nuclear reactor by the MUWC started, while the reactor pressure gradually rose due to the decay heat. On the 14th, the reactor was depressurized by SRV, and the reactor pressure level, which allowed water injection via the MUWC, was maintained. On the 19th, a portable submerged pump started up and SFP cooling by RHR(B) started. On the 20th, RHR (B) cooled the reactor and a cold shutdown was achieved, since which time the SFP and nuclear power reactor have been alternately cooled.

Chapter 4
Overview of Events Occurring at Power Stations Other Than the Fukushima Daiichi Nuclear Power Station

Abstract Chapter 4 provides an overview of events at the Fukushima Daini Nuclear Power Station, the Onagawa Nuclear Power Station, and the Tokai Daini Nuclear Power Station, focusing on factual records.

Keywords Event sequence • Fukushima Daini NPS • Onagawa NPS • Tokai Daini NPS

4.1 The Fukushima Daini Nuclear Power Station

4.1.1 Overview of the Fukushima Daini Nuclear Power Station

The Fukushima Daini Nuclear Power Station of TEPCO (hereinafter the "Fukushima Daini") is located in Naraha and Tomioka towns, Futaba district, Fukushima prefecture and has four BWR plants, with total electric output of 1.1 million kW, which started to run in 1982–1987. The reactor type of Unit 1 is BWR-5 with a Mark II containment vessel while the reactor types of Units 2–4 are BWR-5 with Mark II advanced containment.

4.1.2 Overview of the Earthquake and Tsunami

At 14:46 on March 11, 2011, The Great East Japan Earthquake, the fourth largest in recorded history (M9.0) with strong Level 6 seismic intensity (on the Japan Metrological Agency (JMA) scale) was observed in Naraha and Tomioka towns, Fukushima prefecture. The maximum horizontal acceleration observed at the foundation boards of the plants was 277 Gal at Unit 3 and the maximum vertical acceleration was 305 Gal at Unit 1. Both the observed maximum accelerations were smaller than those of the design basis seismic ground motion Ss in the "Regulatory Guide for Reviewing the Seismic Design of Nuclear Power Reactor Facilities" revised in 2006.

The Fukushima Daiichi Nuclear Accident: Final Report of the AESJ Investigation Committee, DOI 10.1007/978-4-431-55160-7_4

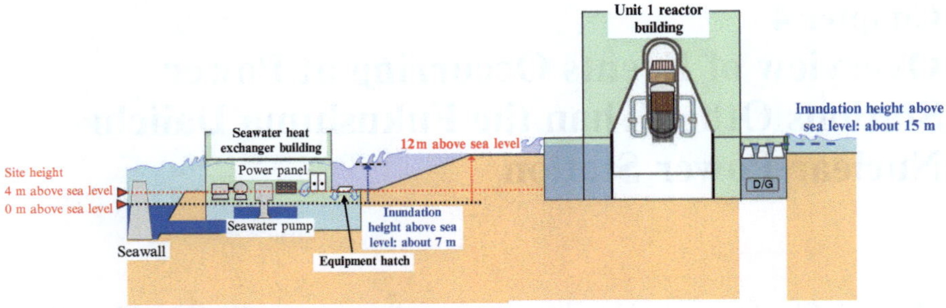

Fig. 4.1 Layout of the Fukushima Daini and the tsunami

After the occurrence of the earthquake, at around 15:30, the tsunami attacked the Fukushima Daini. Subsequent analysis assumed that the offshore tsunami height at the Fukushima Daini was about 9 m, while investigations of the tsunami inundation height show the whole seaside area was inundated (inundation height O.P. about +7 m) at the site height of Onahama port base tide level for construction (O.P.) +4 m (Fig. 4.1).

In addition, the tsunami ran up to the site at O.P. + 12 m along the road from the sea to the Seismic Isolation Building, southeast of the main building site area and proceeded in the direction from Units 1–4, which meant the inundation was deep on the south side of Unit 1, and near the Seismic Isolation Building in particular, reached about 15 m. Conversely, the inundation around Units 2 and 3 was shallower, despite water ingress from Unit 1. No water intruded into the surrounding of Unit 4 building.

4.1.3 Influence of the Seismic Ground Motion and the Tsunami

4.1.3.1 Influence of the Seismic Ground Motion on the Power Station

To investigate the influence of the earthquake, a walkdown (site confirmation) was conducted. The investigation by TEPCO did not uncover any apparent abnormality in critical seismic safety facilities but part of earthquake-proof B and C class equipment, including a desalted water tank, was damaged. In addition, seismic response analysis was performed for representative equipment based on actual seismic ground motion measurements, confirming that the earthquake load exerted on parts was less than the design basis seismic ground motion Ss, or the stress generated in the piping system was below the evaluation standard value.

Conversely, as the disconnecter was damaged at the Shin Fukushima Substation after the earthquake, the 500 kV Tomioka Line No. 2 was disabled. In addition, power to the 66 kV Iwaido Line No. 2 was cut as damage to the lightning arrestor was confirmed. As the 66 kV Iwaido Line No. 1 had been shut down for inspection before the earthquake occurred; only one line of the 500 kV Tomioka Line

No. 1 was available among four lines of external power supplies. In the current seismic design of nuclear power stations, when the design basis seismic ground motion occurred, backup power from an emergency diesel generator (D/G) was assumed. Unlike the Fukushima Daiichi however, at the Fukushima Daini, the inundation of power panels was limited and external power supply was available, albeit only through one line, which was one factor which facilitated the subsequent restoration activities. After the Iwaido Line No. 2 had been restored at around 13:38 on March 12, followed by the Iwaido Line No. 1 at around 5:15 on March 13, the station received power from three lines in total.

4.1.3.2 Influence of the Tsunami on the Power Station

The entrances of the seawater heat exchanger buildings for accommodating large incoming objects on the sea side were destroyed by the tsunami and seawater penetrating inside the buildings except the southern building at Unit 3. Although the doors of the entrance for accommodating large incoming objects opened outward, the wave power of the tsunami destroyed them inwardly. The residual heat removal cooling water system (RHRC) pumps, the emergency equipment cooling water system (EECW) pumps, and the residual heat removal seawater system (RHRS) pumps were all flooded, as well as the power panels of all equipment on the first floor. Under the circumstances, as horizontal RHRC and EECW pumps were shorter than vertical RHRS pumps in terms of the installation height of electric motors, the latter, which were submerged in water, stopped functioning. Further, the basement floors of the seawater heat exchanger buildings were also inundated via equipment hatches and air-conditioning ducts in buildings. Under the circumstances, seawater entering Units 1 and 3 seawater heat exchanger buildings reached the Unit 3 turbine buildings through a concrete trench and inundated the basement floor of the turbine building.

In addition, the tsunami ran up along the road from the southeast side of the main building site to the waste treatment building, the Seismic Isolation Building, and the Unit 1 reactor building. The doors of the waste treatment building and the Seismic Isolation Building were destroyed and inundated with water. Though external power was supplied to the Seismic Isolation Building, power failed due to the tsunami, which also prevented the emergency gas turbine generators from starting up. Under these circumstances, after the tsunami reached the station, the power failure continued at the Seismic Isolation Building until power was restored by installing a temporary cable at around 19:00 on the same day.

Water intruded from the air supply louvers for ventilation and ground equipment hatch in the Unit 1 reactor building. Consequently, three D/Gs set on the second basement of the annex attached to the reactor building stopped functioning due to water damage, as did two out of three emergency M/Cs (high-voltage power panels), although one remained operational.

The inundation around reactor building Units 2–4 was not deep, and no inundation from above-ground opening parts into the reactor buildings was confirmed.

Inundation, however, was recognized in Unit 3 through an underground trench from the seawater heat exchanger building into the reactor building. M/Cs at Units 2–4 remained operational. Considering that one M/C system did not stop functioning at Unit 1 as described above, we can say emergency M/Cs at all units in the Fukushima Daini remained operational in some way. It is assumed that the maintained functions of M/Cs as well as the maintained external power supplies contributed to the effective accident responses. As for Units 2–4, the D/Gs in the annex attached to the reactor building were neither damaged by water nor submerged but because the power panels and motors of the cooling system stopped functioning, all three D/Gs at Unit 2, one of three at Unit 3, and two of three at Unit 4 stopped functioning. Nine of a total twelve D/Gs throughout the Fukushima Daini stopped functioning due to tsunami-related causes.

One notable influence of the tsunami on the equipment was that the soundness of the equipment in the Unit 3 B-system heat exchanger building happened to be maintained because only the entrance for accommodating large objects was not destroyed by the tsunami. Therefore, for Unit 3 alone, TEPCO did not declare a situation that corresponded to the event mentioned in the "Act on Special Measures Concerning Nuclear Emergency Preparedness" (Nuclear Emergency Act) but achieved early cold shutdown according to a normal procedure. For the EECW pumps at Units 2 and 4, A systems were installed on the first floor and B systems on the second floor, and inevitably, only the B systems survived. Conversely, only one of two B-system RHRS pumps at Units 2 and 4 happened to stop functioning though both were located side by side.

4.1.4 Response Before the Arrival of the Tsunami

All Units 1–4 under constant rated thermal power output operation were scrammed by the signal due to the Great East Japan Earthquake, which occurred at 14:46 on March 11. Due to the fall in reactor pressure, the void in the reactor core decreased, which led to the "reactor low water level (L 3)". The lowered reactor water level was restored by supplying water from the reactor feed water system without reaching the level automatically triggering the emergency core cooling system (ECCS) or the reactor core isolation cooling system (RCIC).

4.1.5 Response After the Arrival of the Tsunami

4.1.5.1 Reactor Cooling Responses After the Arrival of the Tsunami

The tsunami attacking the station within about 30 min of the occurrence of the earthquake prevented the emergency equipment cooling pumps from starting up, and the residual heat removal system (RHR) except Unit 3 B-system were no longer unable to dissipate heat from the reactor. Accordingly, at 18:33, TEPCO judged that the situation corresponded to the "loss of reactor heat removal function" event in accordance with Article 10 of the "Act on Special Measures Concerning Nuclear Emergency Preparedness" (Nuclear Emergency Act). Conversely, as for Unit 3, the RHR B-system, which was undamaged by water, could dissipate heat from the reactor, and the reactor went into cold shutdown at 12:15 on March 12.

At Units 1, 2, and 4, which lost the heat removal function due to the tsunami, the main steam isolation valves (MSIV) were manually fully closed, the pressure-reducing operation started with the main steam safety relief valves (SRV), and cooling of the reactors continued via water injection of the reactor core isolation cooling system (RCIC) according to the Emergency Operating Procedures (EOP). Subsequently, after automatic isolation of the RCIC turbine with reduction in steam pressure along with the decrease in reactor pressure, alternate water injection with the make-up water system (condenser) (MUWC), which is accident management (AM) measures, started and the reactor water level was maintained. These flexible responses based on the EOP maintained the reactor cooling under circumstances where the reactor heat removal function had been lost.

4.1.5.2 Cooling of the Containment Vessel Pending Restoration of the Heat Removal Function

During water injection into the nuclear reactor, the S/C water temperature rose and exceeded 100 °C with the RCIC operation and opening of the SRV. Accordingly, at 6:07 on March 12, TEPCO declared the occurrence of the "loss of pressure-suppression function event in accordance with Article 15 of the Nuclear Emergency Act" at all units. To cool the S/C, the containment vessel spray was employed by the MUWC shown in the procedures as well as S/C water injection using the cooling water drain line of the flammable gas control system, which was not usually used, with the wit of operators. These operations temporarily suppressed the increase in the primary containment vessel (PCV) temperature and pressure to obtain sufficient time enough to restore the reactor heat removal function.

Meanwhile, the line was configured for the PCV pressure vent where the reactor heat removal function remained out of action for an extended period. The so-called "feed and breed" line via the MUWC alternate water injection and the PCV pressure vent should be completed with a single opening operation of the outlet valve on the S/C side. As the S/C cooling by the RHR started before the S/C pressure reached the PCV pressure vent exercise pressure, the PCV vent operation was not actually conducted.

4.1.5.3 Restoration Plan Based on the Walkdown

The Fukushima Daini established an on-site organization for nuclear emergency preparedness, headed by the Site Superintendent at the Seismic Isolation Building after the Great East Japan Earthquake and organized a system to provide information, requests for support, and engage in the post-accident restoration. In parallel with the operators' reactor cooling maintenance work after the earthquake and tsunami, the on-site organization for nuclear emergency preparedness planned to determine the level of damage of equipment through site confirmation and prioritize such work.

Still, when planning the site confirmation, there were incessant aftershocks and major tsunami warnings, and work on site without light and with considerable rubble and openings was very dangerous. As no paging system was available as a means of communicating an evacuation signal in the event of a tsunami, nor any PHS in buildings damaged by the tsunami, a disaster restoration group was not immediately dispatched to the site. It was at about 22:00 on March 11 when the procedure for communicating evacuation was established to assign messengers, etc., and when the disaster restoration group with safety apparatus started to check the damaged areas such as the seawater heat exchanger building near the sea.

The on-site organization for nuclear emergency preparedness determined the site situation, receiving reports from the disaster restoration group and decided on a policy to prioritize the emergency equipment cooling pumps with more minor water damage in the seawater heat exchanger building to efficiently restore the heat removal function. Pumps damaged by water were investigated and repaired, and it was decided to replace the damaged motors. In addition, as the power panel that supplied electricity to these pump motors stopped functioning due to the water damage, it was planned that the electricity would be provided to the waste treatment building that was not influenced by the tsunami, namely the power panel in the Unit 3 seawater heat exchanger building or the motors by directly connecting them to high-voltage power supply vehicles.

4.1.5.4 Emergency Material Procurement

The Fukushima Daini asked the On-Site Organization for Nuclear Emergency Preparedness of the TEPCO head office and the TEPCO Kashiwazaki Kariwa Nuclear Power Station (hereinafter the "Kashiwazaki Kariwa") for the emergency procurement of motors, high-voltage power supply vehicles, mobile transformers, and cables, all of which were required to conduct restoration activities, based on the restoration plan developed according to the walkdown results. In response to this request, the On-Site Organization for Nuclear Emergency Preparedness of TEPCO head office and the Kashiwazaki–Kariwa asked the places concerned to confirm whether they were in stock or whether there were any spare materials and equipment and planned to transport materials and equipment with specifications matching those requested by the Fukushima Daini by taking all possible measures, including by air and over land.

The materials and equipment concerned sequentially reached the Fukushima Daini by around 6:00 on March 13. Motors for the EECW(B) and the RHRC(D) at Unit 1 were transported by air by a Self-Defense Force plane from Toshiba Mie Factory to Fukushima Airport and then by land by Self-Defense Forces from the airport to the station. The Unit 4 RHRC(B) motor was transported by a partner's truck from the Kashiwazaki Kariwa. As for high-voltage power supply vehicles, which were requested as electric sources for Unit 1 and 4 EECWs(B), those possessed by the power transmission/distribution sectors of TEPCO branches were taken to the station by branch employees. In addition, the combined total length of temporary cable procured from TEPCO and partners' warehouses for four plants was about 9 km.

The land transportation involved encountered numerous problems and hindrances, which prevented any easy access to the station as follows.

First, numerous depressions, bumps, and other obstructions on the roads disturbed the post-earthquake traffic, which made it important to find out passable by surveying the roads surrounding the station as early as possible. A means of communicating this information to carriers should have been prepared. The best way was to establish a relay exchange point at which to meet carriers and lead them to the station, but this was difficult in the absence of a dedicated group.

Second, there was a need to obtain prior agreement from carriers on transportation to an area possibly contaminated by radioactivity. As many carriers were prevented to enter the area blockaded by the police, power station workers with licenses for heavy vehicles took over the responsibility and drove trucks to the power station. Related measures were also required for delivery without relying on carriers by establishing relay exchange points and a dedicated group.

4.1.5.5 Restoration of the Heat Removal Function and Achievement of Cold Shutdown

The first work to restore the emergency equipment cooling pump in the seawater heat exchanger building involved repairing the access routes to the building. Access to the building was difficult due to scattered drifts caused by the tsunami and the fact that asphalt of roads flowed out, but the workers in charge operated heavy machines, which were also used for other work, to remove drifts, temporarily restored roads with damaged asphalt with gravel, thus securing access routes to the building.

Subsequently, motors for the emergency equipment cooling pump were replaced and a temporary cable to the motor was laid. In particular, a temporary cable with combined length of about 9 km was laid by TEPCO employees and partners' workers, including a total of 200 supporters from the power transmission/distribution sectors, by around 23:30 on March 13. As well as laying the cable, the machine parts of the pump were confirmed, the motor was installed, and pumps, including those in Unit 1, were activated as soon as they were ready from 20:17 on March 13 onward. Subsequently, the RHR pumps (B) were sequentially activated, starting with Unit 1 at 1:24 on March 14, and Unit 4 RHR pumps (B) at 15:42, whereupon the Site Superintendent determined that all units had achieved restoration of the "loss of reactor heat removal function" event in accordance with Article 10 of the Act on Special Measures Concerning Nuclear Emergency Preparedness (Nuclear

Emergency Act). Moreover, to cool reactor water early as well as the S/C, cooling was temporarily conducted with a circulation line which injected water into the reactor from S/C via the RHR heat exchanger (B) by the RHR pump (B) and returned the reactor water to the S/C via the SRV. This lowered the S/C water temperature to below 100 °C; first in Unit 1 at 17:00 on March 14 and finally in Unit 4 at 7:15 on March 15, whereupon the Site Superintendent determined that all units had recovered from the state of the event in accordance with Article 15 of the Nuclear Emergency Act (loss of pressure-suppression function).

4.2 The Onagawa Nuclear Power Station

4.2.1 Overview of the Onagawa Nuclear Power Station

The Onagawa Nuclear Power Station of the Tohoku Electric Power Company, Inc. (hereinafter the "Onagawa") is located in Onagawa town, Oshika district, Miyagi prefecture and Ishinomaki city and has three BWR plants. Unit 1 is BWR-4 with Mark-I containment with rated electric output of 524,000 kW, Units 2 and 3 are BWR-5 with Mark-I advanced containment with rated electric output of 825,000 kW. These plants started to operate during the period 1984–2002.

4.2.2 Overview of the Earthquake and Tsunami

4.2.2.1 Observed Earthquake and Tsunami

Seismic intensity observed in the power station:	Level 6 weak
Earthquake acceleration:	567.5 Gal (seismometers for security check: the second basement of the Unit 1 reactor building) (The maximum previous earthquake acceleration: 251.2 Gal (August 16, 2005))
The maximum tsunami wave height:	About 13 m (value measured by tide gage)
The maximum tsunami arrival time:	At 15:29 on March 11, 2011 (43 min after the earthquake occurrence)

4.2.2.2 Earthquake Observation Record

The maximum acceleration value observed on all floors of Onagawa reactor building Units 1, 2, and 3 in The Great East Japan Earthquake partially exceeded the maximum response acceleration value for the design basis seismic ground motion Ss developed based on the revision of the Regulatory Guide for Reviewing the Seismic Design of Nuclear Power Reactor Facilities (September 2006) but were almost the same.

In addition, the response spectrum of the earthquake observation record of the site ground partially exceeded the response spectrum of the design basis seismic ground motion Ss, including the influence on the ground above the seismometer, although they were almost the same.

A seismic response analysis was conducted based on the earthquake observation record, the deformation of the aseismatic walls of reactor building Units 1, 2 and 3 was evaluated and the shear force exerted on the aseismatic walls on floors confirmed the reactor building functions had been maintained.

4.2.2.3 Tsunami Observation Record

The tsunami height observed by the tide gage in the Onagawa was the work reference level of Onagawa port base tide level for construction (O.P) + about 13 m. It was also confirmed that there were seawater intrusion marks partially on the sea side of the site, and that the tsunami height did not exceed the site height where the main facilities were located (O.P. + about 13.8 m).[1]

4.2.3 *Influence of the Seismic Ground Motion and Tsunami*

4.2.3.1 Collapse of the Unit 1 Heavy Oil Storage Tank

A patrol after the earthquake found that a heavy oil storage tank housing fuel for the backup boiler installed in the seaport had collapsed and heavy oil had spilled on the ocean side of the Unit 1 intake. Accordingly, the heavy oil was absorbed and recovered with oil adsorption mats and countermeasures were taken for the heavy oil diffusion out of the bay with oil fences. When the tanks collapsed, the backup boiler had already shut down and no oil had been supplied.

As this tank was installed at a place O.P. + 2.5 m, lower than the site height (O.P. + 13.8 m) where the main station equipment was installed, it was judged that the tank collapsed due to the tsunami.

[1] O.P. is port base tide level for construction of Onagawa Nuclear Power Station and OP $\pm$ 0 m is the Tokyo Peil (T.P.) −0.74 m. The heights of the tsunami and site are respectively values to be taken into consideration for crustal movements in and around the Onagawa station (– about 1 m) by the Geospatial Information Authority of Japan (GSI) released after the Great East Japan Earthquake. For example, the pre-earthquake site height was O.P. + about 14.8 m, but this report shows the post-earthquake site height: O.P. + about 13.8 m.

4.2.3.2 Fire on the Unit 1 High-Voltage Power Panel (M/C)

After the earthquake occurrence, the fire alarm sounded in the main control room, whereupon the operators headed to confirm the site state and found smoke from the basement of the turbine building at 15:30. Initially, the low visibility due to smoke prevented them from identifying the source, but subsequent site confirmation revealed that the regular system (non-safety grade) M/C on the first basement had been damaged by fire and was emitting smoke. The earthquake-resistant class of the regular system M/C is C class. As the earthquake and tsunami cut off roads in Oshika peninsula, firefighters had trouble reaching the station, and the in-house firefighting team fought the fire and extinguished it at 22:55 on the same day.

The following scenario was provided as a possible cause of the fire: because the circuit breaker lifted at a connecting position in the M/C was largely shaken by the earthquake shock, the disconnecting part of the breaker was damaged, whereupon neighboring structures in the M/C formed a short circuit and other problems; generating sparks which melted cable insulating coating in the M/C and emitted smoke. This fire influenced the functions of ten M/Cs arranged adjacently over a distance of about 10 m. The M/Cs where the fire occurred were replaced with circuit breakers with a horizontal type fixed mechanism and an earthquake-resistant structure.

The loss of regular system M/C function itself does not directly compromise the security of a nuclear power station. However, in case of the Onagawa Unit 1, attention had to be paid to the following three points from a security perspective:

(1) Though the external power supplies themselves remained intact, due to ground faults of the regular system M/Cs (1A), the startup transformers tripped and the external power supplies were lost as a result. After confirming that the startup transformers were not in an abnormal state, they were restored at 2:05 on March 12.
(2) Due to smoke emitted from the fire, it took time to identify where the fire had occurred. In addition, the workers had to temporarily evacuate the turbine building because a carbon dioxide fire extinguishing system was used.
(3) When the fire occurred in the regular M/C (1A), the control cable of the synchronization detection relay connected to the emergency D/G (A) was damaged and a ground fault occurred. Consequently, the synchronization detecting circuit was damaged during a manual start test of the emergency D/G (A) on April 1, leading to an operational shutdown of the emergency D/G (A). In other words, the problem with the regular system caused an indirect ripple effect on the emergency system. In the Onagawa Unit 1, the D/G automatically started up immediately after the earthquake and was in the state of no-load operation. After the startup transformer tripped, the operation state shifted to load operation and power was supplied to the emergency power supply system.

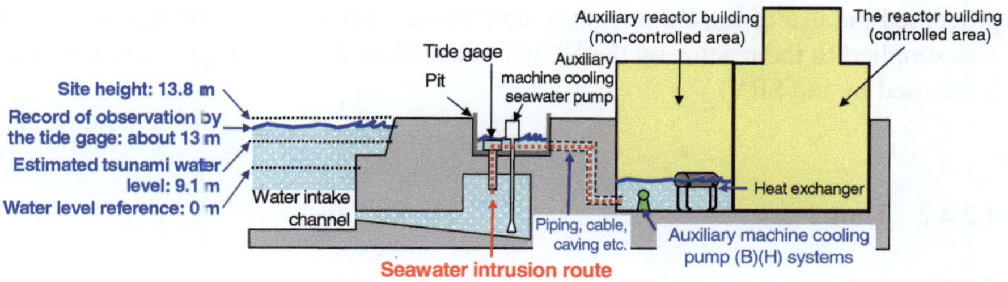

Fig. 4.2 Mechanism of seawater inflow into the auxiliary reactor building of the Onagawa Unit 2. (A tsunami water level of 9.1 m was estimated as the design basis. The Japan Society of Civil Engineers (JSCE) evaluated that the height would be 13.6 m in 2002. As the site height exceeded 13.6 m, it was assumed that no countermeasures would be required.)

4.2.3.3 Inundation of Unit 2 Nuclear Reactor Auxiliary Machine Cooling Water B-System Pump and the High Pressure Core Spray Auxiliary Machine Cooling Water System

A patrol after the earthquake recognized that seawater had flowed into the auxiliary machine cooling system heat exchanger chamber in a non-management zone on the third basement floor of the reactor building and that the nuclear reactor auxiliary machine cooling water B-system pump and the high pressure core spray (HPCS) auxiliary machine cooling system pump had been inundated. Although this stopped the two system pumps operating, as the A system pumps were sound, the nuclear power reactor and fuel pool were cooled without any problem.

It was assumed that the tide level was increased due to tsunami, and seawater had intruded into the auxiliary machine cooling system heat exchanger chamber and others through the pipe-penetrating portion after the upper lid of the tide gage box installed in the seawater pump room had been pushed up, and seawater flowing from there flowed into underpasses including piping through cable trays and the pipe-penetrating portion (see Fig. 4.2). Subsequently, the tide gage box was removed, and a closing plate was installed on the opening.

4.2.4 Response Before the Arrival of the Tsunami

4.2.4.1 Unit 1

At 14:46, severe earthquake tremors were detected, and the reactor automatically shut down. All control rods were normally inserted and reactor subcriticality was confirmed at 15:05. Immediately after the earthquake, external power supplies were secured, but short circuits and ground faults occurred in the regular M/C and the startup transformer, which received the external power supplies, stopped, which prevented the use of the water supply/condensate system pump, whereupon water was supplied to the reactor by the RCIC. In addition, the reactor pressure was also controlled by the SRV.

4.2.4.2 Unit 2

As the reactor was started up at 14:00 on the 11th and was in subcriticality immediately before the earthquake occurrence and the reactor water temperature was less than 100 °C, at 14:49 on March 11, the reactor reached cold shutdown due to "shutdown" operation of the reactor mode switch after the automatic shutdown.

4.2.4.3 Unit 3

Like Unit 1, all control rods were normally inserted due to the automatic shutdown of the reactor and the reactor subcriticality was confirmed at 14:57. Water was supplied to the reactor by the water supply/condensate system after the automatic shutdown of the reactor.

4.2.5 Response After the Arrival of the Tsunami

4.2.5.1 Unit 1

Water was injected to the reactor by the RCIC and subsequently by the control rod drive mechanism (CRD) following depressurization by the SRV. As the water injection by the CRD could maintain the reactor water level, the MUWC was not used to inject the water. Moreover, the reactor was cooled by the RHR and reached cold shutdown status at 0:58 on March 12.

4.2.5.2 Unit 2

Due to the tsunami, the pumps of the reactor auxiliary machine cooling water system (B), reactor auxiliary machine cooling seawater system (B) and high pressure core spray auxiliary machine cooling water system were flooded and stopped functioning. However, as (A) and (C) systems remained operational, it was continuously possible to dissipate the decay heat occurring from the reactor core. Moreover, although the cooling system operational shutdown meant D/G (B) and D/G (H) stopped functioning, D/G (A) remained operational.

4.2.5.3 Unit 3

As the turbine auxiliary machine cooling water pump shut down following seawater intrusion into the seawater pump area due to the tsunami, the reactor feedwater pump, which lost cooling water supply, was manually shut down, and water was supplied by the RCIC. The reactor pressure was controlled by the SRV, and after the nuclear power reactor depressurization, the water was injected by the MUWC. The reactor was cooled by the RHR and reached cold shutdown status at 1:17 on March 12.

4.2.6 Tsunami Countermeasures Before the Accident

4.2.6.1 Site Height Setting

As when constructing Unit 1, it was recognized from the start that tsunami countermeasures would be important when determining the site height, an in-house committee, including external specialists in civil engineering and geophysics, was established in Tohoku Electric Power Co., Inc. and engaged in discussion. It was assumed, according to evaluations based on documentary investigation and a hearing survey at the time, that the height of any tsunami near the power station would be assumed at around 3 m, but following discussions at the in-house committee, the conclusions were summarized as "(1) The site was sufficiently high to take anti-tsunami measures and (2) the height could be around O.P. + 15 m". The committee decided that the height of the first floor of the outdoor important civil engineering structure/the main building should be O.P. + 15 m and the site height should be O.P. + 14.8 m. After applying for the Unit 1 establishment license, the tsunami was evaluated based on the latest knowledge at the time when an application for the Unit 2/3 establishment license was made, and when JSCE established new tsunami evaluation technology, and in all cases, it was confirmed that the estimated tsunami height would be shorter than the site height. Crustal movements associated with the earthquake caused around 1 m of site subsidence and the site height was O.P. + about 13.8 m, but the tsunami (observed height: 13 m) did not exceed the height of the site where the main structures were established.

4.2.6.2 Enhancement of the Tide Embankment

In applying for the establishment license of the Onagawa Unit 2, the predicted tsunami height was reviewed and changed from around 3–9.1 m, using a numerical simulation technique. Thereafter, safety was considered for the slope of the site ground during the backwash of the tsunami, and protective work was carried out by building a 9.7 m high concrete block wall. It is considered that because this work was conducted in advance, the wall would be able to resist not only the first wave of any tsunami but also the second and subsequent waves and remain sound.

The Onagawa station did not install seawater pumps in the harbor zone near the sea surface but dug holes called pits on site about 13 m deep, 13.8 m in height, within which pumps were also installed to prevent water damage. The tsunami would have to exceed the site height to flood the pumps. In The Great East Japan Earthquake, the 13-m high tsunami attacked the station but did not exceed the site height, and the emergency seawater pumps were not submerged. Still, as mentioned before, some pumps stopped functioning due to the inundation.

4.2.6.3 Countermeasures for Backwash

The seawater pump room was designed to maintain a water source in case of backwash for a given period.

The enormous impacts of the tsunami were limited: a collapse of the Unit 1 heavy oil tank and the seawater inflow into the auxiliary reactor building of the Unit 2. However, though the observed tsunami height did not exceed the expected height beforehand, the tsunami overflow led to the loss of the safety system function, which shows the method used to study the seawater leak path should be reconsidered. In particular, various routes were identified as overflows, including trenches, cable tray penetrating portions, piping penetrating portions, sumps, water-tight doors, and elevator shafts. Here, internal overflow PRA and other measures are effective for identifying any leak path.

Incidentally, why did the inundation accident occur only at Unit 2, although it had almost the same design as Unit 3 and the height of both units was sufficient site against the tsunami? Seawater inundated Unit 2 from the tide gage installation zone in the seawater pump room. In applying for the establishment license for Unit 3, the installation of this tide gage was planned to trip the turbine according to the tide, following which, a similar tide gage was installed in Unit 2. A Unit 2 tide gage was additionally installed for the regular system turbine trip in the emergency seawater system area, from the perspective of an installation space. As the tide gage box was pushed up by the water pressure of the anaseism and seawater inundated the auxiliary machine cooling system heat exchanger chamber due to the tide gage opening, the functions of the pumps of the nuclear reactor auxiliary machine cooling water system (B) and the HPCS auxiliary machine cooling system were lost and likewise those of B- and H-system D/Gs. The Onagawa also took countermeasures for backwash from the very start, but failed to ensure sufficient consideration of anaseism pressure when additionally installing a tide gage in Unit 2. As described above, the major site height at Unit 2 exceeded the tsunami height, but the regular system subsequently impacted on the emergency system.

While cooling the D/Gs of Fukushima Daiichi Unit 6, which were additionally installed on a hill due to the installation space and undamaged by water, the tide gage of the Onagawa Unit 2 was inundated because the installation place was inadequate. In both cases, the aim was to improve safety, but these cases with opposite results show the importance of arranging and positioning equipment during safety design and the fact that, during additional installation and backfit, their influences should be more carefully considered.

4.3 The Tokai Daini Nuclear Power Station

4.3.1 Overview of the Tokai Daini Nuclear Power Station

The Tokai Daini Nuclear Power Station (hereinafter the "Tokai Daini") of the Japan Atomic Power Company (hereinafter the "Japan Atomic Power") is located in Tokai village, Ibaraki prefecture. A boiling water reactor (BWR-5) with Mark II containment was adopted and the plant has electric output of 1.1 million kW. Construction was started in 1973 and it entered operation in 1978.

4.3.2 Overview of the Earthquake and Tsunami

The maximum accelerations observed at the foundation boards of the power station's reactor building in The Great East Japan Earthquake were 225 Gal in a horizontal direction (EW) and 189 Gal in a vertical direction, which were smaller than the response value according to the design basis seismic ground motion (Ss) and in the response spectrum, generally smaller than the basis earthquake ground Ss and response spectrum at designing.

In addition, it was confirmed that the maximum water level of the tsunami caused by this earthquake was about Hitachi Port base tide level for construction (H.P.) + 5.5 m[2] (elevation + 4.6 m) at around 16:50 on March 11. During investigations into the inundation height and zone in the power station site, following consideration of the crustal movement based on enhanced accuracy of the leveling and GPS measurement, it was evaluated that the height was H.P. + 5.7 m (elevation + 4.8 m) to H.P. + 6.2 m (elevation + 5.3 m), and that the run-up height was around H.P. + 6.2 m (elevation + 5.3 m).

---

[2] H.P. $\pm$ 0.00 m is Hitachi Port base tide level for construction, which is 0.89 m below of Tokyo Peil (T.P.).

4.3.3 *Influence of the Seismic Ground Motion and Tsunami*

4.3.3.1 Influence of the Seismic Ground Motion on the Power Station

Damage to some equipment, including turbine equipment with low seismic importance (earthquake-resistant classes B and C), were acknowledged as due to this earthquake but not damage to equipment of importance in seismic design (with earthquake-resistant class As/A (New Earthquake-proof Guideline S class).

It was confirmed that following studies by Japan Atomic Power Company concerning the influence of seismic safety of key seismic design equipment using earthquake observation records obtained on floors in reactor buildings, the maximum acceleration in earthquake observation records of nuclear buildings, which are important in seismic design, was below the maximum response acceleration according to the seismic waves for design in the construction (hereinafter "approved design wave") and the design basis seismic ground motion Ss.

In addition, it was confirmed that for equipment and piping important in seismic design, the floor response spectrum in earthquake observation records in reactor buildings exceeded the floor response spectrum in accordance with the approved design wave in some periodic bands (about 0.65–0.9 s) on the second basement through the 6th floor. However, it was below the approved design wave in the natural period in key seismic design for major equipment and piping systems. As the key seismic design equipment and piping systems were designed using tolerance to ensure an elastic state for the approved design wave, it was evaluated that the equipment and piping systems were generally in an elastic state.

4.3.3.2 Influence of the Tsunami on the Power Station

As for the influence of the tsunami, the D/G cooling seawater pump 2C automatically shut down due to the inundation of the north emergency seawater pump room due to the tsunami at 19:52, about 5 h after the earthquake occurrence. D/G and RHR cooling seawater pumps have redundancy and were separately arranged in the south and north emergency seawater pump rooms. It was confirmed that the D/G cooling seawater pump 2C installed in the north room, where seawater intruded, was submerged and automatically shut down.

For the emergency seawater pump rooms, a new sidewall up to H.P. + 7.00 m (elevation + 6.11 m) was established and sealing work (to prevent inundation) on the penetrating portion of the wall was applied to the outside of the existing sidewall raised to H.P. 5.80 m (elevation + 4.91 m) as tsunami countermeasures, and the ground of the north emergency water pump room was easily inundated following watertight work of the cable pit lid buried on the ground (to prevent the lid from rising and making the lid watertight) under construction and because the cable pit perimeter was excavated, and seawater flowed between the newly

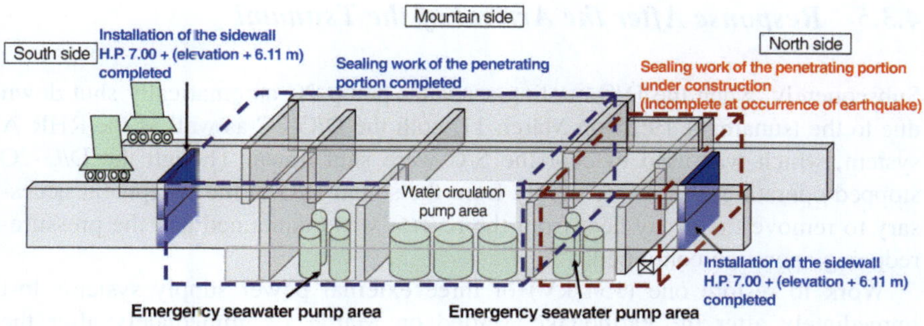

Fig. 4.3 Implementation status of sealing work of the penetrating portion into the emergency seawater pump room

established sidewall and the old partition wall, rode over the previously installed partition wall, and flowed into the north emergency seawater pump area.

Moreover, the inundated D/G cooling seawater pump 2C was restored 10 days after the water damage.

Figure 4.3 shows the implementation status of the sealing work of the penetrating portion into the emergency seawater pump room

4.3.4 *Response Before the Arrival of the Tsunami*

In the Tokai Daini, the main generator automatically stopped immediately after the occurrence of the earthquake because the turbine bearing was severely shaken and the closure of the main steam stop valve triggered the automatic shutdown of the reactor. Moreover, during the shutdown of the power station, three external power supplies, which supplied AC power to the station (two of 275 kV and one of 154 kV (reserved), simultaneously lost functions, but three D/Gs automatically started up and succeeded in supplying the required power to the equipment to safely shut down the station.

The decay heat from the reactor after its automatic shutdown was removed by channeling high pressure steam from the SRV installed on the main steam pipe to the S/C and cooling S/C water with the two RHR systems because the removal by the condenser was not expected when the external power supply system stopped functioning.

As for water injection into the reactor, the trip of the main generator closed the main steam stop valve and the loss of water supply rapidly lowered the reactor water level immediately after the reactor shutdown, and the HPCS and the RCIC automatically started up. Subsequently, the water level and pressure of the reactor pressure vessel (RPV) were continuously adjusted by the RCIC pump and the SRV, and the reactor pressure-reducing operation was conducted.

4.3.5 Response After the Arrival of the Tsunami

Subsequently, when the D/G cooling seawater pump 2C automatically shut down due to the tsunami at 19:25 on March 11, both the D/G-2C as well as the RHR-A system, which was used to cool the S/C were shut down. Though the D/G-2C stopped operating, the power supply from the sound D/G to the equipment necessary to remove the decay heat from the reactor was maintained and the pressure-reducing operation continued.

Work to restore one (154 kV) of three external power supply systems lost immediately after the earthquake, started on March 12 immediately after the earthquake, the station received power at 19:41 on March 13, and the cold shutdown was achieved at 0:40 on March 15 after the functional recovery of the residual heat removal system, which shut down due to the tsunami.

Next, as far as PCV cooling is concerned, external power supplies were lost immediately after the earthquake, but the D/G maintained the required power supply to the equipment to safely shut down the station, while the cooling of the S/C, PCV pressure and temperature were all maintained by the two RHR systems. In addition, immediately after the earthquake, water was injected into the reactor by the HPCS and RCIC pumps. As the pump water source in the early stage was the condensate storage tank (CST), the S/C water level rose up immediately after the earthquake. The issues for the S/P water level included the continuous receipt of water and treatment of received water. Though the RHR pump supplied water to the waste processing system, the processing functions were lost due to the loss of external power supply. Moreover, the power supply interchange with the waste processing system was required to recover the CST water level by injecting water into the reactor immediately after the earthquake. Methods used to secure the power supply interchange with the waste processing system included low-voltage power supply vehicles arranged immediately after the earthquake, air-cooled D/G for Tokai Power Station decommissioning measures, and gas turbine generators for buildings for emergency safety measures (seismic isolation) installed to reflect knowledge acquired in the Niigata prefecture Chuetsu-Oki earthquake. However, considering the equipment layout and the convenience of the power panel connection, the power supply interchange with the waste processing system was conducted via a supply from the gas turbine generator mounted on the building rooftop for emergency safety measures.

In the Tokai Daini, since one of the D/Gs shut down due to the tsunami, one PCV residual heat removal (RHR) system shut down, but it had redundancy and its safety function was maintained. As for the switch to the reactor cold shutdown, considering the fact that power station parameters were maintained in a stable condition and there was hope that external power supplies would be restored in the event of the loss of external power supply and one D/G shutdown, and examining risks in moving to the cold shutdown at an early stage, such as securing a high pressure coolant injection system (HPCI system) to the reactor and the occurrence of secondary trouble in case of the power supply interchange from the sound D/G to the interrupted emergency bus line and by selecting the switch to the reactor shutdown cooling mode when the RHR stopped following the restoration of external power supplies, cold shutdown was achieved at 0:40 on March 15.

In addition, as a result of one D/G cooling seawater pump shutdown, an overview of damage to the equipment in the seawater pump room was checked while a major tsunami warning was issued, and it was confirmed that the north, not south seawater pump room had been inundated. Consequently, the seawater pump room was drained in an effort to prevent the damage spreading.

4.3.6 Tsunami Countermeasures Before the Accident

As for the tsunami evaluation and measures by the Tokai Daini, after the application for reactor establishment license, steps were taken as well as voluntary evaluation as required, based on the knowledge at the time and tsunami evaluation trends in Japan. When a reactor establishment license for the Tokai Daini was applied, there was no clear tsunami standard, and the station was designed based on past documents and peak tide heights in adjacent areas. The estimated tsunami height was H.P. + 2.35 m, the tide level of Kanogawa Typhoon in 1958, which was recorded as the highest in history.

Subsequently, spurred by the occurrence of the tsunami associated with Hokkaido-Nansei-Oki Earthquake in 1993, ministries and agencies proceeded to examine guidelines on tsunami hazard prevention. As the tsunami assessment result of H.P. + 5.3 m was obtained in 1997 as a voluntary review in Japan Atomic Power based on the examination progress, another protective wall of H.P. + 5.8 m was installed in the north emergency seawater pump room. Moreover, in 2002, the Japan Society of Civil Engineers (JSCE) issued the "Tsunami Assessment Method for Nuclear Power Plants in Japan"; according to which the height was H.P. + 5.75 m.

In addition, the tsunami scale used the "Maps of Estimated Tsunami Inundated Area along the Prefectural Coast" released by Ibaraki prefecture in 2007 and reflected in the seismic safety assessment as new knowledge and the tsunami was assessed to obtain the assessment result of H.P. + 6.61 m as well as raising the sidewall of the emergency seawater pump room to H.P. + 7.00 m. This measure was a factor to protect important equipment for safety from the tsunami associated with the Great East Japan Earthquake.

Due to the tsunami, the north seawater pump room was inundated, leading to the automatic shutdown of one D/G seawater pump and manual shutdown of D/G-2C. In addition, one RHR system stopped functioning. Conversely, the south area was not inundated, which meant the station succeeded in an automatic startup after the loss of the external power supply, but one of three D/Gs was lost. Here, the north emergency seawater pump room had different results from the south because the station happened to raise the sidewall of the pump room following the tsunami assessment from Ibaraki prefecture as mentioned above, but the construction work was only completed on the south side, and not the north, from where the water came. The continuous improvement was ongoing in the Tokai Daini resulting in preventing severe accident, and the importance of continuous improvement by incorporating new knowledge was recognized.

4.4 Summary Comparison

Chapter 4 summarized an overview of events in the Fukushima Daini, Onagawa, and Tokai Daini Power Stations. Table 4.1 summarizes the damage conditions of the station equipment and Table 4.2 covers the history of tsunami estimation.

The tsunami and seismic ground motion influenced the safety functions in the Fukushima Daini, Onagawa, and Tokai Daini Power Stations, but the degree of impact was smaller than that in the Fukushima Daiichi. Thanks to restoration measures, including accident management (AM) measures, all plants achieved the cold shutdown. By analyzing and examining the restoration measures in these plants, beneficial knowledge was attained for the AM measures. The following describes the examination perspectives in Chap. 6.

First, it is important to comprehensively consider external events to ensure safety. In designing protection against various external events, there are issues: what hazard should be assumed as a design basis and to what extent should the protection design be prepared. The issue of external events (natural disasters) is handled in Sect. 6.6. It is important for the protection design to "thoroughly consider and take more intensive measures". For example, in the Onagawa, handrail bars were installed in the main control room to enable monitoring and control under a stable condition in case of earthquake, and the seawater pump room was made into pits to accommodate even a major backwash as anti-tsunami measures.

Conversely, the human mindset is limited. It is taken for granted that a proper safety design should be implemented after thoroughly considering what can be anticipated but it is also important to "prepare flexible countermeasures, expecting the unexpected", for example, preparing a portable power supply system. This management issue is discussed again in Sect. 6.5.

Next, "the safety principle, which has long been important, is a priority, as expected".

The top priority is "defence in depth", which will be collectively discussed in Sect. 6.3.

One of the important lessons learned from a bird's eye view of events at the four stations, including the Fukushima Daiichi, is the importance of "continuous improvement". In analyzing the accident, we tend to focus on faults and forget successes. Still, the Fukushima Daiichi Unit 5 reactor, which lost all AC power supplies, avoided the worst case thanks to power supply interchange from Unit 6, which was improved according to the AM. If the earthquake and tsunami had attacked the stations 20 years ago, the consequences would have been even more

Table 4.1 Equipment damage conditions due to earthquake and tsunami at sites and plants of the Fukushima Daiichi, Fukushima Daini, Onagawa, and Tokai Daini

Site	Unit	External power supply[a]	D/G	DC power	Power supply vehicle	Seawater cooling water system[f]	M/C (High-voltage power panel)		P/C (Low-voltage power panel) (Descriptions in brackets indicate systems under construction)		Core damage
							Emergency	Regular	Emergency	Regular	
Fukushima Daiichi	Unit 1	275 kV: × 66 kV: × (seven lines in total)	×[b]	×	Partial use	×	×	×	×	×	Yes
	Unit 2		×[b]	×		×	×	×	2/3	2/4	
	Unit 3		×[b]	O → exhausted		×	×	×	×	×	
	Unit 4		×[b]	×		×	×	×	1/2(1)	1/1(1)	No
	Unit 5		×[b]	O		×	×	×	×	2/7	
	Unit 6		1/3[b]	O		×	O	×	O	×	
Fukushima Daini	Unit 1	500 kV:1/2 66 kV: × (Four lines in total)	×[c]	3/4	Partial use (Secure external power supply & D/G)	×	1/3	O	1/4	O	No
	Unit 2		×[c]	O		×	O	O	2/4	O	
	Unit 3		2/3[c]	O		For D/G(H): O For RHR: 1/2	O	O	3/4	O	
	Unit 4		1/3[c]	O		For D/G(H): O For RHR: ×	O	O	2/4	O	
Onagawa	Unit 1	275 kV:1/4 66 kV: × (Five lines in total)	O	O	(Secure external power supply & D/G)	O	O	1/2	O	O	No
	Unit 2		1/3[d]	O		For D/G(H): × For RHR: 1/2	O	O	O	O	
	Unit 3		O	O		O	O	O	O	O	

(continued)

Table 4.1 (continued)

Site	Unit	External power supply[a]	D/G	DC power	Power supply vehicle	Seawater cooling water system[f]	M/C (High-voltage power panel)		P/C (Low-voltage power panel) (Descriptions in brackets indicate systems under construction)		Core damage
							Emergency	Regular	Emergency	Regular	
Tokai Daini		275 kV: × 154 kV: × (Three lines in total)	2/3[e]	○	Spare secured (D/G secured)	For D/G: 2/3 For RHR: ○	○	○	○	○	No

Some damaged equipment is shown in the "number of sound equipment/total number"

[a]Within one—a few—days, some external power supplies had been restored in the Fukushima Daini, Onagawa, and Tokai Daini Power Stations. One circuit of 66 kV was shut down for inspection in the Fukushima Daini

[b]D/Gs themselves at 1/2 of Units 1 and 4 and 5 were undamaged by water. (The loss was due to the indirect function (submersion of related equipment, including the auxiliary machine cooling system and the M/C)). Units 2, 4, and 6 B systems were air-cooled

[c]In Unit 1, water intruded from the D/G blower air supply opening and others and reached the D/G through the D/G blower. In Units 2–4, water rarely intruded into the annex attached to the reactor building. (The loss was due to the indirect function (submersion of related equipment including the auxiliary machine cooling system and the M/C))

[d]The A system was sound. The nuclear reactor auxiliary machine cooling water B-system pump and the HPCS auxiliary machine cooling water system were lost due to inundation from the opening of the seawater pump room (tide gage)

[e]The seawater pump room was inundated, leading to the automatic shutdown of one D/G seawater pump 2C and manual shutdown of D/G-2C

[f]The seawater system function was lost (including the operational shutdown of the auxiliary machine cooling water system pump)

Seawater pump installation sites: in the Fukushima Daini, the heat exchanger building in the coastal area (inundation from the entrance for accommodating large objects (excluding the Unit 3 south building)), in the Onagawa, the seawater pump room made into pits by digging from the site height (inundation via piping and cable tunnel into the annex attached to the reactor building (non-management zone)), and in the Tokai Daini, the seawater pump room with the sidewall in the coast area was raised up as tsunami countermeasures (inundation via the wall penetrating portion with some work incomplete)

Table 4.2 Details of tsunami estimation at sites of the Fukushima Daiichi, Fukushima Daini, Onagawa, and Tokai Daini

Site	Site height of main building	Application for reactor establishment license	History of estimation after providing the license				
			2002	2007	2007	2009	2011
			Technique by the Japan Society of Civil Engineers (JSCE)	Tsunami estimated by Ibaraki Prefecture	Tsunami estimated by Fukushima Prefecture	The Latest geographical feature and tide level conditions	Tsunami height observation values due to the Great East Japan Earthquake
Fukushima Daiichi	(Units 1–4) O.P. + 10 m[a] (Units 5 and 6) O.P. +13 m	O.P. +3.122 m 1966 (Unit 1)	O.P. + 5.7 m The tsunami where the Fukushima off-shore wave peaked.	O.P. + 4.7 m	O.P. about + 5 m	O.P. + 6.1 m	Tsunami height: O.P. +13.1 m Inundation height: O.P. + 15.5 m
			Measures were taken including the raising of seawater pump	Countermeasures not required	Countermeasures not required	Measures were taken including the raising of seawater pump	
Fukushima Daini	O.P. + 12 m	O.P. +3.122 m 1972 (Unit 1) O.P. + 3.705 m 1978 (Units 3/4)	O.P. + 5.2 m	O.P. + 4.7 m	O.P. + 5 m	O.P. + 5.0 m	Tsunami height: O.P. +7–8 m Inundation height: O.P. + 14.5 m
			Countermeasures were taken such as water tightness of buildings	Countermeasures not required	Countermeasures not required	Countermeasures not required	
Onagawa	O.P. + 14.8 m[a] (+13.8 m[b])	O.P. +2–3 m: 1970 (Unit 1, documentary investigation) O.P. + 9.1 m 1987 (Unit 2, numerical computation)	O.P. + 13.6 m The tsunami where the Sanriku off-shore wave peaked	–	–	–	Tsunami height: O.P. +13.0 m
			Countermeasures not required	–	–	–	

(continued)

Table 4.2 (continued)

Site	Site height of main building	Application for reactor establishment license	History of estimation after providing the license				
			2002	2007	2007	2009	2011
			Technique by the Japan Society of Civil Engineers (JSCE)	Tsunami estimated by Ibaraki Prefecture	Tsunami estimated by Fukushima Prefecture	The Latest geographical feature and tide level conditions	Tsunami height observation values due to the Great East Japan Earthquake
Tokai Daini	H.P. + 8.9 m[c]	H.P. + 2.35 m 1971	H.P. + 5.75 m	H.P. + 6.61 m	–	–	Tsunami height: H.P. + 5.5 m Tsunami height: H.P. + 6.2 m
			Countermeasures not required	Measures were taken including the raising of walls surrounding seawater pump	–	–	

[a]O.P. ± 0.00 m means 0.74 m below Tokyo Peil for Onagawa (Onagawa Port base tide level for construction), and 0.727 m below Tokyo Peil for Fukushima (Onahama Port base tide level for construction)
[b]Considering the ground subsidence due to the earthquake
[c]H.P. ± 0.00 m means 0.89 m below Tokyo Peil for the Tokai Daini (Hitachi Port base tide level for construction)

severe. This is an example of "continuous improvement" serving its purpose. The concept of continuous improvement is reconsidered in Sect. 6.4.

Another important lesson is the principle that "primary responsibility belongs to the reactor licensee" and the importance of "accident response by management". Naturally, the licensee is wholly responsible for the safety design and daily safety management. In addition, responses to an important event are also important. For example, when D/G-2C shut down, the Tokai Daini initially checked the status of other pumps, focusing on the tsunami and other hazards, considered the procedure to prevent the damage spreading and settle events, reducing risks. In retrospect, they did what they should. Such responses were also conducted in other stations as well as the Tokai Daini. The on-site workers were those most familiar with the site, and site confirmation is a rule during earthquakes, tsunamis, and other abnormalities. As abnormality progresses and a higher level of defence in depth is required, the management plays a more significant and demanding role with higher expectations. There is a need to reconfirm that assessing the site state correctly and suitably responding to the same will help prevent any abnormality spreading, reduce its impact, and facilitate convergence in the scope beyond design. The problem of accident management is discussed in Sect. 6.5.

Chapter 5
Off-Site Response

Abstract The off-site response at the accident was one of the key elements of the defence in depth described in the plant design. In Japan, off-site accident responses were based on the assumption of the 1999 JCO accident as the worst accident and nuclear disaster drills are annually implemented. However, off-site support for on-site severe accident measures was scarcely considered. In principle, when an accident occurs, the emergency organization outside the site must assume the role of the fourth stage of defence in depth: “to mitigate the consequences of the accident”. In other words, the off-site organization must offer human and material support for measures taken on-site to mitigate the consequences of the accident. Consequently, off-site accident responses by the Government Nuclear Emergency Response Headquarters and relevant organs were extremely confused. This confusion was exacerbated by the breakdown in the information communication function, due to the impact of the complex disaster. At an early stage, the relevant teams could not communicate with each other at all, and coordination among them was not secured, which caused many problems. For example, during the evacuation, many residents requiring support became victims. Consequently, as the evacuation was conducted before a large emission of radioactive materials, the direct influence of radioactivity could be prevented. However, the core meltdown at Units 1–3 in the Fukushima Daiichi NPP could not be prevented; nor could the radioactive materials emitted from the nuclear power reactors be prevented from polluting the environment and having a profound impact on the society and economy. There were many lessons learned and problems to be solved.

Keywords Environment contamination • Equipment/technical support • Food safety • Integrated management • Off-site response • Residents’ evacuation

The off-site response at the accident was one of the key elements of the defence in depth described in the plant design. In Japan, off-site accident responses were based on the assumption of the 1999 JCO accident as the worst accident and nuclear disaster drills are annually implemented. However, off-site support for on-site severe accident measures was scarcely considered. In principle, when an accident

The Fukushima Daiichi Nuclear Accident: Final Report of the AESJ Investigation Committee, DOI 10.1007/978-4-431-55160-7_5

occurs, the emergency organization outside the site must assume the role of the fourth stage of defence in depth: "to mitigate the consequences of the accident". In other words, the off-site organization must offer human and material support for measures taken on-site to mitigate the consequences of the accident. Consequently, off-site accident responses by the Government Nuclear Emergency Response Headquarters and relevant organs were extremely confused. This confusion was exacerbated by the breakdown in the information communication function, due to the impact of the complex disaster. At an early stage, the relevant teams could not communicate with each other at all, and coordination among them was not secured, which caused many problems. For example, during the evacuation, many residents requiring support became victims. Consequently, as the evacuation was conducted before a large emission of radioactive materials, the direct influence of radioactivity could be prevented. However, the core meltdown at Units 1–3 in the Fukushima Daiichi NPP could not be prevented; nor could the radioactive materials emitted from the nuclear power reactors be prevented from polluting the environment and having a profound impact on the society and economy. There were many lessons learned and problems to be solved.

Figure 5.1 shows an overview of the off-site accident responses when the accident occurred and how the off-site accident responses could be classified into three functions. Namely, the left row shows overall governance in the event of accident function, the middle row the fourth stage of the defence in depth "mitigating the consequences of the accident" function, and the right row, the fifth stage of the defence in depth "emergency measures" function. This chapter shows what activity was actually performed.

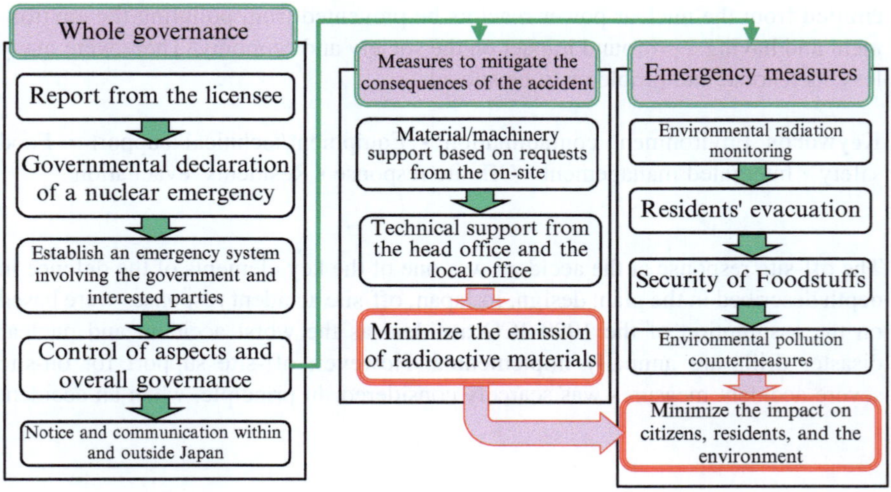

Fig. 5.1 Accident response conducted off-site in the event of a nuclear disaster

5.1 Emergency Response Plan Prior to the Accident

The emergency response plan for nuclear emergency in Japan was based on "Measures related to Nuclear Power Stations to Be Taken for the Time Being" determined in July 1979 by the Central Disaster Management Council. It specified the role of the Government, e.g. in developing an emergency communication system between the national and local governments, organizational expert support systems, including the "Emergency Technical Advisory Body" in case of an emergency at a nuclear power station, and an emergency monitoring system and emergency dispatch system of medical treatment staff. The Nuclear Safety Commission (hereinafter the "NSC") decided on "Emergency Preparedness Guide around Nuclear Power Plant and other facilities" (hereinafter the "emergency preparedness guide") in June 1980, the year after the TMI-2 accident. The "emergency preparedness guide" summarized the review results of technical and specialized matters to focus on specific aspects of a nuclear emergency and to smoothly implement emergency response activities around nuclear power stations.

After the Great Hanshin Awaji Earthquake in January 1995, the Disaster Management Basic Plan based on the Disaster Countermeasures Basic Act (hereinafter the "Basic Act") specifies detailed responses by disaster type, while the "10th Volume of Nuclear Emergency Preparedness" was added to further clarify the responsibilities and roles of agencies and organizations associated with the nuclear emergency preparedness. In response to the criticality accident at the JCO uranium reprocessing facility in Tokaimura (hereinafter the "JCO accident") in 1999, the "Act on Special Measures Concerning Nuclear Emergency Preparedness" (hereinafter the "Nuclear Emergency Act") was newly enacted and promulgated as a special law of the Basic Act and the "Act on the Regulation of Nuclear Source Material, Nuclear Fuel Material and Reactors" (hereinafter the "Nuclear Act"). At the extraordinary Diet where the Nuclear Emergency Act was enacted, related government offices cooperate to establish the budgetary steps and measures necessary to strengthen communication and contact functions, reinforce radiation monitoring, develop off-site centers, materials and equipment for nuclear emergency preparedness, and radiation emergency medical system.

As stated above, the emergency response plan prior to the accident, with the Basic Act and Nuclear Emergency Act at the top, clarified the responsibilities and roles of relevant organs in the Basic Plan and made legislative preparations, but failed to provide off-site support assuming reactor accidents, more specifically, sharing responsibilities and roles among relevant organizations for measures related to the fourth layer of the defence in depth: "measures to mitigate the consequences of the accident" such as containment venting and water injection.

Lessons learned from the accident at the Fukushima Daiichi Nuclear Station identified several issues on the emergency response plan prepared prior to the accident, which are explained in detail in Chap. 6.

5.2 Overview of Emergency Actions Taken in the Event of the Accident

5.2.1 Initial Response Actions During an Emergency

The government started up the Emergency Preparedness Headquarters to respond to the disaster immediately after 14:46 on March 11, 2011, when the earthquake occurred and promptly also started up the emergency response headquarters for this extraordinary and intense disaster at 14:50. As all AC power was lost at Units 1–5 in the Fukushima Daiichi Nuclear Power Station at 15:42 on March 11, the licensee reported the event in accordance with "Article 10 of the Nuclear Emergency Act" to the Nuclear and Industrial Safety Agency (hereinafter the "NISA"). At 16:36, reactor Units 1 and 2 could no longer be cooled, and at 16:45, the licensee reported the occurrence of the "loss of cooling function event in accordance with Article 15" of the Nuclear Emergency Act to the NISA. In response, at 17:45, the NISA started the escalation process according to Article 15, while at 18:22, the Minister of Economy, Trade and Industry escalated the declaration of the nuclear emergency situation to the Prime Minister. At 19:03, the Prime Minister issued a declaration of a nuclear emergency situation and started up the Government Nuclear Emergency Response Headquarters and the local Nuclear Emergency Response Headquarters. Table 5.1 shows what happened from the occurrence of the earthquake through to the startup of the Government Nuclear Emergency Response Headquarters. As shown in Sect. 3.2, the post-accident analysis estimated that the first core damage at Unit 1 started before 19:00 on March 11, which means the core damage had already started when the government headquarters started up.

5.2.2 Urgent Protective Actions for Residents (Evacuation, etc.)

Nuclear emergency drills, which were frequently carried out after the JCO accident, had been adopting a scheme for comparing dose-prediction results from Emergency Response Support System (ERRS) and System for Prediction of Environmental Emergency Dose Information (SPEEDI) with criteria for protective actions such as evacuation and sheltering and defining the area in which evacuation and sheltering should be implemented. However, in the Fukushima Daiichi accident, evacuation and sheltering were not implemented and expanded based on the previous scheme. The evacuation has been implemented in the area beyond the Emergency Planning Zone (EPZ). The earthquake and tsunami caused significant confusion of communicating information to residents and securing transportation, and resulted in a delay of decisions on urgent protective actions and repeated changes of refuges.

Table 5.1 Events concerning the startup of the emergency preparedness organization immediately after the accident

Time	Event	Events concerning the Startup of the Emergency Preparedness Organization.
March 11 14:46	Occurrence of the earthquake	The Ministry of Economy, Trade and Industry (METI) established the earthquake emergency response headquarters (disaster response)
14:50		The government established the emergency response headquarters for extraordinary and intense disaster with the Prime Minister as chief (disaster response)
15:42		TEPCO reported the specific matter in accordance with Article 10 of the Nuclear Emergency Act (the total AC power loss)
After 15:42		The METI established the METI Nuclear Disaster Alert Headquarters in the METI Emergency Response Center (ERC) and its nuclear disaster local alert headquarters in the Off-Site Center (OFC) in Okuma town
16:00		The Nuclear Safety Commission established the Emergency Technical Advisory Body
16:36		The Nuclear Emergency Response Headquarters at the Prime Minister's Office was established and an emergency team was called
At around 16:45		TEPCO reported the specific matter in accordance with Article 15.1 of the Nuclear Emergency Act (inability to conduct water injection of the Emergency Core Cooling System)
At around 16:55		TEPCO said the report of the occurrence of the specific matter at Unit 1 was canceled
17:00–		Upon a request from the Prime Minister, the NISA, emergency team and TEPCO all explained the circumstances
At around 17:12		TEPCO reported the occurrence of the specific matter at Unit 1 again
At around 17:35		The Minister of Economy, Trade and Industry approved the declaration of a nuclear emergency situation
At around 17:42		The Minister of Economy, Trade and Industry reported to the Prime Minister and requested approval to declare a nuclear emergency situation
18:12		The explanatory meeting was suspended because the Prime Minister had to attend a political meeting among leaders of the ruling and opposition parties
19:03		After the political meeting, the explanation resumed. After getting approval from the Prime Minister, a nuclear emergency situation was declared. The Nuclear Emergency Response Headquarters, headed by the Prime Minister, was established in the Prime Minister's office, the local Nuclear Emergency Response Headquarters at the off-site center, and the secretariat of the Nuclear Emergency Response Headquarters at the METI/ERC

5.2.2.1 Urgent Protective Actions After the Accident

At 20:50, the Fukushima Governor independently instructed mayors of Okuma and Futaba towns to evacuate residents living within a 2-km radius of the Fukushima Daiichi Nuclear Power Station before the order by the Government. The 2-km radius was selected according to the zone used in an ordinary nuclear emergency drill (The Government Accident Investigation Committee). At 21:23, the Government ordered residents within a 3-km radius to be evacuated and those within a 3–10-km radius to stay in-house. The 3-km radius was selected, in consideration of the 3–5 km precautionary action zone (PAZ) recommended by the International Atomic Energy Agency (IAEA) associated with precautionary measures before the release, which was considered effective in revising the emergency preparedness guide in 2007. Subsequently, due to the rise in Unit 1 containment vessel pressure and the delay of vent implementation, the evacuation zone was expanded to 10 km at 5:44 on March 12. The Government considered the changes in circumstances and selected a default EPZ in the emergency preparedness guide (The government Accident Investigation Committee). After a hydrogen explosion occurred at Unit 1 at 15:36 on March 12, the evacuation zone was expanded again to residents within a 20-km radius. No clear grounds for the selection of the 20-km radius were shown.

Following events at units such as the explosion of the Unit 3 reactor building at 11:01 on March 14, an explosive event shortly after 6:00 on the March 15, which was assumed to occur near Unit 2 during the accident, damage to the Unit 4 reactor building, and the occurrence of fire, the Government ordered residents within a 20–30-km radius of the Fukushima Daiichi Nuclear Power Station to stay in-house at 11:00 on March 15.

5.2.2.2 Release of Radioactive Materials

At the time of the accident, 24 monitoring posts had been installed in Fukushima prefecture. Excluding the Ohno monitoring post (established about 5 km from the site to the west), they became unavailable due to breaks in communication lines, loss of power, and tsunami outflows. However, portable monitoring posts in seven areas of Fukushima prefecture, the north part (Fukushima city), the central part (Kooriyama city), the southern part (Shirakawa city), Aizu area (Aizu Wakamatsu city), Minami Aizu area (Minami Aizu town), Soso area (Minami Soma city), and Iwaki area (Iwaki city) recorded radiation levels in the surrounding environment. Figure 5.2 shows the time variations of air dose rates in the seven areas in Fukushima prefecture.

The air dose rate rose at 17:46 on March 12 and 20 μSv/h was detected at 21:00 in Minami Soma city, which is located about 24 km away, to the north of the site. This means the radioactive plume caused by the vent of Unit 1 and a subsequent hydrogen explosion in the building was transported by the southerly wind and passed in the vicinity of Minami Soma city. After the plume had passed, the level

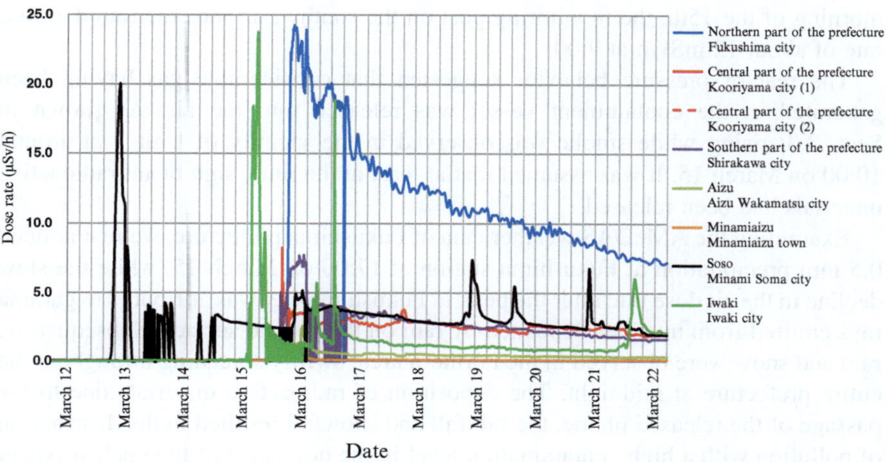

Fig. 5.2 Time variations of air dose rates in seven areas in Fukushima prefecture

declined to less than half. Subsequently, on March 13, it was stable at about 4 μSv/h, which was attributed to the radioactive nuclides depositing on the ground surface. The radioactive plume moved northwards, and a monitoring post in the Tohoku Electric Power Co., Inc., Onagawa Nuclear Power Station showed increases in the air dose at 20:40 and 22:20 on March 12 and 1:50 on March 13, and given a peak dose on March 13 of about 21 μSv/h. The Tohoku Electric Power Co., Inc. notified the relevant organs based on "Article 10 of the Act on Special Measures Concerning Nuclear Emergency Preparedness".

As shown in Sect. 3.3, a reactor core melt occurred at Unit 2 by the evening of March 14, and a subsequent vent operation failed. During this time, the Unit 2 drywell (D/W) pressure was high: over 700 kPa [abs] from the evening on the 14 h through to the morning on March 15, and showed little change. This means part of the steam and hydrogen generated in the reactor leaked from D/W through the reactor building, while the blowout panel of the Unit 2 building was open due to the hydrogen explosion at Unit 1. It was estimated that water vapor, hydrogen, and radioactive materials having leaked in the reactor building were easily released into the environment, which started to increase the on-site radiation level. The released radioactivity was transported to the south in the Hamadori area by the north wind, which was blowing at the time. At 0:00 on March 15, the air dose started to rise in Iwaki city (0.57 μSv/h) and a peak level of 23.72 μSv/h was detected at 4:00. Subsequently, the emitted plume went further southward, the air dose rose at 0:20 in Kitaibaraki city in Ibaraki prefecture (0.144 μSv/h) and a peak dose of 5.575 μSv/h was detected at 5:50. All monitoring stations/posts of the Japan Atomic Energy Agency in Tokaimura also showed a rise in the dose rate before around 1:00 on March 15 and the value peaking shortly after 7:00 on the same day. Subsequently, monitoring post values throughout the Kanto region rose and the radioactive plume seemed to reach Shizuoka prefecture. In addition, due to some events in the early morning of the 15th, the monitoring post on the southwest fence measured a dose rate of about 12 mSv/h at 9:00.

The reactor pressure behavior suggested that considerable gas having been generated in the containment vessel was released into the air. As shown in Sect. 3.3, rising white smoke was observed in the vicinity of Unit 2 at around 10:00 on March 15. It was assumed that at least at the time, significant radioactive materials had been released.

Examining the AMeDAS weather data of Fukushima prefecture, we first noticed 0.5 mm precipitation at Fukushima station at 17:00 on March 15, while the slow decline in the air dose rate after the peak in Fukushima city was attributed to gamma rays emitted from nuclides deposited by rain on the ground surface. Subsequently, rain and snow were observed in the northern area; swiftly spreading throughout the entire prefecture at midnight. The deposition of radioactive materials due to the passage of the released plume, the rainfall and snowfall resulted in the distribution of pollution with a high contamination level in the northwest of the nuclear power station. The Ministry of Education, Culture, Sports, and Science and Technology (MEXT) conducted aerial monitoring within a 100-km range of the Fukushima Daiichi Nuclear Power Station with the cooperation of the U.S. Department of Energy. The radiation dose measurement map created in accordance with the monitoring results and the distribution of the accumulation of Cs-134 and Cs-137 on the ground surface showed non-uniform distributions which were strongly influenced by rainfall when the radioactive plume passed.

5.2.3 Additional Early Protective Actions

5.2.3.1 Implementation of Additional Early Protective Actions

On March 15, high air dose rates were observed throughout the entire area of Fukushima prefecture, particularly in the northwest direction of the Fukushima Daiichi Nuclear Power Station. The air dose rates between 200–300 μSv/h were measured in the vicinity of Namie town, 20 km northwest of the station, at around 21:00. On March 17, the maximum value of 170 μSv/h was observed at point 32 about 30 km northwest of the station. In the vicinity of points 31 and 32 (Tsushima district, Namie town) and point 33 (Iitate village; Warabidaira and Nagadoro districts), where comparatively high air dose rates were observed in local areas, approximately 200 residents remained in their houses. (Subsequently, on April 6, the resident's safety team corrected the number: 128 in Tsushima district, and about 228 in Warabidaira and Nagao districts).

Subsequently, on March 30, the IAEA recommended the Japanese Government to careful assess the situation in the vicinity of Iitate village in the IAEA website of the Fukushima Update Log. The Prime Minister's office examined the possibility for expansion of the evacuation zones and the change of 20–30 km sheltering based on MEXT monitoring data.

On April 22, the Director-General of the Nuclear Emergency Response Headquarters issued an official instruction based on Paragraph 3, Article 20 of the Nuclear Emergency Preparedness Act. It indicated lifting of the stay in-house instruction to residents within a 20–30 km radius of the Fukushima Daiichi station on March 15. It was also ordered that residents in the deliberate evacuation areas should evacuate these areas within about a month as a rule and that those in evacuation prepared areas in case of emergency should prepare for evacuation or stay in-house. In addition, voluntary evacuation continued to be recommended for these areas, in which children, pregnant women, persons requiring care, and inpatients were told not to enter. Then some districts in Date and Minami Soma cities, where it was assumed that the annual integrated dose exceeded 20 mSv, were designated as specific spots recommended for evacuation in June afterwards. It would call their attention to the situation and provide information, and facilitate their evacuation.

5.2.3.2 Estimation of Residents' Dose at an Early Stage of the Accident

Upon a request from the Nuclear Safety Commission on March 25, 1,149 children were taken simple thyroid dose measurements with a NaI scintillation survey meter in Iwaki city, Kawamata town, and Iidate village. The measurement results of 1,080 children were screened on a level of 0.2 μSv/h or less set by the Nuclear Safety Commission (equivalent to 100 mSv of the thyroid dose equivalent to 1-year-old children). Fukushima prefecture conducted the Health Management Survey for the Residents after May. During the basic survey, medical questionnaire sheets were provided to refugees having evacuated within the prefecture from Namie town, Iitate village, and the Yamagiya district of Kawamata town to request their action records and intake situations of meals after March 11, compiled as a priority survey, then sheets were provided to all prefectural residents. Based on the action records and monitoring data, the cumulative external dose was estimated for the 4 month period from March 11 to July 11 in 2011. The collection rate of medical questionnaire sheets was 55 % in the priority survey and 22.9 % for all prefectural residents. The estimated accumulative dose peaked at 25 mSv for 119,450 persons excluding occupationally exposed personnel equivalent to or above 10 mSv for 117, and about 99 % were less than 10 mSv (as of August 31, 2012). In addition, the effective dose of the internal exposure was estimated on whole-body counter for 81,119 residents in the deliberate evacuation areas and Futaba county from June 7, 2011 to September 30, 2012 (assuming the initial acute ingestion on March 12, 2011). Two marked a maximum of 3 mSv, while 81,093 were 11 mSv or less.

5.2.4 Transition to Long-term Protective Actions

On December 26, 2011, the Nuclear Emergency Response Headquarters issued a basic concept for rearranging the restricted areas within a 20 km radius of the Fukushima Daiichi Nuclear Power Station and the deliberate evacuation areas.

Specifically, three areas were defined and response policies for each were shown: the areas where the annual integrated dose would be less than or equal to 20 mSv would be designated as "areas to which evacuation orders are ready to be lifted", the areas where the annual integrated dose would exceed 20 mSv and where residents are ordered to remain evacuated would be designated as "areas in which residents are not permitted to live", and the areas where the annual integrated dose would not be less than 20 mSv within 5 years and the current annual integrated dose would exceeded 50 mSv would be designated as "areas in which residents will face difficulties in returning". Based on this policy, the Nuclear Emergency Response Headquarters consulted and coordinated with Fukushima prefecture, related municipalities and residents, and determined the rearrangement of the areas on March 30, 2012.

5.3 Individual Issues of Emergency Actions

5.3.1 Residents' Evacuation

5.3.1.1 Residents' Evacuation Order

The recommendation or order for residents' evacuation are shown in the BADC and the Nuclear Emergency Act. The Nuclear Emergency Act stipulates that the Chief of the Government Nuclear Emergency Response Headquarters (Prime Minister) should issue residents' evacuation recommendations or orders to be given by mayors to other mayors and governors in areas where emergency response measures should be taken based on the BADC.[1] However, after the startup of the off-site center, this responsibility was going to be delegated to the Director-general of the Local Nuclear Emergency Response Headquarters, who collected the relevant information.[2]

In this accident though, the off-site center was more seriously damaged and, as shown in Table 5.2, could not fully fulfill its function. All the government evacuation orders were issued by the Government Nuclear Emergency Response Headquarters (Prime Minister) in Tokyo. The first evacuation order should have been given as soon as possible based on the accident circumstances, release estimation of radioactive materials and others. However, it was actually only given hours after the start of the reactor core meltdown at Unit 1, an explanation was added that the

[1] Articles 15–2 and 15–3 of the Act on Special Measures Concerning Nuclear Emergency Preparedness.

[2] Articles 15–8 and 15–9 of the Act on Special Measures Concerning Nuclear Emergency Preparedness and "Off-Site Center no Arikata ni Kansuru Kihonntekina Kangaekata ni tsuite Torimatome" (The Overview of Basic Concepts of What Off-Site Center Should Be) the Nuclear and Industrial Safety Agency, August 2012, the last line of p. 12 to the second line of p. 13 http://www.meti.go.jp/press/2012/08/20120831003/20120831003-3.pdf.

Table 5.2 Circumstances of the off-site center

Day	The circumstances of the off-site center (OFC)	Remarks
3/11	14:46 The earthquake occurred.	–
	Power failure → DG startup → 15:23 D/G shutdown → Restoration at around 1:00 on the next morning	–
	15:37 The maximum tsunami reached the station.	The METI alert headquarters was established. (Around 17:00, the Vice Minister, the NISA, and the NSC members departed.)
	The establishment of the local alert headquarters (Around 15:30), the director and three staff members returned from the station. Around 15:45, five staff members of the JNES operation support companies arrived.)	Around 17:35 The Minister of the Economy approved the "declaration of a nuclear emergency situation"
	–	19:03 The "declaration of a nuclear emergency situation" was issued and the Government Nuclear Emergency Response Headquarters was established.
	The local nuclear emergency response headquarters was established (generally managed by the director).	20:55 The Vice President's party of seven persons and two MEXT members moved from the Ministry of Defense by helicopter.
	20:00 Three Futaba police officers joined.	
	Around 21:20 Relocation to the adjacent prefectural nuclear power center.	
	22:40 Three prefectural residents' security group members joined.	
	22:10 The Vice Minister and others arrived at the JASDF Ohtakineyama Sub-Base.	
	Around 23:00 Deputy governors and prefectural staff members arrived.	
	Around 24:00 Arrived at the Environmental Radioactivity Monitoring Center of Fukushima.	
3/12	After OFC D/G was restored, around 3:17, staff members moved to the OFC and at around 5:00 five investigators returned from the Fukushima Daiichi to the OFC. At night on the 11th through the 12th, the police, the Self-Defense Forces, the JAEA, the NIRS, the Nuclear Safety Technology Center, and the Analysis Center joined. The TEPCO Vice President arrived before dawn.	
	06:50–08:00 The local chief and other three attended the visit of the Prime Minister.	
	10:30 The first joint council for nuclear emergency response (determining activity policies, including a grasp of the evacuation situation)	
	18:34 The second joint council for nuclear emergency response (checked on evacuation circumstances ordered in towns as 20-km evacuation expansion)	

(continued)

Table 5.2 (continued)

Day	The circumstances of the off-site center (OFC)	Remarks
3/13	13:30 The third joint council for nuclear emergency response (determined the decontamination screening standard)	
3/14	14:40 The fourth joint council for nuclear emergency response (final council in Okuma town OFC, discussing the situation of residents evacuation completion and others)	
	At around 20:40 The general meeting discussing the progress of consideration of relocation by the local headquarters.	
	At around 22:00 Preparation for relocation to the prefectural government. The outdoor dose was about 1,800 μSv/h.	
3/15	0:10 The OFC outdoor dose was around 100 μSv/h, and intermittent alarm.	
	09:26 An application for approval to relocate the OFC to the prefectural government was submitted to the Minister of Economy, Trade and Industry.	
	At around 11:00 The relocation of the OFC to the prefectural government was started.	

evacuation should be carried out "just in case" because the headquarters did not understand the accident situation, and moreover, the headquarters had to depend on mass media, including TV, to communicate the evacuation order. Consequently, there was a delay in communicating the information on the occurrence of the accident[3] although the municipalities gave evacuation orders quite rapidly and ensured they were made completely known to local residents within a very short time.[4]

5.3.1.2 The Number of Evacuees

In response to the Fukushima Daiichi accident, as shown in Table 5.3, the following evacuation categories were established by the Government Nuclear Emergency Response Headquarters on April 21 and 22 in 2011. An area within a 20-km radius

[3] According to a questionnaire survey by the National Diet of Japan Fukushima Nuclear Accident Independent Investigation Commission, only 20 % or less of residents in five towns surrounding the nuclear power station knew about the occurrence of the accident before the evacuation order before 6:00 on March 12. (p. 356 of the Final Report by the National Diet of Japan Fukushima Nuclear Accident Independent Investigation Commission).

[4] According to a questionnaire survey by the National Diet of Japan Fukushima Nuclear Accident Independent Investigation Commission, 90 % of residents in Futaba, Okuma, and Tomioka towns knew the evacuation order was given three hours after the evacuation order for residents within a 10-km radius issued before 6:00 on March 12. (p. 358 of the Final Report by the National Diet of Japan Fukushima Nuclear Accident Independent Investigation Commission).

Table 5.3 The Number of evacuees from the restricted, deliberate evacuation, and evacuation-prepared areas in case of emergency

	Restricted Area	Deliberate Evacuation Area	The Former Evacuation-Prepared Areas in Case of Emergency	Total	Main Evacuation Destination
Okuma town	11,500	–	–	11,500	Tamura and Aizu Wakamatsu cities and others
Futaba town	6,900	–	–	6,900	Kawamata town and Kazo city of Saitama prefecture
Tomioka town	16,000	–	–	16,000	Kooriyama city and others
Namie town	19,600	1,300	–	20,900	Nihonmatsu city and others
Iidate village	–	6,200	–	6,200	Fukushima city
Katsurao village	300	1,300	–	1,600	Fukushima city, Aizu Bange and Miharu towns, and others
Kawauchi village	400	–	2,500	2,900	Kooriyama city and others
Kawamata town	–	1,300	–	1,300	Kawamata town, Fukushima city and others
Tamura city	400	–	2,100	2,500	Tamura and Kooriyama city and others
Naraha town	7,700	–	50	7,750	Iwaki city, Aizumisato town, and others
Hirono town	–	–	5,100	5,100	Iwaki city and others
Minami Soma city	14,300	10	17,500	31,810	Fukushima and Soma cities, and others
Total	77,100	10,110	27,250	114,460	

from the Fukushima Daiichi was restricted, an area where the annual integration dose may reach 20 mSV (excluding the restricted area) was a deliberate evacuation area, and an area in which sheltering or evacuation is required in the event of an emergency (excluding restricted and deliberate evacuation areas) is an area prepared for evacuation in case of emergency. The evacuation-prepared areas in case of emergency were released on September 30, 2011. According to these evacuation orders, as of November 4, 2011, as shown in Table 5.3, about 114,460 evacuated.

The category of evacuees in Fukushima prefecture by evacuation destination is shown in Table 5.4. At the end of 2012, nearly 40 % of all evacuees (37 %) had evacuated outside Fukushima prefecture.

Table 5.4 Changes to the number of evacuees by evacuation destination

Investigation Time	A	B	C	Sub-total	The number of municipalities with whereabouts known	Evacuees Staying in Other Prefectures	Total
	Shelter (Community Centers, Schools, and Others)	Japanese Inns and Hotels	Houses (including Municipal/ Temporary/Private-sector Houses, and Hospitals)				
12/15/2011	18	22	95,506	95,546	47	59,933	155,439
6/7/2012	0	0	101,320		47	62,084	163,404
12/6/2012	0	0	98,215	98,235	48	57,954	156,189

The Reconstruction Agency web site (Browsed on June 16, 2013)
http://www.reconstruction.go.jp/topics/post.html

5.3.2 *Standard Limits for Radionuclides in Foods*

5.3.2.1 Japanese Regulation Values

As for food and drinking water, measures involving shipments, sales and intake restrictions based on provisional regulation values had been taken since March 17, 2011, but since April 1, 2012, new standard limits have been established and used.

"Index values for radionuclides in foods", as shown in the "Regulatory Guide: Emergency Preparedness for Nuclear Facilities" of the Nuclear Safety Commission of Japan (NSC), were applied to supersede provisional regulation values, which had been in place until March 31, 2012. The guideline level of the Codex Alimentarius international food standards was applied in some cases (provisional regulation values of radioactive iodine in drinking water/milk for infants). In addition, the index of radioactive iodine in fish was not shown, but detected in monitoring. On April 5, 2011, provisional regulation values, to which those in vegetables, excluding edible roots and potatoes, were applied, were added.

Provisional regulation values and new standard limits on and after April 1, 2012 are shown in Tables 5.5 and 5.6. The Task Group on Aspect of the Fukushima Daiichi Nuclear Power Plant Accident of the Nuclear Safety Investigation Committee of Experts of the Atomic Energy Society of Japan (AESJ) suggested ideas on Japanese food regulation in its interim activity report: (1) quickly necessary provision of information on inspection results (2) continuous implementation of internal exposure dose evaluation attributed to food intake, (3) research and study into the transfer of radioactive nuclides to food, and (4) importance of promoting dissemination of the concept of standard values in future.

Table 5.5 Provisional regulation values of radioactive Iodine/Cesium in foods

Nuclides[a]:	Food Category	Provisional regulation value (Bq/kg)
Radioactive iodine	Drinking water	300
	Milk/dairy products[b]:	300
	Vegetables (excluding edible roots and potatoes)	2,000
	Fish	2,000
Radioactive cesium[c]	Drinking water	200
	Milk/dairy products	200
	Vegetables	500
	Grain	500
	Meat, egg, fish, and others	500

[a]Provisional regulation values were separately established for alpha nuclides for uranium, plutonium, and transuranic elements

[b]Instructions were given not to use products with more than 100 Bq/kg for infant-modified milk powder and drinking milk

[c]Regulation values take into account the contribution of radioactive strontium

Table 5.6 New standard limits for radioactive Cesium in foods

Food Category	New standard limits (Bq/kg)
Drinking water	10
Milk	50
Infant foods	50
General foods	100

Note: Standard values take into account the contribution of radioactive strontium

5.3.2.2 Calculation of Standard Limits

These standard values were typically calculated by the following formula:

$$[\text{Standard limit}]\ (\text{Bq/kg}) = [\text{Annual dose}]\ (\text{mSv/year}) / \ [[\text{dose coefficient}]\ (\text{mSv/Bq}) \times [\text{annual intake}]\ (\text{kg/year}) \times [\text{contamination rate}]]$$

The [annual dose], which is equivalent to the exposure dose when a person ingests some food with a concentration of the standard value over the course of a year, is established as a dose requiring the intervention of shipment restrictions, and a certain dose may be allocated to food and radionuclides by type. It is presumed that the provisional regulation value should be 5 mSv/year and a new standard 1 mSv/year.

The [dose coefficient] is the exposure dose per unit of radioactivity taken into the body. The value depends on the radionuclide type, age and gender of the individual. Standard limits can be calculated corresponding to age and gender or the lowest calculated concentration of radionuclides in food can represent the standard limit for all ages and genders. In case of plural radionuclides, standard limits by radionuclide can be calculated according to individual dose coefficients, but in the case of Japanese new standard limits, the dose coefficient is reflected when calculating standard limits, including the dose contribution of radionuclides other than cesium, using the dose coefficient with the contribution to doses added to the dose coefficient of radioactive cesium with the abundance ratio by radionuclide type and transfer rate into food.

The [annual intake] refers to food and likewise, depends on the age/gender of the individual.

The [contamination rate] is the rate of contamination in the food distributed in the market. For new Japanese standard limits, considering the current status of a calorie-based Japanese food self-sufficiency ratio of 39 % (in 2010), targeting 50 % by 2020, 0.5 is used assuming half (1/2) the food distributed in the market is contaminated. The European Community (EC)/European Union (EU) and the current Codex regulation use 0.1, which is a value assuming imports from third countries. The value is also said to be based on statistics of the Food and Agriculture Organization of the United Nations (FAO).

5.3.2.3 Food Guidelines in Europe After Chernobyl NPS Accident

Following the Chernobyl accident, restrictions on radioactive concentrations such as radioactive cesium were imposed on agricultural products from third countries in the European Community. In addition, in the vicinity of the Chernobyl accident reactor, the temporary acceptable level in the event of the accident has been periodically revised as the current updated value. Both these circumstances and Japanese regulatory values are shown in Table 5.7 [1].

After the Fukushima accident, the EU established restrictions on imports of feedstuff and food from Japan in accordance with Japanese regulation limits.

5.3.2.4 Estimation of the Exposure Dose in Japan

When new standard values were discussed, the exposure dose in Japan was estimated [2] (Table 5.8). The estimated value was well below the assumed intervention level of dose of 1 mSv/year: less than 1/10. In addition, even if the provisional regulation value were continuously used, the estimated exposure dose in a median concentration would be only about 1.2 times the case where new standard values were applied, and the annual difference was only 0.008 mSv. From these, we can assume sufficient safety for food conforming to provisional regulation values, but there is uncertainty about actual exposure dose. There is also a need to investigate food contamination and intake states and continuously precede investigation.

5.3.2.5 Summary

The intake/shipment restrictions on food, drinking water, and others are shown below:

- In the early stage of the accident, in terms of intake/shipment restrictions on foods, drinking water, and others, "index values for restrictions on the intake of foods" were applied to provisional regulation values, which were established and operated.
- The standard values replacing the provisional regulation values were studied and applied from April 1, 2012.
- In the new standard values, the standard for radioactive iodine, which had not been detected, was abolished, and the standard values represented by the radioactive cesium were established.
- The estimated value of the actual exposure dose under the operation of food regulatory values was smaller, considering the dose intervention level.
- The Task Group of the AESJ proposed important matters to be conducted in future for the Japanese food regulation.

Table 5.7 Standard limits of radioactive Cesium and others in food in countries (Bq/kg) (UN Chernobyl Forum Expert Group "Environment", Environmental Consequences of the Chernobyl Accident and their Remediation: Created based on Twenty Years of Experience, IAEA (2006), etc.)

Country	Japan		Codex regulation	European Community (EC)	The Union of Soviet Socialist Republics				Russian Federation	Belarus	Ukraine
Effective Date	March 17, 2011	April 1, 2012	1989 and 2006	May 30, 1986	May 06, 1986	May 30, 1986	December 15, 1987	January 22, 1991	2001	1999	1997
Radioactive nuclide	Radioactive cesium[a]:		Radioactive cesium	Radioactive cesium	Iodine-131	β-ray nuclide	Radioactive cesium				
Drinking water	200	10	–	–	3,700	370	18.5	18.5	–	–	–
Milk	200	50	1,000	370	370–3,700	370–3,700	370	370	100	100	100
Infant foods	200	50	1,000	370	–	–	–	–	40–60	37	40
General foods	500	100	1,000	600	18,500–74,000	370–37,000	370–1,850	40–500	40–500	40–500	20–200

[a]Regulation limits take into account the contribution of radioactive strontium

Table 5.8 Effective dose estimated from radioactive Cesium based on new standard limits (All Ages (Average Intake)) (The Radioactive Material Response Working Group in the Food Sanitation Subcommittee of the Pharmaceutical Affairs and Food Sanitation Council report, "Shokuhin chu no Hoshasei Bussitsu ni kakawaru Kikaku Kijun no Settei ni tsuite" (The Establishment of Standards and Criteria on Radioactive Materials in Food) Ministry of Health, Labour and Welfare (Preparation started in December 2011.)

Concentration in Plant	Median Concentration (New Standard Limit)	90 Percentile Value (New Standard Limit)	Median Concentration (If the Provisional Regulation Value is continually used)
Estimation of Exposure Dose (mSv/year)	0.043	0.074	0.051

5.3.3 *Radiation and Exposure Dose Measurements*

5.3.3.1 Status of Environmental Radiation Monitoring

Environmental radiation monitoring should accurately present conditions of radioactivity diffusion caused by an accident to provide appropriate information for residents' evacuation. However, the earthquake and tsunami damaged the functions of necessary equipments in the 3.11 incident. In addition, because of the radioactivity diffusion over broad areas, its distribution had to be measured accurately. The local government is responsible for monitoring in the event of a nuclear disaster. However, in the event of severe accident conditions as the 3.11 incident, a large group of experts from MEXT, the utilities, specified public organizations (the National Institute of Radiological Sciences, the Japan Atomic Energy Agency, etc.) is designated to be dispatched to provide emergency support, and their support played a crucial role during the early stage of the accident.

(1) **Dose measurement in the progress of the accident**

The following shows the status of radiation monitoring conducted with the progress of the accident.

With the earthquake and tsunami of March 11, no fewer than 2,400 workers evacuated from the controlled areas, and they detoured around the exit gate monitor and evacuated with no body surface monitoring. Radiation management staff members were the last to evacuate with the arrival of the tsunami at the station. After 15:50, alarm pocket dosimeters (APDs) were collected from workers who evacuated with no radiation monitoring and the dose recording was continued until 24:00.

Most of the 5,000 or so APDs in the service building (SB) were disabled due to the tsunami. Rented APDs as well as 50 emergency APDs were collected, and by night of the 12th, 320 APDs had been secured. Out of 530 APDs sent from Kashiwazaki Kariwa station, only 30 that were compatible with chargers were used on and after the 12th. All monitoring posts (MPs) in the Fukushima Daiichi site were out of order due to the tsunami. Accordingly, radiation survey

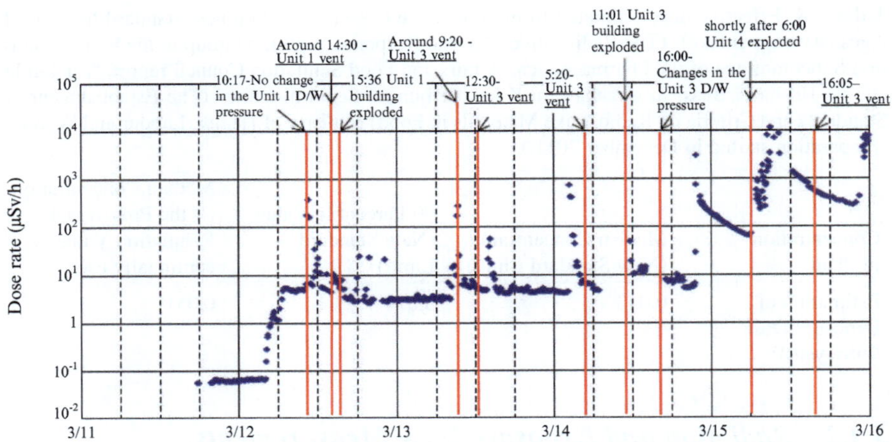

Fig. 5.3 Air dose rate in the vicinity of the front gate of the Fukushima Daiichi NPS (TEPCO, Fukushima Nuclear Accident Analysis Report)

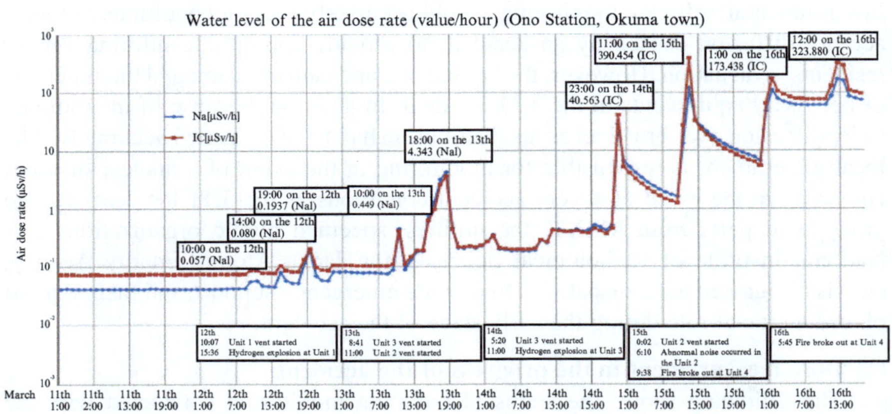

Fig. 5.4 Air dose rate at the monitoring post (Ono station) in Fukushima prefecture (By Fukushima prefecture)

was conducted in the station at 16:30 on the 11th. Figure 5.3 shows data during the initial stage of the accident in the vicinity of the front gate, while Fig. 5.4 shows the monitoring data recorded at Ono Station from March 11 through 16.

As the ventilation stack radiation monitor could not be used, it was difficult to directly evaluate the amount of radiation released.

At 17:19 on the 11th, operators went to the site with a GM survey meter to check the IC water level, the meter scaled out as soon as they opened the reactor building entrance double door, whereupon they gave up checking the level and returned.

At 18:35 on the 11th, work began to arrange an alternate water injection line to the reactor via the core spray systems (hereinafter CS). When it was completed at 20:30, the operators went to the basement of the reactor building with APDs to manually open the valve, and the reading on the APD did not change significantly.

A monitoring car and mini-bus from the Kashiwazaki Kariwa station with a supporting unit of radiation management staff members arrived at 2:49 on the 12th. At the entrance of the Seismic Isolation Building, equipments were lent, radiation survey was conducted, the door opening and closing were managed, and the radiation monitoring of the site was conducted via two monitoring cars, including the existing one. The measured data was released every few hours on the website.

In the staff evacuation to a shelter by bus, contamination inspection was performed by a body surface monitor when leaving the bus.

At 23:00 on the 11th, dose measurement was conducted at the front of the double doors of the reactor building entrance on the first floor of Unit 1 turbine building, recording 1.2 mSv/h at the front of the northern double door on the first floor of the turbine building, and 0.5 mSv/h at the front of the south double door on the first floor of the turbine building. Based on these data, a dose of 300 mSv/h was assumed inside the reactor building, and at 23:05 on the 11th, entry into the reactor building became prohibited. At that time, changes in the dose rates were monitored and marked manually on the arrangement plan of the vicinity of the reactor building amid tension and uncertainties in predicting results of the dose rate, the records of which remain in the seismic isolation building. Accordingly, mistakes in writing down units and numerical values, including neutron measurement results, occurred and the measurement of lower limits was inconsistent. The monitored value in the vicinity of the front gate on the leeward side was 0.060 μSv/h. Given no change in the environmental gamma ray spectrum in prefectural monitoring posts, it was believed that particle and gaseous fission products remained in the containment vessel at this stage. It was 4:04 on the 12th when the leak to the environment was detected. The measurement value of the monitoring car doubled at MP-8 point and at the front gate at 4:05.

5:15 on the 12th	Contacted related parties based on judgment of radioactive material leak off-site.
At 10:17 on the 12th	S/C vent valves were opened.
At 10:40 on the 12th	The radiation level rose in the vicinity of the front gate and the monitoring post 8.
15:36 on the 12th	Hydrogen explosion at Unit 1.
Predawn on the 13th	The judgment criteria of the contamination level was changed from 4 to 40 Bq/cm^2.
5:30–10:50 on the 13th	Neutrons were measured in the vicinity of the front gate though the level barely exceeded detection limit.
14:31 on the 13th	The measurement results were reported: 300 mSV/h, or more on the north side of the Unit 3 reactor building double door and 100 mSv/h on the south side.

(continued)

11:01 on the 14th	Hydrogen explosion at Unit 3.
14:04 on the 14th	The dose limit was raised to 250 mSv.
9:00 on the 14th–13:40 on the 15th	During this time, neutrons were intermittently detected in the vicinity of the front gate.
Shortly after 6:00 on the 15th	Hydrogen explosion at Unit 4.
6:14 on the 15th	The blast sound and vibration, which were initially assumed to have occurred in Unit 2, proved to be due to hydrogen explosion in Unit 4. The sudden pressure drop in the Unit 2 S/C is assumed to have been caused by instrument failures.
9:00 on the 15th	Air dose rate at the front gate was 11.93 mSv/h, the maximum after occurrence of the accident.
23:05 on the 15th	Radiation level exceeding 500 μSv/h (4,548 μSv/h) was measured in the vicinity of the front gate.

(2) **Radiation monitoring plan for the surrounding environment and its progress**

Nuclear Safety Technology Center (NUSTEC) staff members tried to conduct airborne monitoring using a Self-Defense Force helicopter based on studies by MEXT. They tried to meet at the sports park in Rokkasho village, Kamikita district, Aomori prefecture on the afternoon of March 12 only to fail due to differences in timing. Monitoring data at this stage could have been utilized effectively to clarify status of contamination with accident progression and to evacuate residents.

In the evening of March 15, MEXT used monitoring cars to measure air dose rate and recorded a high rate of 330 uSv/h in Namie town.

On March 15, increase in the air dose rates induced by the release from Unit 2 were observed in various locations. It is assumed that the rise in the air dose rates in Kanto neighboring prefectures is attributed to the influences of the release. In addition, significant rise in air dose rate is also observed on March 16.

From March 17 through 19, airborne monitoring was performed by the Aerial Measurement System (AMS) of the U.S. Department of Energy (DOE) (Fig. 5.5). This was conducted over an area of approximately 60-km radius of the Fukushima Daiichi. Consequently, an air dose rate map was obtained, showing a rate of 0.12 mSv/h or more in several areas outside the 20-km radius. This information was communicated to Japan via the Ministry of Foreign Affairs on March 21. Japan and the U.S. planned to perform joint airborne monitoring on and after April 6. Subsequently, following requests from neighboring prefectures, the monitoring scope was widened. At this time, the first through sixth airborne monitoring rounds have been performed within an 80-km range of the Fukushima Daiichi Nuclear Power Station. Conversely, Japan was a little slower in this respect, but under the leadership of MEXT, the Self-Defense Forces performed air sampling over Fukuoka prefecture on and after March 24 to analyze the radioactive concentration, and measured the air dose rate distribution beyond a 30-km radius by loading a radiation measuring

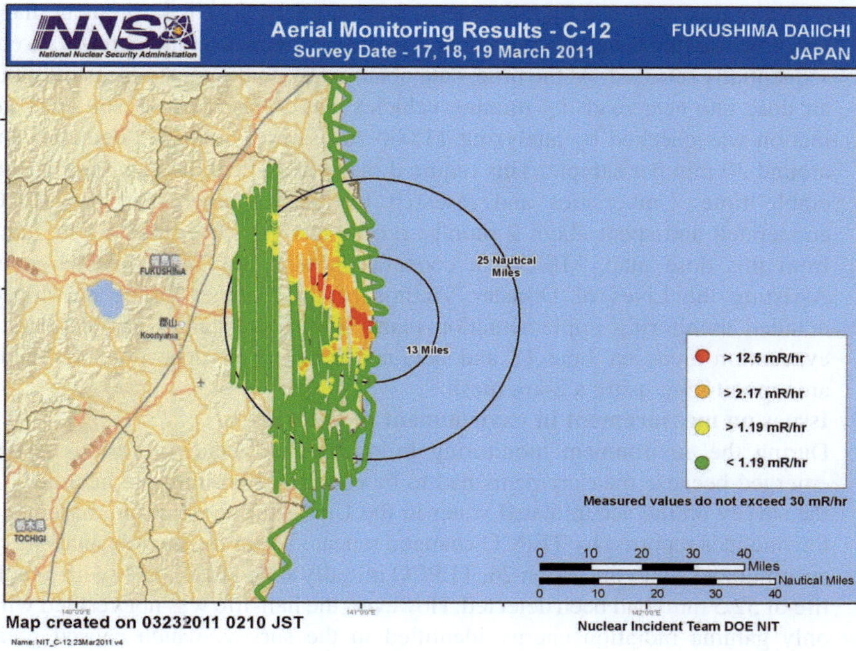

Fig. 5.5 Results of airborne monitoring by the U.S. DOE

instrument of the Nuclear Safety Technology Center (NUSTEC) onto a small aircraft of the Japan Aerospace Exploration Agency (JAXA). Subsequently, the NUSTEC, the Japan Atomic Energy Agency, and the Japan Chemical Analysis Center (JCAC) loaded a measuring instrument on a commercial helicopter, analyzed results at an altitude of 150–300 m and obtained the data such as air dose rates and the amount of cesium deposited from Hokkaido to Tohoku, Kinki, Shikoku, Kyushu, and Okinawa.

Monitoring was also performed in the sea under the leadership of MEXT. On and after March 21, the sea area monitoring was performed by sampling seawater with collaboration from the Japan Coast Guard (JCG).

Fukushima Prefecture Environmental Radiation Monitoring in Mesh Survey Implementation Plan was determined on April 10, based on which an air dose rate investigation was conducted at 2,724 points beyond a 20-km radius in the prefecture on April 12–16 and on 29. The results were compiled as an air dose rate map and released on May 2. Subsequently, on April 18 and 19, MEXT, TEPCO, and the Federation of Electric Power Companies of Japan (FEPC) teamed up to perform monitoring at 128 points within a 20-km radius and released the results on the website.

The Nuclear Emergency Response Headquarters released an environment monitoring enhancement plan on April 22, which stipulated the preparation of air dose rate maps and soil concentration maps of iodine and cesium within the

80-km radius sectioned by 2-km mesh. Subsequently, dose measurement maps and accumulated dose estimation maps based on the measurement results were sequentially released. At the time, data obtained by continuously measuring the air dose rate near roads by running vehicles was released, while soil contamination was checked by analyzing 11,000 soil samples with Ge detectors for around 30 min per sample. This required many analytical devices and considerable time. Universities and research institutions other than the JCAC cooperated and spent about 2 months on the analysis. By matching the data from this dose map, MEXT, in cooperation with the Team in Charge of Assisting the Lives of Disaster Victims of the Cabinet Office developed detailed monitoring implementation plans in restricted areas and deliberate evacuation areas on June 13 and obtained detailed air dose rates via wide area monitoring using a 2-km mesh.

(3) **Issues on measurement in environment monitoring**
During the environment monitoring following this accident, various issues emerged because measurements had to be conducted on multiple nuclides, In the survey on the accumulated water in the Unit 2 turbine building basement, the nuclides reported by TEPCO changed repeatedly. In the nuclide analysis of accumulated water on March 26, TEPCO initially stated that I-134 (with a half-life of 52.5 min) had been detected. However, the half-life was not verified with only gamma radiation energy identified in the survey, which caused great confusion. In addition, there was an error in the results of nuclide analysis of the accumulated water in the Unit 1 turbine basement that sparked doubts over possible re-criticality because CI-38 was detected. This was because the evaluation was based only on an analytical software, and the results on the background discrete variable was mistaken as the peak value. It is also assumed that many problems arose related to the handling of simple measuring devices that later became available. In some cases, measuring apparatus were used for other purposes during emergencies. Due care must be taken in using data obtained in such cases, which must be supplemented with clear, supporting information.

5.3.3.2 Responses to Radiation Exposure (Workers' Dose Management and Surveys on Residents' Dose)

(1) **Response to workers dose**

(a) The dose limit and screening standard in the event of an emergency: the dose limit for regular radiation workers is regulated to not exceed the effective dose of 100 mSv over 5 years and 50 mSv per year. The dose limit for women (excluding those who are diagnosed as unable to achieve pregnancy, and those who have submitted in writing to licensed users or licensees of waste disposal that they have no intention of becoming pregnant), should not exceed 5 mSv over 3 months. In the case of pregnant women, the radiation level from radioisotopes consumed in the body should be 1 mSv, or less until childbirth according to their request.

Conversely, the dose limit should be 100 mSV in the event of an emergency. However, considering the disaster of the Fukushima Daiichi NPS, the President of the National Personnel Authority, the Ministers of Health, Labour and Welfare, and Economy, Trade and Industry consulted with the Radiation Council and determined that the exposure dose limit for emergency work should be 250 mSv on March 14, 2011. The Radiation Council responded the same day that the dose limit of 250 mSv was appropriate; stating that the internationally accepted recommendation value of 500 mSv or less would not cause any acute disorder or late severe disorder, based on "ICRP 2007 recommendation (Pub. 103) no Kokunai Seido nado heno Toriire ni tsuite—Dai Niji Chukan Hokoku—(the Application of 2007 recommendations of the International Commission on Radiological Protection (Publication 103) to the Domestic System - Second Interim Report-)" Subsequently on the same day, the Ministry of Health, Labour and Welfare issued a ministerial order: "Heisei Nijusannen Tohoku chiho Taiheiyo Oki Jishin ni Kiin site Shojita Jitai ni Taiou Suru Tameno Denji Hoshasen Shogai Boshi Kisoku no Tokurei ni Kansuru Shorei" (Ministerial Order on Special Cases of the Ordinance on Prevention of Ionizing Radiation Hazards to Respond to Events Attributable to the Great East Japan Earthquake) (No. 23, the Ministry of Health, Labour and Welfare, 2011) (Notification from the Director-General of the Labor Standards Management Bureau to the Director of the Prefectural Labor Bureau 0315 No. 7). In the ministerial order, the exposure dose limit in the emergency work in Article 7 shifted from the conventional 100 mSv to 250 mSv. As Step 2 in the current road map was completed on December 16, 2011, the ministerial order was promulgated in the gazette to abolish this special provision and came into effect.

The TEPCO Fukushima Nuclear Accident Analysis Report [3] cites that there was no clear standard (screening level) to judge the need for decontamination for on-site workers was established. Therefore, during the initial stage of the accident, 6,000 cpm (GM survey meter) was conservatively adopted from the perspective of internal exposure in accordance with advice from emergency exposure medical experts. On and after April 20, 2011, from the perspective of body decontamination, the screening level was determined as 100,000 cpm in Fukushima prefecture which became the standard value. In addition, on September 16, 2011, the screening level of local residents was reduced from 100,000 to 13,000 cpm and the screening level of workers was established at the same level based on instructions of the local nuclear emergency response headquarters on the basis of ALARA (as low as reasonably achievable).

(b) Status of radiation exposure of workers: The status of radiation exposure of the workers was released in the monthly reports and TEPCO Accident Investigation Report. On July 5, 2013, as the Ministry of Health, Labour and Welfare reconfirmed the validity of internal exposure dose evaluation, TEPCO released results evaluated/modified using the same method.

According to the results, the number of workers who engaged in radiation work from March 2011 to the end of March 2013 was 27,351 (3,710 TEPCO employees and 23,641 subcontractors), among whom 14,055 (about 59 % of all workers) received a dose of less than 10 mSv. The doses received by 3,996 workers (16.9 %) were 10–20 mSv; for 4,233 (17.9 %), 20–50 mSv; for 1,184 (5.0 %), 50–100 mSv; for 138 (0.58 %), 100–150 mSv; for 26 (0.11 %), 150–200 mSv; for 3, 200–250 mSv; and for 6, the dose exceeded 250 mSv. The average dose of workers was 12.48 mSv, with a maximum of 678.80 mSv. After reconfirmation, the number of those exposed to more than 50 mSv increased by 24 (the maximum increase was 48.9 mSv), of whom 6 had a dose exceeding 100 mSv. Conversely, the doses of two decreased before the modification. (The maximum decrease was 9.24 mSv). However, there was no increase in the number of those exposed to more than 150 mSv. All cases of exposure exceeding 250 mSv occurred immediately after the accident in March 2011. On and after April 2011, the doses of most workers were 10 mSv or less. Six workers—one main control room operator and five staff in maintenance operations related to electricity and instrumentation—were exposed to high dose mainly as a result of internal exposure by improper use of protective masks (selection and wearing of the same). According to the report by the Government Accident Investigation Committee, the dose of two people out of the three who might have been exposed to a high dose, was 466 mSv. Two women workers who were not engaged in radiation work were exposed to more than 5 mSv over 3 months, which was the legislative dose limit. They were engaged in refueling fire engines and desk work in the Seismic Isolation Building. Their exposure doses, both external and internal, were respectively 7.49 and 17.55 mSv. It was determined that after March 23, 2011, women would not work in the premises of the Fukushima Daiichi Nuclear Power Station, since which time women have not been exposed to radiation.

(c) Management of radiation exposure to workers: In this accident, while power sources were lost due to the tsunami, dose measuring devices and protective materials and equipment became disabled by inundation or were washed away. The number of personal dosimeters stored in the Seismic Isolation Building was about 320. Until a sufficient number had been secured, representative staff members wore them for performing part of the work according to Article 8–3 provision of the Ordinance on Prevention of Ionizing Radiation Hazards. TEPCO's Fukushima Nuclear Accident Analysis Report states in and after April 2011, that it was ensured that each had a personal dosimeter with the support of other nuclear sites.

(d) Workers' health check/management: On May 17, 2011, the Government Nuclear Emergency Response Headquarters prepared "Immediate Actions to Assist Residents Affected by the Nuclear Accident", which presented the significance of establishing a database to track long-term effects of radiation dose, including the conditions after departure of all workers engaged in emergency work. In June 2011, the Ministry of Health, Labour and Welfare established an "Expert Meeting on the Long-term Healthcare of Workers at the TEPCO Fukushima Daiichi Nuclear Power Plant", after which a report on items required to develop the database and long-term healthcare was compiled in September 2011. The report recommended health guidance after departure by a doctor or public health nurse; healthcare including psychological care; annual general health diagnosis and eye tests (by slit lamp microscope) for those who received dose exceeding 50 mSv; and thyroid inspection, gastric, colon and lung cancer screening, as well as white blood cell count and differential white blood cell count inspections for those who received dose exceeding 100 mSv.

The prescribed healthcare items were to be reviewed after a period of 3 years in view of the advances in medical sciences and changes in inspection methods. In response, TEPCO made decisions to conduct thyroid inspection (thyroid-stimulating hormone (TSH) by drawing blood), free type triiodothyronine (free T3) and free thyroxine (free T4) inspection), cancer screening (stomach, lung, and colon) for workers whose accumulated effective dose exceeded 50 mSv while engaged in specified emergency work, and subsequently worked in TEPCO nuclear power stations up to the end of March 2016. As a result of this test, thorough checkups were to be conducted for those requiring a second inspection (thorough checkup). Moreover, if the thyroid equivalent dose such as accumulated thyroid equivalent doses generated in the specified emergency work and working in TEPCO nuclear power stations up to the end of March 2016 exceeded 100 mSv, thyroid inspection (cervical ultrasonic inspection) would be conducted. The number of those subject to cancer screening was 1,307, and that of thyroid ultrasonic inspection was 1,972 (as of July 22, 2013).

(e) Workers' internal exposure dose measurement: The Japan Atomic Energy Agency (JAEA) conducted detailed measurement of iodine-131 accumulated in thyroid on 560 TEPCO employees engaged in emergency work whose committed effective dose may exceed 20 mSv in a provisional assessment. The maximum effective dose due to internal exposure was 590 mSv, the estimated value for a male worker measured by the JAEA on May 30 (thyroid I-131 residual volume: 9,760 Bq) [4].

(2) **Response to residents' dose**

(a) The Screening standard for residents: The screening level is described in a flowchart of the initial radiation exposure medical care in "Knowledge of Radiation Exposure Medical Care in Emergencies" [5]. The body surface contamination of 40 Bq/cm^2, estimated whole-body dose of 100 mSv, and 13,000 cpm as a GM-type surface contamination survey meter measurement value, which is equal to thyroid iodine-131 quantity of 3 kBq (equivalent to thyroid 100 mSv of iodine-131 for 1-year-old infants) are shown.

Based on these, training and education on the initial radiation exposure medical care in emergencies were provided.

In the accident, screening inspections were performed to evaluate the dose levels of local residents accommodated in shelters and elsewhere, to determine radioactive materials contamination, to measure exposure dose and to take necessary measures. The Ministry of Health, Labour and Welfare issued "Hoshasen no Eikyo ni kansuru Kenko Sodan ni tsuite" (Health Counseling of the Radiation Influence (Request)), dated March 18, 2011 to all prefectures, cities establishing health centers, and the division responsible for regional public health services in special wards. 13,000 cpm was described as an example response in the notice. Conversely, on March 20, 2011, the Nuclear Safety Commission issued "Josen no tame no Screening Level no Henko ni tsuite" (Screening Level Changes for Decontamination). With effectiveness in mind, the provisional value (10,000 cpm) was changed to 1 μSv/h (dose at a place 10 cm away), a standard by the International Atomic Energy Agency (IAEA), and a GM survey meter with a diameter of 5 cm (TGS-136 type) was used so that the measured value would be 100,000 cpm. In response, on March 21, 2011, the Ministry of Health, Labour and Welfare issued "Hoshasen no Eikyo ni kansuru Kenko Sodan ni tsuite (Irai) (Ichibu Shusei oyobi Tsuika)" (Health Counseling of the Radiation Influence (Request) (Partial Revision and addition)), requesting that the division responsible for regional public health services change the screening levels. On March 14, 2011, Fukushima prefecture specified a screening inspection standard for radiation exposure medical care in emergencies, of 100,000 cpm in the case of whole body decontamination. The criteria was based on the treatment of radiation exposure by medical experts dispatched by MEXT, the National Institute of Radiological Sciences, and Fukushima Medical University. If values exceeding 13,000 cpm up to 100,000 cpm were detected, partial wipe-off decontamination should be performed. The screening and decontamination at this level were conducted on evacuated residents during brief visits to their homes.

In view of the confusion related to screening on emergency radiation exposure medical care as part of accident response, the Radiation Exposure Subcommittee of the Special Committee on Emergencies, including nuclear facilities of the Nuclear Safety Commission, recommended organization of the screening system and purposes, and to study related technical issues at the 29th meeting on February 7, 2012.

On September 16, 2011, from the perspective of ALARA (As Low As Reasonably Achievable), the Government Nuclear Emergency Response Headquarters notified Fukushima prefecture and related municipalities to lower the screening level to 13,000 cpm.

As of now, according to the "Nuclear Emergency Response Guidelines" (fully revised on June 5, 2013), the urgent protective measures (OIL4), namely, the beta ray rate of the detector measured at a distance of a few centimeters from the skin was set to 40,000 cpm and the value 1 month later was a standard of 13,000 cpm (the initial value) to conduct decontamination and prevent external exposure from careless oral ingestion and skin pollution. The initial value of 40,000 cpm was an effective level that may be discriminated from background radiation. Considering the accident conditions, however, 100,000 cpm or less can achieve simple decontamination.

(b) Residents' screening results: In Fukushima prefecture, with support from other prefectural medical teams on and after March 13, 2011, emergency exposure screening was conducted at healthcare centers and other mobile and permanent centers. Consequently, from March 13, 2011 to March 13, 2013, 262,366 received the screening test. Following measurements conducted between March 13 through 31, 2011, the number of people whose radiation level exceeded 100,000 cpm were 102.

(c) Residents' internal exposure dose measurement: In Fukushima prefecture, with support from the National Institute of Radiological Sciences, the JAEA, Fukushima prefecture, Minamisoma City General Hospital, Niigata Prefecture Radiation Monitoring Center, Hirosaki University School of Medicine & Hospital and others, internal exposure dose measurement using a whole-body counter, that detects radioactive cesium-137 and -134 in the body was conducted. Children under the age of 18 and pregnant women were prioritized in the measurement. From June 27, 2011 to June 30, 2013, a total of 139,153 received medical checkup. The number of people whose effective dose was 3 mSv were two, 2 mSv were ten, and 1 mSv fourteen. The exposure dose of most residents (99.98 %) were less than 1 mSv. Until the end of January 2012, acute intake scenario (inhalation) was applied, and after this period, continuous intake scenario (oral) was used for estimation. Likewise in future, the internal exposure dose measurement using a whole-body counter is planned to be performed for residents in areas where measurement had not been conducted and for those who evacuated outside Fukushima prefecture. Moreover, Iwaki city, Fukushima city, and others performed the inspection independently.

(d) Reconstruction of the initial internal exposure doses: The National Institute of Radiological Sciences held two international symposiums aiming to reconstruct internal exposure dose in the early stages of the accident in the TEPCO Fukushima Daiichi Nuclear Power Station on July 2012 and January 27 in 2013. During the second symposium, the average exposure dose of residents estimated by combining the dose estimations based on the individual measured values and the air diffusion simulation was generally 10 mSv or less, while the results of accumulated thyroid equivalent doses of 1-year old children in the area surrounding the station was 30 mSv or less in most cases [6, 7].

A draft report by the United Nations Scientific Committee on the Effects of Atomic Radiation (UNSCEAR) in May 2013 showed the accumulated thyroid equivalent dose of adults and children in Fukushima prefecture. This draft report stated the dose of adults in the evacuation zones was 8–24 mSv, that of 1-year children in the evacuation zones was 20–82 mSv, and that of 1-year-old children outside the zones was 33–66 mSv.

(e) Residents' healthcare management: Fukushima prefecture conducted "Health Management Survey for the Residents" to resolve their concern over health by radiation contamination of the Fukushima Daiichi Nuclear Power Station accident and to promote healthcare in future. The Health Management Survey was consigned to, and led by Fukushima Medical University with support from other prefectures, municipalities, the Council of Radiation Effects Research Organizations, relevant ministries, the Japan Medical Association (JMA), universities, academic societies all over Japan, and included items as exposure dose estimation, healthcare of evacuated residents, and regular checkups for all residents in Fukushima Prefecture over the future. All healthcare data are expected to be compiled into databases and continually controlled. Survey items include basic behavior survey for external exposure dose estimation, thyroid inspection, medical checkup, mental health check, lifestyle survey, and survey on pregnant women and others.

(i) Basic survey: As of March 31, 2013, 481,423 (23.4 %) out of 2,056,994 survey respondents had responded to medical interview sheets (including behavior records) distributed by Fukushima prefecture. As of the end of January, 394,369 (82.7 % of responses) external exposure dose estimation had been completed, and 410,539 had been notified of results. The number of temporary sojourners who submitted medical interview sheets was 2,064 and the estimation of 1,589 had been completed. As for the effective dose estimation, 420,543 (411,922 excluding those engaged in radiation work) were estimated. The number of those whose effective dose was less than 1 mSv was 271,822 (94.9 %). less than 1–2 mSv is 119,018 (4.7 %), and less than 2–3 mSv is 18,589 (0.1 %). The maximum value was 25 mSv, of a resident in the Soso area.

(ii) Thyroid inspection: Thyroid ultrasound inspection started in October 2011 for residents under 18 years old in Fukushima prefecture at the time of the accident as part of long-term health monitoring of children based on the results of the Chernobyl accident health effects. Likewise in future, those under the age of 20 would undergo the inspection every 2 years and subsequently, every 5 years at inspection facilities in Fukushima prefecture. As of March 31, 2013, the number of those undergoing the first thyroid inspection with confirmed results was 40,302 in 2011 and 134,074 in 2012. The number of people who had 5.0-mm or less tubercles or cysts (A2 judgment) was 14.427 (35.8 %) in 2011 and 59,746 (44.6 %) in 2012, the number of those who had 5.1-mm or more tubercles, or 20.1-mm or more cysts (B judgment) was 205 (0.5 %) in 2011 and 934 (0.7 %) in 2012. In addition, the number of those immediately requiring a second inspection (C judgment) was one. The 11th Health Management Survey Committee reported on June 5, 2013 that 12 persons who underwent the second inspection had thyroid cancer (early-stage papilla cancer), and one had a benign node.

(3) **Summary**

In the accident at TEPCO's Fukushima Daiichi Nuclear Power Plant, radiation was often manually controlled, and issues involving erroneous descriptions and entry mistakes occurred. There is a need to fully examine the workers' and local residents' dose management in the future and the dose/radiation management system construction. In addition, it is also important to release data on the radiation levels of workers who were engaged in initial emergency activities such as the Self-Defense Forces, the police, and the fire fighting force.

One-tenth of surface contamination density (nuclides that do not release alpha ray: 40 Bq/cm^2) was the regulated value for the transfer of properties from controlled areas of nuclear power station. However, no clear screening standard in emergencies had been established. In addition, although screening standard of local residents were developed and used in previous training, the grounds on the values were not explained sufficiently to disaster prevention staff members. There were also changes in these values, etc., which caused confusion.

Few workers were exposed to more than 100 mSv and many workers received lower dose. Health checks over the long term must be provided for those exposed to more than 50 mSv.

The majority of the exposure dose of local residents showed less than 1 mSv. However, Fukushima prefecture intends to conduct long-term healthcare survey in an integrated manner including behavior surveys, internal exposure dose measurement by the whole-body counter, a lifestyle-related disease checkup, and mental healthcare.

5.3.4 Environmental Pollution by Radioactive Material and Decontamination

5.3.4.1 Pollution of Soil and Water Environments by Radioactive Material

(1) **Occurrence of pollution**

Due to the vent operation in Units 1 and 3, the explosion of the reactor building Units 1 and 3, and the damage to the Unit 2 containment vessel following the accident at the Fukushima Daiichi Nuclear Power Station, a large quantity of radioactive iodine and cesium was released into the environment, resulting in radioactive materials polluting a wide area of East Japan, mainly Fukushima

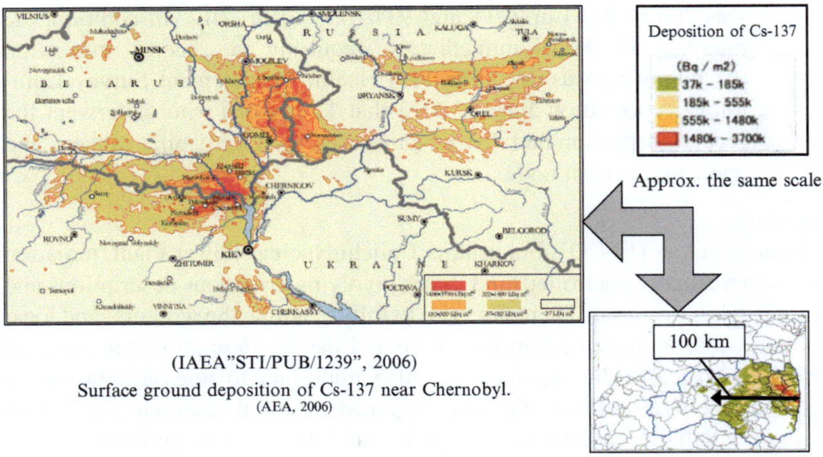

(IAEA"STI/PUB/1239", 2006)
Surface ground deposition of Cs-137 near Chernobyl.
(AEA, 2006)

Surface ground deposition of Cs-137 near Fukushima-1
(Edited from MEXT data)

Fig. 5.6 Comparison of areas contaminated by the accidents of Chernobyl and Fukushima Daiichi

prefecture. March 15 and 16, 2011 in particular saw the amount of radioactive material released peak, which resulted in high environmental pollution to the northwest of Fukuoka prefecture, namely the Nakadori area. In addition, some materials reached Abiko, Kashiwa and Nagareyama cities, among others, and monitoring [8] revealed individually contaminated hot spots within the same.

(2) **Characteristics of contaminated areas**
Comparing the areas contaminated by the accident in Fukushima prefecture with those contaminated by the explosion accident involving Unit 4 at the Chernobyl Power Station in April 1986, we notice the area contaminated by cesium-137 in this accident was a fraction of the contaminated areas in the Chernobyl accident [9] as shown in Fig. 5.6. In addition, in the Chernobyl accident, the explosion of the reactor core widely scattered radionuclides, including plutonium and other nuclear fuel materials, which were contained in the reactor core. In this accident, the core melted down due to the high temperature, but the core itself did not explode. Accordingly, cesium and iodine, which are relatively volatile, accounted for the majority of radionuclides. As for other radionuclides, strontium-90 was confirmed in the area contaminated in this accident [10], but as the measured strontium-90 concentration was lower than cesium by several orders of magnitude, cesium removal was the basic method used to reduce the air dose rate. In addition, one of the characteristics of the area contaminated in this accident is the fact that forest comprises 70 % or more of Fukushima prefecture.

(3) **Estimation of the decontamination zone**
To cope with this pollution and reduce the air dose rate, the Government strove to limit the annual additional exposure dose to 1 mSv long term. According to

Table 5.9 Area by division of radioactive Cesium concentration in the agricultural land soil in Fukushima prefecture (estimated values) (Secretariat of Agriculture, Forestry and Fisheries Research Council (AFFRC), "Nochi Dojo no Hoshasei Bussitu Nodo Bunpu Map Kanren Chosa Kenkyu Hokokusho (Dai 3 Hen)" (Research Study Report Related to Map of Radioactive Materials Concentration in Agricultural Soil) (Part 3) (March 2012))

Radioactive Cesium Concentration (Bq/kg)	Rice Paddy (ha)	Dry Field (ha)
0–1,000	59,942	22,022
1,000–5,000	39,164	14,658
5,000–10,000	1,958	796
10,000–25,000	2,575	751
25,000–	1,646	581

documents submitted to the Committee on Environment Recovery of the Ministry of the Environment, the estimated areas subject to decontamination in the event of wide area decontamination of areas where the additional annual exposure dose is 5 mSv or more and spot decontamination of those where it is 1–5 mSv (excluding forest) were 51 km^2 of the building lot, requiring wide area decontamination, 13 km^2 of the arterial traffic lot, 349 km^2 of agricultural land and 23 km^2 of other lots. In addition, 642 km^2 required spot decontamination. The areas were calculated in case where 10–100 % of forests exceeding 5 mSv were widely decontaminated. For example, in the case of 100 % decontamination of the forest, the estimated area was 2,419 km^2, in combination of the area mentioned above.

(4) **Soil contamination**

Monitoring [11] was conducted to measure the soil contamination. Table 5.9 shows the areas of rice paddies and dry fields per concentration division in Fukushima prefecture. Observing the table, we can see a total of 62,129 ha (about 621 km^2) is 1,000 or more Bq/kg rice paddies, while dry fields make up no less than 8,307 ha (about 83 km^2). Deposited cesium was also repeatedly measured. It emerged that in case of untilled ground, 90 % or more remained in a surface layer at a depth of 5-cm or less [12]. It also emerged that past tests had shown cesium firmly adhered to clay particles in the soil and hardly dissolved into water.

The Clean-up Subcommittee of the Atomic Energy Society of Japan (AESJ) conducted a decontamination test by stirring soil puddling [13–15]. After putting the water drained by the test quietly for a given period, the radioactive concentration of the filtered water was measured, and the result showed that the radioactive concentration in water was less than the detection limit. This showed, based on soil decontamination, the removal of microscopic particles absorbing radioactive materials was effective.

(5) **Water pollution**

As wide-area water pollution, the environmental pollution of rivers and lakes used for tap water, etc. was assumed, and radiation monitoring of bodies of water, water bottom, etc. was conducted. In the early stages after the accident, iodine-131 was detected in tap water and elsewhere. The Nuclear Safety

Commission cited "indicators of food and drink intake restrictions and temporary restrictions according to the Food Sanitation Act" as the concentration target of radioactive materials in the tap water,

The Ministry of Health, Labour and Welfare compiled the report on June 2011, saying "unless a large amount of radioactive materials was released again, it is less likely that restrictions on the intake of tap water will be required" and showed new standard values for radioactive materials in food (enacted on April 1, 2012).

5.3.4.2 Decontamination

(1) **The enforcement system and framework of decontamination**

As shown in Sect. 5.3.4.1, the Government aimed to reduce the additional annual exposure dose to 1 mSv long term and established 100 Bq/kg for common foods as the new standard values for radioactive materials in foods on April 1, 2012. (See Sect. 5.3.2 for details.) To do so, as well as residential and agricultural land, a wide living space (portions of public facilities, roads, and forests) had to be decontaminated. In 2011, mainly the Cabinet Office promoted the decontamination model project and commissioned the evaluation of decontamination technologies in residential land and elsewhere to the Japan Atomic Energy Agency. In addition, the "Act on Special Measures Concerning Response to Environmental Contamination by Radioactive Material Released from the Accident of the Nuclear Power Station Caused by the Great East Japan Earthquake, which occurred on March 11, 2011" (Law No. 110, August 30, 2011) (hereinafter the "Act on Special Measures" concerning the Handling of Radioactive Pollution was enacted, stipulating the system and standards to decontaminate and handle waste polluted by radioactive materials attributed to the accident at the Fukushima Daiichi Nuclear Power Station. The act specified the special decontamination area (formerly restricted and former deliberate evacuation areas) and the intensive contamination survey area. The former would be decontaminated directly by the Government, and the latter, where the additional annual exposure dose exceeded 1 mSv, would be by municipalities. (However, if the land and its construction were controlled by Japan, a prefecture, a municipality, and by that specified in the Ordinance of the Ministry of the Environment, they would take measures such as decontamination, and the agricultural land could be decontaminated by the prefecture at the request of the municipality.

For this reason, the Ministry of the Environment established the Ministry of the Environment Fukushima Office for Environmental Restoration in January 1, 2012 to develop a decontamination plan and decontamination project for the national land and helped the municipality prepare said decontamination plan. In addition, the Ministry of the Environment developed a new decontamination process sheet by the area under evacuation order, and has been conducting/planning the full-fledged decontamination of residential land in zones where plans to cancel evacuation directives had been prepared since the first quarter of 2012. (In many public facilities, including schools and government offices, decontamination was conducted as a model project.) Conversely, municipalities developed decontamination plans with priority and feasibility in mind and based on the Act on Special Measures concerning the Handling of Radioactive Pollution and actual circumstances of areas and developed decontamination implementation plan. If this plan is carried out, the municipality should select a proper method from the decontamination methods in the guideline relevant to decontamination released by the Ministry of the Environment in December 2011. The guidelines provided specific measures to investigate contaminated points (determination of measurement points and actual measurement) and decontamination of roofs, gutters, ditches, exterior walls, garden trees, walls, fences, benches, play equipment, etc. as decontamination measures for structures including buildings. In addition, decontamination methods for ditches, paved surfaces, unpaved roads, and others were conducted as road decontamination. Moreover, as soil decontamination measures, decontamination methods for schoolyards, gardens, parks, and agricultural land as well as plant decontamination measures, specific decontamination methods for lawns, street trees and other trees in the living space, and forests were conducted.

(2) **Decontamination technologies**

As for the decontamination technologies, with reference to the decontamination technologies adopted in the Chernobyl accident, from 2011 to 2012, the Cabinet Office commissioned the Japan Atomic Energy Agency to perform a model project [11] for building lots and surrounding forests exposed to high doses of 20 mSv/year or more. In this model project, the municipalities were divided into three groups to verify the decontamination technologies: Group A (Minami Soma, Kawamata and Namie towns, and Iidate village), Group B (Tamura city, Futaba and Tomoka towns, and Katsurao village), and Group C (Hirono, Okuma and Naraha towns, and Kawauchi village). The results of the model project showed that the surface contamination density-reducing effect differed quite significantly, even though the object was the same; significantly dependent on the material, the surface, adhesion states and other factors. In addition, the cesium gamma radiation had an impact at a radius of a few meters from the source. For effective decontamination, both a wider area as well as the point should be decontaminated. Accordingly, the effect would differ significantly, even if the same decontamination technology were applied. In selecting a suitable decontamination technology, it is important to decide on a case-by-case basis to determine proper decontamination methods and points, etc.

The AFFRC developed "practical technologies to promote new agricultural, forestry, and fishery policies" and "technologies to remove/reduce radioactive materials of facilities surrounding forests and agricultural land" in three areas in 2011: decontamination technologies in agricultural land, separation/removal technologies for radioactive cesium and volume-reduction technologies for contaminated rice straw, pasture and similar, and technologies to suppress the migration of radioactive cesium. As for issues of decontamination and volume-reduction, verification tests were continued or implemented as new issues in 2012. The Forestry and Forest Products Research Institute under the AFFRC Forestry Agency conducted "forestry decontamination verification tests by removing undergrowth and fallen leaves in conifer and deciduous broad-leaved forests", successfully reducing the air dose rate by about 60–70 % compared to before the decontamination by removing undergrowth and fallen leaves from forests. In addition, as a new decontamination technology, the Cabinet Office publicly solicited projects (by consigning it to the Japan Atomic Energy Agency) [16] and selected a total of 25 issues as technologies to enhance the efficiency of contamination work, technologies to compact decontaminated materials such as soil, shipment and temporary storage of removed materials, decontamination support, and other related technologies, and tested their effectiveness. Moreover, in Fukushima prefecture, in 2011, as a demonstration (publicly solicited) project of decontamination technologies in Fukushima prefecture [17] the effectiveness of technologies covering a total of 19 issues were tested in two fields of structures (roof, roof terrace, walls, underside, etc.) for decontamination and soil volume-reduction technologies (excluding agricultural land) respectively. In 2012, new issues were solicited and the project was continued. These included not only on-site decontamination but also, as the progress of decontamination to date reveals, issues of handling considerable organic waste produced by decontamination such as contaminated soil and plants. In other words, it was clear that reducing the concentration of radioactive materials by decontamination and compacting organic matter by burning would be important issues in the near future, and these issues were investigated to determine effective technologies for such measures. Various organizations such as universities, research institutions, and private companies other than those described here developed and verified decontamination and volume-reduction technologies.

5.4 Radioactive Material Release and INES Evaluation

5.4.1 Estimated Amount of Radioactive Material Release

5.4.1.1 Release to the Atmosphere and Time Sequence

To reduce the uncertainty of the amount of evaluated fission products (FP) release discharged into the atmosphere during the accident, the dose rate should be used at a point as close to the origin as possible, such as the main ventilation stack monitor. In the accident at the Fukushima Daiichi, the monitors attached to plant facilities could not be used because they had no electrical power supply and likewise the monitoring posts for the same reason. However, TEPCO measured dose rates in the power station by moving monitoring cars (Figs. 5.7 and 5.8), based on which the FP amount of release was evaluated [18]. This evaluation method requires repeated

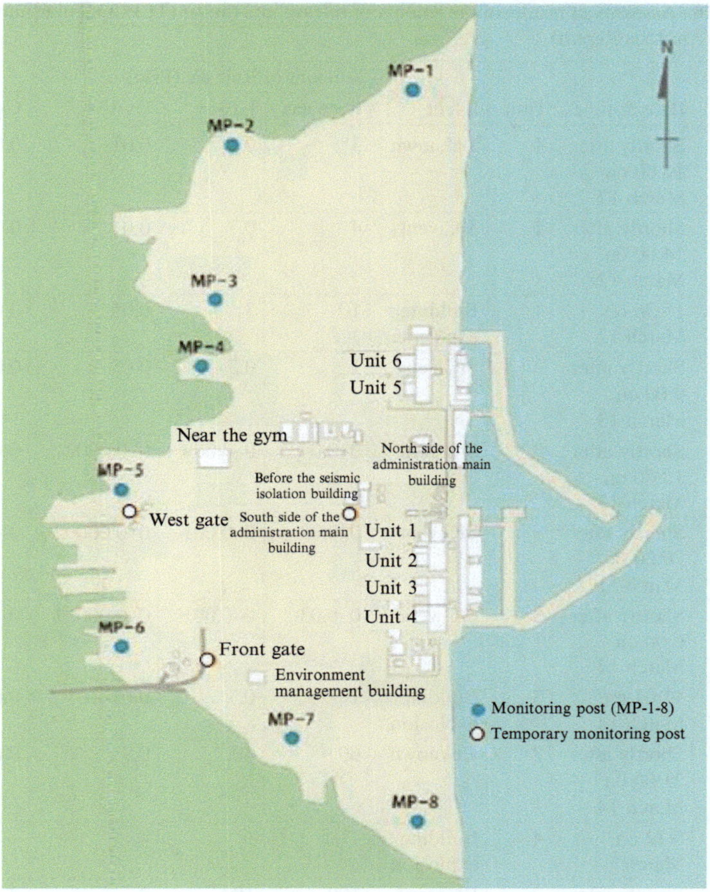

Fig. 5.7 Monitoring position (TEPCO, Fukushima Nuclear Accident Analysis Report)

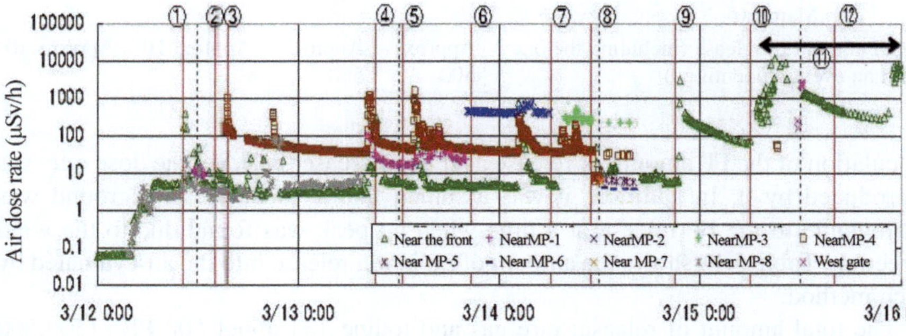

Fig. 5.8 Monitoring data (TEPCO, Fukushima Nuclear Accident Analysis Report)

Table 5.10 Assessment results of the amount of release into the air (TEPCO, Fukushima Nuclear Accident Analysis Report)

Serial Number	Date & time	Unit	Event	Amount of Release (PBq)			
				Rare gas	I-131	Cs-134	Cs-137
[1]	Shortly after 10:00 on March 12	1	Unknown	3	0.5	0.01	0.008
[2]	Shortly after 14:00 on March 12	1	S/C vent	4	0.7	0.01	0.01
[3]	15:36 on March 12	1	Building explosion	10	3	0.05	0.04
[4]	Shortly after 9:00 on March 13	3	S/C vent	1	0.3	0.05	0.03
[5]	Shortly after 12:00 on March 13	3	S/C vent	0–0.04	0–0.009	0–0.0002	0–0.0001
[6]	Shortly after 20:00 on March 11	3	S/C vent	0–0.03	0–0.001	0–0.00002	0–0.00002
[7]	Shortly after 6:00 on March 14	3	S/C vent	0–0.03	0–0.001	0–0.00002	0–0.00002
[8]	11:01 on March 14	3	Building explosion	1	0.7	0.01	0.009
[9]	Shortly after 21:00 on March 14	2	Unknown	60	40	0.9	0.6
[10]	6:12 on March 5	4	Building explosion	–	–	–	–
[11]	Shortly after 7:00 on March 15	2	Release from the building	100	100	2	2
[12]	Shortly after 16:00 on March 15	3	S/C vent	0–0.03	0–0.001	0–0.00002	0–0.00002
Total amount of release (including the one with an event unidentified)				Approx. 500	Approx. 500	Approx. 10	Approx. 10

calculation of the FP amount of release as if the increase (peak) of the dose rate was reproduced by it. In addition, it was assumed that 1 % of the background was attributable to the FP release at a time when no peak was found due to the wind direction. Table 5.10 shows the amount of radiation release into the air evaluated by such method.

The total amount of release: rare gas and iodine-131 about 500 PBq (500,000 TBq), respectively, cesium-134 and -137 about 10 PBq (10,000 TBq). It was the release from Unit 2 after 7:00 on March 15 that contributed most to the total amount

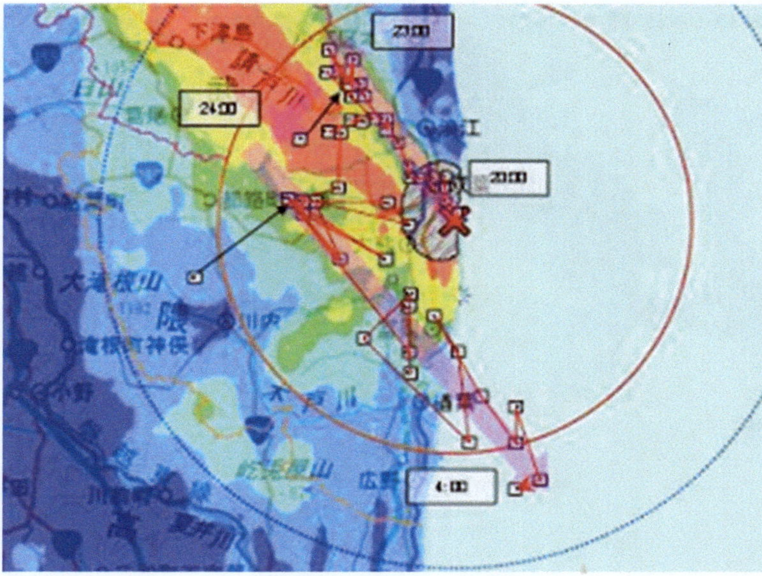

Fig. 5.9 Path of the plume released at 20:00 on March 15 (TEPCO, Fukushima Nuclear Accident Analysis Report)

of release. The FP release by the containment vessel vent was significantly smaller than the total amount of release and was not assumed to be a primary factor in the peripheral soil contamination.

It seemed that on March 15, Unit 2 continuously released FP. On that date, the wind blew toward the northwest (Fig. 5.9). The rain cloud was observed in the area to the northwest of the station at night (Fig. 5.10) and it seemed there was rainfall. Contaminated zones with high doses northwest of the station were identified, and it seemed that the main cause was FP release from Unit 2 on March 15.

As described below, the Nuclear and Industrial Safety Agency and Japan Atomic Energy Agency made a similar assessment and their results for the amount of cesium were almost the same as that of TEPCO. As for the amount of iodine, however, TEPCO's result was about three times larger. TEPCO assumed constant release rates of rare gas, iodine, and cesium, which might be influential.

5.4.1.2 Amount of Release into the Ocean

In evaluating the amount of release into the ocean, TEPCO evaluated the reproducible FP amount of release based on the radioactivity concentration measured in the ocean in the vicinity of outlets [18].

The main causes of marine pollution were the release from the vicinity of intake screen of Units 2 and 3, the release of low-density contaminated water in the Radioactive Waste Treatment Building, the release of low-concentration

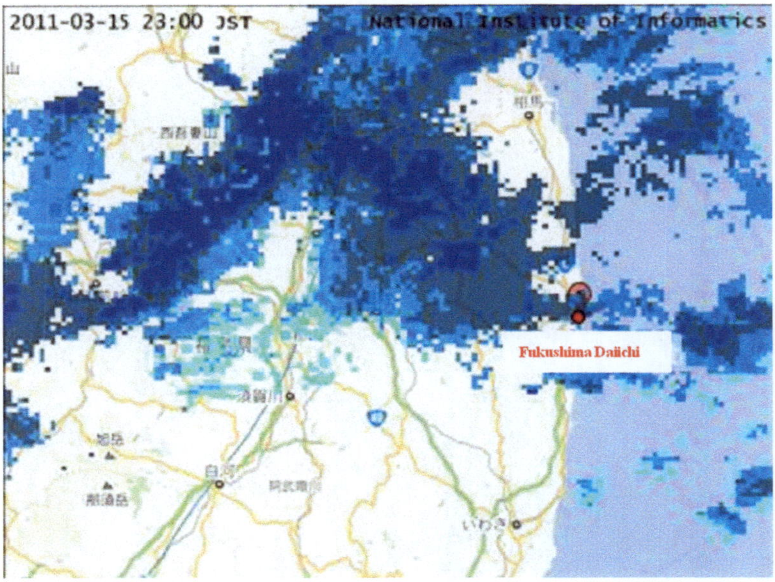

Fig. 5.10 Rain cloud radar map at 23:00 on March 15 (TEPCO, Fukushima Nuclear Accident Analysis Report)

Table 5.11 Assessment results of the amount of release into the ocean (PBq) (TEPCO, Fukushima Nuclear Accident Analysis Report)

Nuclide	Total Amount	March 26–31	April 1–June 30	July 1–September 30	Remarks
I-131	11	6.1	4.9	5.7×10^{-6}	Including direct leak (2.8) (April 1–6, April 4–10, May 10–11)
Cs-134	3.5	1.3	2.2 (1.26 + 0.94)	1.9×10^{-2}	Including direct leak (0.94) (April 1–6, April 4–10, May 10–11)
Cs-137	3.6	1.3	2.2 (1.26 + 0.95)	2.2×10^{-2}	Including direct leak (0.94) (April 1–6, April 4–10, May 10–11)

subsurface water accumulated in the Units 5 and 6 sub-drain pits, and FP falling down from the air and the inflow from rainwater as well.

As a result of evaluation, the amounts of release were: about 11 PBq (11,000 TBq) of iodine-131, about 3.5 PBq (3,500 TBq) and about 3.6 PBq (3,600 TBq) of cesium-134 and -137 respectively (Tables 5.5, 5.6, 5.7, 5.8, 5.9, 5.10, and 5.11). As shown in Fig. 5.11, the radiation flow into the ocean was greatly reduced by the end of April.

Japan Atomic Energy Agency and others conducted similar evaluations and showed numerically similar results.

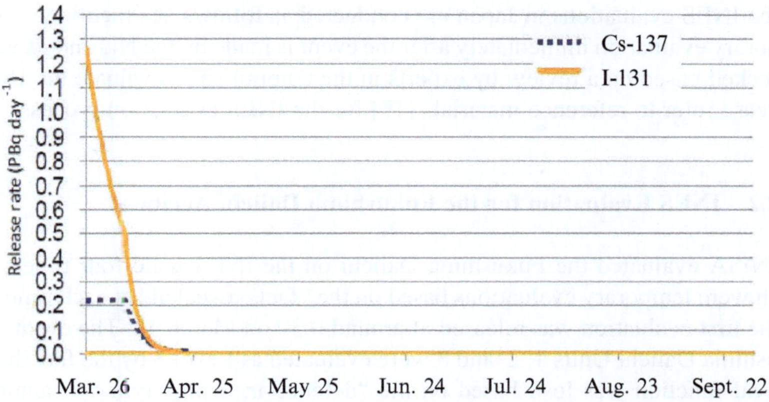

Fig. 5.11 Changes of rate of release into the ocean (TEPCO, Fukushima Nuclear Accident Analysis Report)

5.4.2 INES Evaluation

5.4.2.1 Overview of the INES Evaluation

The International Nuclear Event Scale (INES) is a system to inform of safety-significance of accidents/events caused by the utilization of nuclear power or radiation to the general public in the country promptly and of major items to the International Atomic Energy Agency (IAEA), where the safety-significance is evaluated based on an internationally unified standard (scale).

INES scale is classified as follows: starting from the scale without any safety-significance and Level 0, then Levels 1–3 as incident and Levels 4–7 as accident. The level on the scale is determined by three criteria: "people and environment" (damage caused by radioactive materials to the general public and workers), "radiological barriers and controls at facility" (degree of loss for the containment function of radioactive materials), and "defence in depth" (operability of the safety function and frequency of initiating events). Moreover, the detailed "INES User's Manual" is also prepared to allow any country to evaluate similarly. At present, the 2008 version [19] is used.

INES participating countries have INES national officers (NO) respectively. When notified of the occurrence of an event by the licensee, the regulatory authority immediately evaluates the safety-significance based on INES, an internationally common scale, and like the magnitude and seismic intensity in the event of an earthquake, announces "this is the event of a certain level". At this point it is recognized as a "temporary evaluation". The NO should also notify the IAEA of a level 2 or higher event. The IAEA staff in charge of INES notifies each country's NO of a summary of the event and the INES evaluation result through NEWS, the information system operated by the IAEA.

The INES evaluations in Japan are conducted as follows. As mentioned before, temporary evaluation immediately after the event is made by the NO and its validity is checked based on a review by experts in the committee to evaluate the INES.

Please refer to reference materials [19] for the INES evaluation process.

5.4.2.2 INES Evaluation for the Fukushima Daiichi Accident

The NISA evaluated the Fukushima Daiichi on the INES scale four times; all of which were temporary evaluations based on the "facts revealed" at each time point.

The first evaluation was released at around 0:30 on March 12. The events at the Fukushima Daiichi Units 1, 2, and 3 were evaluated as Level 3 by the fact that heat removal function was lost, based on the "defence in depth" criterion among the previously mentioned three criteria. Similar evaluations were conducted for the Fukushima Daini Units 1, 2, and 4.

The second evaluation was conducted in the evening of March 12. The event at the Fukushima Daiichi Unit 1 was reevaluated as Level 4 based on the criterion of "radiological barriers and controls at facility" by the fact that the radiation level on the Fukushima Daiichi site boundary rose at 16:17, which was judged to be due to the radiation released from Unit 1.

The third evaluation was conducted on March 18 [20]. The events at the Fukushima Daiichi Units 1, 2, and 3 were reevaluated as Level 5 based on the criterion of "radiological barriers and controls at facility". As the following situations were observed, it was judged that a core meltdown was likely to occur:

(1) There was a time when the reactor water level was below the top of active fuel, and the temperature of the fuel seemed to rise.
(2) Hydrogen combustion seemed to cause explosion.
(3) The radiation level rose both inside and outside the site.

The fourth evaluation was conducted on April 12 [21]. As described above, the INES prepared the criterion of "people and environment", as one of three criteria. This is based on the following concepts of human exposure dose and the release of radioactive materials into the environment.

"How many persons were exposed and how seriously" is the simplest scale. However, suppose it is the only basis for evaluation and that disaster prevention measures were conducted effectively, the exposure dose would be so small that the evaluation result would not reveal the severity of the accident, despite the serious damage of the facility. Accordingly, "how much radioactive material was released into the environment" was introduced as the criterion. This criterion is used to judge the event at Level 4 or higher.

In April, the Nuclear and Industrial Safety Agency evaluated the amount of radioactivity released into the air of 370,000 TBq in terms of iodine-131 as a trial calculation using the conversion factor shown in the INES User's Manual based on the analytical results of the Fukushima Daiichi Units 1, 2, and 3 reactors conducted by the JNES. In addition, by back-calculation from monitoring data, the Nuclear

Safety Commission evaluated the total release from the Fukushima Daiichi into the air of 630,000 TBq in terms of iodine-131.

Significant uncertainty was remained in these results, however both of which exceeded the criterion of Level 7 of "an amount of release of radioactive material is more than several tens of thousands TBq in terms of iodine-131". Therefore, the NISA rated the events as Level 7 for whole Fukushima Daiichi and published the result, while the INES evaluation had conducted for each unit according to the INES User's Manual until March 18.

The INES evaluation of the Fukushima Daiichi remained temporary as of the end of March 2013. It seems the amounts of release of radioactive materials into the air, as estimated by the NISA, NSC, TEPCO and some research institutes, show the evaluation of whole Fukushima Daiichi is remained at Level 7.

5.4.2.3 Issues on the INES Evaluation and International Discussion

The Fukushima Daiichi was the first INES level 7 event after the start of INES and was unprecedented: the INES evaluation was conducted under the situation progressing and in emergency where sufficient information could not be obtained.

The following issues of INES evaluation of the Fukushima Daiichi accident were highlighted:

(1) INES aimed to "immediately inform the significance of the accident to the general public". Due to the loss of functions for plant parameters and radiation monitoring, it was difficult to estimate the progress of the accident and the status of the release of radioactive materials.
(2) Consequently, the accident was rated at Level 5 one week after the start of the accident and Level 7 one month after, which was far from an "immediate notice".
(3) INES evaluation was conducted each time based on "facts with high reliability". However, the changes of evaluation results according to the progress of accident invited the criticism that they tried to hide the significance of the accident.
(4) Some criticized the evaluation of Level 7, the same level as the Chernobyl accident, when the amount of release of radioactive materials in Fukushima Daiichi was smaller than that of Chernobyl accident by one digit.
(5) As INES evaluation had been conducted on a unit-by-unit basis, the issue is the way how to evaluate similar simultaneous multi-unit accidents, in addition to the accidents in several NPSs in the same region.

Doubts were raised concerning the method used to estimate the amount of release of radioactive materials by the NISA. As described above, the NISA made a trial calculation of the amount of release using the severe accident analysis code. At the time when the trial calculation was conducted, it was difficult to expect an accurate analysis for each unit.

According to the result of the amount of release evaluated by TEPCO shown in Sect. 5.4.1, it was the release from Unit 2 after 7:00 on March 15 that contributed most to the total amount of release, which seemed due to a leak from the drywell (D/W).

On the other hand, as at Unit 2, "an impulsive sound was heard in the vicinity of the suppression chamber (S/C) at around 6:00–6:10 on March 15 and almost simultaneously the Unit 2 S/C pressure was lowered", most of the persons concerned at the time erroneously guessed "significant break occurred in the S/C" (Refer to Sect. 3.3). In this case, it was assumed that radioactive materials leaked from the S/C.

In estimating with calculation codes, the difference in assumptions of the leak source, the drywell or the S/C, results in a big difference. This is because depending on whether the scrubbing effect by the S/C should be considered or not, there may be a double-digit difference in the amount of release. In estimating the amount of release for INES evaluation, the results of environment monitoring data should be prioritized, and the calculation code results should remain as a reference.

Following the Fukushima Daiichi accident, doubts were raised in global society, namely: "whether INES is effective as a means of conveying information in emergencies". Under the circumstances, Director General Amano referred the task of enhancing INES effectiveness as a means of conveying information to the INES advisory committee (INES-AC), during the IAEA ministerial meeting in June 2011.

Subsequently, the INES-AC spent 1 year considering the same. At the meeting, first of all, "whom INES was for" was discussed. The conclusion was INES could not be regarded as the means of conveying necessary information to the general public who might be influenced in emergencies but rather the significance of the accident to the general public who would not be influenced. While the INES-AC had prepared the "Use of INES" guidance, the "additional guidance" was also prepared to add INES evaluations during the severe accident and guide to convey information. The new guidance shows that INES evaluation should be conducted when an event such as the occurrence of a reactor core meltdown is revealed, gaining considerable credibility, but in the case such event progressing, great caution should be paid to the means of announcing the evaluation, and that one event should not be excessively evaluated with INES. In response to criticism as to why the Fukushima Daiichi accident was rated at the same Level 7 as Chernobyl, seven levels are unchanged, that means Level 7 shows an extremely serious accident and subdividing it is not meaningful.

5.5 Communication After the Accident

From March 11 through 15, 2011, the main actors managing the accident in the Fukushima Daiichi Nuclear Power Station were the Prime Minister's official residence, governmental agencies including the Nuclear and Industrial Safety

Agency, TEPCO including the Tokyo head office and the Fukushima Daiichi Nuclear Power Station, municipalities near the station, and others. At the time however, there was a remarkable lack of communication internally, among the actors, and between actors and the general public.

Section 5.5.1 of this chapter introduces major examples of miscommunication related to the actors involved, followed by the activities to convey information by AESJ in Sect. 5.5.2. These evaluations are described in Sect. 6.13.

5.5.1 Miscommunication Related to the Actors Involved

(1) **The Government and TEPCO**

Before the accident, communication channels in emergencies between the Prime Minister's official residence/the Nuclear and Industrial Safety Agency and nuclear licensees had been prepared. However, there was a huge need for information immediately after the accident this time, therefore communication channels prepared could not satisfy the need, and new channels were added on an ad hoc basis.

For example, the TEPCO head office and Fukushima Daiichi Nuclear Power Station used a video-conferencing system to share information in real time, but the Prime Minister's official residence and the Nuclear and Industrial Safety Agency could not use it until March 15. Accordingly, the Emergency Response Center (ERC) established in the Ministry of Economy, Trade and Industry collected information on plant parameters by TEPCO staff members calling the TEPCO head office by mobile phone and orally communicating. At the time, Prime Minister Kan was on the fifth floor of the Prime Minister's official residence. TEPCO executives there collected information by mobile phone and communicated it to the Prime Minister. Moreover, mobile phones could not be used on the basement floor where the crisis management center was located.

Consequently, the information obtained by the Prime Minister's official residence and NISA from TEPCO was delayed and fragmented one. The extent of information sharing among the three parties was insufficient in terms of both quality and quantity.

(2) **The Government, municipalities, and evacuees**

Communication between the Government and municipalities or evacuees were neither proper nor sufficient.

Fukushima prefecture issued an evacuation order to residents within a 2-km radius of Fukushima Daiichi by its own judgment at 20:50 on March 11, while the Government issued an evacuation order to residents within a 3-km radius at 21:23, 33 min after 20:50. However, this evacuation order by the Government was not notified to Fukushima prefecture in advance. The orders themselves were communicated to the target residents relatively swiftly via various tools such as emergency radio and television. However, when an evacuation order was issued to residents within a 10-km radius at 5:44 on March 12, only 20 % of residents were aware that an accident had occurred. And there was no information at the time of evacuation that it would take a long time to return home upon completion of evacuation. In addition, some residents evacuated to areas which were found out to be affected by high doses.

(3) **Government and the General Public**
Some cases of the information disclosure by the Government in emergencies were improper.

For example, when radioactive materials caused by the accident were detected in food products immediately after the accident, then Chief Cabinet Secretary Yukio Edano frequently repeated the remark: "(the amounts detected in food products) would not immediately affect the human body and health". This remark was interpreted in various ways.

As for the core meltdown, NISA Deputy Director General Koichiro Nakamura explained the possibility at the press conference at 14:00 on March 12. Immediately after the conference, the Prime Minister's official residence requested that the Prime Minister's official residence be notified of the details of NISA press conference in advance, since which time the NISA continued to avoid the expression "core meltdown". On April 10, the NISA decided to use the term "fuel pellet meltdown" and notified TEPCO of the decision. It was May 16, 2011 that the Government formally used the term "core meltdown".

5.5.2 Communications to the Public by AESJ

(1) **Activities of the board of directors, committee, and subcommittees**
On March 18, 2011, Atomic Energy Society of Japan (AESJ) released "Kokumin no Minasama he - Tohoku Chiho Taiheiyo Oki Jishin niokeru Genshiryoku Saigai nitsuite-" (To the People of Japan—Nuclear Disaster in the Great East Japan Earthquake) on its website. On March 17, AESJ established an e-mail address "Q and A" to respond to questions from citizens and had responded to 100 or more questions by December 2012. In terms of content, about 50 questions related to radioactive materials and radiation, about 60 related to the reactor, and about 20 related to the activities of AESJ.

In addition, AESJ established Committee for Investigation of Nuclear Safety in April 2011, under which three subcommittees were also established: Technical Analysis Subcommittee, Task Group on Radiological Aspects, and the Cleanup Subcommittee. The Committee for Investigation of Nuclear Safety was inherited to AESJ Investigation Committee on the Nuclear Accident at the Fukushima Daiichi Nuclear Power Plant in August 2012. In addition, in June 2012, AESJ established the "Fukushima Special Project" to support the environmental restoration in Fukushima. Besides, AESJ subcommittees, committees, and division engaged in activities related to the accident, and the Nuclear Safety Division held a "Seminar on the Accident at the Fukushima Daiichi

Nuclear Power Station". The results of these activities were published in annual meetings and symposiums.

(2) **Activities of Team 110**

Team 110 to provide easy-to-understand explanations of abnormal events at nuclear facilities, introduces nuclear experts as commentators upon request from mass media and municipalities when an accident occurred, such as the release of radioactive materials in nuclear facilities. Since its establishment in 2010, the response to the accident at the Fukushima Daiichi Nuclear Power Station was the first activity. Following the accident, AESJ received many inquiries from newspapers, TV stations, network information agencies, foreign media, and others. As ten commentators, who registered themselves in advance, were unable to accept all requests, the team selected new commentators from the technical supporter list for help. The actual number of requests handled by Team 110 was 352 from March 2011 to December 2012.

Team 110 was established based on an assumption of accidents equivalent to the JCO accident. Immediately after the accident, they were unable to satisfy media needs in terms of both quality and quantity. Under the circumstances, commentators contributed to some degree by providing easy-to-understand explanations on expert knowledge, but the effect achieved was limited. In addition, their explanation to media was mainly based on their personal viewpoints, rather than AESJ's message.

(3) **Information through "AtomoΣ: journal of Atomic Energy Society of Japan"**

The contents of "AtomoΣ", the monthly journal issued by AESJ, was changed in the May 2011 number. The journal mainly deals with accidents and the countermeasures and more than half the articles have been about the accidents. (See http://www.aesj.or.jp/atomos/tachiyomi/mihon.html for details)

5.6 Off-site Support Activities

Before the accident, no response to a severe accident had been assumed, but the related parties judged situations flexibly in providing various off-site supports for equipment and materials: batteries, power supplies, equipment for water injection, fuel, radiation management/protective articles, food and beverages, and others (commodities, clothes, bedclothes, living water, toilets, etc.).

5.6.1 Actual Conditions of the Off-site Distribution

The earthquake and tsunami resulted in events such as an inability to deliver equipment and materials to the required site due to several distribution obstructions such as road damage (closures to traffic, traffic jams, etc.) over a wide area from the

Tohoku district to the Kanto district, the deterioration of the communication environment, and outdoor contamination around the Fukushima Daiichi site due to the explosions at Units 1, 3, and 4.

As the hydrogen explosion occurred at the Fukushima Daiichi Unit 1 on March 12, the dose in J village became high and there was concern that it might be subject to an evacuation order. On March 12, TEPCO decided to use the Onahama Call Center in Iwaki city as a base for receiving equipment and materials. However, because the Onahama Call Center and surrounding suffered from the earthquake and tsunami, there were many unknown senders/destinations of materials and equipment, and there was insufficient preparation to receive materials, therefore, the received materials, if any, could not be organized but were forced to be simply stacked up in order of acceptance, the settlement/inventory management at the call center was confusing and chaotic. Moreover, the radiation level in the surrounding areas rose further on March 15, and the transport to Fukushima Daiichi became difficult. Therefore, the transport from the outside was stopped at Onahama Call Center and elsewhere, and TEPCO employees and other supporters took over the transport from there to Fukushima Daiichi. However, because of insufficient road information due to a lack of communication means, insufficient experience of driving large vehicles, wearing full-face masks, a lack of sense of geography, goods delivery places changed repeatedly, a shortage of unloading heavy machinery or a shortage of heavy machinery operators, and goods in unexpected places, long delays in transport and non-delivery of goods occurred frequently, which made it difficult for goods to reach the station in time.

5.6.2 Status of Securing Materials and Equipment

(1) **Securing batteries**
The batteries to supply DC power are essential in the event of accidents for plant monitoring, depressurization, water injection and cooling. Before the accident, no spare was required. From the evening of March 11, when the DC power had been lost, the Nuclear Emergency Response Headquarters at the Fukushima Daiichi strove to secure batteries, and the TEPCO head office also made every effort to gather batteries regardless of the specification. There are three main ways used to secure batteries: gathered and purchased by the station, and transferred within TEPCO.

(2) **Support from the TEPCO head office, etc.**
Understanding an overview of the damage of the electric facility, the TEPCO head office prioritized the order of vehicle batteries, which were easily carried around. From midnight on March 11 to the morning of the 12th, it ordered 1,000 units of 12 V vehicle batteries, but permission to use highways was not smoothly obtained and it was around 0:00 on March 14 when they were delivered to Onahama Call Center. By 21:00 on March 14, the head office support staff members transported about 320 battery units using two heavy

trucks to the Fukushima Daiichi, whereupon transportation was suspended due to the deterioration of the site environment and resumed from March 17. Moreover, the Kashiwazaki Kariwa Nuclear Power Station bought twenty 12 V vehicle batteries in Kashiwazaki city and sent them with support staff to the Fukushima Daiichi on March 14.

(3) **Use within TEPCO**
The TEPCO head office succeeded in urging thermal power stations and branches in TEPCO to supply various types of batteries they possessed. The Hirono Thermal Power Station removed fifty 2 V batteries (12.5 kg/unit) from itself, which were brought to Fukushima Daiichi by the Self-Defense Forces helicopter at around 1:20 on March 12 and used to start up the Unit 1 diesel-driven fire pump and restore the Unit 3 reactor water level measurement system. Besides, additional 2 V batteries were delivered by the Kawasaki Thermal Power Station and Tokyo branch.

(4) **Procurement from outside by the Fukushima Daiichi**
The residents' evacuation made it impossible to procure batteries from shops around the station, hence the station tried to procure vehicle batteries in distant cities, including Iwaki city. While a series of major aftershocks occurred, it was difficult to find open shops due to large-scaled landslides and obstacles caused by the earthquake and the tsunami blocking National Route 6 and other highways at many points, in addition to a traffic jam caused by the vehicles of residents evacuating and the damage of Iwaki city by the earthquake. However, on this occasion, eight 12 V vehicle batteries were purchased and used in the control rooms of Units 3 and 4. Because the office building was seriously damaged and big aftershocks occurred frequently, the building was closed for the time being and the safe was inaccessible, hence staff members' money was temporarily borrowed and used for the purchase. In addition, five 12 V batteries were removed from buses and vehicles for business use on the station, and brought to the main control rooms of Units 1 and 2 on the night of March 11, whereupon part of the reactor water level measurement systems became active. Twenty batteries were also secured from individual staff members' commuting vehicles; ten of which were connected in series to operate the main stream safety relief valves to depressurize the reactors of Units 2 and 3.

5.6.3 Securing Power Supply Vehicles

The Fukushima Daiichi judged that early restoration of the power supplies was difficult and aimed to restore the power supply using mobile power supply vehicles. Three potential suppliers were considered; TEPCO itself, other utilities, and the Self-Defense Forces.

The total number of TEPCO power supply vehicles was fifteen, including both high- and low-voltage power supply vehicles, while the total number of other utilities and the Self-Defense Forces was seven. Because of the damage to the roads and traffic jams, it took longer time to transport them. At around 17:50 on March 11, though the TEPCO Head Office asked the Self-Defense Forces to transport power supply vehicles by helicopter, air transportation was abandoned as the vehicles were too heavy. The power supply vehicles sequentially arrived at the station by land on and after March 12 after all. Subsequently, cable laying routes was considered, cables were prepared, rubble was removed, and cable was laid. On and after March 13 power was partially restored via the power supply vehicles. However, the influence of the explosion of the reactor building and troubles, including cable damage, hindered the power supply recovery. As TEPCO power supply vehicles arrived early, those from the other suppliers were not used.

5.6.4 Securing Fire Engines

At the Fukushima Daiichi, injecting water into the reactor using fire engines was considered, in addition to injecting water via fire extinguishing piping prepared as a part of accident management measures. Accordingly the local headquarters arranged for not only fire engines possessed by the station but also additional ones. One of the three fire engines possessed by Fukushima Daiichi was out of order and another was at the side of Units 5 and 6 blocked off by rubble, which meant only one was available. On March 13, as rubble was removed, one fire engine became available and was used to inject water into the reactor henceforth. A total of 12 off-site fire engines were provided: seven from the Kashiwazaki Kariwa Nuclear Power Station and the TEPCO thermal power station on Tokyo Bay, and five from other utilities, the Government, the Self-Defense Forces, and others. These fire engines sequentially arrived at the station on and after the morning of March 12 and were used to transfer water and seawater to the fire-protecting water tanks, that were the water source of cooling water to the reactor, and to inject them into the reactor directly.

Besides, following this accident, various types of support activities were provided from off-site: taking pictures of the accident scenes by remote-control helicopter and removal of rubble from areas of high radiation by robot. Conversely however, many issues were revealed. For example, robots, which were developed for emergencies, could not be used because they had not been maintained for a long time.

Issues for future off-site support are shown in Chap. 6.

References

1. UN Chernobyl Forum Expert Group “Environment”, *Environmental Consequences of the Chernobyl Accident and Their Remediation: Twenty Years of Experience* (IAEA, 2006), Austria, p. 71
2. Ministry of Health, Labour and Welfare, The Radioactive Material Response Working Group in the Food Sanitation Subcommittee of the Pharmaceutical Affairs and Food Sanitation Council report, “Shokuhin chu no Hoshasei Bussitsu ni kakawaru Kikaku Kijun no Settei ni

tsuite (The Establishment of Standards and Criteria on Radioactive Materials in Food)" (Ministry of Health, Labour and Welfare, 2011), Japan, p. 10

3. TEPCO, *Fukushima Nuclear Accident Analysis Report* (TEPCO, 2012), Japan
4. O. Kurihara, K. Kanai, T. Nakagawa et al., J. Nucl. Sci. Technol. **50**(2), 122–129 (2013)
5. Public Interest Incorporated Foundation Nuclear Safety Research Association (NSRA), *Kinkyuji Hibaku Iryo no Chishiki (Knowledge of Radiation Exposure Medical Care in Emergencies)* (Public Interest Incorporated Foundation Nuclear Safety Research Association (NSRA), 2003), Japan
6. National Institute of Radiological Sciences, NIRS-M-252 (Proceedings: The 1st. NIRS Symposium on Reconstruction of Early Internal Dose in the TEPCO Fukushima Daiichi Nuclear Power Station Accident) (2012)
7. National Institute of Radiological Sciences, *Dai Ni Kai Kokusai Sinpojium Tokyo Denryoku Fukushima Daiichi Gensiryoku Hatudensho Jiko ni okeru Shoki Naibu Hibaku Senryo no Saikochiku (The Second International Symposium Reconstruction of Initial Internal Exposure Doses in TEPCO's Fukushima Daiichi Nuclear Power Station Accident), Summary Collections* (National Institute of Radiological Sciences, 2013), Japan
8. The Ministry of Education, Culture, Sports, and Science and Technology, Dai Goji Kokuki Monitoring no Sokutei Kekka, and (2) Fukushima Daiichi Genshiryoku Hatsudensho kara 80 km Kengai no Kokuki Monitoring no Sokutei Kekka ni tsuite (The Fifth Airborne Monitoring Measurement Results and (2) Measurement Results of Airborne Monitoring beyond a 80 km Radius of the Fukushima Daiichi Nuclear Power Station) (2012)
9. IAEA, *Environmental Consequences of the Chernobyl Accident and their Remediation: Twenty Years of Experience*, STI/PUB/1239, (IAEA, 2006), Austria
10. The Ministry of Education, Culture, Sports, and Science and Technology, Monbu Kagaku-sho niyoru, (1) Ganma-sen Hoshutsu Kakushu no Bunseki Kekka, oyobi (2) Strontium 89, 90 no Bunseki Kekka (Dai Niji Bunpu Jokyo Chosa) ni tsuite (Analytical Results of Gamma Ray Emitting Nuclides and (2) Analysis Results of Strontium-89 and Strontium-90 (the Second Distribution State Survey) by MEXT (2012)
11. Secretariat of Agriculture, Forestry and Fisheries Research Council (AFFRC), *Nochi Dojo no Hoshasei Bussitsu Nodo Bunpu Map Kanren Chosa Kenkyu Hokokusho (Dai 3 Hen) (Research Study Report Related to Map of Radioactive Materials Concentration in Agricultural Soil)* (Part 3) (2012)
12. Japan Atomic Energy Agency, Fukushima Daiichi Genshiryoku Hatsudensho Jiko ni Matsuwaru Hinan Kuiki nado ni okeru Josen Jissho Gyomu Hokokusho (Report on Decontamination Validation Report in the Evacuation Zones related to the Accident at the Fukushima Daiichi Nuclear Power Station) (the Cabinet Office's Commissioned Project) (2012)
13. N. Sato et al., A Test case of clean-up of a radioactivity-contaminated paddy field, 1; Entire scheme and outline of the field, 033, the 2012 autumn meeting of the Atomic Energy Society of Japan (AESJ)
14. M. Saso et al., A Test case of clean-up of a radioactivity-contaminated paddy field, 2; Evaluation of radiation measurement and the contamination status, 034, the preliminary draft of the 2012 autumn meeting of the Atomic Energy Society of Japan (AESJ)
15. T. Jintoku et al., A Test case of clean-up of a radioactivity-contaminated paddy field (3) Evaluation of rough stirring and the decontamination effects, 035, the preliminary draft of the 2012 Autumn meeting of the Atomic Energy Society of Japan (AESJ)
16. Japan Atomic Energy Agency, Heisei 24 nendo Josen Gijutsu Hyoka nado Gyomu Hokokusho (the 2012 Report of Decontamination Technologies Evaluation) (2012)
17. Living Environment Division of Fukushima Prefecture, *Heisei 23 nendo Fukushima ken Josen Gijutsu Jissho Jigyo Jicchi Shiken Kekka (the 2011 Results of Decontamination Technologies Demonstration Project Field Test in Fukushima pref.)* (2012)
18. TEPCO, *Fukushima Daiichi Genshiryoku Hatsudensho Jiko ni okeru Hoshasei Busshitsu no Taiki chu heno Hoshuturyo no Suitei ni tsuite* (*Estimated Radioactive Material Release into the Air in the Accident at the Fukushima Daiichi Nuclear Power Station*) (TEPCO, 2012), Japan

19. IAEA, INES, *The International Nuclear and Radiological Event Scale* – User's Manual, 2008 Edition, (IAEA, 2009), Austria
20. The Nuclear and Industrial Safety Agency, Tohoku Chiho Taiheiyo-oki Jishin niyoru Fukushima Daiichi Genshiryoku Hatsudensho oyobi Fukushima Daini Genshiryoku Hatsudensho no Jiko Trouble ni taisuru INES (Kokusai Genshiryoku Hoshasen Jisho Hyoka Shakudo) no Tekiyo ni tsuite (Application of INES (the International Nuclear and Radiological Event Scale) to the Accidents/Troubles at the Fukushima Daiichi Nuclear Power Station and the Fukushima Daini Nuclear Power Station due to the Great East Japan Earthquake) (2011)
21. The Nuclear and Industrial Safety Agency, Tohoku Chiho Taiheiyo-oki Jishin niyoru Fukushima Daiichi Genshiryoku Hatsudensho no Jiko Trouble ni taisuru INES (Kokusai Genshiryoku Hoshasen Jisho Hyoka Shakudo) no Tekiyo ni tsuite (Application of INES (the International Nuclear and Radiological Event Scale) to the Accidents/Troubles at the Fukushima Daiichi Nuclear Power Station due to the Great East Japan Earthquake) (2011)

Chapter 6
Accident Analysis and Issues

Abstract Chapters 2–5 presented an outline of Nuclear Power Station, an overview of the accident at the Fukushima Daiichi NPS, an overview of the events that took place at nuclear power stations other than the Fukushima Daiichi, and responses to the accident outside the power station based on fact. In keeping with these facts, Chap. 6 analyzes and evaluates the accident from various perspectives. First, the items of the analysis in Chap. 6 are explained, the accident progression behavior is simulated, and the findings from the evaluation of the releases of radioactive materials during the accident are presented in Sect. 6.1. Second, each item is analyzed and evaluated and the relevant issues are addressed in Sect. 6.2 and subsequent sections.

Keywords Accident analysis • Accident management • Defence in depth • Emergency preparedness • Environment remediation • External events • Human resources in nuclear field

6.1 Overview of Accident Analysis

The accident progression in the first around 100 h (4–5 days), during which time most radioactive materials were released into the external environment, is important to be clarified in the accident analysis. Subsequently, the focus is shifted to the stable cooling of fuel debris, movement of radioactive materials in reactor pressure vessel and containment vessel, their behavior in terms of the environmental release, and the reduction of the release. Finally the ongoing decommissioning of reactors requires various factors to be taken into consideration. The conditions and measures taken, from a few days after the accident to date, vary significantly depending on how the accident progressing in the first few days. It is crucial, therefore, to understand the accident progression over the first few days, and estimate the reactor condition by simulation in the situation of inability to observe the inside of the reactor, and clarify the releases of radioactive materials into the environment.

In Sect. 6.1.1, first the items required to consider the nuclear plant equipment and management from the standpoint of nuclear safety are extracted. Subsequently, issues are clarified from the accident progression at Units 1–4 and the actual condition of their nuclear disasters, and from the events that took place at nuclear

The Fukushima Daiichi Nuclear Accident: Final Report of the AESJ Investigation Committee, DOI 10.1007/978-4-431-55160-7_6

power plants other than the Fukushima Daiichi. Then technical and investigative areas (items) related to these issues are presented. These items are analyzed and evaluated in detail in Sect. 6.2 and subsequent sections. In Sect. 6.1.2, the accident progression behavior simulated by the severe accident simulation code SAMPSON is described. The SAMPSON code is used because it can represent events more accurately than other codes using a mechanistic model, and it was used independently by the Accident Investigation Committee of AESJ. In Sect. 6.1.3, the release of radioactive materials into the environment (atmosphere and sea) is explained primarily in the first few days. For the release of radioactive materials into the atmosphere in particular, monitoring data is compared with analytical results using the MELCOR code to demonstrate the capability of severe accident analysis to indicate most of actual releases into the environment.

6.1.1 Items in the Analysis

Chapter 6 analyzes and evaluates the Fukushima Daiichi accident to identify the problems behind it. Items subject to analysis and evaluation were selected using the following two methods to avoid omissions:

- Selecting items necessary when considering power plant equipment and management in terms of nuclear safety
- Clarifying issues from accident progression at Units 1 to 4 (Chap. 3), the status of accompanying nuclear disasters (Chap. 5), events in power stations other than the Fukushima Daiichi NPS (Chap. 4), and selecting technical and investigative areas (items) in which these issues are included

These items are examined independently in separate sections from Sect. 6.2.

6.1.1.1 Nuclear Power Plant Equipment and Management

The items listed below are chosen by the Accident Investigation Committee of AESJ which are necessary when considering nuclear power plant equipment and management in terms of nuclear safety.

- Concept of nuclear safety
- Internal/external events
- Defence in depth
- Plant design, accident management
- Disaster preparedness: Emergency preparedness and response, environmental restoration and decontamination
- Simulation analysis
- Human resources and human factors
- Radiation and radiation monitoring
- Nuclear security, physical protection and safeguards

- Safety regulations
- Relationship with international society
- Information dissemination
- Other

6.1.1.2 Accident Progression at Units 1 to 3, etc.

The issues conduced from the accident progression at Units 1 to 3, and technical and investigative areas (items) in which these issues are included are clarified (see Table 6.1).

(1) **Accident progression in Unit 1**

(a) At 14:46 on March 11, 2011, an earthquake occurred and Unit 1 in operation automatically shut down. When external power was lost, two emergency diesel generators (D/Gs) automatically started to supply power. The isolation condenser (IC) maintained the plant in hot standby state, and neither plant parameters nor visual inspection of the reactor building showed any abnormality.

(b) With the onslaught of the tsunami at around 15:30, all D/Gs, DC power and seawater pumps stopped functioning, causing a station blackout, while the monitoring and control instruments in the central control room also lost their function (Issue A).

(c) Operation of IC: The IC system was the only option available to cool the reactor in the state of the plant following the tsunami.
The tsunami caused a station blackout, and the interlock mechanism on the isolation valves in the IC system activated to close almost fully, putting the IC system into a near loss of cooling function (Issue B).

(d) Core damage: After the IC system failed to function, the reactor pressure was controlled by the main steam safety-relief valve (SRV), but the water level of the reactor gradually decreased due to the SRV operation, and dropped below the top of the active fuel (TAF) past 18:00. Core damage presumably started before 19:00.
The molten core is believed to have damaged the RPV before 2:30 on March 12 when the dry well (D/W) pressure almost reached the reactor pressure, 0.8 MPa [gage] (Issue C).

(e) Alternative water injection: At around 4:00 on March 12, water injection into the reactor was started from an alternative injection line using a fire engine (Issue D).

(f) PCV venting: After 14:00 on March 12, PCV venting was realized as PCV pressure was confirmed to drop when the S/C vent valve was opened with a temporary compressor (Issue E).

(g) Explosion in the reactor building: At 15:36 on March 12, a hydrogen explosion occurred in the reactor building.
It is estimated that hydrogen generated inside the reactor leaked out via the PCV flange to the reactor building (Issue F).

Table 6.1 Issues and related technical and investigative areas on accidents at Units 1 to 3

	Issue	Technical/investigative area (item)
A	Tsunami countermeasures	External events
	Beyond severe accident assumption	Accident management
	Loss of instrumentation and power systems	Accident management
B	Loss of IC function due to interlock operation to close valve	Plant design
	Opportunities of IC design review according to information from abroad	Plant design
		Relationship with international society
	Failure to inform plant management of IC operation state at an early stage	Human resources/factors
		Information dissemination
C	Failure to catch the internal state of the reactor (instrumentation systems)	Accident management
D	Difficulty in depressurizing RPV with SRV	Accident management
	Failure to inject water with an alternative injection system due to low pressure	Accident management
E	Difficulty in configuring the PCV venting line	Accident management
	Confused instructions concerning the evacuation of local people	Emergency preparedness and response
F	Loss of PCV containment function	Accident management
	Hydrogen accumulation inside reactor buildings beyond assumption	Accident management
G	Beyond design basis earthquake motion	External events
H	Delay in confirming the operation of RCIC system	Accident management
I	Work disturbed by explosion at Unit 3 (problem of multiple plants at the same site)	Accident management
J	Difficulty in configuring the PCV venting line	Accident management
K	Tsunami countermeasures	External events
	Beyond severe accident assumption	Accident management
	Loss of all AC power	Accident management
L	Automatic stop of the RCIC system due to failures	Plant design
M	Manual stop of the HPCI system appropriate?	Human resources/factors
	Failure to depressurize RPV with SRV immediately	Accident management
	Failure to inject water with D/DFP due to low pressure	Accident management
N	How to design piping junction	Plant design
O	Accident responses mainly taken outside the power station	Radiation and radiation monitoring
		Environmental restoration and decontamination
		Emergency preparedness and response
		Relationship with international society
P	Events in nuclear power stations other than Fukushima Daiichi	Accident management

(2) **Accident progression in Unit 2**

(a) At 14:46 on March 11, 2011, an earthquake occurred and Unit 2 in operation automatically shut down. When external power was lost, two emergency diesel generators (D/Gs) automatically started to supply power. SRVs for reactor depressurization and the reactor core isolation system (RCIC) for water injection to cool the reactor maintained the plant in a hot standby state (Issue G).
(b) When the tsunami struck at around 15:30, all D/Gs, DC power and seawater pumps stopped functioning, causing a station blackout. The monitoring and control instruments in the central control room also lost their function (Issue A).
(c) Operation of RCIC: Before DC power was lost due to the tsunami (15:39), the operator had manually activated the RCIC, but neither controlling nor monitoring was possible. The reactor water level was kept high and the reactor pressure maintained at around 6 MPa [gage] until late morning of March 14 (Issue H).
(d) Alternative water injection: Preparation of an alternative injection line (fire extinguishing system—condensate water makeup system—low-pressure core injection system) was underway to prepare for failure of the RCIC, but it was around 19:57 on March 14 when water injection using fire engines finally got underway. As the water injection into the reactor was delayed, the reactor water level dropped below TAF at around 17:00 on March 14, and presumably core damage started around 19:20 (Issues C, D and I).
(e) PCV venting: PCV venting was ultimately considered unsuccessful. The D/W pressure dropped to 0.155 MPa [abs] at around 11:25 on March 15, which was considered attributable to gas leaks from the PCV head flange (Issues I and J).

(3) **Accident progression in Unit 3**

(a) At 14:46 on March 11, 2011, an earthquake occurred and Unit 3 in operation automatically shut down. When external power was lost, two emergency diesel generators (D/Gs) automatically started to supply power. SRVs for reactor depressurization and the reactor core isolation system (RCIC) for water injection to cool the reactor maintained the plant in a hot standby state (Issue G).
(b) With the tsunami striking at around 15:30, D/Gs and seawater pumps stopped functioning, but DC power was retained. Accordingly, the reactor state could be monitored in the central control room and the RCIC and HPCI (high-pressure core injection system) were possible to activate (Issue K).
(c) Operation of the RCIC system: The RCIC was manually activated at 16:03 on March 11 and continuously injected water into the reactor, but automatically stopped at around 11:36 on March 12 because of failures (Issue L).

(d) Operation of the HPCI system: After the RCIC stopped, the reactor water level started decreasing and reached L2 at around 12:35 on March 12 and the HPCI automatically started. The reactor pressure was maintained at between 0.8 and 1.0 MPa [gage] by the operation of HPCI system.
At around 02:42 on March 13 the HPCI was stopped, and SRVs open was planned to depressurize the reactor for switching water injection to D/DFP (diesel-driven fire pump), but this attempt did not succeed (Issue M).
(e) Alternative water injection: While depressurization work with SRVs continued, the reactor pressure started decreasing at around 09:08 on March 13, and reached 0.350 MPa [gage]. Water injection into the reactor using a fire engine was started at around 09:25.
As the alternative water injection was delayed, the reactor water level dropped below TAF past 09:00, and caused core damage at around 10:40, according to the analysis (Issue C).
(f) PCV venting: PCV venting was prepared, and at around 09:20 on March 13, the D/W pressure decreased probably due to the rupture disk being triggered to vent the PCV (Issue J).
(g) Explosion in the reactor building: At 11:01 on March 14, a hydrogen explosion occurred in the reactor building of Unit 3.
It is estimated that hydrogen was generated in the reactor and leaked out directly from the seal of the PCV joint into the reactor building (Issue F).

(4) **Event progression in Unit 4**

(a) At 14:46 on March 11, 2011, an earthquake occurred. Unit 4 was under outage for periodic inspection, and when the AC power was lost, one of two emergency diesel generators (D/Gs) automatically started to supply power (the other was under periodic inspection). All fuel had been relocated to the SFP, which was cooled by the residual heat removal (RHR) system. Cooling of the SFP, however, stopped when all external power was lost.
(b) With the tsunami striking at around 15:30, D/G, DC power supply and seawater pumps were disabled, causing a station blackout. The SFP lost its cooling and water supply functions. The exposure of fuel in the SFP was estimated to take place around the end of March, accordingly countermeasures to Units 1 to 3 were prioritized, during which no significant problem on the SFP emerged (Issue A).
(c) At 06:12 on March 15, a hydrogen explosion occurred in the reactor building at Unit 4. This was possibly attributable to PCV venting in Unit 3. The hydrogen gas of Unit 3 flowed through the junction of Unit 3 and Unit 4 ventilation lines to the reactor building of Unit 4 (Issue N).

(5) **Accident responses mainly taken outside the power station**
Japan lagged behind international standards provided by the IAEA in preparedness and response to emergencies.

There are problems in terms of environmental monitoring such as SPEEDI and wide-area monitoring.

There are also many problems concerning the distribution, dissemination and disclosure of information, particularly inadequacy in terms of the international distribution of information.

Environmental restoration and restoration of contaminated areas are also critical issues hereafter (Issue O).

(6) **Events in nuclear power stations other than Fukushima Daiichi Units 1 to 4**
The analysis of nuclear power stations other than Fukushima Daiichi, where the worst accident cases could be prevented, revealed the importance of restoring failed equipment, accident management (AM) and continuous improvement (Issue P).

(a) Fukushima Daini Units 1, 2 and 4 maintained the reactor water level by injecting water from the MUWC (make up water condensate) system provided as a means of AM. The cooling system of the RHR system, disabled by the tsunami, was restored by replacing motors, and cold shutdown was achieved.
(b) Fukushima Daiichi Unit 5 prevented the worst case scenario by sharing power with Unit 6, which had been provided as AM.
(c) The floodwall, the height of which had been raised, protected the safety function of Tokai Daini.

(7) **Summary of the analysis and evaluation items from accident progression, etc.**
The items subject to analysis and evaluation, selected from the above-mentioned accident progression and other facts, are listed below:

- External events
- Plant design, accident management
- Emergency preparedness and response
- Human resources and human factors
- Radiation and radiation monitoring
- Environmental restoration and decontamination
- Relationship with international society
- Information dissemination

6.1.1.3 Conclusion

The items subject to analysis and evaluation in the subsequent sections were selected through two methods. Table 6.2 lists these items and the sections to cover.

Note that internal events are discussed as part of the plant design, not independent topic in this chapter. Safety regulations will be analyzed in Chap. 7.

These items are correlated and can be summarized as a systematic diagram as shown in Fig. 6.1.

Table 6.2 Sections of Chap. 6

Section	Title
6.2	Concept of Nuclear Safety
6.3	Defence in Depth
6.4	Plant Design
6.5	Accident Management
6.6	External Events
6.7	Radiation Monitoring and Environment Remediation Activities
6.8	Simulation Analysis
6.9	Emergency Preparedness and Response
6.10	Nuclear Security, Physical Protection, and Safeguards
6.11	Human Resources and Human Factors
6.12	Relationship with International Society
6.13	Information Dissemination

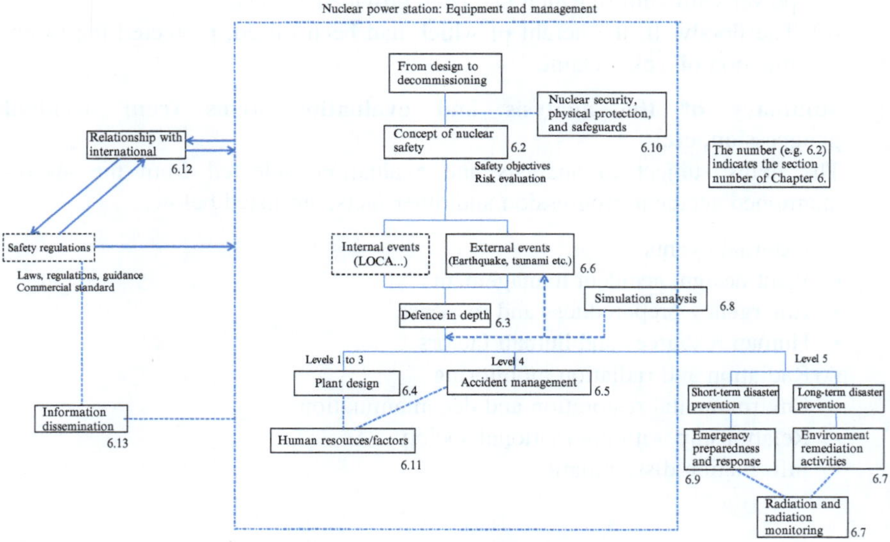

Fig. 6.1 Correlation of items regarding plant equipment and management

6.1.2 Evaluation of Accident Progression Behavior

The events in the RPVs of the Fukushima Daiichi NPS Units 1 to 3 were analyzed using severe accident analysis code SAMPSON, and measures for preventing meltdown were investigated.

6.1.2.1 Features of SAMPSON

SAMPSON is a severe accident analysis code owned by the Institute of Applied Energy (IAE) having the following features:

(1) The code is configured with mechanistic models that indicate various phenomena from the reactor scram to PCV damage with elaborated mathematic representation, and explains physical phenomena for the analytical results.
(2) While mechanistic models are used, none of the user-tuning parameters are incorporated In other words, there is no intention to adjust analytical values to measured values with the parameters.
(3) Because of the complex and lengthy numerical calculations, it is not suitable for analyses such as sensitivity analysis which require numerous calculations.

6.1.2.2 Outline of Plant Event Progression

Most of the equipment in nuclear power plants requires electricity (at least DC power and mostly AC power) to ensure plant safety. The post-earthquake state of power supplies at the Fukushima Daiichi NPS was as follows:

(1) Period from earthquake to the tsunami: Off-site power was lost, mainly because the power transmission towers collapsed following landslides after the earthquake. Immediately after the loss of off-site power, emergency diesel generators (D/Gs) started operating and power was supplied until the tsunami struck. All data indicating the plant state was recorded.
(2) After the tsunami struck: D/Gs and metal-clad switchgear (M/C) were inundated and their functions lost [station blackout event]. At Units 1 and 2, the batteries were also inundated and DC power was lost [loss of all AC and DC power event]. The battery at Unit 3 remained active and was used to operate major valves until it dried up. After DC power was lost, the on-site operators used potable batteries to measure major plant data intermittently.

Details of the plant operation during the initial stage of the accident were not altogether clear, but TEPCO's voluntary interview with operators and confirmation of equipment operation revealed a considerable portion of the operation. Data were added or updated periodically, and released on the website (https://fdada.info/) together with associated measurements. The data was also used to set conditions in this analysis.

The operational state of individual units indicated below is mainly for cooling of fuel, and plant operations relating to containment vessels (PCVs) and cooling of suppression pools are omitted.

Table 6.3 Major operations at Unit 1

Time	Duration after scram	Major event	Time	Duration after scram	Major event
March 11		Earthquake	March 12		
14:46	00 h 01 min	Automatic reactor scram	1:05	09 h 44 min	D/W pressure 0.6 MPa
14:47	00 h 01 min	Automatic closure of main steam isolation valve	2:30	11 h 44 min	D/W pressure 0.84 MPa
14:52	00 h 06 min	Automatic start of IC-A and IC-B	2:45	11 h 59 min	RPV pressure 0.9 MPa
15:03	00 h 17 min	Manual stop of IC-A and IC-B	5:46	15 h 0 min	Start of fresh water injection into RPV (total amount of discharged water from the pump up to 14:53: 80 m^3)
15:17	00 h 31 min	Manual start of IC-A	6:00	15 h 14 min	D/W pressure 0.74 MPa
15:19	00 h 33 m	Manual stop of IC-A	14:00	23 h 14 m	W/W vent (till 14:11)
15:24	00 h 38 min	Manual start of IC-A	14:53	24 h 07 min	Stop of freshwater injection into RPV
15:26	00 h 40 min	Manual stop of IC-A	15:36	24 h 50 min	Hydrogen explosion in RB
15:32	00 h 46 min	Manual start of IC-A	19:04	28 h 18 min	Start of seawater injection into RPV
15:34	00 h 48 min	Manual stop of IC-A			
15:37	00 h 51 min	Loss of all AC and DC power (tsunami)			
18:18	03 h 32 min	Attempt to open the IC-A valve			
18:25	03 h 39 min	Attempt to open the IC-A valve			
20:07	05 h 21 min	RPV pressure 7.0 MPa			
21:51	07 h 05 min	RB dose rate 288 mSv/h			

Note: *RPV* reactor pressure vessel, *D/W* dry well, *RB* reactor building, *W/W* wet well

6.1.2.3 Analysis on Unit 1

(1) **Analysis conditions**

Table 6.3 indicates the main operational state during the period from the occurrence of the earthquake to seawater injection with fire engines. All events until the tsunami struck were within the design assumptions and the isolation condensers (ICs) were intermittently operated. The sound associated with a massive steam discharge from the IC shell side was confirmed in the central

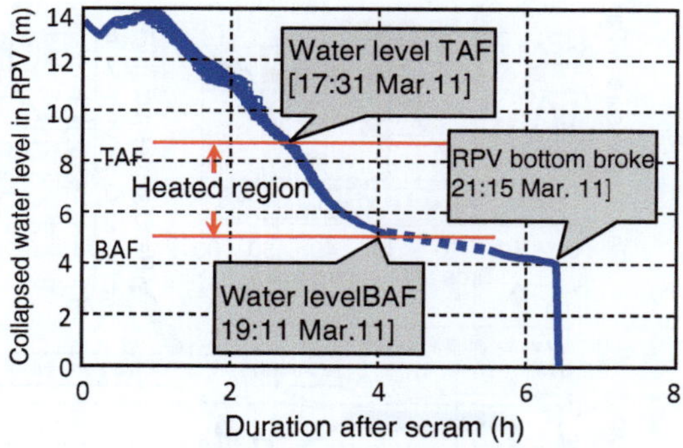

Fig. 6.2 Changes in water level in RPV (Unit 1)

control room while the IC was operating, which meant the operation continued. Following the loss of all AC and DC power, the instrumentation in the central control room showed signs of partial battery recovery after the seawater retreated, and the operators tried to open the valve to resume the IC (18:18 on March 11). At the time, the operators heard steam generation and saw steam escaping over the reactor building, but it was unlikely that the operation to open the valve in the central control room under the condition of loss of all A and DC power caused the valve to open. It is rationally concluded that the IC stopped operating after the loss of all AC and DC power. While the IC operated intermittently, any abnormal rise in reactor pressure was prevented, and the safety relief valve (SRV) did not open. On March 12, alternative water injection was continued for about 9 h from 05:46, and the recorded total amount of water discharged from the pump was 80 m^3, but the actual amount of water injected into the reactor was still unclear because of the considerable leakage from the branch lines in the pump piping system. The IC was the only reactor cooling system that worked in Unit 1 until the start of the alternative water injection.

(2) **Analytical result**

Figure 6.2 shows the analytical result of the water level transient in the RPV. The key events are shown in the hatched boxes. The water level in the figure indicates the collapsed water level and the two-phase (boiling) level inside the reactor is much higher than the collapsed water level. The IC operated for 48 min after the scram, whereupon no steam was released from the SRV, and the water level deviated but did not decrease. After the IC stopped, the safety valve function of the SRV (spring-loaded opening/closing) functioned to release steam in the RPV intermittently, and the water level started decreasing. At 17:31 on March 11 (2 h 45 min after the scram), the collapsed water level dropped to the top of the active fuel (TAF). Since water in the reactor core boiled due to the decay heat, the boiling water level reached TAF at 10 min later, or at 17:41 on March 11, and the fuel temperature started rising.

Figure 6.3 shows the transient of the reactor pressure. The reactor pressure before the tsunami, during which plant parameters were recorded, fluctuated according to the intermittent operation of the IC, and the analytical result

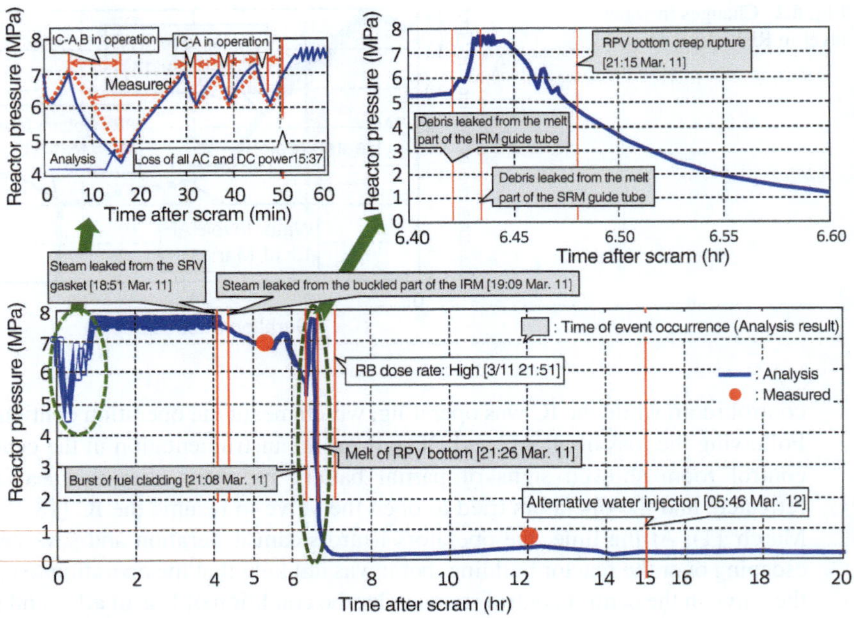

Fig. 6.3 Changes in reactor pressure (Unit 1)

showed fairly good agreement with the measured data. After the tsunami, that caused the loss of all AC and DC power, and the IC was disabled, the reactor pressure rose due to the decay heat generation, and the SRV intermittently operated to release excess steam to the pressure suppression pool to maintain the reactor pressure at around 7.5 MPa. This state lasted for about 3 h, during which the water level gradually declined as the excess steam flowed out, as shown in Fig. 6.2. The characteristics of Unit 1 were (a) after the water level dropped below TAF, steam in the reactor was also heated with the overheat of fuel, and if the SRV gasket temperature exceeded the maximum design value (450 °C) when the high-temperature steam passed through the SRV, the sealing capability would be deteriorated, and the steam would escape from the deteriorated parts of the gasket directly into the dry well (D/W) of the PCV, (b) if the steam temperature in the reactor exceeded 1,027 °C (1,300 K), part of the in-core monitor guide tubes would buckle, and the steam would flow out from the cracks directly into the dry well. In this analysis, the steam that passed through the SRV reached the maximum design temperature 450 °C (723 K) at 18:51 on March 11, which resulted in the start of steam leakage from the gasket. The amount of leak from the gasket was too small to drastically decrease the reactor pressure, but stopped flow out of steam from the SRV.

As shown in Fig. 6.4, guide tubes for the control rods and in-core monitors penetrate the lower head wall of the RPV. Since the bottom ends of some of the in-core monitor guide tubes, such as the source range monitoring system (SRM)

Fig. 6.4 Pressure boundary at the lower part of the RPV

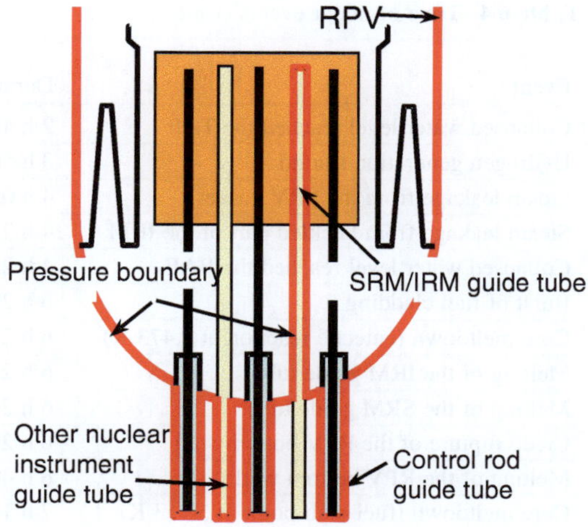

and intermediate range monitoring system (IRM), are open to the drywell, the pressure boundary of these tubes is inside the RPV. The pressure difference between the inside and the outside of these tubes is always about 7 MPa under the normal operating condition. The material of these tubes has enough strength to keep their integrity under the normal operating condition. However, if the reactor temperature rises under severe accident conditions, the material strength would be deteriorated, which would result in buckling under the 7 MPa pressure difference. In the present analysis, steam in the reactor reached 1,027 °C (1,300 K) at 19:09 on March 11, and the buckling of the IRM guide tube occurred, which resulted in direct steam leakage into the drywell from the cracks caused by the buckling. Subsequently, the SRM guide tube buckled, and a total of 12 IRM/SRM guide tubes broke one after the other, accelerating depressurization of the reactor. Accordingly, the means to remove decay heat was lost after the IC stopped, and reactor coolant leaked from the SRV gasket and the IRM/SRM guide tubes. The reactor core thus overheated without coolant, and finally the creep rupture of the RPV bottom wall occurred after a fuel meltdown.

Figure 6.3 also shows the measured values of the RPV pressure. After the loss of all AC and DC power, the reactor pressure could be measured at only two points during the period indicated in the figure, which shows the depressurization of the RPV during this period. The analytical result showed the depressurization at the time between the two measured points.

Table 6.4 lists the analyzed time of occurrence of the main events. Hydrogen was generated by the reaction of in-core metal materials (mainly zirconium) and water or steam. The time for hydrogen generation in the table stands for the time at which hydrogen was generated in order of a gram per second (with the fuel surface temperature at 750 °C). Hydrogen generation reached the order of 10 g/s at 19:23 on March 11 (4 h 37 min after the scram, and with a fuel surface temperature of 1,400 °C). With eutectic reaction of UO_2 and zirconium

Table 6.4 Time of major events (Unit 1)

Event	Duration after scram	Time of occurrence (March 11)
Collapsed water level reached the TAF	2 h 45 min	17:31
Hydrogen generation started	3 h 50 min	18:36
Steam leakage from the SRV gasket	4 h 05 min	18:51
Steam leakage from buckled part of the IRM	4 h 23 min	19:09
Collapsed water level reached the BAF.	4 h 25 min	19:11
Burst of fuel cladding	6 h 22 min	21:08
Core meltdown (eutectic reaction at 2,473 K)	6 h 25 min	21:11
Melting of the IRM guide tube	6 h 25 min	21:11
Melting of the SRM guide tube	6 h 26 min	21:12
Creep rupture of the RPV bottom wall	6 h 29 min	21:15
Melting of the RPV bottom wall	6 h 40 min	21:26
Core meltdown (fuel melt point at 3.113 K)	7 h 13 min	21:59

(i.e. both dissolving simultaneously at a lower temperature than their specific melting point, despite these elements being mixed, and the reaction was assumed to start at 2,473 K), the fuel and fuel cladding started melting at 21:11 on March 11 (6 h 25 min after the scram). The molten material ("corium" or the mixture of UO_2 fuel, zirconium alloy, steel, and—in some cases—molten control materials) fell into the lower plenum of the RPV, and heated the through hole sections of the SRM/IRM guide tubes. Since the wall thickness of these tubes is thin (about 3 mm), the tubes would melt in little more than 10 s when in contact with high-temperature corium (the tube's melting point is about 1,700 K), which caused the corium to drop into the drywell (D/W). Shortly afterward, the RPV bottom wall was breached by creep deformation (a phenomenon of material deformation depending on temperature and added stress). At 21:51 on March 11, the dose rate of the operating floor in the reactor building had already increased to 288 mSv/h. The burst of the fuel cladding tubes and the melting of the fuels and the claddings caused the fission product release to the coolant (mainly steam), which passed through the deteriorated portion of the SRV gasket and buckled parts of the SRM/IRM guide tubes into the drywell, and moreover leaked directly into the drywell from the damaged portions of the RPV bottom wall, and then the fission products in the drywell leaked into the reactor building. The fuel melt continued without coolant, resulting in the melt of the RPV bottom wall itself at 21:26 on March 11 (4 h 37 min after the scram). The UO_2 temperature continued increasing over the eutectic temperature and reached the melting point (3,110 K) at 21:59 on March 11. Alternative water injection was started about 7 1/2 h after the core meltdown caused by eutectic reaction or creep damage at the bottom of the RPV.

Table 6.5 shows the results of the status of core meltdown, the amount of hydrogen generated, and other data at 5 h 45 min after the alternative water injection was started (20 h 45 min after the scram, or at 11:31 on March 12). As

Table 6.5 Summary of analysis (Unit 1)

Item	Result
Ratio of UO_2 melt to the total initially loaded	38.50 %
Ratio of melts of core materials to the total[a]	58.50 %
Amount of hydrogen generated in the reactor core	686 kg
Cesium released from fuel	61 kg (72 %)[b]
Iodine released from fuel	4.9 kg (72&)[b]
Damage to bottom of the RPV	Yes

[a]Total amount of fuel, steel, control materials and zircaloy in the reactor core
[b]Percentages in parentheses denote the proportions of cesium and iodine released to coolant after the scram to ones contained in the fuel at the time of scram

the bottom of the RPV had already been breached when the alternative water injection was started, the injected water into the core would flow down into the D/W through the broken walls to cool the corium on the D/W floor, while part of the injected water cooled the unmelted fuel and corium remaining in the RPV, and the steam generated in this process continued to cool the reactor. The time after water injection, 5 h 45 min, was chosen because the core and the corium on the D/W floor were supposed to be stably cooled with the alternative water injection.

At 11:31 on March 12, or 20 h 45 min after the scram, 38.5 % of UO_2 fuel had melted. Except for the fuel, the materials of the core structures (steel, zircaloy, control materials) had lower melting points than the fuel, meaning the proportion of melts of the core structures (including the fuel) increased to 58.5 %. Most of the corium dropped to the drywell floor from the broken bottom of the RPV, during which 72 % of cesium and iodine included in the fuel at the time of scram was released into the coolant, and 28 % remained in the fuel (unmelted fuel and corium). At the time, the fuel (unmelted fuel and corium) has been cooled stably with the "circulating injection cooling system", but the water refreshing capability (removal of cesium) of the accumulated water processing facility in the system began declining in mid-2012, and the water quality (concentration of ^{137}Cs) has changed little since then (http://www.tepco.co.jp/nu/fukushima-np/roadmap/images/d131128_06-j.pdf). Additional supply of cesium due to diffusion from highly contaminated sources such as components in the drywell and the suppression pool and elution from the fuel (unmelted fuel and corium) may be the cause of degradation.

As shown in Table 6.3, freshwater injection was started at 05:46 on March 12 from fire engines, considering the leak from the branch (leak path) on the piping connecting the fire engine and RPV, the actual amount of water injected in the reactor could be smaller than the water discharged from the fire engines. To compound things, the water injection was stopped about 9 h later at 14:53 on March 12. Seawater injection was started about 4 h later, but was also suspended twice. Seawater injection was finally stabilized at 20:00 on March 14, or 77 h 14 min after the scram. During the suspension of alternative water injection, unmelted fuel could heat up to increase the amount of corium. This long-term analysis is currently underway, and the result may change the outcome listed in Table 6.5.

Table 6.6 Major operations at Unit 2

Time	Time after scram	Major event	Time	Time after scram	Major event
March 11			March 14		
14:46		Earthquake	9:00	66 h 13 min	RCIC stopped
14:47	0	Automatic reactor scram	18:02	75 h 15 min	Manual opening of a SRV
14:50	00 h 03 min	Manual start of RCIC	19:54	77 h 07 min	Seawater injection
14:51	00 h 04 min	Trip of RCIC [L 8]	21:20	78 h 33 min	Manual opening of another SRV
15:02	00 h 15 min	Manual start of RCIC	23:00	80 h 13 min	A SRV closed
15:28	00 h 41 min	Trip of RCIC [L 8]	23:25	80 h 38 min	Manual opening of another SRV
15:39	00 h 52 min	Manual start of RCIC	March 15		
15:41	00 h 54 min	Loss of all AC and DC power (tsunami)	2:22	83 h 35 min	Manual opening of another SRV
March 12			6:14	87 h 27 min	Abnormal sound and vibration (maybe due to hydrogen explosion at Unit 4?)
4:20	13 h 33 min	RCIC water source switched from CST to S/P			

Notes: *CST* condensate storage tank, *S/P* suppression pool

6.1.2.4 Analysis on Unit 2

(1) **Analysis conditions**

Table 6.6 shows the major operations within the first few days of the accident. Following the scram, operators activated the reactor core isolation cooling system (RCIC), a cooling system with a pump driven by the steam turbine. When power is normally supplied, the system valves automatically open with a low water level (L 2) signal, and close with a high water level (L 8) signal from the reactor. Operators first manually activated the RCIC in the central control room at 3 min after the scram, and when the system automatically stopped with the high water level (L 8) signal, manually started again the system. The operation was repeated until the loss of all AC and DC power due to the tsunami. At 15:39 on March 11, operators restarted the RCIC, and the loss of all AC and DC power occurred 2 min later, during which the RCIC valves opened, and the turbine continued to run. The high water level (L 8) signal could not be transmitted to stop the RCIC, resulting in continued water injection with the RCIC for about 65 h after the loss of all AC and DC power.

After the RCIC stopped, an attempt to reduce reactor pressure and inject water from fire engines was made. To depressurize the reactor, however, the SRVs had to be opened, which is normally done using nitrogen pressure and

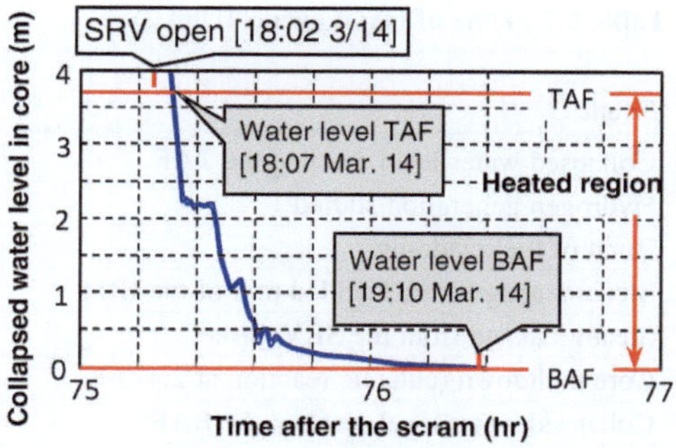

Fig. 6.5 Changes in core water level (unit 2)

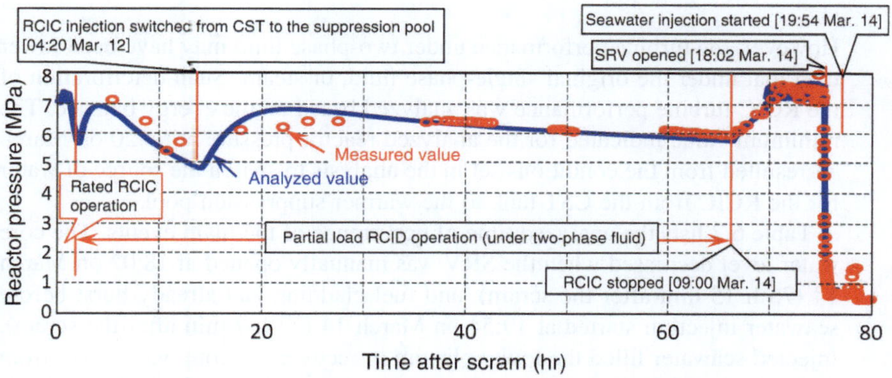

Fig. 6.6 Changes in reactor pressure (Unit 2)

means electric power is needed to operate the valve of the nitrogen supply pipe. Portable batteries were brought to the site to operate the valve, but the work was only accomplished 9 h after the RCIC had stopped, and after the SRVs were opened, 1 h 52 min elapsed until fire engines started injecting water.

(2) **Analytical results**

Figure 6.5 shows the analytical results of the collapsed water level transient in the RPV. While the RCIC was in operation, the water level changed little. When the RCIC stopped and the RPV pressure rose, a safety relief valve (SRV) operated to gradually reduce the water level. Subsequently, when the SRVs were opened to depressurize the reactor, the water level dropped immediately. The core was heated totally without coolant for 44 min from the water level below BAF to seawater injection.

Figure 6.6 shows the transient of the reactor pressure. Water injection from the RCIC was continued even after the loss of all AC and DC power, but as a result of the water level continuing to rise due to the absence of the L 8 signal, two-phase fluid, or a mixture of water and steam, flowed into the RCIC turbine. The RCIC is originally designed to inject sufficient water to remove the decay heat, and reactor pressure should be reduced when the RCIC is activated.

Table 6.7 Time of major events (Unit 2)

Event	Time after scram	Time of occurrence (March 14)
Collapsed water level reached the TAF	75 h 31 min	18:18
Hydrogen generation started	76 h 35 min	19:22
Burst of fuel cladding	76 h 58 min	19:45
Steam leakage from buckled part of the IRM	77 h 08 min	19:55
Steam leakage from the SRV gasket	77 h 35 min	20:22
Core meltdown (eutectic reaction at 2,473 K)	77 h 38 min	20:25
Collapsed water level reached the BAF	77 h 41 min	20:41
Creep rupture of the RPV bottom wall	81 h 28 min	24:15:00

However, the turbine performance under two-phase fluid may have been lower than that under the original single-phase fluid, or steam. Such deterioration of the RCIC turbine performance was analyzed based on the energy balance. The minimum value indicated for the analyzed reactor pressure at 04:20 on March 12 resulted from the conditions set in the analysis to switch the source of water for the RCIC from the CST tank to the warmer suppression pool.

Table 6.7 lists the analyzed time of occurrence of the main events. The core water level decreased when the SRV was manually opened at 18:02 on March 14 (75 h 15 min after the scram), and fuel cladding had already burst before seawater injection started at 19:54 on March 14 (77 h 7 min after the scram). Injected seawater filled the lower plenum to recover the core water level from the bottom, which significantly delayed effective cooling of the fuel. Similar to Unit 1, water leaked from the branch of the piping system connecting the seawater injection pump (fire engine) to the reactor, and only part of the water from the pump reached inside the reactor. This analysis estimated about 30 % of water discharged from the fire engine reached the reactor. Almost all water flowing in the RPV evaporated by decay heat, thus hardly contributing to help recover the reactor water level. Accordingly, the fuel temperature continued to rise, even after seawater injection, causing a core meltdown due to eutectic reaction, and 1 h 35 min after the start of seawater injection, creep rupture of the RPV bottom wall occurred.

Table 6.8 shows the results of the status of core meltdown, the amount of hydrogen generated and other data about 9 h after the start of seawater injection (at 04:49, March 15, 86 h 02 min after scram). Although only a part of the discharged water from the fire engine reached the reactor, melting was judged to stop 9 h later because water injection continued, unlike discontinuation at Unit 1, and the physical quantities obtained in the analysis became nearly constant.

As shown in Table 6.8, 20.8 % of UO_2 fuel had melted at 86 h 2 min after the reactor scram, or at 04:49 on March 15. The percentage of melts of the core structures including fuel increased to 28.1 %. Most of the corium dropped to the D/W floor from broken bottom of the RPV, during which 46 % of cesium and

Table 6.8 Summary of analysis (Unit 2)

Item	Result
Ratio of UO_2 melt to the total initially loaded	20.8 %
Ratio of melts of core materials to the total[a]	28.1 %
Amount of hydrogen generated in the reactor core	711 kg
Cesium released from fuel	65 kg (46 %)[b]
Iodine released from fuel	5.2 kg (46 %)[b]
Damage to bottom of the RPV	Yes

[a]Total amount of fuel, steel, control materials and zircaloy in the reactor core
[b]Percentages in parentheses denote the proportions of cesium and iodine released to coolant after the scram to ones contained in the fuel at the time of scram

iodine included in the fuel at the time of scram was released into the coolant, and 54 % remained in fuel (unmelted fuel and corium). Stable cooling has also been continued at Unit 2 with the "circulating injection cooling system", but similar to Unit 1 in (3) [2], the water refreshing capability deteriorated in mid-2012, and the concentration of ^{137}Cs has changed little since then, which suggests that fission products are still discharged by diffusion from the fuel (unmelted fuel and corium) etc.

6.1.2.5 Analysis on Unit 3

(1) **Analysis conditions**
Table 6.9 shows the major operations within the first few days after the accident. The RCIC and HPCI (high pressure coolant injection system) were activated after the accident. The HPCI is also designed to automatically start and stop depending on the water level signals as the RCIC. As the battery was available even after the tsunami struck, the operators could control the amount of water injected in from the RCIC and HPCI. After the reactor scram, the operators manually started the RCIC, which then automatically tripped with the high water level (L 8) signal, whereupon the operators manually started it. However, the RCIC did not restart after tripping at 11:36 on March 12. Then the HPCI automatically started with the low water level (L 2) signal about an hour after the RCIC last tripped, but operators manually stopped the HPCI at 02:42 on March 13. Operators then attempted to open SRVs to allow water injection into the reactor by fire engines, but the automatic depressurization system (ADS) had already automatically activated. Steam discharged with the ADS, which is equivalent to six opened SRVs in capacity, caused a sudden decrease in reactor pressure and core water level. Core cooling was stopped for 6 h and 43 min after the HPCI was stopped and water injection was started by fire engines.

(2) **Analytical results**
Figure 6.7 shows the transient of the reactor pressure, and Fig. 6.8 the collapsed water level transient in the RPV.

Table 6.9 Major operations at Unit 3

Time	Time after scram	Major event	Time	Time after scram	Major event
March 11			March 13		
14:46		Earthquake	2:42	35 h 55 min	HPCI stopped
14:47	0	Automatic reactor scram	9:08	42 h 21 min	ADS activated
15:05	00 h 18 min	Manual start of RCIC	9:20	42 h 33 min	W/W gas venting started
15:25	00 h 38 min	RCIC tripped [L 8]	9:25	42 h 38 min	Freshwater injection by fire engines
15:38	00 h 51 min	Station blackout [tsunami]	11:17	44 h 30 min	W/W venting valve closed, and opened several times
16:03	00 h 16 min	Manual start of RCIC	12:20	45 h 33 min	Freshwater injection stopped (depletion)
March 12			13:12	46 h 25 min	Seawater injection by fire engines
11:36	20 h 49 min	RCIC tripped			
12:35	21 h 48 min	Automatic start of HPCI [L 2]			

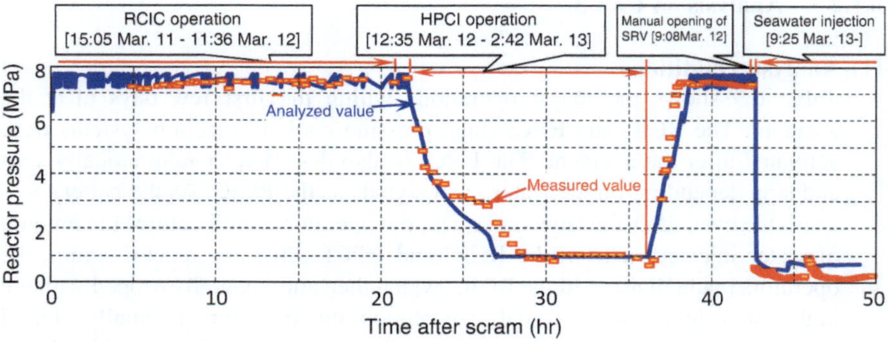

Fig. 6.7 Changes in reactor pressure (Unit 3)

While the RCIC was in operation, reactor pressure was maintained at just over 7 MPa by operators controlling the flow rate and repeatedly starting and stopping the RCIC, while the RPV water level also remained about the same. When the RCIC tripped at 11:36 on March 12, however, the water level decreased and the HPCI automatically started with a low water level (L 2) signal generated about an hour later. With the HPCI capacity exceeding the RCIC, the water level showed signs of recovering, but the reactor pressure dropped to 1 MPa. As steam pressure for driving the HPCI turbine reduced, the amount of injection water decreased, and the RPV water level gradually declined. When the HPCI stopped at 02:42 on March 13, the RPV water level

Fig. 6.8 Changes in water level in RPV (Unit 3)

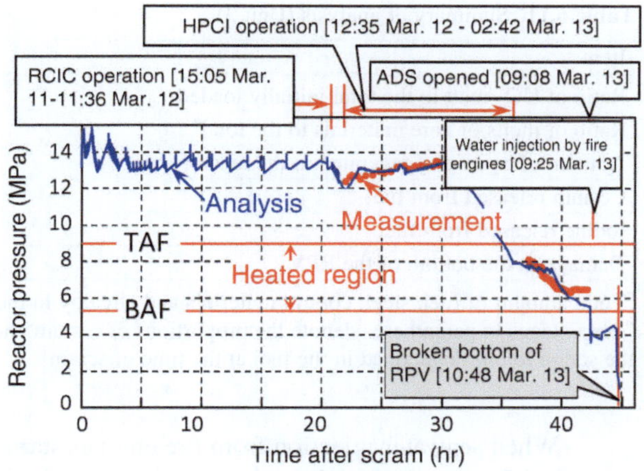

Table 6.10 Time of major events (Unit 3)

Event	Time after scram	Time of occurrence (March 14)
Collapsed water level reached the TAF	34 h 54 min	1:41
Collapsed water level reached the BAF	42 h 22 min	9:09
Burst of fuel cladding	43 h 17 min	10:04
Core meltdown (eutectic reaction at 2,473 K)	43 h 48 min	10:35
Melting of the RPV bottom wall	44 h 01 min	10:48
Core meltdown (fuel melt point at 3,113 K)	44 h 24 min	11:01

had already dropped just below TAF, and the fuel started overheating. At 09:08 on March 13, SRVs were opened to start depressurization, but the water level decreased almost to BAF. Water injection by fire engines started 18 min after the SRVs were opened, but little water reached inside the reactor due to leaks from the branch piping system. The water injected into the reactor filled the lower plenum to recover the core water level from the bottom, which significantly delayed core cooling, and caused overheating of the core. The reactor core started melting and the bottom of the RPV was breached at 11:37 on March 13.

Table 6.10 lists the analyzed time of occurrence of the main events. While the HPCI operated at around 1 MPa, which is the lower operation limit, steam was discharged into the suppression pool through the HPCI turbine. The water flow rate injected by the HPCI decreased with declining performance due to low-pressure operation around 1 MPa, and it became smaller than extracted steam flow rate from the RPV to the HPCI turbine, resulting in the decrease of the reactor water level while the HPCI was in operation. Seawater injection by fire engines started at 09:25 on March 13, 42 h 38 min after the scram, but the core water level had already decreased below the BAF.

Table 6.11 Summary of analysis (Unit 3)

Item	Result
Ratio of UO_2 melt to the total initially loaded	24.9 %
Ratio of melts of core materials to the total[a]	38.7 %
Amount of hydrogen generated in the reactor core	562 kg
Cesium released from fuel	61 kg (39 %)[b]
Iodine released from fuel	4.9 kg (39 %)[b]
Damage to the bottom of the RPV	Yes

[a]Total amount of fuel, steel, control materials and zircaloy in the reactor
[b]Percentages in parentheses denote the proportions of cesium and iodine released to coolant after the scram to ones contained in the fuel at the time of scram

When seawater injection from fire engines started at Unit 3, similar to Units 1 1nd 2, water leaked from the branch of the piping system connecting the fire engines to the reactor and only part of the water from the pump reached inside the reactor. This analysis estimated about 60 % of seawater is charged from the fire engine reached the reactor and 40 % leaked from the branches. Similar to Unit 2, the injected seawater filled the lower plenum to recover the core water level from the bottom, which significantly delayed the effective cooling of fuel. Consequently, burst of the fuel cladding, core meltdown by the eutectic reaction and melt of the RPV bottom wall took place successively after seawater injection.

Table 6.11 shows the results of the status of core meltdown, amounts of hydrogen generated, and other data after the temperature of UO_2 fuel reached its melt point, 3.113 K. The flow rate of seawater into the RPV was twice at Unit 2 on the average, and the melting behavior had almost stopped by this time. 24.9 % of UO_2 fuel and 38.7 % of the core structures (including fuel) melted; most of which dropped on the D/W floor. The total amount of hydrogen generated in the core was 562 kg, during which 39 % of cesium and iodine included in the fuel at the time of scram was released into the coolant, and 61 % remained in the fuel (unmelted fuel and corium). Similar to Units 1 and 2, fission products still appeared to be released into the cooling water in the "circulating injection cooling system".

Seawater injection from fire engines was also suspended at Unit 3, whereupon unmelted fuel could probably heat up to increase the amount of corium. This long-term analysis is currently underway, and the result may change the outcome listed in Table 6.11.

6.1.2.6 Summary

Accident progressions from the beginning to water injection into reactors by fire engines at the Fukushima Daiichi NPS Units 1 to 3 were analyzed, focusing on the behavior of the reactor pressure vessels. Models of characteristic in each reactor unit, which had not been considered in the original SAMPSON code, were created during this analysis, the results of which are summarized below.

(1) The ratio of UO_2 melt to the total initially loaded was 38.5 % at Unit 1, 20.8 % at Unit 2, and 24.9 % at Unit 3. The amount of hydrogen generated in the cores was 686, 711 and 562 kg, respectively.
(2) The ratio of melts of core materials (including fuel) to the total was 58.5 % at Unit 1, 28.1 % at Unit 2, and 38.7 % at Unit 3.
(3) The RPV bottom wall was breached in all units, and most of the corium dropped to the drywell.
(4) Cesium and iodine included in the fuel at the time of scram was partly released into the coolant as the fuel overheated and melted. The release ratios were 72 % at Unit 1, 46 % at Unit 2 and 39 % at Unit 3.

6.1.2.7 Was There Any Means to Prevent Core Meltdown?

Since nobody had considered that measures against the loss of all AC and DC power were actually required in Japan up to the Fukushima Daiichi NPS accident, the reactor personnel had never got relevant education and never received hands-on trainings. Emergency manuals had, of course, been provided, but they were premised on that the DC power, at least, was available. Under such circumstances, the reactor personnel at the Fukushima Daiichi NPS site had to determine what to do then and there, surely making their best efforts at the time of the accident. However, reviewing the accident with more advanced insights into the phenomena through the simulation as mentioned above, suggests that there were indeed means to prevent core meltdown using systems and equipment available at the time. It would have been difficult to ask the reactor personnel to take the steps listed below at the time of the accident, but they are useful as a reference to prevent the recurrence of similar accidents in future:

(1) **Unit 1**

(a) Result: Core meltdown could be prevented by continuous isolation condenser (IC) operation and water supply to the IC tank from fire engines.
(b) Consideration: Responses to the tsunami which caused the station blackout and the loss of DC power are discussed below.
The actual state of Unit 1: Two ICs shut down after intermittent operation (valves of IC system closed), the SRV intermittently operated to cause steam in the reactor to flow out, and i) the two phase (boiling) level dropped to TAF at around 17:41 on March 11, causing overheating of the fuel, ii) the maximum temperature of the fuel rods reached 750 °C at 18:36 on March 11, whereupon hydrogen generation significantly increased (no burst of the fuel cladding yet).

If ICs could be reactivated within a few hours after the tsunami struck (the time of (i) or (ii) above), overheating of the fuel and significant hydrogen generation would be preventable. The approach to achieve this is shown below.

(i) After the tsunami, prioritize the procurement of mobile (in-vehicle) batteries or other portable DC power supply to operate IC system valves, or restarting ICs. (Actually, at the site, top priority was given to AC power supply vehicles, but restoration of all AC power required the installation of a switchgear. Early restoration of AC power should be considered impractical.)

(ii) When ICs start operating, reactor pressure settles below the SRV operating pressure, preventing release of steam in the reactor, and the amount of coolant in the reactor can be kept constant. On this assumption and based on the balance between decay heat generation and the quantity of heat required to evaporate water in the IC tank, restart one of two ICs 2.5 h after the tsunami. When this IC loses cooling function after operation for a little over 6 h (meanwhile, all water in the IC tank evaporates), start the second IC to continue removing decay heat for another seven plus hours. Namely, by activating two ICs, one at a time, the removal of decay heat can be continued for at least 16 h (decay heat reduces over time after the scram, namely, the time of the first IC operation is shorter than the second IC). Water injection from fire engines was actually started at 05:46 on March 12 (15 h after the scram). When the fire engine injected water in the IC tank, the removal of decay heat could be continued for a prolonged period (indeed, a line for injecting water in the IC tank was provided as a means of accident management, but was not used in the accident). Early restarting of ICs prevents the release of fission products, which increases radiation in the turbine and reactor buildings, and thus allows reactor personnel to access these buildings.

(iii) The next best means is direct injection of water from fire engines to reactors. In this case, there are some important things to bear in mind: Use a core spray system as the piping system for fire engines, promptly inject sufficient cooling water directly over the fuel, and close all valves of the branch piping to prevent leaks. Omission of the last process may result in insufficient cooling, causing increased reactor pressure that exceeds the discharge pressure of fire engines, and preventing water from being injected into the reactor.

(2) **Units 2 and 3**

(a) Result: Core meltdown could be prevented by depressurization of the reactors as early as possible, immediately starting water injection from fire engines via a core spray system, and closing all valves of branch pipes.

(b) Consideration: At Unit 2, reactor pressure increased after the RCIC shut down, and intermittent operation of SRVs caused release of steam in the reactor. When water injection started, the reactor water level had gone below BAF, and the fuel cladding had burst (see Table 6.7). At Unit 3, intermittent operation of SRVs after the HPCI shutdown caused the reactor water level to drop near BAF (see Table 6.10). Namely, both Units 2 and 3 had already overheated when SRVs were opened for depressurization. As water injected from fire engines filled the lower plenum first, and effective reactor cooling was delayed, part of the discharged water leaked from the branch. The means to solve these problems and prevent core meltdown involved shortening the time of steam release caused by intermittent operation of SRVs as much as possible, and depressurization by opening SRVs and water injection from fire engines needed to be carried out immediately after the RCIC or HPCI shut down as a series of means.

(i) Open SRVs for depressurization immediately after the RCIC or HPCI shut down, and promptly start water injection from fire engines. There are some important things to bear in mind: (a) Use a core spray system as the piping system for fire engines and inject sufficient cooling water directly on the top of fuel (eliminating delay) and (b) close all valves of the branch piping to prevent leaks. For example, even if a delay of 18 min at Unit 3 from the SRV open to water injection from fire engines occurs, burst of fuel cladding could be prevented.

(ii) It is difficult to predict when the RCIC or HPCI trips, but measurement of RPV pressure increase may be used for evaluation. When the RCIC or HPCI tripped, reactor pressure started rising up to 7 MPa causing intermittent operation of SRVs to keep the RPV pressure constant at about 7 MPa. In this case, promptly open SRVs for early depressurization.

(iii) DC power is required to open SRVs. Provide sufficient portable batteries or similar means while the RCIC or HPCI is in operation, and after the fire engines are ready, check the activation of SRVs even before tripping.

6.1.3 Evaluation of Radioactive Material Release

The objective of nuclear safety is to protect people and the environment from the harmful influence of radiation. To achieve this, it is important to relate the release of radioactive materials to events which may directly cause such release, and verify consistency with analytical results based on the progression of accident. This facilitates efforts preparing measures to mitigate the release of radioactive materials. By analyzing relations between releases of radioactive materials and environmental contamination, effective disaster prevention measures (including post-accident responses) can be developed. Before addressing this issue, however, we must understand the behavior of radioactive materials in the reactor and leakage paths, as well as environmental behavior, including ground contamination, in the event that radioactive materials are released to the atmosphere.

Accordingly, evaluation of releases based on simulation of severe accidents, review of accident progression scenarios, evaluation of release amounts, measures to reduce releases, and issues for the future are discussed here respectively.

6.1.3.1 Evaluation of Releases Based on the Simulation of Severe Accidents

Simulation-based evaluation is effective for the judgment of probability of accident progression scenarios, and assessment of the forms and amounts of release. Accordingly it is important to confirm the association between simulation results using analysis codes and actual releases of radioactive materials to verify consistency between analysis according to accident progression and actual releases.

To reproduce the progress of accidents, simulations using various analysis codes including MAAP, MELCOR, SAMPSON and THALES have been performed; most of which are designed to simulate damage to reactor core, pressure vessel and containment vessel, and changes in pressure, temperature and water level due to deterioration or loss of the cooling function. Some are intended to simulate the behavior of radioactive materials in the reactor core and release into the environment. Leak point and parameter settings based on findings in the chemical behavior of radioactive materials are required for these analysis codes, however such information is often incomplete. Accordingly analysis codes should be used understanding their limit.

Releases of radioactive materials into the environment and the impact on the environment are usually evaluated in following four steps: (1) nuclide and element inventories in fuel at the accident, (2) releases of radioactive materials from fuel according to the presumed accident progression scenario, (3) behavior of radioactive materials released from fuel in reactor pressure vessel, containment vessel and reactor building, and leak from them, (4) behavior of radioactive materials in the environment. In many cases, however, the behavior in step (3) cannot be traced completely, and is often assumed from their chemical properties. Figure 6.9 shows the nuclide and element inventories 1 day after the scram at Unit 1. The inventories of Units 2 and 3 are about 1.5 times the inventories of Unit 1. In step (2), radioactive materials released from the fuel can be calculated according to the temperature of the fuel at the accident using the CORSOR model, etc. In step (3), parameter settings are based on the chemical properties of groups containing similar elements. Table 6.12 shows the groups of elements used in MELCOR and THALES2 codes; typical elements of which are listed in the table.

Of the analysis codes above-mentioned, the MELCOR code computes the release of radioactive materials in addition to the accident progression and these analytical results are used for discussions in the subsequent paragraphs. According to the conditions of a recent analysis by the Japan Nuclear Energy Safety Organization (JNES), leaks are caused due to overheating, and potentially from safety relief valves (SRV), source range monitors (SRM), intermediate region monitors

Calculation of inventory at Unit 1 using ORIGEN2.2
- Library: ORIGEN2.2 with BWRUS.lib
- Burnup: 26 GWd/t (1,000 days) with 460 MWe (heat efficiency: 30%)
- Fuel (UO_2): 75 t
- Composition: U-238: 96%, U-235:4%

	Unit 1	Unit 2	Unit 3
Output (MWe)	460	784	784
Fuel assembly	400	548	548
Mean burnup (GWd/t)	26	23	22
(Output) x (burnup) (normal)	1	1.5	1.5

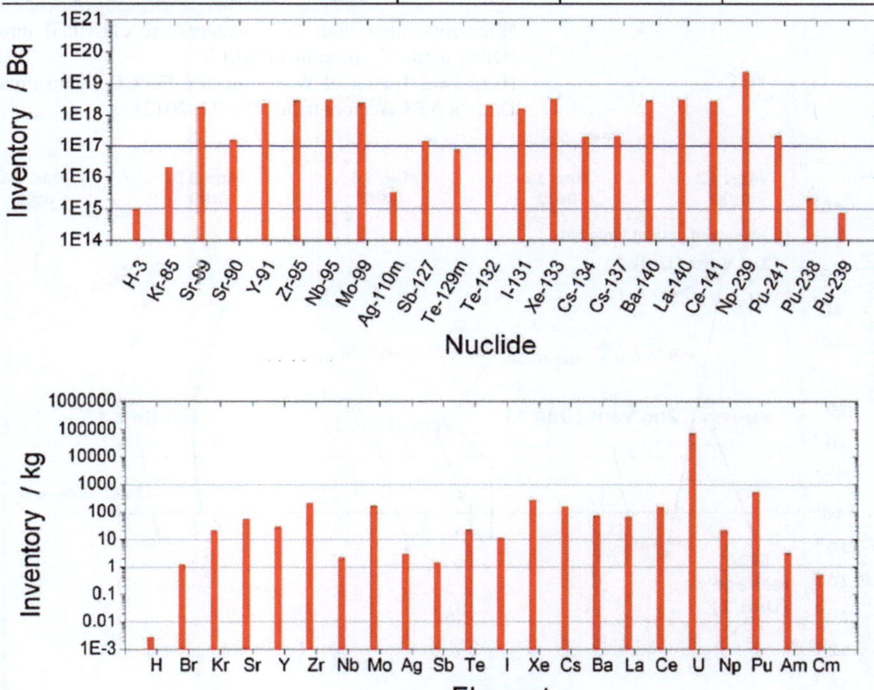

The release and diffusion behavior of radioactive materials from fuel strongly depends on their chemical form, while the total stoichiometric volume of individual elements, including nonradioactive nuclides, is required to estimate to determine their chemical form.

Fig. 6.9 Estimated inventories of nuclides and elements at Unit 1 (1 day after scram)

Table 6.12 Grouping of elements to evaluate release behavior of radioactive materials in MELCOR and THALES2

	MELCOR	THALES2
1.	Xe	Xe
2.	CsI	CsI
3.	CsOH	CsOH
4.	Te	Te
5.	Ba	Sr
6.	Ru	Ru
7.	Ce	Ce
8.	La	Other aerosol[a]
9.	Mo	
10.	U	
11.	Sn	

[a]Elements other than 1–7 elements are classified into 'Other aerosol' group in THALES2
(Ishikawa, Technical Workshop for TEPCO Fukushima Daiichi NPS Accident; July 23–24, 2012)

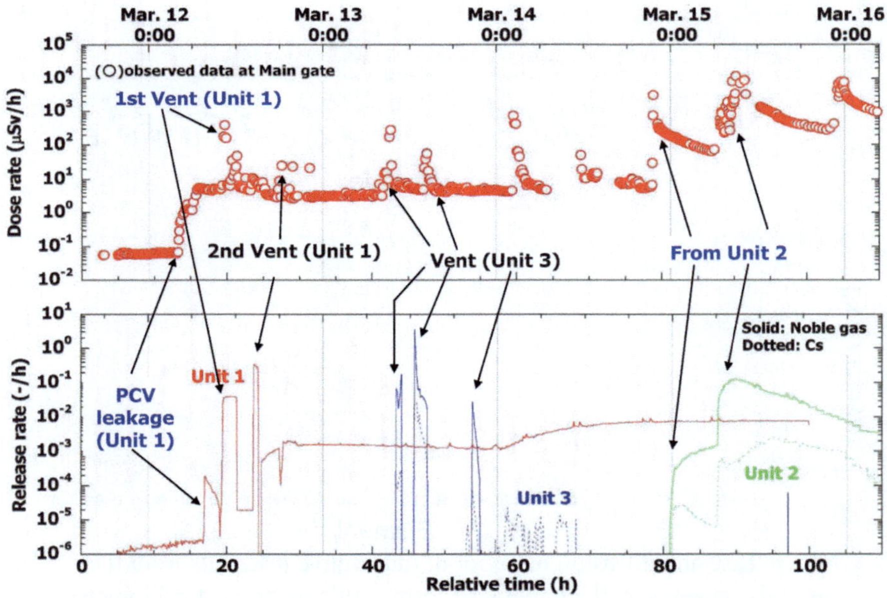

Fig. 6.10 Estimated amount of release in MELCOR and comparison with monitoring post data (Hoshi, Technical Workshop for the TEPCO Fukushima Daiichi NPS Accident, July 23–24, 2012)

(IRM), and traversing in-core probe (TIP) in the reactor pressure vessel, and from the top flange packing and equipment hatch in the containment vessel. Assumption of thermal stratification of the suppression chamber is one of the features in this simulation. However, molten core-concrete interactions (MCCI) are not taken into consideration. Figure 6.10 shows a comparison of the estimated release volume of

Table 6.13 MELCOR evaluation of release rate (Unit 2)

Element group	Release rate (–)	Element group	Release rate (–)
Noble gas	9.6×10^{-1}	Ba	4.7×10^{-3}
CsI	9.7×10^{-2}	Ru	2.1×10^{-8}
Cs	2.6×10^{-2}	Ce	1.0×10^{-9}
Te	5.4×10^{-2}	La	1.9×10^{-6}

(Hoshi, Technical Workshop for the TEPCO Fukushima Daiichi NPS Accident, July 23–24, 2012)

Table 6.14 MELCOR evaluation on environment release (Bq)

Nuclide	Unit 1	Unit 2	Unit 3
Xe-133	1.6×10^{18}	3.3×10^{18}	4.3×10^{18}
I-131	4.8×10^{16}	1.9×10^{17}	1.4×10^{16}–1.0×10^{17}
Cs-134	1.2×10^{15}	7.1×10^{15}	2.22×10^{13}–6.7×10^{15}
Cs-137	9.7×10^{14}	6.3×10^{15}	1.3×10^{12}–5.8×10^{15}
Sr-89	6.9×10^{14}	1.2×10^{16}	4.5×10^{13}–2.2×10^{14}
Ba-140	1.0×10^{15}	1.9×10^{16}	2.7×10^{14}–3.55×10^{14}
Te-132	4.6×10^{16}	8.3×10^{16}	2.8×10^{16}–3.3×10^{16}
Ru-103	8.8×10^{7}	6.8×10^{10}	3.2×10^{9}–4.0×10^{9}
Pu-241	6.3×10^{6}	3.0×10^{8}	3.00×10^{5}–2.6×10^{7}
Cm-242	2.4×10^{8}	7.5×10^{9}	2.1×10^{9}–6.7×10^{9}

Nobel gases and Cs in MELCOR with monitoring data (at the front gate of Fukushima Daiichi NPS). The result of MELCOR appears to explain changes in monitoring data up to noon on March 15, suggesting the validity of the accident progression scenario. Tables 6.13 and 6.14 show the evaluation on environment release from each unit and release rate of Unit 2. The broad range for Unit 3 results from the changes in leak area of PCV and amounts of external water injection.

The evaluation results show the following:

- Elements with higher volatility are generally linked to larger release rates.
- The release at Unit 2 is large, but the release at Unit 3 could have been as large as that at Unit 2, suggesting releases of radioactive materials may be significantly reduced with sufficient external water injection.
- The chemical forms and in-core behavior of I and Cs, which are critical for environmental evaluation, must be studied in more detail.
- The amounts of Sr and Pu released into the environment are small, but their release behavior must be studied in more detail in addition to measuring contamination in and around the site.
- Significant contamination of areas northwest of Fukushima Daiichi NPS on March 15 can be estimated from release evaluation and weather condition, namely, a significant release of 131-I, 134-Cs and 137-Cs from Unit 2 from late evening of March 14 to the following day, and northwesterly wind and rain as the causes.

Accordingly, the simulation results of the accident progression scenario show the following:

- Release of radioactive materials (time, nuclides and quantity) can be reproduced with suitable assumptions, which suggests the validity of the accident progression scenario. However, this may be applied to releases up until around March 15. More detailed analysis is required for subsequent releases.
- Many correlations are found between venting and monitoring data. Data that does not show correlations may indicate a functional deterioration of the RPV and PCV containment capability. Such analytical processes could be key in estimating the causes of release. This analysis examined correlations with data on the monitoring post at the front gate of Fukushima Daiichi, but when data in other monitoring posts are examined, much more findings can be obtained.
- Significant releases from late evening of March 14 to daytime of 15 were considered attributable to continuous leak from the PCV of Unit 2, which resulted in significant contamination in the northwest regions of Fukushima Daiichi. A continuous release resulting in serious environmental contamination clearly underlines the importance of controlled release e.g. by venting.

6.1.3.2 Evaluation on Releases into the Ocean

Releases into the ocean involve the deposition of radioactive materials released into the atmosphere and the direct release from the power station. Figure 6.11 shows the amount of material released into the atmosphere from March 12 to March 20, as estimated from the 134-Cs concentration of surface seawater by the Japan Atomic Energy Agency (JAEA). Evaluation is based on the quantity of radioactive materials diffused into the ocean area at the accident, which fell over the ocean as either

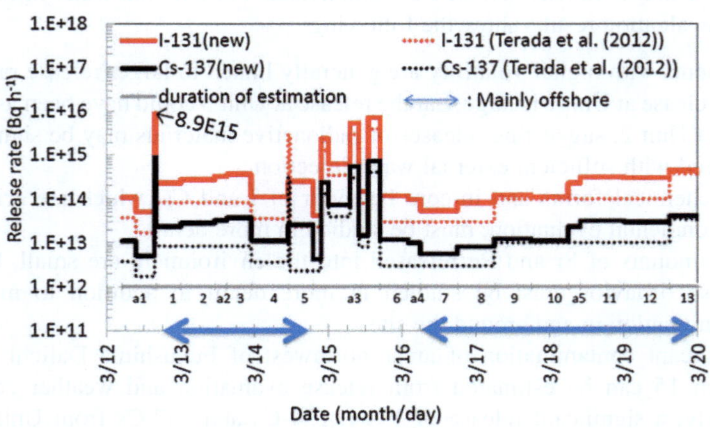

Fig. 6.11 Atmospheric releases estimated from the 134-Cs concentration in surface seawater (JAEA) (Chino, material for the 18th JAEC Special Meeting)

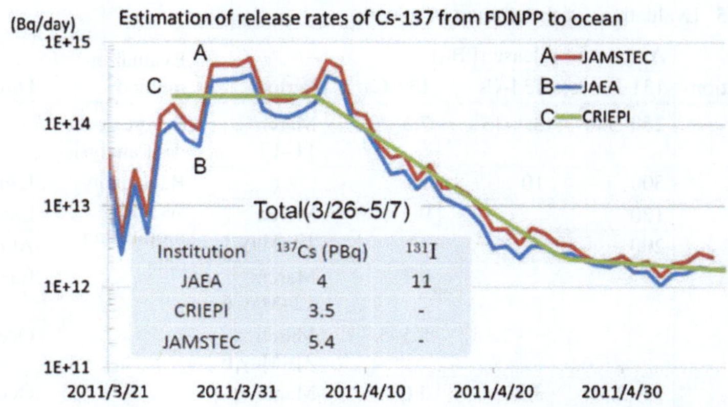

Fig. 6.12 Evaluations of 137-Cs released to ocean by three institutions (Chino, material for the 18th JAEC Special Meeting)

dry or wet depositions. Significant releases were found at noon on March 12 and from late evening of March 14 to daytime of 15, which corresponds to the data in Fig. 6.10. Figure 6.12 shows releases into the ocean from March 21 to April 30. The amount of releases into the ocean can be estimated in ocean simulations. The evaluation results of the Japan Agency for Marine-Earth Science and Technology (JAMSTEC), Japan Atomic Energy Agency (JAEA) and Central Research Institute of Electric Power Industry (CRIEPI) are approximately consistent. According to the CRIEPI evaluation, direct releases into the ocean up until September 2011 account for about 1/4 of the total volume of releases into the ocean.

6.1.3.3 Evaluation of Radioactive Material Release

Table 6.15 shows the amount of released radioactive materials evaluated by several organizations. Evaluations following severe accident analysis and back analyses using monitoring data approximately coincide. In terms of the total volume of released radioactive materials, comparison of estimated values based on the accident progression scenario with those based on the actual monitoring data in the vicinity of the power station will confirm oversights, if any, in the accident progression scenario. The total volume of releases should be reflected in environmental restoration activities and future disaster prevention plans.

6.1.3.4 Measures to Reduce Radioactive Material Release

In light of the above evaluations on releases of radioactive materials, an approach is taken to sustain the integrity and containment capabilities of various types of equipment to reduce releases from the leak path. This includes protecting fuel

Table 6.15 Evaluations of the amount of release

Organization	Amount of release (PBq) 131-I	134-Cs	137-Cs	Period	Evaluation method	Direction
JNES	250–340	8.3–15	7.3–13	March 11–17	Severe accident analysis	
TEPCO	500	10	10		Back analysis with monitoring data	Land
JAEA	120		9	March 12–May 1		Land
	200		13			Atmosphere
JAMSTEC			9.7	March 12–May 6		Land
			5.5–5.7	March 21–May 6		Ocean
CRIEPI	11 (2.8)	3.5 (0.94)	3.6 (0.94)	March 26–September 30		Ocean

Note: Numbers in parentheses indicate direct releases to the ocean (From the JNES report, etc.)

and fuel cladding from melting, and sustaining the integrity of RPV containment capability (control rod driving system, safety relief valves (SRVs), instrument systems, various piping systems), PCV containment capability (venting system, flange packing, instrument systems, various piping systems), and reactor building containment capability (structure, ducts, ventilation systems, exhaust systems (relation with hydrogen explosion). Severe accident countermeasures (software, hardware and management), protection of fuel cladding from overheating, and improvements in the containment capabilities of RPV, PCV, and reactor building are related to defence in depth, plant design and accident management.

6.1.3.5 Issues for the Future

First of all, advanced simulation is required. For this purpose, grouping of elements and parameter settings, including the behavior of iodine such as organic iodine, cesium and aerosol, to simulate releases must be improved based on experimental data and so forth. There is also a need for elaborated models that clarify the relations between physical behavior such as entrainment and the behavior of nuclides. The release path from the RPV to the environment via the PCV and reactor building must also be examined in more detail hereafter. Field investigations in decommissioning processes will provide more information.

Improvements for the analytical method, diffusion behavior, and the density of data regardless of locations are issues for the future in back analysis with monitoring data.

Releases after March 16 need to be explained differently by other methods than used up to March 16. Changes in the pressure of the suppression pool (S/P) and drywell (D/W) are no longer useful and evaluation of the evaporation behavior based on temperature changes of various locations is required. In the longer term, releases from fuel debris and radioactive materials remaining in various locations must be taken into consideration, which are related to the changes in the concentration of radioactive materials in contaminated water. Accurate evaluation of releases of radioactive materials into the atmosphere and ocean from immediately after the accident to date is important as reducing releases in future.

6.1.3.6 Conclusion

A comparison of simulation results with release data confirmed the validity of the accident progression scenario that allows the reproduction of releases of radioactive materials (time, nuclides and quantities) for the first few days from the accident by setting adequate assumptions. This gives a certain value to accident progression scenario under circumstances where directly observing the inside of the reactor is impossible.

Significant contamination of the northwest zone caused by a large amount of releases at Unit 2 due to the damage of the containment vessel, and the seriousness of the environmental contamination caused by continuous releases suggest the importance of controlled releases such as venting in severe accidents. Furthermore it is important to understand the release paths and reinforce the containment function. Consideration of high-temperature degradation of parts such as flange packing and in-core instrument tube is also required.

In terms of reinforcing measures to protect people and the environment, more advanced simulation, and continued study of various issues concerning source term assessments are also required.

6.2 Concept of Nuclear Safety

Criteria of judgment are essential to analyze accidents and investigate causes and countermeasures. If the accident is caused by a certain technical problem, the analysis may be based on the existing technical standards and guidelines, but if the accident subject to analysis presents doubt about the validity of these standards, we must revisit the most basic concept of nuclear safety.

This section, therefore, outlines the concept of nuclear safety, and technical approaches and methods based on this concept, and considers their relations to accidents.

INSAG-12 [1] published by the IAEA based on experience of ensuring nuclear safety in various countries and Safety Fundamentals-1 (SF-1) [2] which extensively addresses the goals of nuclear safety explained in INSAG-12 can be referred to as the basic principle of nuclear safety. SF-1 documents a superordinate safety philosophy, which integrates the nuclear safety standards developed by the IAEA for individual technical areas as the nuclear safety standard system. It was agreed after a decade of discussions by experienced experts of various countries.

Such superordinate safety philosophy has not been introduced in the corresponding regulations in Japan. However, recognizing the importance of presenting the basic concept of nuclear safety, the Nuclear Safety Commission started discussing the documentation of basic principles of nuclear safety in February 2011. Unfortunately, this movement was discontinued after the Fukushima Daiichi accident, but the Atomic Energy Society of Japan (AESJ) restarted the discussion and developed the basic principles of nuclear safety [3].

The basic principles of nuclear safety are outlined and their correspondence with accidents is discussed in Sect. 6.2.1.

As the Fukushima Daiichi accident at is clearly shown, nuclear energy accompanies risks. Understanding the conventional risk assessment and awareness of risks in regulatory bodies is important for considering nuclear safety in future. Risk assessment and utilization of risk information at nuclear power stations are outlined in Sect. 6.2.2.

Since utilization of nuclear energy accompanies risks, the extent to which the society accepts risks, in other words, quantitative targets for "how safe is safe enough?" and social acceptance of these targets are necessary. The safety targets and risk reduction are outlined in Sect. 6.2.3.

In Sect. 6.2.4, technical aspects to ensure nuclear safety are outlined. Of the topics discussed in Sect. 6.2.4, critical issues that emerged in the light of the Fukushima Daiichi accident, such as defence in depth, severe accident management, and nuclear emergency readiness, are explained separately.

In Sect. 6.2.5, relations between security and nuclear safety are discussed.

6.2.1 Basic Principles of Nuclear Safety

The basic objective of nuclear safety is to "protect the people and the environment from nuclear power reactor facilities and harmful effect of radiation caused by the activities in these facilities." The basic principles to achieve this goal can be expanded in terms of "who (subject)," "for what (object)" and "how (means)."

The subject can be considered as the basic principles concerning "responsibility and management." This is summarized in SF-1 to three principles:

Principle 1: Responsibility for safety
Principle 2: Role of government
Principle 3: Leadership and management for safety

In AESJ basic safety principle to five principles:

Principle 1: Responsibility for safety
Principle 2: Roles of the Government
Principle 3: Roles of regulatory bodies
Principle 4: Leadership and management for safety
Principle 5: Development of safety culture

In these principles, the organizations and people responsible for safety, roles of safety-related authorities, leadership to be exercised, management to be achieved, and safety culture on which nuclear safety is depend are explained below SF-1 does not define safety culture as an explicit principle, but INSAG-12, which is the predecessor of SF-1, did do so. One of the lessons learned from the Fukushima Daiichi accident is the insufficiency of the safety culture, which is defined as one of the AESJ basic safety principles. Note that "responsibility" as defined under the AESJ basic safety principles is such that a broad range of organizations and individuals relating to nuclear safety fulfill their responsibilities by doing their jobs under circumstances in which their capability is supposed to be exerted given the operators of nuclear power reactor facilities have the largest responsibility to ensure safety.

The "object" is to "protect the people and the environment from radiation risks." This is summarized in SF-1 to:

Principle 4: Justification of facilities and activities
Principle 5: Optimization of protection
Principle 6: Limitations of risks to individuals
Principle 7: Protection of present and future generation

In AESJ basic safety principles to:

Principle 6: Justification of nuclear power reactor facilities and activities
Principle 7: Limitations of risks to the environment and continued efforts

As nuclear facilities are the source of potential risks, there is a need to compare a broad range of risks and benefits of nuclear facilities to explain the validity of nuclear facilities in rational ways and reduce the related risks (as low as reasonably achievable, ALARA). Obtaining the latest insight and continuous improvements, namely, continuous risk reduction, is one of the major lessons from the Fukushima Daiichi accident. The AESJ basic safety principles therefore clearly indicate continuous risk reduction based on the ALARA rule.

The "mean" is to "prevent actual radiation risks," namely, accident prevention and emergency response. This is summarized in SF-1 to:

Principle 8: Prevention of accidents
Principle 9: Emergency preparedness and response
Principle 10: Protective action to reduce existing or unregulated radiation risks

In AESJ basic safety principles to:

Principle 8: Prevention of accidents and reduction of effects
Principle 9: Emergency preparedness and response
Principle 10: Nuclear security measures to reduce existing or unregulated radiation risks

These principles cover subjects such as the prevention of accidents based on defence in depth, mitigation of radiation effects in the event of an accident, accident management during an accident and preparation of emergency response plans.

As explained above, while technical and hardware aspects, such as accident prevention and effect mitigation, tend to be focused on protecting people and the environment, which is the goal of nuclear safety, software aspects including management, leadership, safety culture and accident management are essential. Heavy dependence on hardware aspects in Japan was pointed out as a lesson learned from the Fukushima Daiichi accident. The AESJ basic safety principles may be useful as an indicator to avoid these pitfalls.

6.2.2 *Risk Assessment and Utilization of Risk Information*

6.2.2.1 Background

The Nuclear Safety Commission[1] explained about the probabilistic risk assessment (PRA), "global nuclear safety stakeholders recognized the importance of reducing risks that lead to serious core damage beyond the design basis assumption (severe accident) in nuclear power generation facilities as a valuable lesson from the experience of accidents at TMI and Chernobyl. Accordingly, a probabilistic safety assessment (PSA) technique was developed to evaluate probabilities of disasters caused by failure of multiple safety devices during malfunction or wrong operation of equipment and the significance of effects, and quantify the risks of severe accidents". It continued, "this technique is used to assess the risks of severe accidents in nuclear power reactor facilities in Japan. Consequently, it is judged that the level of severe accident risk management in nuclear power reactor facilities in Japan is equivalent to international standards". The safety of nuclear power reactor facilities in Japan was declared via a risk assessment.

Various countries use various information obtained from the risk assessment for improving safety, increasing efficiency in operation and maintenance, reducing exposure to radiation, improving the capacity factor, and providing logical rules.[2] For instance, in the U.S., the government issued a public statement concerning PRA utilization in 1995 to promote the PRA technique in all nuclear regulatory activities targeting (1) improvements in safety decision-making, (2) effective use of NRC resources, and (3) reduction in unproductive load of operators. Specifically included are the elimination of unnecessary maintenance in the current regulations and proposals for new regulatory requirements using PRA, representation of actual conditions in PRA as far as possible, disclosure for public review, application of safety goals and additional targets in appropriate consideration of uncertainties of

---

[1] Interim Report of the Special Committee on Safety Goals for Nuclear Safety Installations in Japan, Commission," Special Committee on Safety Goals, Nuclear Safety Commission, December 2003.

[2] "Scientific and reasonable rules" is a typical representation, but "reasonable" often elicits misunderstanding. "logical" is therefore used in this paragraph as rules that are logically structured based on certain concept and principle.

PRA, and so on. The framework of risk information utilization was determined in a couple of years. For instance, regulations and guidelines encouraging the use of risk information were developed in order in 1998, and reactor monitoring processes that evaluate plant operation according to plant performance indexes based on risk information, and regulating the plants based on the result were conducted in 1999. In 1997, a plant-specific PRA (IPE program)[3] for internal events was developed, and in 2002, a similar program for external events (IPEEE program)[4] was also developed.

In Japan, regulations and guidelines have also been developed. These include a decision document of the Nuclear Safety Commission concerning accident management (AM) and regular safety review requested by the Ministry of International Trade and Industry in 1992, AM development report in 2002, policies of introducing safety goals (stated above) and risk information utilization regulation by the Nuclear Safety Commission in 2003, performance goals set by the Nuclear Safety Commission, and basic policies of risk information utilization and the immediate implementation plan by the Nuclear and Industrial Safety Agency (NISA) in 2005, and the risk information utilization guideline and PSA quality guideline by NISA in 2006. Around the same time, organizations and systems were also streamlined, including the change in the parent body of the Nuclear Safety Commission to the Cabinet Office to increase its independence (2001), establishment of the Nuclear and Industrial Safety Agency (2001), and establishment of the Japan Nuclear Energy Safety Organization (2003). However, successive incidents of broken pipes of the residual heat removal system at the Hamaoka NPS reactor Unit 1 (November 2001), dishonest acts of TEPCO including the falsification of data (2002), improper construction work of the Rokkasho Reprocessing Plant (2003), and breakage of pipes in the secondary side piping at the Mihama NPS reactor Unit 3 (August 2004) caused nuclear power plants and transport of used fuel to the reprocessing plant to stop for an extended period. This is a valuable lesson concerning the difficulty in the stable use of nuclear energy if the security of nuclear power reactor facilities by the licensees and safety regulations enforced by the government does not gain public trust.

A picture of effective and efficient regulations based on plant-specific risk assessment was not implemented due to an increase in the safety and capacity factor of power plants by utilizing risk information. The use of probabilistic risk assessment on internal events was limited except for the representation of high safety levels of nuclear power plants in Japan. Efforts to promote risk assessments of external events made little progress. In 2006, the Regulatory Guide for

[3] NUREG-1560, Individual Plant Examination Program: Perspectives on Reactor Safety and Plant Performance, December (1997).

[4] NUREG-1742, Perspectives Gained from the Individual Plant Examination of External Events (IPEEE) Program, April 2002.

Reviewing Seismic Design of Nuclear Power Reactor Facilities of the Nuclear Safety Commission was eventually amended to recognize and minimize residual risk as much as possible. The AESJ developed a practice standard of seismic PRA in 2007, but in effect, did not address risk assessments for external events including tsunamis.

6.2.2.2 Risk Assessment and Risk Information Utilization

What would be the results of the Fukushima Daiichi accidents if a risk assessment had been widely accepted in Japan and was used as a general means for confirming safety?

The Regulatory Guide for Reviewing Seismic Design of Nuclear Power Reactor Facilities requires active faults which are proved to have moved from 120,000 to 130,000 years ago to the present to be taken into consideration. The frequency of the Jogan Earthquake and Tsunami off the coast of Sanriku in 869 is considered to be more than once every 1,000 years in a geological assessment. The tsunami allegedly took the lives of about 1,000 people. First of all, we must consider the possibility of having employed incorrect frequencies of earthquakes and accompanying tsunamis. Except for Japan, many counties adopt the requirement of considering natural phenomena in 10^{-3}–10^{-4}/years, based on the assumption that the frequency shorter than this period is far beyond the experience of human societies, and thus practical and effective measures cannot be taken. However, when considering the impact of accident, taking appropriate safety margin makes sense in some cases. It is regrettable that detailed impact and event progress were not studied in Japan. Some also pointed out that we lacked a perspective of balanced and systematic risk analysis on various natural phenomena.

Probabilistic risk assessment was conducted for internal events. Since other aspects such as risk information utilization were not promoted, skeptic began doubting the meaning and role of PRA. This may be why level 2 severe accident analysis and level 3 off-site consequences analysis have never been conducted, resulting in poor assessment on severe accidents and off-site consequences, and also discouraging studies on severe accident and resource investment.

In Japan, no programs for external events like IPEEE employed by the U.S. Nuclear Regulatory Commission were developed. One of the excuses is the fact that methods of assessing external events were still immature, or there was no reliable data. This means, prematurity of PRA of external events was concluded due to the low reliability of evaluation results. In fact, evaluation methods should be improved not only for external events, but even for internal events, and continuous collection of data is required. Waiting for the maturity of evaluation methods degraded the meaning and role of the assessment.

The mindset must stem from comprehension that the purpose of using PRA was to prove that "risks are very low at nuclear power plants which are already safe". If so, immature method of assessment and incomplete data justify the omission of PRA.

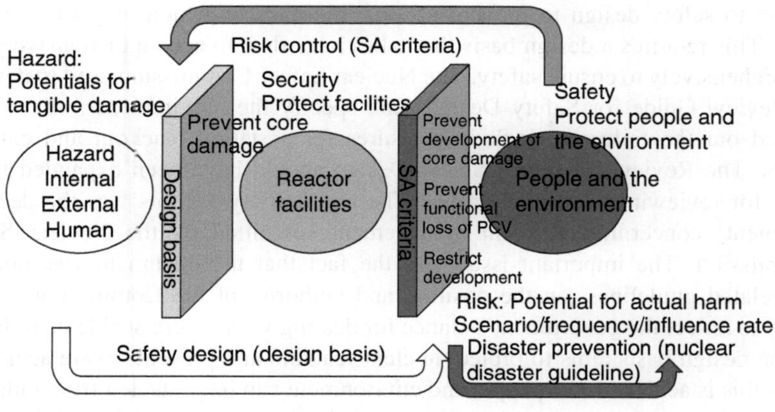

Fig. 6.13 Relations of hazards with design and SA criteria

According to the interim report of safety goals (stated above) by the Nuclear Safety Commission, "risk assessment on severe accidents has been conducted in Japan, and the result showed that severe accident risk reduction level in nuclear power reactor facilities in Japan could be judged to bear comparison with the international standard". However, this risk reduction level was only based on the result of level 1 PRA.

Certain very low probability plant states that are beyond design basis accident conditions and which may arise owing to multiple failures of safety systems leading to significant core degradation may jeopardize the integrity of many or all of the barriers to the release of radioactive material. These event sequences are called severe accidents (IAEA[5]). The report is, however, inconsistent in that while it says it "recognized the importance of reducing risks of events leading to serious reactor core damage beyond the design assumption (severe accident)," the frequency of reactor core damage was cited to judgment to bear comparison with international standard. Even more extraordinary is that they only considered internal events.

For external events, a comprehensive and systematic approach must be developed by extracting as many natural phenomena and human events in the past as possible, deciding whether they are taken into account in terms of the frequency of occurrence in power plants, physical distance, temporal allowance, and influence rate, and selecting an appropriate evaluation method by listing available methods. This approach is shown in Fig. 6.13.

Hazards that must be considered in nuclear power reactor facilities include internal events, external events and human events, each having numerous subgroups. Hazards are possibility to become the hazard to actual harm. Engineering facilities are, in addition to ordinary design for implementing their primary roles,

[5] IAEA, Severe Accident Management.

subject to safety design to prevent all possible hazards which may cause actual harm. This requires a design basis that takes hazards into account extensively and comprehensively to ensure safety. The Nuclear Safety Commission used to develop the Review Guide for Safety Design[6] and specify the requirements, but sources pointed out the failure to include measures for a station blackout and external events. The Review Guide for Safety Design should have been expanded to the guide for reviewing severe accidents. The key activity for this was the decision document[7] concerning accident management submitted by the Nuclear Safety Commission. The important issue was the fact that the document was unclear, like related guidelines, on the position and authority of the Commission, which seems to reflect the government's stance for dealing with severe accidents in Japan.

The design basis aims to protect nuclear reactors and prevent severe accidents. When this is achieved, the people and environment can be protected from radiation hazards. Even though events exceeding the design basis may take place and reactors cannot be protected, extensive releases of radioactive materials to the premises can be stopped by preventing such events progressing or escalating to severe accidents, avoiding the loss of containment function, and reducing leaks of radioactive materials. These are basic countermeasures for severe accidents, targeting efforts to prevent the releases of radioactive materials to the premises. Therefore, countermeasures are required to consider not only the progression of events leading to severe accidents but also to the manner and scale of releases, as indicated by the arrow from "People and the Environment" to "Reactor facilities" in Fig. 6.13.

Based on this approach, nuclear disaster measures for releases into the external environment, namely, nuclear emergency preparedness guidelines should be provided. The risks in this case are potential hazards causing tangible damage to people and the environment. They must be restrained by referring to qualitative and qualitative safety goals.

Level 1 PRA evaluates the performance of protecting nuclear power reactor facilities from hazards, and quantifies the frequency of reactor core damage, accident sequence, and core damage state. Level 2 PRA evaluates the performance of preventing releases of radioactive materials from damaged core, and quantifies the frequency of the functional loss of PCV, as well as the form and amounts of releases of radioactive materials. Level 3 PRA evaluates effects of released radioactive materials on the lives, health and assets of the people and the environment. A wide range of external events must be taken into account in these evaluations. All levels of PRA including level 3 should be conducted.

Even if the methods and integration were incomplete, comprehensive risk assessment could lead to an event progression scenario like the Fukushima Daiichi accident. If it resulted in a low frequency of occurrence, it would still be possible and make sense to take accident prevention measures, confirm their effectiveness with a risk assessment, and conduct suitable education and training to make them

[6] Review Guide for Safety Design of Nuclear Power Reactors, Nuclear Safety Commission.

[7] Accident Management, Nuclear Safety Commission, 1992 (revised in 1997, abolished in 2010)

effective when considering the probability of mechanical failures caused by a common event, difficulties in preventing the escalation to a severe accident, and simplified distinction of equipment to be protected.

6.2.3 Safety Goals and Risk Reduction

6.2.3.1 Background

Japan used to depend on three levels of protection in defence in depth, prevention of incidence, prevention of expansion of incidence, and mitigation of effects of accident, to ensure the safety of nuclear reactors (Sect. 6.3). Preparations for design basis accidents were employed based on the fundamental concept of design basis external events (earthquakes, tsunami) by structuring the safety critical systems according to single failure criteria.

After the TMI accident in the U.S., and the Chernobyl accident in former Soviet Union, however, many countries took measures against beyond design basis accidents as severe accidents. In Japan, the then Nuclear Safety Commission discussed the matter, and in May 1992, voluntarily integrated an effective accident management for licensees of reactor operation concerning "events beyond design basis accidents," which is comparable with level 4 of defence in depth, and strongly recommended the suitable application of the same in preparation for emergencies. To prevent an event escalating to a severe accident, and mitigating the same, the Commission recommended the effective use of potentially useful and new equipment provided for severe accidents, in addition to the existing safety margin and equipment and facilities assumed in the original safety design. The application was, however, left to the discretion of licensees and the Ministry of International Trade and Industry (later to the Nuclear and Industrial Safety Agency), with the submission of a report.

The Ministry of International Trade and Industry promoted severe accident measures through administrative guidance according to its policy, and requested a progress report to licensees, who interpreted the accident management as voluntary measures and did not clearly define the accidents potentially leading to "events beyond the design basis accident" or severe accidents as control subjects. This policy was by no means a delay from world trends at that time. For example, it required a new plant to take measures in the design phase. While many countries made it mandatory later, Japan left it to the discretion of licensees.

6.2.3.2 Defence in Depth and Reduction of Severe Accident Risk

The interim report on safety goals [4] issued by the Nuclear Safety Commission in 2003 addressed the following in its introduction: "The business activities utilizing nuclear energy are beneficial in that energy sources for all mankind, including our descendants, while they have a wide range of radiation application, including medical, industrial and agricultural sectors, but the existence of potential risks of accidents accompanying the extensive diffusion of radioactive materials which may largely affect our health and social environment is undeniable". To reduce these risks, a concept of multiple protection[8] was designated as basic safety measures for nuclear reactor facilities. It consists of three steps of prevention, namely, prevention of occurrence of incidents, prevention of escalation to accidents, and prevention of significant releases of radioactive materials.

The Nuclear Regulation Authority Establishment Act, approved in June 2012 after the Fukushima Daiichi accident, includes extensive amendments to safety regulations that assumed large-scale natural disasters and terrorism. The act also requires measures to prevent escalation to beyond design basis accident and mitigate radiation effects in case of severe accidents. The three-step safety measures were eventually developed to level 4 of defence in depth, and the reduction of severe accident risks to people and social environment was clearly defined in the framework of nuclear regulations.

The magnitude of tsunami following the Tohoku District-off the Pacific Ocean Earthquake far exceeded the design basis assumption. Nuclear power facilities with potential risks resulting in serious effects on the health of people and social environment must not be excluded from risk management as "beyond assumption," and give grave consequences to the people and environment in the vicinity of site boundary with radioactive materials. Nuclear regulations, therefore, require fundamental safety goals to evaluate risk management and risk reduction. It is essential to apply new insights, research outcomes and operational experience in Japan and abroad to risk management and nuclear regulations.

6.2.3.3 Risk Reduction Targets (Safety Goals)

Adequate risk management is carried out according to defence in depth to reduce risks that exert large effects on the people and environment. The way to evaluate the state of risk management has been an issue internationally discussed as "How safe is safe enough?" Qualitative safety targets with probabilistic values have been adopted in many countries, and increasingly used as a supplement of deterministic safety regulations. As mentioned above, the former Nuclear Safety Commission also proposed safety goals in Japan. This is because the Commission considered that, as a certain level of risk reduction that is achieved by activities pursuant to nuclear safety regulations, safety ensuring activities would be more effective if safety goals were set based on the concept of probabilistic risk and used to assess the safety regulatory activities.

---

[8] "Multiple defence" is the term used in the original text, but "defence in depth" is used in this document to avoid misunderstanding that defence is achieved by redundant installation of engineering facilities.

Safety goals quantitatively define the degree of risk or how low the probability of occurrence which the safety regulatory activities of the government require the licensees to include in their risk management. Setting safety goals based on public risks will help increase the transparency of regulatory activities, predictability, rationality and consistency, allowing the effective and efficient exchange of opinion as to what the nuclear regulatory activities of the government including the development of regulations and standards should be.

Qualitative safety goals indicate the level of risk reduction, and "the probability of public health hazard due to the release of radiations and radioactive rand materials in the utilization of nuclear energy should be limited to a level at which significant increases in health risk in daily lives are unlikely". Quantitative goals represent the level of safety, and should measure the attainment of qualitative goals objectively. The Commission suggested that the average acute mortality risk of individuals living in areas near the site border, and the average mortality risk of individuals living in areas at a certain distance from the site border, by cancer potentially developed by accident initiating radiation exposures should not exceed the order of 1/1,000,000 per year (10^{-6}/people-year). Quantitative safety goals are used to determine the depth and breadth of safety ensuring activities at nuclear power facilities. Therefore, the probability of occurrence of serious accidents inherent in each nuclear facility and relevant to quantitative goals for that facility were set as performance goal, for which the core damage frequency was set as 10^{-4}/rector-year, and PCV breakage frequency as 10^{-5}/rector-year in the accident scenario including internal and external initiating events (intended/manmade events are excluded). These numerical values are required to indicate rational and viable risk reduction planned and implemented, and whether the frequencies mentioned above are lower than target values is not important.

The Nuclear Regulation Authority made it clear to address the performance goals from the day it was founded, and announced the following values in April 2013 after having investigated the performance goals in various countries.[9]

Core damage frequency (CDF)	10^{-4}/reactor-year
Confinement failure frequency (controlled release) (CFF-1)	10^{-5}/reactor-year
Confinement failure frequency (uncontrolled release) (CFF-2)	10^{-6}/reactor-year
Amounts of released radioactive materials	100 TBq (Cs-137)

CFF-1 is assumed to include controlled releases using filtered vent or other means, and CFF-2 uncontrolled releases. 100 TBq is set as the upper limit for the amounts of releases.

Quantitative safety goals and accompanying performance goals are presented above. Qualitative safety goals are the basis for the acceptable risk level that can be shared by us. Sufficient discussion on these targets is important, but omitted from the objectives set by the Nuclear Regulatory Authority. Sharing efforts to reduce

[9] http://www.nsr.go.jp/committee/kisei/h24fy/data/0032_10.pdf.

risks accompanying nuclear energy utilization to the public and the environment as much as reasonably achievable, and the importance of setting risk reduction levels needs safety goals to be accepted broadly by the society and esteemed by stakeholders. A broad array of dialogue with the public must be continued in every step of developing and applying safety goals for purposes of safety goals, details and applicable laws and regulations.

In the Fukushima Daiichi accident, abundant radioactive materials were released into the environment, contaminated the ground in vast areas, and forced residents in affected areas to bear an enormous burden of prolonged evacuation. Such a long-term evacuation clearly indicates a serious social risk. The report by the Nuclear Safety Commission states on social risks; "the consequences of a serious accident not only result in radiation-related health impact to the public, but also restraints of the living environment due to ground contamination, which is a social impact. Compared with the direct influence of radiation on individual's health due to an accident, the social impact is difficult to quantify, and a targeted risk reduction level is not sufficiently discussed. For this reason, the proposal does not include objectives with these attributes. Of course, this is not the decision of this advisory committee on the social impact as insignificant." The Nuclear Safety Fundamentals of the Atomic Energy Society of Japan [3] focusing on social risks include the limitation of releases of radioactive materials that cause serious ground contamination as one of the risk reduction factors. Having experienced the Chernobyl accident, major European countries set safety goals containing the restriction of radiation releases as one of their performance goals. It is insufficient for safety goals to include only public health risk, but how to confront social risks is important.

6.2.4 Safety of Nuclear Power Generation and Mechanism of Ensuring Safety

6.2.4.1 Meaning of Nuclear Safety

Radioactive materials (materials holding radioactivity) stored in nuclear facilities including nuclear power plants present "potential danger" of releasing radioactivity. "Nuclear safety" is to prevent it from becoming serious, which is the fundamental of "nuclear safety." In other words, the goal of nuclear safety is to prevent hazards due to radioactive rays and materials in nuclear facilities, and continue activities to ensure safety in all phases from design, manufacture, construction to operation management and maintenance. In the operation of nuclear power plants, consideration should be given to the potential effects of operation on living, society, economy and the environment. The concept that this is part of nuclear safety has been gradually understood. In this sense, it is important to start public discussions about "what is nuclear safety" or "ensuring nuclear safety" and what to do for achieving it, including the process of discussion, and gain public acceptance as well as listening to public opinion and mutual understanding.

The accident at the Fukushima Daiichi NPS (Fukushima Daiichi accident) caused by the Great East Japan Earthquake on March 11, 2011 showed us that the danger of nuclear power generation in Japan was not hypothetical, but real. Since then, various opinions have been presented to ensure nuclear safety. This section will analyze the technical aspects of ensuring nuclear safety in nuclear power plants.

6.2.4.2 Technologies to Ensure Safety

(1) **Inherent safety**

The nuclear fuel in light-water reactors contains about 2–4 % of U-235 that contributes to nuclear reaction, and the rest of fuel is U-238. When the reactor output increases, the fuel temperature rises, and the thermal motion of uranium atoms promotes the neutron absorption of U-238, and fission chain reaction is decreased. Because the fuel of light-water reactors is mostly composed of U-238 as explained above, chain reaction of nuclear fission is less active with increasing temperature. A large-scale explosive reaction, therefore, will never take place. If output increases and the cooling water becomes hot during normal operation, bubbles are generated in the water (or its density decreases) to decrease the fission chain reaction. The physical mechanism of decreasing the fission chain reaction with increased core output is called a self-regulating characteristic, which is a fundamental requirement in the safety design of reactors.

When the earthquake took place, the nuclear reaction was stopped at the Fukushima Daiichi NPS, and assumed not to have escalated to criticality when considering the post-accident conditions.

(2) **Basic safety design**

The basic safety design of nuclear reactor facilities involves implementing measures to prevent incidents and accidents, escalation and mitigation of effects through multiple barriers, graded approach,[10] and single failure criterion based on internationally accepted principles, defence in depth.

To achieve this, it is important to install high quality equipment and facilities, implement reliable operation and control, and involve all stakeholders, including licensees, equipment manufacturers and regulatory personnel, ranging from design to control, manufacture, construction, operation and maintenance, as well as safety regulations to share safety awareness, and ensure nuclear safety.

In the Fukushima Daiichi accident, a graded approach based on defence in depth presumably did not work well, leading to the accident. For detailed analysis, see Sects. 6.3, 6.4, 6.5 and 6.6.

[10] For example, classification by significance of risk and a relevant response is selected to achieve goals.

(3) **Basic control of radioactive materials**
Radioactive materials and radiations emitted from radioactive materials are the sources of nuclear hazards. Basically they are contained in physical barriers and areas when using radioactive materials. The degree of confinement and control depends on the risks of the radioactive materials used.

In nuclear power plants, confinement of radioactive materials, which are basically enclosed in the PCV or prevented from leaking, is one of the principles used to ensure safety. In the Fukushima Daiichi accident, radioactive material leaked from locations hard to estimate as the accident progressed. Conflicting situations of confinement and core cooling resulted in a failure to cool the reactor core, core meltdown, and releases of significant amounts of radioactive materials into the atmosphere.

(4) **Specific design**
The safety of reactors is ensured by preventing reactivity initiated accidents, keeping the integrity of fuel assemblies and continuing the confinement of radioactive materials. The basic measures to achieve this are:

(a) Defence in depth: Multi-level protection having different effects. The concept involves (i) developing design policy to prevent abnormal conditions, (ii) implementing it to prevent defects in normal operation from developing to abnormal incidents or accidents, and (iii) preventing escalation of accidents, mitigating unfavorable effects, taking post-accident measures, and preventing diffusion of radioactive material.
(b) Multiple barriers: Preventing or reducing leaks and outflow of radioactive materials using multiple physical barriers, including fuel pellets, fuel claddings, reactor pressure vessels, containment vessels and reactor buildings.
(c) Site isolation: Radiation hazards should be prevented by installing the source of radiation at a certain distance from the public (separation distance). In the Fukushima Daiichi accident, neither defence in depth nor multiple-barrier was not fully effective. In contrast, site isolation was effective to some extent though there were issues to be discussed on the requirement of isolation.

(5) **Design basis events**
The events considered in the safety design and assessment of nuclear reactor facilities are called "design basis events". These are conditions that confirm the normal operation and safety of equipment and machines in the design of facilities and equipment, and the events assumed to be caused by various equipment and system failures or incorrect operation. In addition, the design is required so that the safety of nuclear reactor facilities is violated in an assessment that assumes failures of safety mechanisms or power loss.

Design basis events should not cause serious damage to the reactor and secondary damage causing abnormal state in their progression. The barrier design must be valid for the diffusion of radioactive materials, and reactor does not result in "nuclear accident".

(6) **Accident management (AM) and severe accident management (SAM)**
Events beyond the design basis must be handled by every possible means of accident management (AM) based on the reliability of equipment and operators, effects of external events and frequency of occurrence, in the form consistent with performance goals described in Sect. 6.2.3. The beyond design basis events cannot be completely eliminated. These events are assumed to take place due to multiple failures, abnormal events, and multiple incorrect operations, and though the probability may be extremely low, once they happen, they may cause serious core damage, and extensive release of radioactive materials, generally called severe accidents (SA). For example, a loss-of-coolant accident (LOCA) with the loss of external power and emergency power, resulting failure of ECCS operation and failure of residual heat removal, which may develop to core meltdown, penetration of the RPV, overpressure or overheating damage to the PCV, and in some cases, hydrogen explosions and finally the release of a large amount of radioactive materials.
Prevention of severe accidents is essential in terms of ensuring nuclear safety. Before the Fukushima Daiichi accident, beyond design basis event "would not happen," but measures were taken "to make sure" but they were not complete. Assuming various events, and determining relevant measures are crucial to ensure safety.

6.2.4.3 Safety Measures for Potentially Serious Accidents

One of the lessons learned from the Fukushima Daiichi accident is that the safety of people and the environment must be ensured if a potentially serious accident emerges, that is, actually happens. This is the severe accident response inside the site, and the nuclear disaster readiness outside the site.

It is important to always consider the response to the fourth level of defence in depth in nuclear safety, prevention of accidents beyond design events, i.e. severe accidents, and effect mitigation, and the fifth level, disaster prevention, and build a system to conduct PDCA. Considering their importance, these issues are discussed separately in Sects. 6.5 and 6.9.

6.2.5 Relationship Between Nuclear Safety and Nuclear Security

The Fukushima Daiichi accident suggests the potential for similar serious effects on society by terrorism in nuclear power facilities. Accordingly, licensees and regulatory agencies must strengthen their efforts, not only for safety but also nuclear security assuming that terrorism to nuclear power facilities can happen, and take effective countermeasures through mutual cooperation.

Licensees must be aware of their prime responsibility to ensure safety and nuclear security, and related ministries and government agencies, including individuals of these bodies, foster organizational culture (safety culture), be aware of their responsibilities, continuously review and improve measures, especially taking their responsibility extensively to avoid oversight between mutual measures.

6.2.5.1 Common Elements

Nuclear safety and nuclear security have many common elements. Both protect nuclear power facilities with the ultimate goal of protecting the people, society and the environment. Namely, their basic objective is the same, i.e., protection of the people, society and the environment. Regardless of the initial cause, whether safety or security issues, the radiation risk imposed on the people, society and the environment is the same as is the concept of achieving their basic objectives.

Both safety and security are achieved according to the policy of defence in depth in principle, and composed of a number of defence levels. The basic characteristics of these levels are also the same in safety and security. Both require to give top priority to prevention, then early detection of abnormal condition and prompt response to prevent damage. The next effective action is mitigation, and finally, the development of an extensive emergency plan in the case of failure of prevention, protection and mitigation.

An example of complementary nuclear security measures incorporated in nuclear safety measures is the Extensive Damage Mitigating Guidelines in the U.S., which is intended to maintain or restore core cooling, confinement, and spent fuel pool cooling capabilities under the circumstances associated with loss of large areas of the plant due to explosions or fire caused by airplane crash or similar matter, issued based on 10CFR50.54 (hh) after terrorist attacks in the U.S. on September 11, 2001.

This is a provision for mitigating the impact of an accident or fault, and counter security failure potentially causing radiation risk, but also useful for a large-scale accidental event accompanying core melt due to external events such as the one at the Fukushima Daiichi accident.

6.2.5.2 Conflicting Elements

While having many common elements with nuclear safety, events relating to nuclear security and safeguard, such as attacks on nuclear power facilities, theft of nuclear materials, sabotage or intended release of radioactive materials into the environment, are derived from intentionally "intellectual" or "deliberate" deeds with the initial intention of avoiding protective measures. Nuclear security measures against risks caused by malicious actions (e.g. nuclear terrorism) willfully conducted with the intention of doing so require totally different approaches from nuclear safety measures, which respond to risks of natural phenomena, malfunction of facilities or equipment, other internal events or disorders, or unintended events cause by human errors. We must consider nuclear safety and nuclear security as basically different systems when taking relevant measures.

Internal threat measures such as management and reliability confirmation of those who handle classified information concerning nuclear materials and nuclear power reactor facilities, and those who access critical facilities and equipment are effective to limit access to the source of radioactive risks and reduce risks. However, setting delay barriers for nuclear security could limit the scope of employees to promptly access the facilities to deal with safety-related events, or evaluate in an emergency.

When safety or security facilities, which must work all the time, are inspected or maintained, an alternate means is required. For example, the surveillance function, used for security, may be disabled when power to the relevant area is shut off for maintenance. In this case, a compensating security means must be used, and coordination of safety and security functions is required.

Quality assurance is required for both safety and security. For example, management of classified information is especially important for security. In contrast, transparency or accountability is essential for safety-related matters and quality assurance must take both into account.

The government must directly engage in identifying threats, and support counter-terrorism activity. For this reason, the role of the government in security is not the same as its role in abnormal events in safety, and its involvement also varies. While the security authority is the only organization countering armed terrorist organizations, and expertise on engineering, machinery or maintenance is emphasized for persons relating to security, skills varying from general employees in power plants, who are nearly ordinary workers, are required for them.

6.2.5.3 Nuclear Safety Measures and Nuclear Security Measures

As stated above, each of nuclear safety or nuclear security is not included in the other, and safety measures and security measures are either conflicting or complementary. For these reasons, measures are planned and implemented comprehensively so as not to make light of security for implementing safety and vice versa. This indicates that the goal is to maximize the protection of the public, asset, society and the environment by improving and reinforcing interfaces between safety and security. The organizations responsible for safety and security must strive to promote information sharing and exchange of opinions, produce as much synergy as possible in two different areas, and not make light of each other.

6.3 Defence in Depth

The report of the Japanese government submitted to the Ministerial Conference of the International Atomic Energy Agency (IAEA) described how the Fukushima Daiichi accident resulted in a severe accident, shaking up public trust in nuclear safety, warning people in nuclear society against over-reliance on nuclear safety, and highlighting the importance of extensively learning lessons from the accident. It classified these lessons to five categories with the prospect that "defence in depth is the fundamental concept of ensuring nuclear safety". A report of the Near Term Task Force of the U.S. Nuclear Regulatory Commission (NRC) that presented recommendations for nuclear safety in the twenty-first century [5] states that "the defence in depth concept has functioned well for the NRC and licensees, and remains valuable. However, it was neither used consistently, nor provided with guidelines for the extent of application of the defence in depth. Risk assessment provides valuable and practical insights into potential exposure (public exposure). Coupled with other technical analyses, it brings information to determine appropriate defence in depth implementations". Safety Report Series No.46 [6] of the International Atomic Energy Agency cites, "defence in depth is a comprehensive safety approach developed by nuclear experts to ensure with high certainty that the public and the environment are protected from any hazard relating to nuclear power generation".

These reports indicate a common global perspective that defence in depth is the most important and fundamental approach or concept for nuclear safety. It suggests adequate combinations of various provisions that are practical and effective for ensuring safety. The significance and role of the defence in depth concept do not and will not change. In the light of the defence in depth, lessons are being learned from the Fukushima Daiichi accident at present. At the same time, in the light of the Fukushima Daiichi accident, some key insights on the defence in depth are about to be unraveled. The defence in depth and the risk assessment are conjointly applied to obtain useful insights for ensuring safety. Defence in depth is a concept and approach of protecting the public and the environment in nuclear safety.

From a historical perspective, the importance of defence in depth was recognized after the Three Mile Island accident in the U.S. when some pointed out that safety could not be ensured with the design basis protection alone. In the present section, we discuss the need to evaluate the effectiveness of defence in depth, points to pursue the defence in depth application, and what to be achieved to ensure safety in response to the Fukushima Daiichi accident.

Section 6.3.1 gives an overview of the history of defence in depth understanding and application in Japan. Section 6.3.2 identifies insufficiencies and implementation of defence in depth to ensure safety to date in terms of initiating events, design and maintenance of safety equipment, and accident management based on the lessons-learned from the Fukushima Daiichi accident. Section 6.3.3 addresses the defence in depth concept in relation to the objectives of the protection, and whether safety goals can be achieved by providing various measures based on defence in depth.

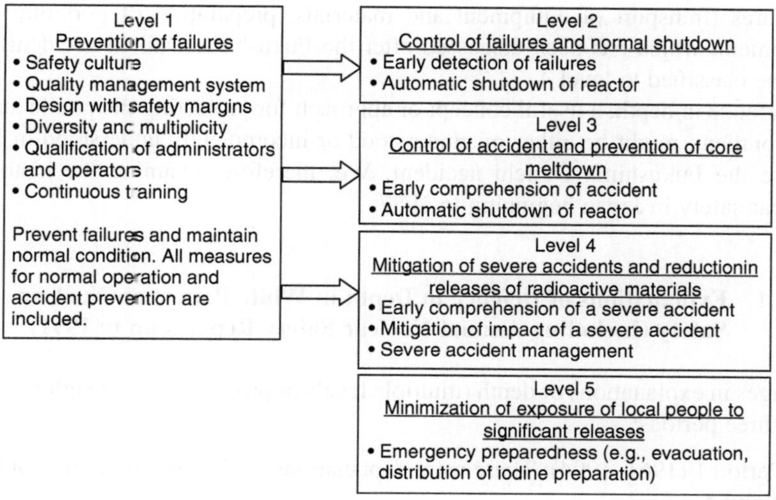

Fig. 6.14 Overview of defence in depth

6.3.1 Defence in Depth Perception in Japan

Defence in depth was considered as a fundamental principle of safety from the beginning of the development of nuclear power generation technology in Japan, and technologically achieving this principle has been targeted for safety study and design. The defence in depth for nuclear power plants generally comprises five levels (see Fig. 6.14):

Level 1: Prevention of abnormal operation and failures (provision of avoiding disturbances in plants)

Level 2: Control of abnormal operation and failures (provision of reducing impact of disturbances on equipment)

Level 3: Mitigation of accidents (provision of preventing core damage under serious conditions of equipment)

Level 4: Control of beyond-design-basis accidents (provision of preventing significant releases of radioactive materials into the environment in the event of core damage)

Level 5: Protection of the public and the environment (provision to reduce public exposure in case of significant releases of radioactive materials)

Note that the "accident" means "design basis accident (DBA)". On the other hand, "Beyond DBA (BDBA)" ranges from incidents exceeding the design basis but not resulting in core damage to severe core damage (severe accident). Including BDBAs, situations preceding to the severe accident can be classified to level 3, and severe accidents to level 4. In all cases, level 4 includes beyond-design-basis conditions, where the focus is placed on accident management. Off-site BDBA measures (transport of equipment and materials, preparation of portable safety equipment, which has been discussed after the Fukushima Daiichi accident) may also be classified to level 4.

Defence in depth, a useful concept or approach for protecting the public and the environment, might be either misunderstood or incompletely understood in Japan before the Fukushima Daiichi accident. We, therefore, examine the history of nuclear safety in Japan beginning in 1957.

6.3.1.1 Explanations of Defence in Depth in White Papers on Nuclear Safety (Including Annual Nuclear Safety Reports up to 1991)

Changes in explanations of depth (multiple levels of protection) are roughly divided into three periods:

(1) Period 1 (1961–1994): There were explanations for levels 1 to 3, but not levels 4 and 5.
(2) Period 2 (1995–2002): Explanations were changed every year.
   - 1995: Potential of severe accident (SA) is inconceivably low (Chairman: Yasumasa Togo)
   - 1997: SA measures are provided by licensees on a voluntary basis (Chairman: Yasumasa Togo)
   - 1998: Measures are required on the premises of the occurrence of severe accidents (Chairman: Kazuo Sato)
   - 2000: No one can say "absolutely safe" (levels 4 and 5 were explained for the first time (Chairman: Kazuo Sato)
   - 2002: Explanations of levels 4 and 5, and the need for accident management (AM) (Chairman: Shojiro Matsuura)
   - 2003–2004: No explanations for levels 4 and 5, explanations for levels 1 to 3 again (Chairman: Shojiro Matsuura)
(3) Period 3 (2005–): No explanations of defence in depth at all (Chairman: Atsuyuki Suzuki, Haruki Madarame)

6.3.1.2 Understanding of Defence in Depth and Safety Assurance

The concept of defence in depth has been used in the nuclear energy industry, but as shown in the history in the preceding paragraph, it consisted of levels 1 to 3, namely, limited to safety measures within the design basis. The Chernobyl accident drastically changed the notion of safety with the events far beyond-design-basis assumptions. The concept of defence in depth was clearly designated as the international safety standard for the first time in 1996; mainly by European countries which were severely affected by the Chernobyl accident. Level 4 for mitigating the impact of beyond-design-basis events or SAs, and level 5 for disaster prevention in the event of SA were added to levels 1 to 3 which covers the design basis requirements, and the current defence in depth principles (INSAG-10) [7] were developed.

It was discussed in Japan for some time, but legal provisions that premise no severe accident (no problem for prevention of disasters caused by reactors) were inconsistent with the concept of level 4. Hence it was decided to leave whether to use level 4 provisions to voluntary activities of licensees, and this was formally described in the white paper on nuclear safety in 1997. In 2000 and 2002, the concept of defence in depth, levels 1 to 5 as defined in INSAG-10 was explained in detail in the White Papers on Nuclear Safety for the first time in Japan. However, the descriptions on the levels 4 and 5 disappeared in 2003 and thereafter as in the white papers before 1998. There might be a reason to change the situation during this period.

In April 2006, the Nuclear Safety Commission (Chairman: Atsuyuki Suzuki) started amending domestic guidelines in line with international safety standards, but the Nuclear and Industrial Safety Agency (NISA) requested to cancel it, and in May 2006, the chairman of NISA wrote to the chairman of NSC requesting "Do not wake a sleeping lion". This was recorded in the minutes, and the NSC (Chairman: Haruki Madarame) later announced it.

It was most regrettable that the implementation of SA was explained in the white paper on nuclear safety in 2000 and 2002, but later withdrawn. Defence in depth for achieving goals of nuclear safety seemed to be interpreted as the explanation of absolute safety. The highly independent Nuclear Regulation Authority, founded as an Article 3 (of the National Government Organization Law) based commission in September 2012 as part of nuclear regulation reform, is expected to understand the defence in depth and tie it to adequate safety ensurances.

6.3.1.3 Documentation of the Role of Defence in Depth

Defence in depth is the most fundamental safety logic for nuclear safety. It is obvious with the statement "defence in depth is the primary means of preventing and mitigating the impact of accidents" in Principle 8 "Prevention of accidents" of the Basic Safety Principles (SF-1) [8] of the IAEA, which is ranked the highest in 132 safety standards. It is the crucial accident prevention concept. For this reason, major countries utilizing nuclear energy give the regulatory documents covering the application of defence in depth the highest priority in safety regulation documents. Unfortunately, there are no such regulation documents in Japan. The Atomic Energy Society of Japan addressed the concept of the IAEA's defence in depth in detail in the Fundamental Concept on Nuclear Safety published in November 2012 (Commentary 15). The explanatory material provided by the Nuclear Regulation Authority states that the new regulations and standards are based on the concept of defence in depth. It is expected to be designated one of high-priority regulation documents of the NRA, Japan.

6.3.2 *Analysis of Defence in Depth in the Light of the Fukushima Daiichi NPS Accident*

6.3.2.1 Relations Between Ensuring Safety and Defence in Depth

The purpose of nuclear safety is to protect residents in vicinity to nuclear reactor facilities from radiological hazards. It is based on the concept of defence in depth, for providing multi-level safety measures in general, and "independent effectiveness" in particular when safety measures in each level are concerned. The reason for providing multiple levels of defence in safety measures is that preparations are to be provided for uncertainties of failure or loss of function in one level, regardless of how robust the level is implemented. The reason for pursuing independent effectiveness without excessively relying on specific levels is that ensuring safety is difficult if the levels fail to function. "Independent effectiveness" involves achieving overall protection across all levels of defence, even if one or more levels fail to work. The thickness and number of levels depend on the nature of the hazard and the degree of uncertainty.

Ensuring safety not only includes requirements for facility design, but also adequate control and management in daily facility operation and in accidents. Whether safety is sufficiently ensured can be confirmed by means such as probabilistic risk assessment (PRA). The concept of defence in depth and the methodology of risk assessment are essential for ensuring safety. Adequate and effective improvements in safety may be possible if the PRA suggests improvements in equipment or operational procedures effective for reducing risks, and enabling the development of more suitable defence in depth.

In the Fukushima Daiichi accident, natural phenomena exceeding the design basis conditions caused multiple damage to equipment with safety function, which, in turn, caused core meltdown, and triggered the releases of significant amounts of radioactive materials. Considering these facts, the failure of defence in depth based measures to function properly for external events such as natural phenomena is discussed below.

6.3.2.2 Specifically Important Issues in the Fukushima Daiichi Accident

The issues derived from the Fukushima Daiichi accident relate to almost all levels of defence. Issues in individual level are discussed first of all.

Level 1 is to prevent abnormal operations and failures that may initiate an accident sequence. Design must include sufficient safety margins for maneuver based on proved techniques, including defence against individual events such as earthquakes and flying objects if necessary. In addition, the facility is maintained based on high quality control systems. However, a large number of safety systems lost their capabilities due to a single external event, the tsunami, in the Fukushima Daiichi accident.

Level 2 is to detect anomalies and failures immediately to avoid escalation to an accident. For example, when an operation parameter exceeds the tolerance level, the relevant signal is transmitted for the automatic insertion of control rods to shutdown the reactor. In the Fukushima Daiichi accident, the reactors were confirmed as automatically shutdown immediately after the earthquake. That is, the earthquake caused the loss of external power supply, classified as a postulated transient change, but the reactors automatically shut down with signals from the seismometers, and all emergency diesel generators (D/Gs) were activated, putting reactor facilities under control.

Level 3 is to prepare for potential accidents and mitigate their impact, for example, installation of the emergency core cooling system (ECCS) which could prevent the reactor core from being heated up if the circulated cooling water has leaked from the broken pipe of the reactor cooling system, a robust and airtight primary containment vessel (PCV) to prevent the release of radioactive materials into the environment, and a PCV cooling system to prevent failure of the PCV due to internal pressure rise. However, though there were some differences among the reactor units, all of them lost AC power, seawater cooling systems, and even DC power in the Fukushima Daiichi accident, and most of the safety systems did not work, resulting in the core meltdown of three units. The requirements of the Safety Design Guideline by the Nuclear Safety Commission concerning the station blackout, "prolonged SBO need not be taken into account" were subject to harsh criticism.

Level 4 is to prevent a design basis accident from escalating to an SA (the phase 1 AM), and mitigate the impact of SA (the phase 2 AM). However, many of such AM activities required power supply, and the on-site work was extremely difficult because of the significant impacts of the earthquake and tsunami on plant system, and environmental conditions of SA, especially high radioactivity.

Level 5 is to protect the public and the environment. Besides the fact that emergency planning in Japan was not effective, the accident revealed some problems in inter-organizational communication. It was also revealed that the lessons-learned from the criticality accident by the JCO in 1999, such as solicitude toward vulnerable people, had remained unresolved.

Of numerous problems identified from the Fukushima Daiichi accident, the following three are particularly important:

(1) **Defence against external events, particularly natural phenomena**

Prevention of component failures is classified to Level 1 of the defence in depth. "IV. General reactor facilities (Provisions 1 to 10)" in the Safety Design Guideline of the former Nuclear Safety Commission requires safety-related structures, systems and components (SSCs) to have high reliability. Provisions 2 to 5 require natural phenomena, manmade external hazards, internal flying objects and fire to be considered in the design

However, defence against external events, particularly natural phenomena was not fully covered for adequate protection. While a wide range of design basis hazards should have been included in the plant design according to the influence of events even if the occurrence frequency is low. Tsunami had not been considered in the design appropriately. A safety design is required to be conformed to the characteristics of individual initiating events. For earthquakes, an earthquake-resistant design with a sufficient safety margin is required for each SSC. For tsunamis, suitable tide walls, watertight buildings, and substantial measures are required.

(2) **Reliability of accident management**

Level 4 of the defence in depth is the AM for severe accidents. Impacts of beyond-design-basis events should have been mitigated by the AM to prevent severe accidents, but the earthquake and tsunami made the operation of predefined equipment and procedures for AM extremely difficult in the Fukushima Daiichi accident because common cause failure of major safety-related SSCs. In addition, consideration was incomplete for the potential internal and external environmental conditions resulting from severe accidents caused by earthquakes and tsunami, which impeded access to facilities and transport and the installation of the required equipment and materials. Required information sharing and effective cooperation were also disturbed because of communication failure.

Due to poor reliability and effectiveness, accident management measures provided for severe accidents was not useful in practice under the accident conditions. In AM for beyond-design-basis natural disasters, measures that take situations unique to natural phenomena into consideration are also essential.

(3) **Flexible response to beyond-design-basis events**

The Fukushima Daiichi accident reminded us the problem of adequate defence in depth implementation for low-frequency/high-consequence natural phenomena. As mentioned above, nuclear safety is achieved technically based on defence in depth, but consideration and evaluation of the effectiveness of the defence in depth for events caused by external factors appeared insufficient. It should be noted that the equipment inside the plant alone cannot be enough for effective management and control for extreme natural events. For example, an earthquake far exceeding the design basis assumption may cause the loss of safety function of SSCs in the reactor building due to a common cause. Even if watertightness of buildings is reinforced to prevent the impact of the tsunami, the loss of function may also be induced by a common cause if the water level exceeds the critical height. It is clear from the mitigation actions taken in nuclear power plants other than Fukushima Daiichi that flexible approaches are effective for preparing for low-frequency and high-consequence events with uncertainties. There was no idea of systematic preparation for flexible responses to events of beyond-design-basis assumptions in Japan. Even if appropriate safety measures covering every possible condition were implemented in advance, extra measures such as transportable safety equipment should have been provided to avoid unexpected worst scenario. Uncertainties do exist in anything deemed perfect, which is the premise and practice in the U.S., but not in Japan.

6.3.2.3 Safety Assurance by Defence in Depth

The risk of severe accidents that exceed design basis has been recognized, and accident management measures have been developed. It was obvious that defence against natural phenomena, particularly tsunamis, was insufficient in the Fukushima Daiichi accident, and accident management was not effective. Natural phenomena exceeding design conditions may result in the failures of a number of SSCs simultaneously due to a common cause. This apparently indicates even if the frequency is low, once an event takes place, the consequence may be serious. In the Fukushima Daiichi NPS, many safety-related SSCs lost their function due to the tsunami far exceeding the design level, which resulted in core meltdown and the significant releases of radioactive materials into the environment.

Risk analysis experts have recognized the risk of severe accidents resulting from natural phenomena is more significant than that of internal events. Discussions have, however, been limited to the enforcement of the design basis to respond to external events, leaving many unsolved issues such as diversification of water injection and cooling systems, effectiveness of accident management, and implementation of emergency preparedness. Because of excessive emphasis on preventing severe accidents and over confidence in the credibility of this approach, the defence in depth turned to be ineffective immediately after a severe accident occurrence. In other words, the requirement for independent effectiveness that inhibited excessive dependence on a specific level was missing.

Measures taken at the Onagawa NPS and Tokai Daini NPS, such as the elevation of the NPS site determined with cognition of tsunami risks, and review of the height of the tsunami as required according to new evidence and knowledge, helped prevent severe accidents. Continuous efforts to improve safety (effective application of new insights and operational experience) would help make the defence in depth more effective.

Another important problem identified is the lack of communication between experts for sharing common views for tsunami risks; based on which the design basis tsunami was assumed. This is probably because there were no shared goals to ensure nuclear safety. The defence in depth levels are not mutually exclusive. The safety experts should have explained the design basis tsunami hazards quantitatively and have obtained in-depth information on the height of tsunamis from tsunami experts. They should have, based on the shared common information, determined or amended the design basis.

6.3.2.4 Severe Accident Measures and Ensuring Their Effectiveness

Consideration should be given to implement safety measures based on the defence in depth philosophy. Even though the design basis is exceeded resulting in a severe accident, it would be impossible to consider all possible scenarios. In this case, a flexible and general response such as the use of transportable equipment is effective, while organizations and those in charge must have responsibility, suitable judgment and leadership to implement diversified accident management.

It is also revealed that many difficulties are involved in accident management when dealing with actual severe accidents. For example, there were constraints of time and space for working in high radiation conditions, and operators in the control rooms were forced into uncomfortable situations. Some of the accident management operations were difficult to be conducted under certain accident conditions such as the recovery of the power supply is impossible, for example.

To allow the defence in depth to function effectively, flexibly integrating a number of elements, such as measures to ensure effective accident management, and continuous review of them, are required in addition to the on-site and off-site, multi-level and multiple phased safety assurance.

6.3.3 Defence in Depth Deepening and Future Steps

6.3.3.1 Preparation for Beyond-Design-Basis Events

In the report submitted by the Japanese government to the Ministerial Conference of the IAEA, the importance of defence in depth in ensuring safety was stressed, and "five groups of lessons to be learned were presented with the prospect that defence in depth is the crucial principle for ensuring nuclear safety". These groups indicate three elements (prevention, mitigation and emergency response) of defence in depth as lessons to be learned. The first group of lessons deals with preventive measures for severe accidents, the second group of lessons provides mitigation to severe accidents, and the third provides emergency responses to nuclear disasters. It is noted that a lesson in the fifth group on commitment to safety culture declares "in future, returning to the origin of the concept that defence in depth is essential for ensuring nuclear safety, nuclear safety operators undertake implementation of safety culture by continuously learning expertise on safety, and reviewing improvements in vulnerabilities in nuclear safety and enhancement of safety".

The NISA suggested that "based on defence in depth, safety should be ensured by evaluating the design basis events, and based on the premise that beyond assumption incidents could happen, strict 'denial of previous level' (deny the effectiveness of required measures to prevent an accident, and conducting subsequent measures on the premise that an accident could happen) should be applied".

The Safety Report Series No.46 [6] of the IAEA states "Defence in depth is a comprehensive and safe approach developed by nuclear experts, ensuring protection of the public and the environment from any hazard relating to nuclear power generation with a high credibility".

The U.S. NRC presented 12 recommendations [5]. The report contains Chap. 4 "Safety Through Defence in depth". Sect. 4.1 "Ensuring Protection from External Events", Sect. 4.2 "Mitigation", and Sect. 4.3 "Emergency Readiness". It also describes in Chap. 3 "Regulatory Framework for the 21st Century" for Recommendation 1: "However, the Task Force also concludes that a more balanced application of the Commission's defence in depth philosophy using risk insights would provide an enhanced regulatory framework that is logical, systematic, coherent, and better understood". The importance of the defence in depth was reevaluated in the U.S. as lessons-learned from the TMI-2 accident. The report pointed out that ensuring safety is insufficient if it is solely based on design basis events.

The way to systemize safety-related SSCs based on design basis is to provide suitable means for the predefined group of design basis events. It allows effective and adequate provisions and assures high levels of safety. Quality and effectiveness of risk management, however, depend on the framework used to determine the selection of design basis events. Our experience indicates that every severe accident in the past was resulted from unexpected incidents that were not postulated in the design basis. The loss of safety function due to multiple failures associated with human factors or external events are the causes. This indicates that high levels of safety cannot be achieved by preparing for predefined design basis events alone. Incompleteness uncertainties in the postulated design basis events must be taken into consideration. The importance of creating the framework for design basis event and requirements for preparing for beyond-design-basis events must be equally emphasized. If uncertainties in design basis events are appropriately handled, nuclear safety can be more adequate and reliable through appropriate risk management.

6.3.3.2 Concept of Defence in Depth

The concept of defence in depth was originated from the policy of providing multiple physical barriers to prevent the release of radioactive materials into the environment. This is understood as more general multiple structure (so-called multiple levels of barriers) comprising physical barriers and auxiliary provisions. Defence in depth ensures high levels of safety with sufficient margin; assuming equipment failures and human errors are unavoidable.

The number of levels of defence in depth is often subject to discussion. According to INSAG-12 [8], defence in depth is implemented to compensate for potential human and mechanical failures including successive barriers preventing the release of radioactive material to the environment. It comprises two phases; (1) prevention of accidents, and (2) mitigation of the impact of accidents to prevent more serious situations. The actual structure of defence in depth must comprise a number of levels of protection including multiple barriers according to the progression of events.

Defence in depth is the fundamental concept of safety assurance, and covers a wide range of contents. Defence in depth must be clear and comprehensive so that the degree of safety assurance can be identified. Defence in depth requires providing generality for application to every activity for ensuring safety.

According to INSAG-12, the reliability of defence in depth is improved by applying defence in depth to each level of defence to prevent damage to the barrier, which makes hierarchical defence in depth effective. It may also be reasonable to divide the goal of (2) (prevention of impact of accidents to prevent more serious situations) into two to make three fundamental elements ((1) prevention of accidents, (2) mitigation of the impact of accidents to prevent the release of radioactive materials, and (3) mitigation of the impact of releases to protect public exposure) for defence in depth aiming to protect the public and the environment. This may be interpreted that reducing public radiation exposure is apparently the safety purpose, and to perform the followings: to identify threats for the protection, to prevent the occurrence via technology and safety regulations (engineering provisions), to reduce threats by mitigating their impacts if the engineering provisions do not work (provision of accident management), and to prevent serious damage to the protected if the impact reduction does not work perfectly (provision of disaster preparation).

What to be protected must be defined first in defence in depth. It is to prevent significant harmful effects of radiation on the public (health and living of the people). For this purpose, what are necessary are the safety design of nuclear power plants, safety assurance activities such as developing accident management measures, and preparations/responses to prevent serious impact on the public when safety assurance activities are not effective. Defence in depth is a philosophy of implementing effective preparations for uncertainties in all safety assurance activities concerning the attainment of safety goals.

6.3.3.3 Provisions to Implement Defence in Depth

Provisions play an important role in the concept of defence in depth. Defence in depth is achieved with adequate arrangement of provisions. There are also provisions to prevent the occurrence of serious accidents, provisions to mitigate the impacts of severe accidents and provisions to protect the public and the environment in the event of significant releases of radioactive materials into the environment. These provisions may be arranged in a hierarchical manner depending on the objectives. For example, to prevent the occurrence of severe accidents, (1) preventing the occurrence of disturbance (abnormal state) that may affect the facilities, (2) detecting disturbances, preventing their escalation by activating reactor protection systems and mitigating impacts, and (3) installing engineering safety features to prevent serious damage to reactors when reactor protection and mitigation failed. Using defence in depth in this way, the high reliability of barriers that prevent severe accidents can be incorporated into the design.

All activities for ensuring safety, including those for organizations, activities and facilities, are based on multiple levels of related "preparations". Multiple levels of defence are a fundamental feature of defence in depth to avoid failures of facilities to cause extensive damage to the public and society.

The "provision" is an important concept containing wide-ranging implications. According to Safety Report Series 46 [5], the provision is a means practiced in design and operation, such as inherent plant characteristics, safety margins, system design facilities and operational procedures, which help prevent the occurrence of relevant mechanisms. INSAG-3 uses the term as "provision" of design, and explains the "provision" to mitigate impacts of severe accident may extend the concept of defence in depth beyond the accident prevention. Accident management, engineering safety facilities and off-site measures are three provisions of mitigating the impact of severe accident".

Japan is targeting the nuclear regulation standards will be among the world's highest. Hence we should understand defence in depth profoundly as the crucial fundamental principle for safety assurance. The basics of defence in depth should be respected to establish safety logics and approaches. Consistent development of elements at various levels, including objectives of utilizing nuclear energy, safety goals, safety principles, safety design basis and guidelines, and practical design methods, is required to ensure safety effectively. The relations of these elements to the concept of the defence in depth, and conformity of safety design requirements with the concept of the defence in depth must be confirmed for each element.

As stated in Sect. 6.3.1, defence in depth is often considered to be composed of five levels. Levels 1 to 3 represent intact reactor core situations or consistency with design basis and can be integrated into a single level to prevent the occurrence of severe accidents. In other words, it is a defence level of the confinement of radioactive materials inside the reactor core. Of the five levels in Fig. 6.14, 1 to 3 are based on the provision of engineering design, level 4 on the provision of accident management, and level 5 on the provision of emergency preparations. These three types of provisions must be included in safety design in a balanced manner. Safety assessment involves quantitatively evaluating the achievement level of these three provisions. We must define the relevant safety regulations and safety requirements.

Imperfect accident management and emergency preparedness were pointed out in the Fukushima Daiichi accident. The criticism focused on defence in depth comprising only three levels, levels 1 to 3, in Japan. In reality, however, accident management (level 4) and emergency planning (level 5) have also been implemented in Japan. The root of the problem is the absence of a thorough strategy to activate levels 4 and 5 when levels 1 to 3 are no longer effective, because of excessive dependence on level 3; assuming that accidents exceeding design basis were unlikely. The false belief in achieving the goals of defence in depth stems from the grounds that engineering safety equipment had been installed to prevent reactor core damage, the PCV could prevent the release of radioactive materials into the environment, or the safety design had been enhanced. The lack of independent effectiveness (as mentioned above, which means effectiveness not excessively dependent on a particular level), and excessive dependence on level 3 which impeded the balance of defence in depth cannot be denied.

6.3.3.4 Lessons on Defence in Depth

The only explanation for defence in depth in the report of the Government Accident Investigation Committee [12] is the excerpt from the IAEA International Expert Mission report [10], “inadequate defence in depth against tsunami disasters was pointed out”. The report of the Diet Accident Investigation Committee [11] merely stated “introduction of regulations for reducing accident risks is slow, and Japan lagged behind in the international standards in term of nonconformity with the concept of level 5 in defence in depth”, not in the text, but in the summary as the organizational problem of the parties relating to the accident in Part 5. As described in the beginning of this chapter, many pages of the IAEA Ministerial Conference report and the U.S. NRC report were allocated to discussions concerning the reason that the defence in depth to ensure safety was not effective in the Fukushima Daiichi accident. The problems in understanding of defence in depth and its application should be identified to achieve high quality and robust safety in the light of lessons-learned from the accident.

Imperfect defence in depth and nonconformity with the concept of level 5 in defence in depth are reviewed in this paragraph. The purpose of nuclear safety regulations is to protect the public and the environment. In this regard, prevention of severe accidents comes first and every practical measure must be taken. On the contrary, we understand the ultimate safety goal is prevention of severe accidents in Japan. It seems the prevention of abnormal conditions, mitigation of impacts and prevention of accident, and the use of engineering safety features in the design basis accident were applied in this context. A principle that the public safety is ensured if a severe accident is prevented. The misperception results in nonconformity with the concept of defence in depth with five levels, as was revealed by the consequences of the Fukushima Daiichi accident. The defence in depth provides effective provisions against uncertainties. All levels of defence should be focused on in a balanced manner, and remain effective as a whole.

The IAEA Ministerial Conference report and NRC report describe lessons-learned for three categories each of which corresponds to three fundamental levels of defence in depth, namely, prevention of severe accidents, mitigation of impacts and the release of radioactive materials, and emergency plans and response, as the categories of lessons. Mitigation of impacts and prevention of the significant release of radioactive materials into the environment (level 4), and emergency response in association with significant releases (level 5) are to be enforced and issues to be discussed in more detail in future. As already mentioned, flexible accident management is effective in level 4. At the same time, according to new insights and operational experience, the design basis and relevant requirements should be enhanced. Considering the relation to the design basis events defined with postulated scenarios in mind, design enhancement and flexible responses should be arranged in a balanced manner to respond to severe accidents.

Defence in depth is a concept of ensuring safety, and is generally applicable to external and man-made events. To confirm the effectiveness of the defence in depth for these events, deterministic or probabilistic risk assessment is required. The former is a methodology for clarifying safety margins, and the latter for clarifying uncertainties, essential to evaluate the effectiveness of defence in depth, i.e. the comprehensive provisions for uncertainties.

It is required to understand the defence in depth and evaluate the level of achievement of safety continuously to prevent the recurrence of serious situations in which exert grave impacts on the public and the environment as in the Fukushima Daiichi accident. Evaluating the effectiveness of prevention of severe accidents, mitigation of impacts and confinement of severe accidents to the site, and emergency response in the event of escalation outside is required to prevent significant adverse effects on the public. Steady implementation of these measures leads to continuous improvements in safety.

6.4 Plant Design

Nuclear power generation systems are designed, constructed and operated according to specification requirements like other commercial products. Their reliability and safety is ensured by theories, laws, regulations and standards, with the underlying fundamental approach on "nuclear safety" and "defence in depth" concept. Organizations must assume their roles based on the awareness on the significance of "nuclear safety" and "defence in depth" concept.

Nuclear power generating system is designed to ensure safety under operating conditions assumed at the time point in the design, by conforming to performance requirements and by evaluating integrity of SSCs through analysis assessments. SSCs are manufactured and established based on this framework. Plan on operation and maintenance must be developed, and consolidated into design documents including drawings and procedure manuals with consideration given to maintaining consistency. The analysis on Fukushima Daiichi Nuclear Power Plant accident was developed from these perspectives.

6.4.1 Analysis on Design

6.4.1.1 Design Basis, Design Basis Events, and Basic Approach on Design

The framework on the design of nuclear power plant comprises of safety design pertaining to system design and reactor core design. "Safety design" is a concept integrating the entire framework on design that is intended to ensure nuclear safety by the application of all levels of "defence in depth" while "facility design" involves logical configuration of plant facilities based on this concept. Safety design has so far evolved over levels 1 to 3 of defence in depth for ensuring safety, governed by the key element of design basis and corresponding design basis events. Facility design concerns levels 1 to 3 of defence in depth intended to ensure and maintain nuclear safety by establishing a comprehensive plant system consistent with safety design and equipped with SSCs with adequate specifications.

A general approach in coping with various events, in particular, natural hazards as external events has not yet been set forth. Facility design criteria on seismic ground motion has been specified by national standards, to which commercial standards have developed relevant detailed assessment methods. Design standards have been considered in this way—natural hazards have been conservatively dealt with, within the range of scientific assumption based on past records. A quantitative design basis was provided to maintain structural integrity under a condition slightly exceeding the standard, with consideration given to uncertainties. On the basis of this approach, sufficient margin was provided in the design for manufacturing and constructing plants. However, only earthquake hazard was taken into account in the design, but not other natural hazards such as tornados, volcanic eruptions, and meteorite falls. In most cases, evaluations on other natural hazards based on these types of data had not been conducted. Analysis techniques for tsunami events had just been formulated, and by the application of state-of-the-art technologies, progress was being made on the reevaluation of tsunami consequences on nuclear power plants.

6.4.1.2 Discussion Points on Design in the TEPCO Fukushima Daiichi Accident

Discussion points on design related to the Fukushima Daiichi Accident have been analyzed and countermeasures are shown as follows.

(1) **Discussion points and analysis on safety design**

The design of nuclear power plants in Japan mainly focused on establishing three elements of IAEA's defence in depth, "shutdown", "cooling", and "containment". These had been appropriately achieved before the arrival of tsunami that caused the Fukushima Daiichi accident. However, the three elements in the design had not been sufficiently established to cope with station blackout conditions caused by the arrival of the tsunami. The consequence of the tsunami-caused station blackout was the loss of the "cooling" and "containment" elements that must be maintained. One of the important lessons learned from the accident is that power supply and power system are critical elements in ensuring safety against tsunami events.

Safety design of nuclear power plants is applied to system design, where the requirements in ensuring functionality is integrated with those for ensuring safety of functions related to the systems. System design embodies design on mechanical systems, piping systems, instruments and control and electric systems, and integrates all associated components. One of the major underlying causes of the Fukushima Daiichi Accident was the lack of an integrated assessment on system design including power supply and electrical systems necessary for ensuring safety during abnormal and accident conditions.

The accident was initiated by the loss of all AC power and electric systems, which hindered support by auxiliary systems and portable equipments. Failure to maintain instrumentation and electric systems hampered the use of equipments that can be activated by open and close manipulation of the valves, which led to the severe accident. The isolation condenser (IC) which does not require power sources is activated for removing heat and cooling the reactor in the event of emergency of loss of power by opening and closing valves. However, because of the conflicting requirements on isolating systems to protect pipes from breakage, and that for core cooling, and in the absence of strategy on prioritizing isolation or cooling, isolation was selected in the end, and cooling, or IC function was not effectively utilized.

Depressurization of BWRs requires the activation of the main steam safety-relief valve of the reactor pressure vessel, or venting of the containment vessel. Given the lack of emergency measures in accident management, neither power supply nor orifice gas required for activation of valves were available, which made situations worse. While the safety design of nuclear power plants have been subject to periodic safety review (PSR) for incorporating the latest technologies and enhancing safety of nuclear plants, it was not effectively utilized. It is essential to incorporate the approach on "system safety" in the safety assessment for evaluating important functions in the safety design in the future. The concept on defence in depth must be fully integrated in the safety design, where a framework of defence in depth applied consistently and coherently over design is appropriately established.

(2) **Discussion points and analysis on facility design**

Facility design has focused on ensuring safety by means of robust facilities to protect against design basis internal events such as loss of coolant accident (LOCA). The management of external events has focused on seismic ground motion. Tsunami events had only been dealt with as an earthquake subordinate event and not considered in the regulatory guideline. Further, because accurate evaluation of risk associated with natural hazards is difficult, risk assessment on the consequences of natural hazards on nuclear disasters had not been sufficiently considered. Residual risk assessment on seismic ground motion had just started at the time of the accident, and thus, efforts in developing latest expertise and state-of-the-art technologies must be promoted.

Design safety in the period preceding the accident was not founded on the premise of events exceeding design basis. However, consequences of the Fukushima Daiichi accident brought to light the significance of considering, and developing measures against beyond design basis events together with consideration of design basis events.

The method in managing beyond design basis events is risk assessment applied to seismic ground motion assessment, which aims to improve safety reliability by assessing "residual risk" of beyond design basis occurrences, and implementing measures to reduce such risk.

The concept of defence in depth is applied to events exceeding the design basis, which is called accident management (AM). Provision of facilities and components incorporating diversity and independence in design are provided against beyond design basis events falling in the region of accident management. In many cases, SSCs that are not used under normal conditions are accommodated in the accident management of abnormal occurrences and accident conditions, for which training and drills must be provided.

In principle, facilities are designed, manufactured and constructed to meet the design basis according to specification requirements. However, the occurrence of Fukushima Daiichi NPS accident has created circumstances on the need to consider design basis extension to cope with abnormal incidents and accident conditions. The scope on facility design is an issue to be discussed in the future, where a design structure corresponding to each level of defence should be given consideration.

6.4.1.3 Summary Conclusion

Design, based on anticipated operation at the time point in design, has so far consisted of graphical presentation of facilities that meet design requirements and performance specification requirements, and was developed to the extent covering operational procedure. Based on this, analysis for verifying the integrity of facilities was conducted to ensure "nuclear safety," based on which facilities for nuclear power generation systems were manufactured and constructed. Attention was not paid to "beyond assumption events" containing temporal elements. Design is an important element of nuclear safety. Safety design founded on defence in depth concept must be established for ensuring nuclear safety. Safety design must be optimized by the integration of defence in depth levels of 1 to 5. The interaction between the region of facility design of levels 1 to 3, and those of management of levels 4 and 5 must be coordinated to establish a plant system integrating all functions in order to realize a framework for ensuring safety.

The three aspects of (1) total system, or integrated plant system comprising the grounds, buildings, components, piping, electric and instrumentation systems, etc.; (2) a total process, or comprehensive safety over the entire plant process, on a safety framework under normal operation, accident conditions and emergency response; and (3) total management, or optimizing management, facilities and operation that take into account the entire realm of defence in depth of both tangible and intangible aspects, must be coordinated and integrated to reinforce total design, symbolizing comprehensive efforts in optimizing plant safety.

The issue on how the concept of resilience [1][11] may be incorporated into design should be addressed in the future.

[11] Resilience: Methods and capacity of restoring the required function to safety assurance level as a system to respond to an incident or accident.

6.4.2 *System Safety in Plant Design*

6.4.2.1 Beyond Design Basis Conditions in Safety Design and Facility Design

(1) **Defence in depth and the role of design before 3.11**

In the design of a nuclear power plant before the 3.11 accident, beyond-design-basis events were considered as part of safety assessment in safety design as follows: First of all, plant siting was assessed before the development of plant and facility design. Defence in depth, or "preclusion of preceding defence level" used in Japan, was applied to evaluate the validity of siting based on accident scenario of extensive radioactive material release from the containment vessel. This was, however, not a practical scenario that presented the occurrence of a severe accident and its progression. Assessment in the realm of defence in depth level 5 is evaluating dose in the site boundary under severe accident conditions. The accident is evaluated according to the accident sequence, not a scenario.

Performance requirements for ensuring safety must be shown clearly in the safety design in the realm of D-I-D level 1 to level 4. Safety design of levels 1 to 3 associated with production and manufacturing, and safety design associated with plant operation including accident management in the event of failure (malfunctioning) of safety systems caused by design basis accident, had each been assumed by plant manufacturers and utilities.

Facility design ensures safety associated with D-I-D levels 1 to 3, focusing on the structural design of goods produced that meet safety design requirements.

Defence in depth for nuclear power plants is generally structured in five levels (see Sect. 6.3):

Level 1: Prevention of abnormal operation and failures
Level 2: Control of abnormal operation and failures
Level 3: Mitigation of accident
Level 4: Onsite response to beyond design basis accident
Level 5: Protection of the public and the environment

"Accident" stands for design basis accident (DBA).

(2) **Issues on safety design**

The importance of design lies in safety design that maintains the integrity of the whole, as well as enhancing protection levels. As Fukushima Daiichi accident indicates, the approach on safety in Japan had focused on facility design of D-I-D levels 1 to 3. Even beyond-design-basis events was thought to be controllable with adequate facility management.

Accident analysis suggests the need for safety assessment of a nuclear power plant as a total system in design. In the light of the Fukushima Daiichi accident, accidents causing not only single equipment failure or functional damage, but also multiple failures or functional damage, failures or loss of function caused by common factors, and effects of system affecting another systems or propagation of failures or loss of function should be taken into consideration. The accident suggests the importance of "system safety" where the functions/roles of equipments, pipes, electrical and instrumentation systems are coordinated over the entirety of plant system for the achievement of allocated functions of each system, under the concept of "system safety" [12].

(3) **Issues on facility design**

In facility design, the design conditions of facilities, including those for emergencies, were determined by assuming LOCA (loss of coolant accident) caused by pipe breakage, based on design basis to ensure "shutdown", "cooling" and "containment" for levels 1 to 3 of defence in depth.

However, for events exceeding design basis, or severe accidents, response varies depending on the type of event and circumstances. This is why scenarios are important for beyond design basis events falling in the region of defence in depth level 4. As many scenarios as possible should be developed, together with countermeasures. This is because the conditions of each accident scenario differ, particularly, those initiated by external events, from their emersion to progression, which must be flexibly dealt with. Accordingly, it is essential to create accident scenarios over the broad range and to develop measures. A mechanism should be laid out for continuous development of scenarios and measures, which will lead to reducing "unexpected" occurrences. As well, there are scenarios that are difficult to create even by applying current knowledge. Based on the understanding of these conditions, a framework on a more adequate and systematic response may be established by standardizing, re-evaluating and modifying SSCs and procedures. Although this extends to the realm of management, measures that are linked to facility design should be discussed.

Plant design is formulated on the basis of design basis conditions (accidents), to which design rules as redundancy and diversity is applied to SSCs for the effective management of various events. Design basis concept so far has included a broad spectrum of events and challenges against design basis, on the premise that beyond design basis events do not occur. In view of the Three-Mile Island and Chernobyl accidents, there had been a strong focus on internal events underlined by a strong belief in the design integrity of SSCs and the unlikelihood of the occurrence of beyond design basis events.

Whereas, in dealing with external events (natural hazards), under the geophysical conditions of frequent earthquakes, Japan has conducted numerous studies and developed measures against earthquakes from the very early stages of introduction of nuclear power generation. Lessons learned from the 1995 Great Hanshin Awaji Earthquake and state-of-the-art expertise were incorporated into the revised Seismic design guidelines in 2006. Back-checks were conducted at each plant with necessary reinforcements made. In the revised guidelines, seismic design basis was reevaluated and measures against beyond design basis events developed; risk assessment as a means for assessing plant safety was highlighted, which resulted in encouraging the voluntary initiatives of the operators in "conducting "residual risk" assessment", and in the establishment of PRA assessment technique. Kashiwazaki Kariya Nuclear Power Plant experienced beyond design basis seismic motion in the Chuetsu Offshore Earthquake in 2007. However, because the back-check process was already in place, sufficient margin was ensured and key SSCs maintained structural integrity. Consequently, the case led to prompting seismic back-checks in all nuclear power plants in Japan.

(4) **Issues on accident management**

Facilities must meet the design criteria such as seismic resistance standards; however, as shown by the March 11 Earthquake and Tsunami, even rare events with very small occurrence frequency/probability may lead to beyond design basis conditions, if they occur. Hence, it is important that provisions are made for incidents exceeding the design basis by identifying functions required for ensuring nuclear safety over the entirety of the plant system flexibly based on a broad range of scenarios on failures in nuclear power plants, which is called accident management.

Natural hazards is difficult to predict. It is also impossible to manage all circumstances, the responses which may end up with the same results as doing nothing at all. Thus, accident management trainings and drills based on assumptions on various situations to flexibly deal with beyond design basis events should be provided.

6.4.2.2 Plant Systems Comprised of Functions Required for Ensuring Nuclear Safety and Role of Defence in Depth

Table 6.16 lists the functions essential for nuclear power plants. They include the boundary, cooling and control functions, as well as the power supply function.

Nuclear power plants are designed to ensure coordinated interaction between these functions, which, in turn, ensures the integrity and safety of facilities whatever the design basis event is.

Whereas, for beyond-design-basis events, because of the broad scope in responses depending on the type of event or conditions, development of scenarios is essential. Responses/measures should be developed on the basis of these scenarios. Subsequently, SSCs and accident management procedures should be standardized to establish a systematic accident management framework. Table 6.17 shows the relationship between defence in depth and key safety functions. Each function has a sub-function; the roles of which are associated with each levels of defence in depth. The table also shows examples of back-up—functions when a function is lost. As shown by the relationship between these functions, the preconditions, or the fundamental driving source of all functions is power supply. Facilities that do not require power supply are extremely rare. Obviously, power supply must be backed up by alternative sources for which interactions and the correlation between associated functions should be clarified. Functional configuration of power supply,

Table 6.16 Critical safety functions and associated structures, systems & components

Function	Sub-function	Structures, Systems, and Components	
		PWR	BWR
Boundary examples	Reactor coolant pressure boundary	Equipment and piping composing boundary (small caliber pipes and equipment for instrumentation, etc. excluded)	Equipment and piping composing boundary (small caliber pipes and equipment for instrumentation, etc. excluded)
	Reactor coolant pressure boundary overpressure prevention	Pressurizer safety valve (open function)	SRV safety valve
	Radioactive material containment, radiation shielding, and release reduction (1)	PCV, annulus, PCV separation valve, PCV spray system, annulus air recycling system	PCV, PCV separation valve, PCV spray cooling system, FCS
	Radioactive material containment, radiation shielding, and release reduction (2)	Safety accessory air cleaning system, combustible gas concentration control system	R/B, SGTS, emergency re-circulating gas processing system (related system) stack (SGTS exhaust pipe support)
Cooling examples	Reactor morphology maintaining	Core support structure (except for fuel)	Core support structure, fuel assembly (except for fuel)
	Heat removal after reactor shutdown	Residual heat removal system: Residual heat removal system, aux. feed-water system, and main steam-feed water system, main steam safety valve, main steam escape valve (manual escape) up to SG 2nd-side separation valve	Residual heat removal system: RHR system, RCIC system, HPCS system, SRV (escape valve function), automatic pressure reducing valve, (manual escape)
	Reactor cooling	Emergency reactor cooling system: Low-pressure injection system, high-pressure injection system, accumulator injection system	ECCS: RHR system, HPCS system, LPCS system, ADS
Control examples	Excess reactivity supply prevention	Control rod drive unit pressure Housing	CR coupling
	emergency reactor shutdown	Reactor shutdown system in control rod system	Scram
	Subcriticality control (1)	Reactor shutdown system	CR/CRD system
	Subcriticality control (2)	Reactor shutdown system	Safety protection system

(continued)

Table 6.16 (continued)

Function	Sub-function	Structures, Systems, and Components	
		PWR	BWR
Other examples	Signal generator for operating engineering safety facilities and reactor shutdown system	Safety protection system	Safety protection system
	Safety critical related function	– Emergency onsite power supply – Control room and shielding – Ventilation & air-conditioning system – Reactor aux. cooling water system – DC power supply system – Instrumentation air system	– Emergency onsite power supply system (related systems), DG fuel transportation system, DG cooling system – Control room & shield, emergency ventilation & air-conditioning system – Emergency aux. cooling water system – DC power supply system

Table 6.17 Relationship between safety functions and defence in depth

Defence in depth	Boundary function	Cooling function	Control function	Other (common)
Level 1	• Reactor coolant pressure boundary • Housing of reactor coolant • Storage of radioactive material and not directly connected to reactor coolant pressure boundary • Prevention of radioactive material release • Maintaining of reactor coolant • Radioactive material storage • Prevention of fission product diffusion to reactor coolant	• Core morphology control • **Normal state reactor cooling** ↓	• Prevention of excess reactivity application • Reactor coolant circulation	• Safe handling of fuel • **Power supply (excluding power for emergency use)** • Plant measurement and control (1) (2) (3) (excluding safety protection systems) • Plant operation support (1) (2) • Reactor coolant purification • Sustainment of commonly shared functions of power supply and activation signal systems is important.

(continued)

Table 6.17 (continued)

Defence in depth	Boundary function	Cooling function	Control function	Other (common)
Level 2	• Prevention overpressure of reactor pressure boundary • Blowout of safety and relief valves	• **Heat removal after reactor shutdown** ↓	• **Heat removal after reactor shutdown** ↓	*Common in levels 2 and 3* • Generation of operation signals to engineering safety facilities and reactor shutdown system
		• Safe shutdown from outside control room ↓	• **Subcriticality control** (**control rod system**) ↓	• Critical for safety (1) (emergency onsite power supply) • Critical for safety (2) (control room)
		• Mitigation of reactor pressure increase ↓	• **Subcriticality control** (SLCS) ↓	• Critical for safety (3) (reactor aux. cooling water system) • Critical for safety (4) (DC power system)
		• Reactor coolant supply	• Power output increase control	• Understanding of plant state under accident conditions
Level 3	• Radioactive material containment, radiation shielding and release reduction (PCV) • Radioactive material containment, radiation shielding and release reduction (R/B)	• **Reactor cooling** • Water make-up for fuel pool ↓	↓	
Levels 4 and 5	• Severe accident management (PCV venting)	• **Severe accident management** (**water supply system**) • **Operation during severe accidents** (**fire extinguisher system**)	• **Severe accident management** (**standby liquid control system**)	• Requirements for emergency responses and understanding of emergency state (1) (2) (3)

which must be maintained for defence in depth levels 1 to 3 is a matter to be dealt with separately.

Events exceeding the scope of level 3 defence in depth fall in the region of severe accidents, or defence in depth level 4, to which provisions against various unanticipated circumstances should be arranged. Not only consideration of measures based on extensive scenarios focusing on the hardware aspects of key design basis SSCs, but measures emphasizing human factors and intangible aspects, including the utilization of all available SSCs is the key to accident management in this region. For example, it is essential that the valves may be opened and closed manually in the event of an SBO.

6.4.2.3 Points of Discussions on the Fukushima Daiichi Accident

(1) **Response to beyond design basis events**
The impact of seismic ground motion on the structural integrity of plant facilities is one of the key issues. The seismic ground motion at some units of the Kashiwazaki Kariwa NPS exceeded the design basis by three times during the Chuetsu Offshore Earthquake. However, the facilities maintained integrity, verifying the plant was equipped with sufficient margin against design seismic motion. Subsequently, design seismic standards was tightened for all nuclear power plants in Japan. However, the back checking on structural integrity that was being conducted on plants showed sufficient margin against the revised criteria. It is assumed that there was more margin with regard to functional integrity of the plants.

The significance of "beyond design basis" must be reconsidered. Ground motion acceleration which slightly exceeded the design basis seismic ground motion was observed at many nuclear power stations along the Pacific coast during the Great East Japan Earthquake. The impact was too small to be reflected on observation data on plant state. Analysis results and plant records suggest that the integrity of power stations were maintained. For beyond design basis seismic ground motion, more extensive assessment methods including the introduction of base isolators should be taken into account in future, instead of simply reviewing reference values for acceleration response.

The criteria on "the time point that the design basis is exceeded" for implementing accident management should be clearly shown.

(2) **Response to natural hazards by design**
The magnitude of the tsunami in the Great East Japan Earthquake was simply unforeseen and outside the range of assumptions.

Predicting a tsunami that occurs in the order of once in a millennium is extremely difficult. Tsunami-resistant design is a combination of structural design for buildings and facilities, and electric and instrumentation design based on extensive evaluation of various phenomena, including earth crustal movement, wave generation, propagation and run-up, and flood. Tsunami magnitude has been simulated in various ways, and various hypotheses presented. Even if design basis criteria on tsunami was defined, there is no guarantee that a tsunami exceeding such criteria would not occur. Although the conditions of earthquake and tsunami exceeding assumptions differ, management of beyond design tsunami should be developed. Consideration should be extended to beyond design basis conditions of other natural disasters as well.

(3) **Accident management and emergency preparedness**
All power sources and cooling safety systems were lost due to beyond design basis conditions, which was pointed out as inadequacies in managing events in the region of defence in depth level 4 based on various scenarios, or the so-called accident management. Accident management requires the utilization of all available resources, materials, and human knowledge and wisdom. Not many surveys on the specific development of accident sequences have been conducted. Hence, emergency preparedness and trainings by developing as many scenarios as possible is very important. Compared to emergency response in the Fukushima Daiichi accident, preparedness and response to terrorist attacks developed in the U.S. after 9.11 have been reported as more effective. It may be necessary to investigate how the measures were developed.

(4) **Failsafe system and robustness[12] in design**
The consequences of Fukushima Daiichi accident was the significant radioactive material release from the containment, involving two issues on the use of IC and PCV venting. The IC (isolation condenser) is used for reactor cooling in the event of AC power failure. It was a failsafe design, where on the detection of a pipe rupture signal, the isolation valve is closed. The isolation valves were installed on both sides of the PCV wall on the pipe that traversed the PCV. If the AC power is lost, there is no way that the valve inside the PCV will open. At the time point that DC power to the instrumentation control system is lost, the valve close signal is issued even though there was no piping rupture, and the DC power that was still active was supplied. As a result, although designed to maintain "as-is", or valve open state under loss of power conditions, the valve close signal was issued. The design related to nuclear safety which is ensured by "shutdown", "cooling" and finally "containment" should be reevaluated.

PCV venting is designed to be activated at the time point that the highest, or the predetermined pressure is exceeded, involving the breaking of the rupture disk to release pressure. At the same time, the PCV vent valve must also be manually operated to ensure safety. However, because of the difficulties in, and the delayed opening of the vent valve, this is assumed to have led to the subsequent hydrogen explosions and the escalation of the accident.

Normally, there is a certain amount of leakage from the PCV, and thus, sealing is not a strict requirement. Robustness relates to the facility as well as in the management. The robustness of containment venting is maintained by the redundancy in arranging a number of exhaust systems. However, with regard to

[12] Flexibility of responses.

the timing that the vent valves are opened, no redundancy or diversity is provided for ensuring robustness, but is managed solely by prioritizing the maintenance of the boundary functions, or containment. Accordingly, facility robustness was ineffective under these circumstances.

Safety design goals should be defined by identifying the requirements on nuclear safety, failsafe design and robustness.

6.4.2.4 New Approach on Design

Design so far has emphasized nuclear safety by warranting the integrity of design basis, based on safety design requirements applied to the facilities. However, there is a significant limitation on this approach.

The approach on new design is optimization by means of common assessment index by integrating all levels of defence, including detection of normal/abnormal operations, response to failures, beyond-design-basis events, and emergency (disaster preparedness) into safety design, combined with a system ensuring safety in post accident management, resilience, and restoration.

Management governs facility design and disaster management. Plant safety cannot be ensured by design alone; a new conceptual framework for ensuring nuclear safety focusing on functionality that extend over, or are governed by different criteria scales should be established.

Design plays an essential role in ensuring nuclear safety. Nuclear safety must be founded on safety design incorporating defence in depth Level 1 to Level 5 of the IAEA's defence in depth concept are integrated into safety design and optimized—Level 1 to Level 3, realm governed by facility design and Level 4 and 5, governed by management must be coordinated for ensuring the effectiveness over the entirety of the plant system comprising of functions. Figure 6.15 shows the relationship between IAEA's defence in depth concept and design basis management [13].

Nuclear power generation system design must be comprehensive and comprise of total system—integrated plant system comprising the grounds, buildings, components, piping, electric and instrumentation systems, etc.; total process—a safety framework over the entire process under normal operation, accident conditions and emergency response; and total management—optimizing management, facilities and operation that take into account the entire realm of defence in depth of both tangible & intangible aspects.

6.4.3 Discussion Points on the Isolation Condenser (IC)

6.4.3.1 Introduction

The isolation condenser (IC) is installed only at the Fukushima Daiichi NPS reactor Unit 1 and Tsuruga NPS reactor Unit 1. These are the first Mark I boiling water

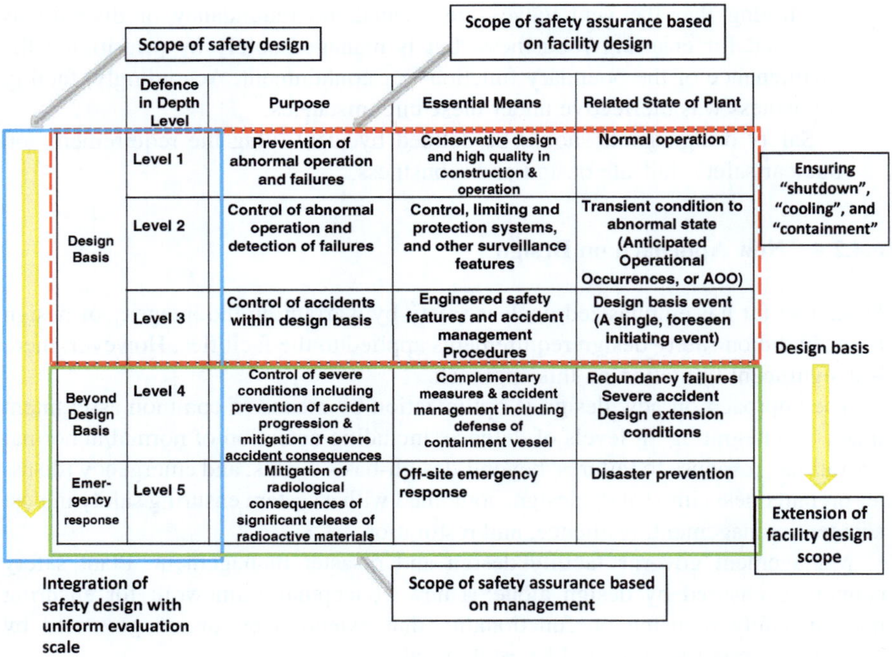

Fig. 6.15 Relationship between IAEA's defence in depth concept and design basis management

reactors (BWR) to be installed the IC in Japan. The IC system containing the IC is not the emergency core cooling system (ECCS) that is activated in the event of loss of core coolant, but a heat removal equipment utilized to condense steam generated in the reactor when the condenser in the turbine system is not available, and considered as a "safety equipment" [3].[13]

Immediately after the earthquake, the IC system was activated in Unit 1 and decreased reactor pressure from 7 to 4 MPa within about 15 min.

The IC system was the only heat removal system available among the systems expected to operate during station blackout in Unit 1, and if this system had operated successfully, accident responses would have been more effective. Issues on design in the IC system of Unit 1 is analyzed as follows.

[13] According to the classification of importance, the IC as the "safety equipment" is "classified as (1) the building, system or equipment used to shut down the reactor in an emergency, remove residual heat, prevent overpressure at the reactor coolant boundary, and mitigate significant impact of radiation on local residents," and one of residual heat removal systems in BWR, specifically as "(4) the building, system or equipment to remove residual heat after reactor shutdown."

6.4.3.2 IC System

The IC system takes steam in the reactor into the isolation condenser (IC), removes heat, condenses the steam, and returns the condensed water to the reactor by natural circulation. It is driven by water, not a pump (see Fig. 2.3). The IC stores water to maintain reactor cooling for approximately 8 h after isolation. It comprises of two systems, (A) and (B), and they operate together as required to double the cooling capacity.

The IC system is mainly equipped with isolation valves, and is designed as follows:

- Isolation valves are installed one each at the inside and outside of the PCV on the steam outlet line, and one each at the inside and outside of the return line. All are electric-powered gate valves (MO valves).
- The isolation valve inside the PCV is driven by AC power and valve outside the PCV by DC power.
- The valves are all opened except for the valve outside the return line during normal plant operation (standby state).
- The IC system is activated when the normally closed isolation valve outside the return line is opened. This valve is either fully open or fully closed. When the IC system is used to adjust the amount of heat removed from the reactor, this valve is repeatedly opened or closed (to start and stop the IC system).

In Unit 1, an elbow flowmeter was installed on the steam line and return line inside the PCV. When the pipe of the IC system broke, the water flow became stronger than under normal conditions. The IC system was isolated when the elbow flowmeter detected the pipe rupture. If the power (DC) for the rupture detection circuit of the elbow flowmeter is lost, an isolation valve close signal is generated as in the case of pipe rupture. The circuit may be designed "to open the isolation valve after confirming normal operation," because the IC system is not an emergency safety system, but whether this logic was included in the design phase cannot be confirmed without an investigation.

6.4.3.3 Operation of the IC System

The operation of the IC system during the accident is described below (according to the report of the government accident investigation committee).

- The outer isolation valves of the IC system were closed and inner valves half opened following the station blackout due to the earthquake on March 11, from 15:37 to 15:42 (interlock operation after station blackout).
- Operators could not confirm the system status because the indicators in the control room were disabled after the tsunami.

- At 16:42 to 17:30, the reactor water level was indicated in the central control room, and operators found a little decrease. At 17:15, the water level was judged to reach TAF within 1 h.
- At 18:18, the indicators in the central control room were temporarily restored, and operators found the outer valves (2A and 3A) of the IC (A) system were closed. The state of the inner isolation valves (1A and 4A) were not shown on the indicators. 2A and 3A were opened using DC power, and the operation of the IC system was confirmed with the exhaust line (nicknamed as pig's nose) on the cooling water side through the reactor building. A small amount of steam was confirmed briefly, but soon disappeared. The IC system was judged to stop functioning, and 3A was closed (18:25).
- At around 21:30, 3A was reopened, and steam release sound was heard for only a very short while.
- The reactor water level at Unit 1 dropped below TAF past 18:00, and core damage was assumed to start before 19:00.

6.4.3.4 Analysis on Operation of the IC System

The analysis on the IC system (A) operation (operator manipulation) on March 11th is shown below.

(1) **System status before Tsunami**
The IC system was confirmed as on standby state ready to be activated before the tsunami struck. Figure 6.16 shows the state of the IC system. In this figure, the layout image of the piping system is used in a vertical direction. The condensation line indicated by a solid line was laid out at an incline, in which steam was assumed to accumulate as condensed water, and the water

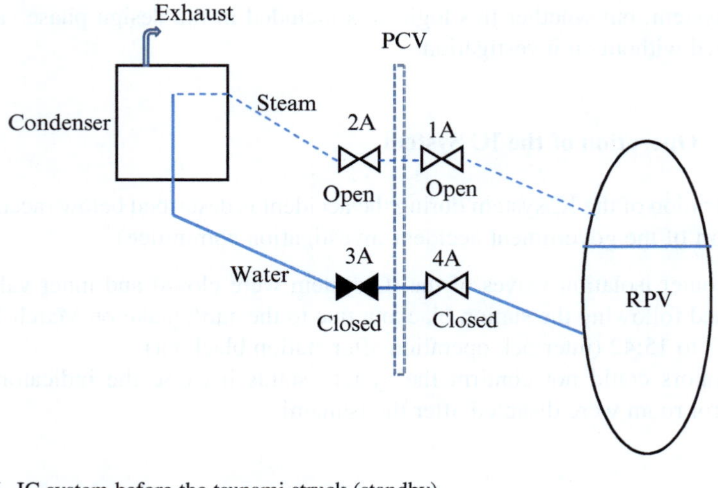

Fig. 6.16 IC system before the tsunami struck (standby)

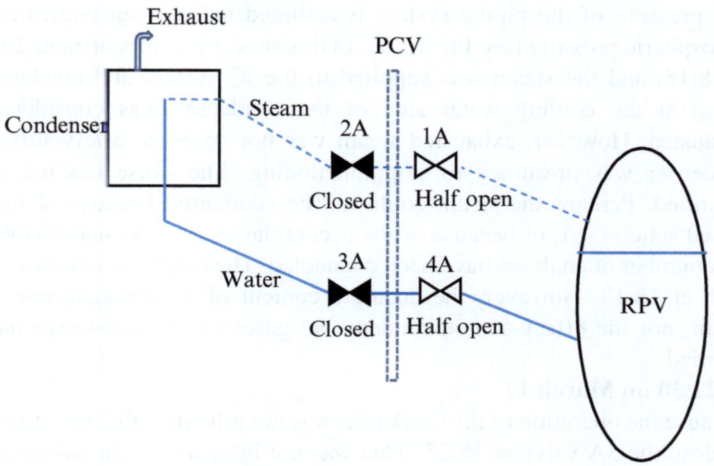

Fig. 6.17 IC system after the tsunami struck

temperature was assumed to be about the ambient temperature around the piping. The steam line indicated by dotted line is connected to the condenser by a riser pipe. The steam inside the pipe was condensed when the pipe was cooled in the PCV or reactor building, and the condensed water returned to the RPV along the downward pipe. This supplies steam in the pipe, regulating the steam line temperature to almost the RPV temperature (about 280 °C). It should be noted that if non-condensable gases were frequently generated in the reactor, these gases would have flown in and accumulated inside the IC system, damaging the IC heat removal function.

(2) **System status after the Tsunami on March 11**

Tsunami caused a station blackout, and interlock started working after failure of control power. The 2A and 3A valves were fully closed and the 1A and 4A valves were half open but the degree of the opening was not confirmed (see Fig. 6.17). The DC power for driving valves was lost sufficiently after the loss of DC power for control, and 2A and 3A could be driven to fully close state, while there was little time lag from the loss of DC power for control to the loss of AC power for driving 1A and 4A inside the PCV, and operation was aborted before valve close indicators were activated.

According to the investigation report of TEPCO (May 10, 2013), almost close state of inner valves 1A and 4A in the (A) system, and almost open state of inner valves 1B and 4B in the (B) system were confirmed. Accordingly, it is assumed that the inner valves in the (B) system was completely closed only for a short period of time, and were in a half open, or near fully open state.

(3) **At 18:18 on March 11**

When the valves of 2A and 3A were completely closed due to the tsunami, the high-pressure and high-temperature steam between 2A and the condenser was isolated, gradually cooled in the ambience, and condensed to water.

The pressure of the piping system is assumed to be about equivalent to the atmospheric pressure (see Fig. 6.17). In this state, operators opened 2A and 3A at 18:18, and the steam was supplied to the IC system and condensed. The steam at the cooling water side of the condenser was considered to be exhausted. However, exhausted steam was not observed afterwards, and the condenser was presumed to stop functioning. The cause has not yet been identified. Perhaps the steam could not be condensed because of the almost closed state of 4A, or because of the accumulation of non-condensable gases.

A number of analyses have been conducted. The reactor water level was near TAF at 18:18. However, the accurate content of non-condensable gases in steam, nor the effect of non-condensable gases on the condenser have been clarified.

(4) **At 21:30 on March 11**

Because the operation of the condenser was not effective, the operators decided to close the 3A valve at 18:25. This was not informed to the seismic-isolated emergency headquarters. There was, therefore, a different understanding on the status of the IC system between the operating staff and headquarters. The staff re-opened the 3A valve at 21:30. However, steam release sound issued from the "pig's nose" was heard only for a short period of time. The conditions of the IC system at the this time point could be as follows:

Core damage at Unit 1 is considered to have started before 19:00. When the 3A valve was opened at 21:30, the reactor water level is assumed to have dropped below the shroud (see Fig. 6.18). With the progression of core damage large amount of non-condensable gases may have been generated by water-zirconium reaction. The water in the return line is the steam condensed in the steam line. If sufficient amount of steam is not supplied, accumulated water may have decreased, although the amount of residual condensed water has not been clarified. When the 3A valve was opened at 21:30, the accumulated water

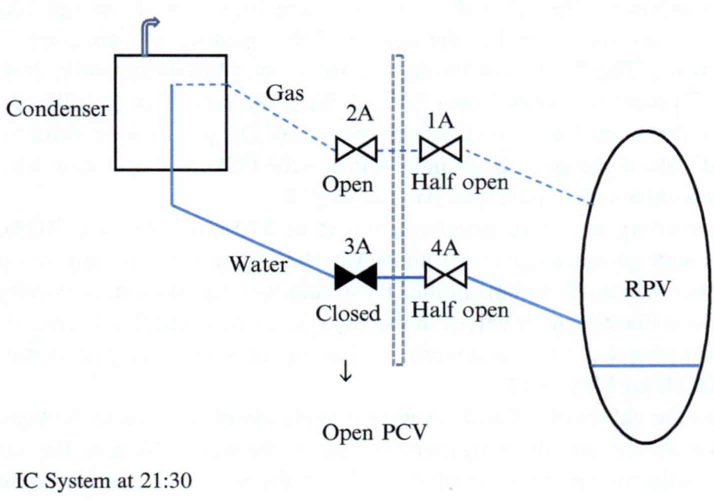

Fig. 6.18 IC System at 21:30

was discharged to the RPV by gravity through a narrow opening of the 4A valve. At the same time, gases containing large amount of non-condensable gases were supplied to the steam line, and was cooled by the condenser. However, because non-condensable gases do not transform into water, the IC system pipes became filled with non-condensable gas, and cooling of the condenser could not be maintained.

6.4.3.5 Analysis on Functional Maintenance of the IC System

If the IC system had operated normally at Unit 1, heat removal of the reactor core could have been maintained for an extended period via continuous supply of cooling water to the IC system condenser and fuel damage consequences could have been forestalled. The analyses on key points in maintaining the IC system are shown below.

(1) **If IC of the (B) system had been activated**
The situation might have been different if the IC system had functioned successfully after the tsunami struck. The IC system was a dual system, but only the (A) system was activated during the accident. The inner isolation valves of the (B) system (1B, 4B) were thought to be almost fully open state. If the following procedure was taken under this condition, the situation would have turned out much better:

- The work procedure clearly states that two systems must be used for heat removal in an emergency.
- After the tsunami struck on March 11, reactor instrumentation system was disabled, and the reactor state and IC system operation could not be identified. With event in progress, facilities with the immediately required cooling functions were limited. Hence, in management, efforts should be extended for operating available facilities, i.e., the IC system.
- If both IC (A) and (B) systems were activated soon after the tsunami (for example, manually open 3A and 3B as well as 2A and 2B on the site), the IC (B) system could have operated normally to continue to remove heat in the reactor.

Details on the use of the IC system could have been discussed thoroughly before the accident, though this may have been difficult, depending on the situation at the site.

(2) **US B5b**
B5b, developed by the U.S. NRC in 2001 in the light of 9.11 experiences, contains recommendations for improving the IC system. There is a strong implication that if these recommendations had been applied in Japan, the IC system could have been used in the accident, which is discussed below. It should be noted that the licensee did not have the information on B5b before the Fukushima Daiichi accident, and did not know about the recommendations.

As indicated in the analysis in Sect. 6.4.3.4, manual open of the almost closed isolation valves inside the PCV was required to restore the IC system. Six requirements for IC operation in Phases 2 and 3 cited in B5b are as follows.

(a) Provide a procedure/guidance that describes the plant-specific steps necessary to start and operate RCIC or the Isolation Condenser without AC or DC power.
(b) Indicate the locations of major systems to be manually operated.
(c) Compute the time the IC endures without feed-water.
(d) Indicate instruments that require no electricity to measure the IC shell water level.
(e) Ensure means for feeding water to the IC shell in the event of power failure (e.g., portable pumps).
(f) Use onsite instruments to monitor reactor water level (e.g., existing instruments not requiring electricity).

Of the above B5b recommendations, a. relates to the functional failure of the IC system. The activation of the IC system, and IC valve operation onsite under power failure conditions is designated in a. Therefore, a. is important to achieve the operation in (1). This is related to manual operation for opening the outer isolation valves (3A, 3B) on the condenser return line, required when the system is activated, and suggest that the manual operation of 3A/3B on site should be included in the procedure.

As already described, a tool (rod, etc.) was required to open the inner isolation valves from outside of the PCV to restore the IC system, but recommendations on the use of such tool is not clearly designated in B5b.

However, a further review may suggest the following:

- Scenario on AC power availability and DC power failure, because "conditions on the loss of DC or AC power source" should be given consideration.
- The IC systems under the above power supply conditions is that the interlock works to completely close the inner valves.
- Therefore, manual opening of the outer and inner isolation valves from outside the PCV are required to activate the IC system.
- A tool (rod, etc.) is required for opening inner valves from outside the PCV. If B5b recommendations were applied, a tool (rod, etc.) would need to have been installed to restore the IC system in the accident.

(3) **Design obligations and remodeling proposal by manufacturers**

GE of the U.S. designed the reactor facilities of Unit 1. GE makes remodeling proposals for BWR plants as required, and provides the information to the user, or TEPCO in this case. There was no such information related to the IC system, indicating no proposals had been made by GE.

Remodeling based on B5b could be implemented independently by the electric power companies. In the case remodeling is made as part of safety measures, it must be made public.

6.4.3.6 Issues Identified and Actions Required

The issues and solutions related to the use of IC system in the accident are discussed as follows:

(1) **Consideration of accident management for all facilities**
The transmission of the automatic close signal to the PCV isolation valves of the IC system in the event of loss of control power for the rupture detection circuit mounted on these valves should be considered as part of accident management. The key concern is maintaining functionality of the systems under any circumstances.

For instance, isolation valves mounted on a safety-critical system may be "fail as is (FAI)". The two PCV isolation valves may be installed outside the PCV, and not one inside and the other outside.

(2) **Check-ups & tests on facilities related to safety (IC system)**
The IC system, which is designed according to the same criteria as safety systems needs to be tested regularly. However in reality, tests had been conducted only on valve open and close, and not on the operation of the IC system. Hence, the operators had no experiences on what happens when the IC system is activated. A test procedure on the IC technology which will be used in the future should be examined.

(3) **Coordinated approach on PCV isolation and design of safety systems**
In terms of safety systems, remotely controlled components such as electric valves should not be installed inside the PCV. All safety systems and components should be installed outside the PCV so that they can be repaired promptly or operated as required in the event of a severe accident.

At present, all ECCSs except for the RCIC of ABWR are installed outside the PCV. Because the RCIC was formerly considered as a general system, but became classified as an ECCS in ABWR, an isolation valve on the turbine steam line is installed inside the PCV. A valve is also installed inside the RCIC system. The valves on the steam line of the RCIC system are designed under the same concept as with the isolation valves in the IC system. However, the RCIC is not designed to close the isolation valves when the rupture detection system control power is lost. Since the design on the RCIC system is newer than the IC system, the design features of the IC system shared with the RCIC system should have been modified with the revision in the design of the RCIC system.

6.4.3.7 Summary Conclusion

In the Fukushima Daiichi accident, there were high expectations on the effective performance of the IC system in reactor core cooling. Hydrogen generated due to core damage in Unit 1 resulted in a hydrogen explosion, followed by core melt-down, hydrogen explosion at other units, and consequently to extensive radioactive materials release. In this regard, the effective operation of the IC system is a crucial issue. All power was lost by the tsunami. With the loss of AC power, a valve close signal was transmitted to PCV isolation valves of the IC system, and the valves were closed until the power driving valves were cut off. Consequently, the isolation valves of the (A) system inside the PCV is assumed to have been closed or near closed, and the IC system lost cooling function.

If the IC system had operated, it might have functioned as the sole heat removal system and facilitated more effective accident responses. If there had been more awareness on the significance of the IC system, then both (A) and (B) systems would have been operated for better consequences.

When the RCIC system with similar features as the IC system was changed, the revision should have been reflected on the IC system also.

If the licensee obtained B5b with recommendations on the improvement of the IC system, and did something based on these recommendations, for instance, improvements and accident management, the IC system would have performed effectively in the Fukushima Daiichi accident.

What should be done to operate necessary systems under any situation must always be considered for continuous improvements.

6.4.4 Materials and Structural Integrity

6.4.4.1 Objectives and Method of Investigation

The Fukushima Daiichi accident was unprecedented from the aspects of reactor core material, which resulted in significant core damage and seawater injection for core cooling. A detailed on-site investigation is impossible at present, and far from the conclusion in terms of material technology. It is important to analyze the accident progression using as much information as possible from the perspective of material science and material engineering, and to identify the issues at present and in the future as a means of ensuring post-accident stability (safety).

Phenomena associated with structural materials and their effects during the Fukushima Daiichi accident were assessed based on various simulations, experiments and investigations conducted after the accident, events in other nuclear reactors, insights obtained from Unit 2 of Three Mile Island NPP, and other data, and the results obtained to date were summarized.

It was found that most phenomena were qualitatively clarified to some extent, however, some were not understood at all. The unknown segment caused significant discrepancy in the accident analysis results, which must be given consideration in future accident investigations and severe accident management planning.

Issues related to material science/technology aspects that must be addressed were extracted from the simulation results of material behavior during the accident and after cold shutdown. Although the situation was not the same as the Fukushima Daiichi NPS insights obtained from the seawater inflow event at Unit 5 of Hamaoka NPS has been shown.

6.4.4.2 Summary of Events from Material Perspectives

Materials affected with accident progression, and specific features of the changes in these materials have been organized, from which potential issues on items with not much accumulated insights has been extracted as follows. The accident progression is classified into the following phase: (1) before core damage, (2) from core damage to RPV damage, (3) from RPV damage to loss of PCV, (4) from loss of PCV to cold shutdown, and (5) spent fuel pools (SFPs). Table 6.18 shows issues related to material technology aspects based on the discussions by the Safety Measure Enhancement Technology Investigation Special Committee of AESJ (http://www.aesj.or.jp/special/senmon.html). Some of the issues in the table have already been substantiated by accumulated research and expertise, however, most are predictions based on the lessons learned from the TMI-2 accident, and require verification. The events and issues concerning fuel and molten fuel are covered in Chap. 9, and hence, omitted in this chapter, except for reactor core material and related items shown in the list.

A number of research projects have been established to date, and various simulations, experiments, and assessments have been conducted on the accident. These projects were classified as follows according to Table 6.18: (1) evaluation of the direct effects of the earthquake, (2) overview of study on simulations on accident behaviors such as fuel melt and PCV damage, and (3) seawater inflow event which reactor core materials have never undergone. The direct effects of the earthquake in (1) were studied in relation to the degradation event in the Public Hearing for Aging Degradation held by the former Nuclear and Industrial Safety Agency. Although no special concerns were identified in this public hearing, verification by on-site investigations is required. A number of studies on the effects of seawater that have not been clarified are conducted for the event in (3). In addition, the seawater influx event into the Hamaoka NPS reactor Unit 5 in May 2011 (http://www.chuden.co.jp/energy/hamaoka/hama_info/hinf_topics/__icsFiles/afieldfile/2011/12/06/hyoukakentouiinkaigiji.pdf, http://www.nsr.go.jp/archive/nisa/oshirase/2012/09/240914-2.html) presented similar situation as at the Fukushima Daiichi NPS where seawater mixed with reactor cooling water and the concentration of chloride ions became higher than usual. This event was therefore considered to offer critical information for understanding potential events in fuel and construction materials in the saline water environment. The summary is provided in the next section. In addition to the discussions at the Corrosion Countermeasure Investigation Committee of the Fukushima Daiichi NPS, investigations into the effects of saline water included studies on the corrosion induced by high-temperature seawater prior to cold shutdown, corrosion in seawater after cold shutdown, and investigations into the events at the Hamaoka NPS. Based on these results, issues on (2) and (3) were identified.

Table 6.18 Potential material related issues in damaged reactors

	Issues on material technology
(1) Before core damage	[Cladding tube] Post-accident material research; sampling and analysis for clarifying progression of events (temperature history, etc.); clarification of fuel history at the time of accident (irradiation degradation, oxidation, hydrogenation, internal pressure of fuel rod, amounts of accumulated FP, etc.) and damage; effects of intensive hydrogen embrittlement near damaged area and progression on damage to the entire fuel rods; and clarification of effects of water quality and cladding.
	[Fuel assembly] Evaluation of cladding temperatures after fuel exposure (history and maximum value); hydrogen induced gas-phase corrosion and hydrogenation; evaluation of swollen cladding due to creep deformation of assembly; and effects of flow path blockade and heat transfer deterioration due to salt deposition, etc.
	[Reactor] Vapor-phase corrosion of RPV and core materials; evaluation of high-temperature corrosion; and evaluation of effects of fuel releases due to damage on pressure boundary
	[RPV] Database for high-temperature properties of reactor materials
	[Piping] Investigation into effects of the earthquake
(2) From core damage to RPV damage	[Fuel] Clarification of behavior under conditions exceeding the performance evaluation standard for emergency core cooling system (1,200 °C, 15 % ECR) and meltdown (limits of temperature and time of cladding containment, creep rupture under steam corrosion conditions)
	[Core internals] Development of multi-component system phase diagram in high-temperature steam environment and transitional non-equilibrium condition, in particular, of the reaction of materials such as stainless steel, zircaloy, control materials and fuel; and clarification of control rod drop event and fuel melting in terms of materials
	[RPV] Identification of RPV damage features; high-temperature damage of gaskets; interaction between molten core and RPV; clarification of damage to lower head of RPV based on insights from the TMI-2 accident.; and high-temperature seawater steam corrosion
	[Piping] Effects of the earthquake
(3) From RPV damage to loss of PCV function	[Transition of FP] Clarification of FP transition behavior from molten fuel to water phase (FPs are assumed to have moved from the damaged core to outside RPV via the injected water)
	[PCV] Behavior of excessive temperature protection seal materials; and relations between jet impingement induced corrosion (a phenomenon of corrosion of structures hit by molten jets) and PCV damage
	[Concrete] Effects of gas generation and water immersion, increases in convection heat transfer rate, and flow property of molten debris due to erosion of concrete
	[Cooling] Thermal interaction between molten core property and cooling water in pedestal

(continued)

Table 6.18 (continued)

	Issues on material technology
(4) From loss of PCV function and stop	[Molten fuel] Evaluation of seawater corrosion; evaluation of long-term cooling of fuel after LOCA (no standard in Japan)
	[Structural materials] Evaluation of seawater corrosion
	[Concrete] Evaluation of seawater erosion and corrosion
(5) SFP	[Fuel] Evaluation of fuel rod damage behavior (ballooning, diffusion behavior of FP, etc.) and seawater corrosion; investigation and measures for preventing damage to spent fuel due to rubble when taking it out; and handling of fuel exceeding limits of present reprocessing facilities
	[Rack] Evaluation of seawater corrosion
(6) Whole unit	[Accident analysis] Remote sampling at accident plant and analysis and testing

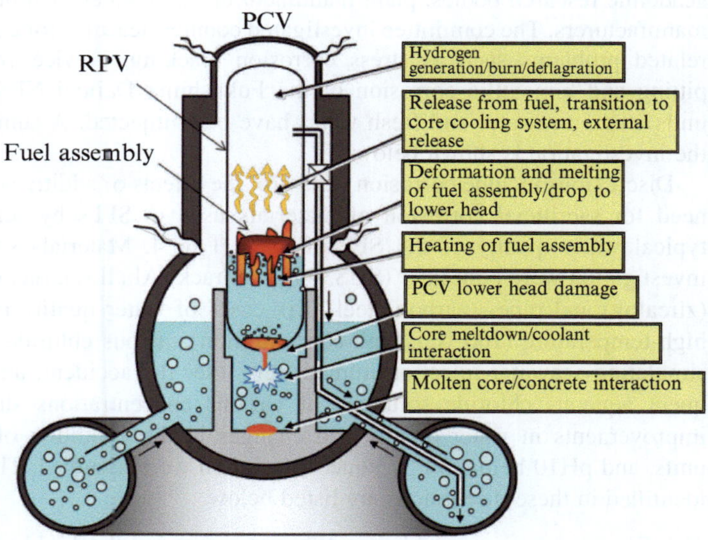

Fig. 6.19 Potential phenomena in the reactor core during a severe accident

6.4.4.3 Current Investigation of Events

(1) **Material behavior during accident**
It was estimated that water injection was restarted about 6–14 h after fuel assemblies had been exposed from coolant in the Fukushima Daiichi accident. During this period, meltdown of a large amount of fuel at Units 1 to 3 was highly likely, and part of the molten fuel is assumed to have melted through the RPV wall and dropped into the PCV. Various phenomena may take place in the reactor during a severe accident (Fig. 6.19). However, in-core conditions such as fuel debris distribution and reactor core damage have not been clarified as of

present. New insights on these conditions are critical to estimate fuel material behavior during the accident (e.g., maximum temperature and cooling speed of the damaged fuel). In-core conditions, including fuel debris distribution need to be understood for effective and accurate removal of fuel debris.

(2) **Influence of seawater inflow**

(a) Discussions at the Fukushima Daiichi NPS Corrosion Countermeasure Investigation Committee: Fresh water and seawater were injected into the reactors of Units 1 to 3 of the Fukushima Daiichi NPS to cool the reactor cores. Fresh water was injected into the SFP at Unit 1, and seawater and fresh water were injected into the SFP at Units 2 to 4 during the accident. The Central Research Institute of Electric Power Industry (CRIEPI) launched the Fukushima Daiichi NPS Corrosion Countermeasure Investigation Committee (http://criepi.denken.or.jp/result/pub/annual/FY2011/P96-P97_kiban8.pdf) comprised of academic experts, and members of academic research bodies, plant manufacturers, and water treatment plant manufacturers. The committee investigated countermeasures for corrosion-related problems such as stress corrosion cracking, crevice corrosion, pitting and bimetallic corrosion of the Fukushima Daiichi NPS reactor units where seawater and fresh water have been injected. A summary of the investigation is shown below.

Discussions include corrosion behavior, the effects of additives, and the need for sacrificial corrosion of materials used in SFPs by setting the typical water quality of the SFP of Units 1 to 4. Materials subject to investigation were pool liner (SUS304), fuel rack (Al alloy), fuel cladding (zircaloy) and pipes (carbon steel). Six cases of water quality including high-temperature (100 °C), high-concentration aqueous chloride solution simulating seawater injection immediately after the accident, and subsequent aqueous chloride solutions of various concentrations simulating improvements in water quality and changes in the condition of reactor units, and pH10 below 50 °C aqueous solution were assumed. The issues identified in these discussions are listed below.

(i) Stress corrosion cracking and crevice corrosion are highly likely to have been generated in the pool liner of Units 2 to 4. The progression rate could reach several tens of millimeters per year in a high-concentration aqueous chloride solution condition after the accident. There was no possibility of local corrosion or stress corrosion cracking after water quality was improved, and there is a strong possibility that the progression itself stopped. Voluntary stop of progression of local corrosion is, however, difficult if the effects of microbes are significant.

(ii) In Units 1 and 3, concrete structures were broken by hydrogen explosions and lumps of concrete mixed in water increased its pH. Regarding fuel racks made of Al alloy, uniform corrosion is likely under the chloride solution condition, but the amount is in the order of tens of millimeters per year in the assumed pH range (around pH 10). The effects of bimetallic corrosion in the area in contact with pool liner largely depend on water quality (electric conductivity) and tend to be accelerated by the effects of microbes.

(iii) Local corrosion of fuel cladding is unlikely, but crevice corrosion in high-temperature, high-concentration aqueous chloride solution immediately after the accident is undeniable.

(iv) Uniform corrosion on carbon steels occur under all conditions in the order of 1–2 mm per year at most. Local corrosion is likely in low-concentration aqueous chloride solution with relatively high pH.

(v) Simulated results of this corrosion behavior suggests measures such as the reduction of oxidiants, controlling microbial effects and monitoring potential, etc., to deter the corrosion of SFPs.

(b) Understanding of the effects of seawater injection on RPVs and fuel assemblies, etc.: Fuel assemblies in the SFPs, and fuel debris in reactors are exposed to water containing seawater components and high radiation. Under such extraordinary conditions, removal of fuel assemblies and fuel debris and subsequent long-term storage require surveys on material degradation and damage to develop preventive and mitigative measures. The RPVs and PCVs in which seawater has been injected are assumed to be exposed to diluted seawater conditions for an extended period in the future, and their structural strength may decrease due to corrosion of materials.

(c) Effects of damage to condenser tubes at the Hamaoka NPS Unit 5

(i) Outline of events and responses Hamaoka NPS Unit 5 (rated output: 1,380 MkW) of the Chubu Electric Power was suspended at 10:15 on May 14, 2011. After it reached sub-criticality at around 13:00 the same day while being depressurized, seawater flowed into the reactor water from the damaged narrow tubes of the condenser past 17:00. The reactor water was immediately diluted with accumulated water in the suppression chamber, cleaned with the reactor coolant purification system, and desalinated. The water in the turbine system was drained to the condenser hot well, and replaced with desalinated water. The chloride ion concentration of the reactor water rose to 400 ppm immediately after the event, and continued to exceed 100 ppm for about 70 h afterwards. Details of this event including the cause of damage to the condenser tubes were recorded (http://www.chuden.co.jp/energy/hamaoka/hama_info/hinf_topics/__icsFiles/afieldfile/2011/12/06/hyoukakentouiinkaigiji.pdf, http://www.nsr.go.jp/archive/nisa/oshirase/2012/09/240914-2.html).

The Chubu Electric Power confirmed the integrity of all seawater inundated facilities through mechanical and system inspections, and replaced or overhauled components as required. This was the first reactor seawater inundation event in Japan, other than the 3.11 incident at the Fukushima Daiichi NPS. The results of adequate surveys and measures taken would help to understand material behavior in the damaged reactors in the Fukushima Daiichi NPS, as well as instrumental in the responses to similar accidents in the future.

In subsequent inspections completed in 2012 at the Hamaoka NPS, degradation of the reactor and components was identified, including partial corrosion and tarnish of the weld of austenitic stainless steel liner of the reactor and other parts, which were confirmed not to cause leakage or breakage. In addition, adherence of oxidation products mainly ferrioxide was also observed on pumps, heat exchangers and valves during inspections. The amount of adherence was slightly more than during normal inspections, however, removal was possible.

Typical steels composing the seawater inundated facilities were tested to evaluate corrosion impact. The facilities subject to evaluation and test results are summarized as follows.

The reactor and the turbine system inundated by seawater were simulated (seawater concentration, temperature and soaking time). Local corrosion was observed on stainless steels but no stress corrosion cracking was found. Corrosion of the adhesion enhancement portion, the nitrided portion and crevice corrosion of superimposed part of stainless steels were also observed.

Evaluation of fuel includes appearance inspection, tensile test and sectional metallographic examination conducted by simulating seawater inundation using a simulant fuel. Appearance inspection, hydrogen analysis, tensile test and sectional metallographic examination were also conducted simulating seawater inundation and associated operations using irradiated fuel. The test results confirmed fuel integrity with limited seawater impacts such as minor tarnish on the surface and corrosion on the plenum.

Reactor water quality was also improved and maintained after inflow of seawater. This suggests that the event did not affect withdrawal and transfer of the fuel to the SFP.

(ii) Applying insights to the Fukushima Daiichi NPS accident The Hamaoka Unit 5 was exposed to high-concentration (400 ppm) chloride ions of approximately 240 °C after seawater flooded into the reactor from the damaged condenser tubes. However, this was only for a short while, and the corrosion was halted by the dilution and purification of reactor water. Although it is not necessarily appropriate to apply the insights obtained in the event at the Hamaoka NPS to the Fukushima Daiichi NPS because of the different conditions, due attention must be paid to the corrosion of the nitride portion of stainless steels obtained from the results of component tests.

Leaks were found at the corroded weld of the recirculation pipe (carbon steel pipe) for the condensate return pump used to supply water to remove saline matter from the water containing seawater components in the hot well of the condenser. Corrosion of the weld was presumably generated in a corrosive environment inside the pipe due to the circulation of water containing seawater components. Corrosion spread across the pipe and concentrated on the weld that is prone to corrosion compared with the base material. This resulted from continuous operation of the condensate return pump constantly supplied seawater components and dissolved oxygen which accelerated corrosion.

Local corrosion is not likely on the carbon steel pipe which uniformly corrodes in general, and the weld is especially prone to corrode.

(d) Evaluation of crevice corrosion behavior of stainless steels in diluted seawater: When seawater flows into the condenser of LWR, corrosion may be induced by chloride ions in the condensate and feed-water system depending on the amount of inflow. Chloride ions may induce local corrosion such as crevice corrosion on stainless steels, used for condensate and feed-water system. Numerous insights have been accumulated on the effects of chloride ions on stainless steels, however, not necessarily sufficient data on the corrosion behavior of stainless steels used in the LWR in chloride ion concentration and temperatures of the condensate/feed-water system under extensive seawater inundation conditions have been collected.

On the basis of chloride ion concentrations of steel grades assumed to be used for the reactor core and turbine system, the crevice corrosion potential was obtained according to JIS [14], and compared with corrosion potential [15] to identify corrosion conditions. Crevice corrosion was observed on SUS304L, SUS316L and SCS19A at 50 °C when the chloride ion concentration was around 500 ppm or more, but uniform corrosion was significant when the chloride ion concentration was less than 500 ppm. Crevice corrosion on SUS403 was observed when the chloride ion concentration was approximately 15 ppm, or more. This is probably because SUS403 contains smaller quantities of chloride-resistant additive elements than other steels. These insights may be used as indices to estimate material corrosion behavior or the purification of affected systems in the event of seawater inundation.

The generation and progression of corrosion were evaluated in controlled-potential corrosion tests [16]. In the temperature range between room temperature and 100 °C at a relatively high chloride ion concentration (6,000 ppm), crevice corrosion occurred within comparatively short time, and shortened with the increase in temperature. The depth of crevice corrosion was found to increase at about the 0.5th of the corrosion progression time. Various forms of crevice corrosion are assumed in the reactors in the accident, including significant progression of crevice corrosion on these steels.

6.4.4.4 Summary

Potential issues were identified in terms of material technology with the progression of various events during the accident and after cold shutdown (Table 6.18). Considering the results of various discussions, the direct impact of the earthquake may be small. However, this awaits to be verified by on-site investigations. Insights on the seawater inflow event have been obtained from reactors other than those in the Fukushima Daiichi NPS. Though not specified in this section,, there are a number of issues on the maintenance of specific reactor facilities after cold shutdown in terms of material technology, such as the long-term cooling and seismic resistance of post-LOCA reactor core associated with the corrosion of damaged reactors, SFPs, waste water tanks, and decontamination equipments.

6.4.5 *Ageing Degradation*

6.4.5.1 Introduction

All six reactor units of the Fukushima Daiichi NPS had been in operation for more than 30 years before being hit by the tsunami in the Great East Japan Earthquake. A technical assessment had been conducted on these plants in terms of structural degradation for another long-term operation following 30 years of operation. By this technical assessment, their structural integrity for continuous long-term operation had been confirmed.

Future analyses will clarify the causes of the accident. Plant life and ageing management is discussed here.

6.4.5.2 Discussion Points on Plant Ageing

(1) **Aging in terms of structural integrity**
On what ground are plants declared termination of service life period?

(a) Design life and actual life: Life period in design, or "design life" is defined and designated in which the period of operation is assumed and a plant is designed to ensure the operation of critical systems, equipment and structures within this period. In reality, however, all processes (including design, manufacturing and inspection) have some margin, and the actual life for which the required functions are available far exceeds the design life.

(b) Life extension by replacement: A nuclear power plant comprises some tens of thousands of components and millions of parts, many of which are regularly replaced or upgraded with newly developed products. They are renewed before the plant comes to the end of its life. In addition, general structures start showing various symptoms of degradation after 30 or 40 years of operation. Before they reach the end of their service life, measures are taken through analytical assessment in the design phase, inspections from start of operation, and maintenance. This is achieved through the continuous effort of operators in collecting nonconformity information worldwide utilizing cutting-edge findings to predict degradation, and conduct inspections for application to actual operation.

(2) **Plant life period and application of changes in safety concept**
Nuclear power plants in Japan undergo regular inspections almost every year, a safety review every 10 years, and a technical assessment every 10 years after 30 years of operation. Thirty years of service (operation) is granted with the approval of nuclear reactor establishment license on the preconditions of these foregoing examinations. The technical assessment confirms the integrity of facilities for extended long-term operation after 30 years.

However, this mechanism did not take into consideration changes in the safety concept with time, which is as important as structural integrity. This issue could jeopardize even the most advanced plants. Excessive emphasis has been placed on structural integrity, including aging degradation and reduction of failures to the exclusion of changes in the safety concept, application of new insights, and f improvement efforts.

Numerous insights obtained from the TMI-2 and Chernobyl accidents have been incorporated into safety assessments worldwide, which have been revised and updated as necessary. Particularly in the U.S., safety provisions against beyond design basis conditions, including counter-terrorism was re-evaluated thoroughly after the September 11, 2001. This was a significant change in the approach on ensuring safety made as a result of accumulated insight. The application of such insight is critical for the effective implementation and fulfillment of the required safety functions.

(3) **Plant life period and application of changes in assessment criteria**
The majority of plant components and equipment are replaceable. In strict technical terms, even the RPV can be replaced. Assessment should not only focus on structural degradation, but must be based on the latest insights to reduce risk consequences on the entire system, to ensure nuclear safety.

Although tsunami assumption was enforced and the height was raised in tsunami assessment, the development and implementation of measures was delayed. Regardless of whatever redundant standby equipments are provided, the consequences of tsunami inundation may not be deterred because the emergency diesel generators were designed to be installed in the basement of the reactor building on the ocean side. The Fukushima Daiichi accident revealed the need to prepare for and respond to small probability events with high consequences as the tsunami. The integrity of the plant, systems and equipment must be maintained under any circumstances. In the event of emergency, if functional integrity of alternative equipments and systems are maintained, there are no issues. With consideration given to the consequences of combined disasters by earthquake and tsunami events, and based on the approach on the latest design and degradation management, framework for maintaining functionality of the entire plant system should be established. It is the role of the regulatory body in establishing such framework, and their efforts must be extended to reduce risk involving nuclear power plants in all aspects. This has been achieved in many countries by means of regulatory safety reviews which are the key to long-term operation.

6.4.5.3 Plant Operation and Duration of Life

As described above, the determination process on plant service life is complex. The extension of operation period has been approved for many nuclear power plants worldwide on such basis.

According to the IAEA data, nuclear power plants operating for more than 40 years account for 5 % of all nuclear power plants worldwide. Operating license of 40 years is granted, and application for extension of 20 years is approved in the U.S. The NRC has already approved 60-year operation for more than 70 nuclear power plants in US. More recently, preparation is in progress to revise the approval to 80 years operation. In countries including France, the U.K., and Spain, there are no regulated restrictions on the period of operation. In France, licensees must provide an aging degradation control program and certification of appropriateness for continuous operation of plants in operation for over 30 years. Continued operation is also approved every 10 years through regular safety reviews in the U.K. These suggest that the operating life of nuclear power plants is not limited to 40 years. Nuclear power plants are managed under adequate aging degradation control programs in these countries, and the IAEA has established a database for common degradation events (International Generic Aging Lessons Learned: IGALL).

6.4.5.4 Important Perspective on Nuclear Power Plant Ageing Management

The key point in the operating life of nuclear power plants is that it is not simply defined by service life period of the hardware aspects, as, including equipment, structures and materials. Perspective on the entirety of the plant system with view to ensuring nuclear safety is required. Periodic safety reviews (PSR) are effective for verifying whether the application of the safety concept is appropriate, and whether the safety standards are updated with the latest insights, or not. Application of changes in the safety concept, adoption of new safety concept, and the back-fitting of changes in safety standard will be difficult without quantification of nuclear safety concept in some way or other. Obsolescence of the system can be prevented, and the loss of "life", or loss of functionality for ensuring safety can be e avoided by examining changes in the safety concept and criteria, and introducing an assessment system. New insights in the design phase go out of date over time. Old insights must make way for new insights. This is also about "life period."

6.4.5.5 Conclusion

Most systems and equipment at nuclear power plants are replaced with the latest models. Some of the key hardware are said to have a life beyond a century. Nuclear power plant ageing management do not simply involve structures and hardware, but must be determined with view to various aspects.

6.5 Accident Management

Issues identified from the Fukushima Daiichi accident and corresponding measures are discussed in terms of accident management in this section. Accident management stands for taking a series of measures through the effective utilization of extra functions assumed in the safety margin included in the design or safety design, or independent equipment installed in preparation for such event that may exceed the design basis and cause serious damage to the reactor core or spent fuel pool. The introduction of accident management used to be at the discretion of licensees, but has been legalized for complete application according to new regulatory standards since July 2013.

The important perspective of improvements in accident management, in the light of the Fukushima Daiichi accident, is that the role of severe accident management is not an extension of the preceding three levels, but independent effectiveness which is the key perspective of defence in depth (see Sect. 6.3.3). Assessment of a severe accident through the design method according to levels 1 to 3 could see the same mistakes as in the Fukushima Daiichi accident repeated. Measures based on reference scenarios and reference events designated in levels 1 to 3 are insufficient, or wrong. Namely, measures (management) should be developed in a different perspective from level 3 (design) (see Fig. 6.20).

In addition it is important to build consistent and integrated safety measures according to the concept of defence in depth based on risk assessment, namely, the

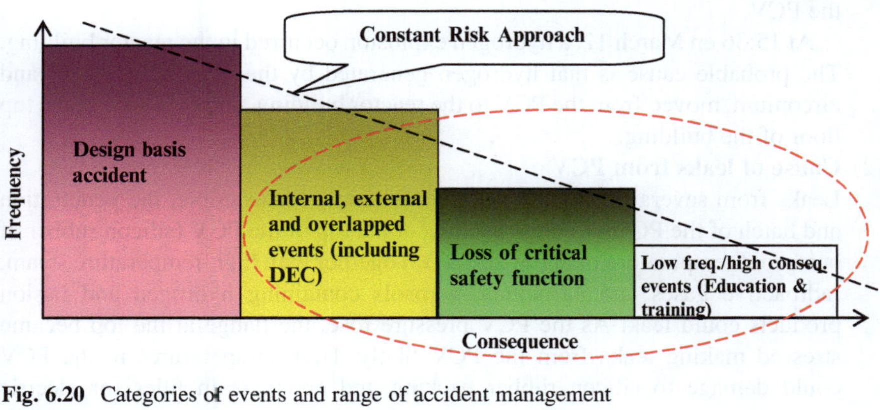

Fig. 6.20 Categories of events and range of accident management

importance of safety determined by the frequency, effect and scenario, for all potential initiating events (including multiple failures and overlapped events) without ignoring beyond design basis events.

Examples of accident management in terms of the equipment for defence in depth for natural disasters (earthquakes, tsunamis), and accident management for multiple reactors in the same site are discussed in detail based on the Fukushima Daiichi accident. Discussions are based not only on new regulatory standards, but on a broader perspective using hardware and software.

6.5.1 Radioactive Material Containment Function of Primary Containment Vessel

6.5.1.1 Overpressure and Overheat Caused Damage to PCV, and Countermeasures

(1) **Locations of leak**

One of the characteristics of data on reactor pressure, water level and PCV pressure at the Fukushima Daiichi Unit 1 was a rise in reactor water level past 21:00 on March 11 despite the failure to inject cooling water into the reactor. This indicates an abnormal condition, for example, high temperatures in the PCV caused water in the reference head tube of one of two differential pressure gauge tube exposing in the drywell (D/W) of the PCV, to start evaporating due to overheated gases (see Sect. 6.5.2). The PCV was probably heated up by high-temperature steam that leaked from the broken neutron counting instrument inside the RPV or the broken bolt or gasket of main steam safety relief valve (see Sect. 6.1.2).

The PCV pressure became almost the same as the RPV pressure before 3:00 on March 12 due to damage to RPV. Continuous discharge of energy from the reactor core could have caused the PCV pressure to rise, but maintained to 0.75 MPa [abs] past 12:00 on March 12, probably resulting from leaks from the PCV.

At 15:36 on March 12, a hydrogen explosion occurred in the reactor building. The probable cause is that hydrogen generated by the reaction of water and zirconium, moved from the PCV to the reactor building, and exploded at the top floor of the building.

(2) **Cause of leaks from PCV**

Leaks from several locations could be assumed, for example, the penetration and hatch of the PCV or flange packing at the top of the PCV (silicon rubber of heatproof temperature of about 200 °C). Together with high-temperature steam, radioactive gases and particulate aerosols containing hydrogen and fission products could leak. As the PCV pressure rose, the flange at the top became stressed making leaks from the PCV likely. High temperatures in the PCV could damage to silicon rubber packing and epoxy resin filler for electric penetration, and cause leaks from these locations. Field investigation should clarify the locations of leaks in the future.

(3) **Locations of contaminated water leaks**
Contaminated water leaks from the broken part at the lower part of the PCV to outside have spread to a wide area through the trenches to turbine building. Leaks from the lower part of the PCV or the part of venting pipe connecting the drywell and suppression chamber at Unit 1 was confirmed by the TEPCO investigation in November 2013. Hereafter investigations identifying leak locations need to be continued widely to reduce the generation of contaminated water and stop leaks to prepare for removal of fuel debris.

6.5.1.2 Countermeasures for Venting

(1) **Venting situation at Fukushima Daiichi**
A reinforced vent from PCV was installed at Unit 1, but the vent valves could not be opened remotely due to the loss of control power, and could not be opened by manual operation fast enough to prevent the PCV pressure from rising and partly causing leaks of hydrogen, thereby hydrogen explosion. Similarly, PCV venting at Units 2 and 3 did not proceed smoothly.

(2) **Situations at overseas**
A wide range of releases of radioactive materials at the Chernobyl accident significantly affected nearby European countries. After the accident, the filtered vent has been installed at almost all nuclear power plants in Europe. In France, a filter unit consisting of water and gravel is installed inside an upside-down bowl shaped container of 8 m in diameter and 4 m in height. In addition, many catalytic recombiners are installed inside the PCV to counter hydrogen. Two filtered vent systems each having 50 % capacity are installed at the Laibstadt NPS (BWR) in Switzerland. SBO is highly likely to cause a severe accident. In this case, a rupture disk blows out at about 0.3 MPa to start venting automatically.

(3) **Lessons from the Fukushima Daiichi accident and countermeasures**
Despite of a rupture disk installed on the reinforced vent system of the PCV as measures for accident management, the PCV pressure rose to an abnormal level because the vent system was designed to switch several valves but failed. This is a lesson learned from the accident. If the vent system was designed to be operable during the accident, early venting could have avoided overpressure caused damage to the PCV and hydrogen explosions, largely reducing releases.

After the Fukushima Daiichi accident, new regulatory standards were established to obligate licensees the installation of filtered vent in Japan. Operating filtered vent effectively, countermeasures for preventing overheat caused damage to the PCV, such as the enhancement of PCV spray or the suppression chamber (S/C) cooling system, are required. A connected water supply inlet for fire-fighting and dedicated piping are installed for direct water injection to D/W or S/C from the fire engine in the event that PCV spray with engineering safety facilities is unusable. Water injection to reactor well can prevent damage to top flange packing of the PCV due to overtemperature.

As for improvements in the certainty of operation and operability of existing reinforced vent systems and new filtered vent systems, the auxiliary air cylinder and the power supply to open air operation valve must be ensured. Auxiliary systems and equipment to make use of these system must also be taken into account.

(4) **Installation of filtered vent and effects**

The most effective way to reduce influences of venting on the external environment is to install a filtered vent, which largely reduces releases of radioactive materials such as cesium and iodine.

Looking at the trend overseas, filtered vent having higher decontamination factor has been used in France since the Fukushima Daiichi accident, and installation of filtered vent in all nuclear power stations has been determined in Russia.

6.5.2 Reactor Instrumentation Systems (Reactor Water Level Instrumentation)

Various instrumentation systems, essential for the operation of the reactor, are installed around the RPV and PCV. Reactor water level, reactor pressure, D/W pressure and S/C pressure are mainly measured. Many of these instrumentation systems were lost in the Fukushima Daiichi accident, which made determining the plant conditions extremely difficult.

Of these instrumentation systems, the reactor water level instrumentation system, which made the especially large impact to the accident, is discussed here to reveal its mechanism and the problems identified in the accident.

6.5.2.1 Importance of Reactor Water Level Instrumentation

The core cooling water for boiling water reactors (BWRs) usually maintains its water level at about 5.3 m above the top of active fuel (TAF) during normal operation. When the water level falls due to a pipe rupture, the reactor water level instrumentation system detects it and an emergency core cooling system (ECCS) is required to start operating. The reactor water level instrumentation system is a key system for BWRs.

6.5.2.2 Mechanism of Reactor Water Level Instrumentation

A differential pressure detection system is used to measure the reactor water level as the difference in pressure. The differential pressure transmitter detects a differential pressure signal, and changes the deflection of its diaphragm when the

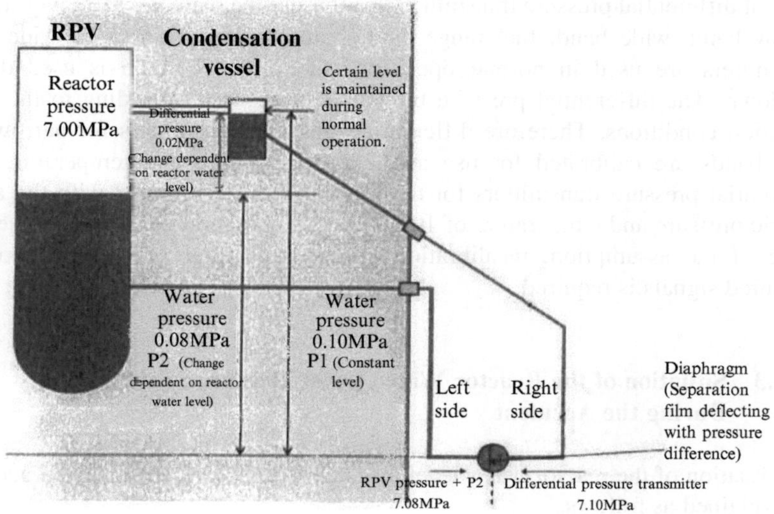

Fig. 6.21 Reactor water level measurement (Created based on TEPCO data)

differential pressure changes. The change in deflection is converted to an electric signal in the semiconductor of the transmitter, and the output is indicated in the central control room.

The differential pressure transmitter connects to a reference water level container called the condensation vessel that connects to the RPV, and the tap of the reactor near the bottom of the reactor fuel (Fig. 6.21). The pressures are transmitted through these instrument pipes to the differential pressure transmitter, which detects a difference in pressures of these pipes. The pressure from the reference water level container is the sum of the reactor pressure and the hydraulic head pressure from the container to the differential pressure transmitter. The pressure from the reactor side tap is the sum of the reactor pressure and the hydraulic head pressure of the reactor water level. The reactor pressure is added to both pressures, namely the differential pressure is the difference between the hydraulic head pressures from the container water level to the transmitter, and the hydraulic head pressure of reactor water level.

The reference water level container always contains the steam from the reactor, which is condensed due to the difference in temperatures between the container and reactor. The condensation is continuously carried out during plant normal operation, and the water level of the container is maintained, thereby its hydraulic head pressure is also maintained constantly. The pressure from the reactor side tap changes with the change in the reactor water level, therefore the change in differential pressure transmitter shows the change of the reactor water level.

The measuring conditions of the reactor water level, including the range of measurement, temperature, and pressure, vary depending on the plant operation conditions, except for the measurement using pressure difference. There are several kinds of differential pressure transmitters to measure the reactor water level such as narrow band, wide band, fuel range, and shutdown, etc. Narrow or wide band transmitters are used in normal operation, and the fuel range is used during shutdown. The differential pressure transmitters are used according to the plant operation conditions. Therefore differential pressure transmitters for narrow and wide bands are calibrated for use at the rated pressure and temperature, and differential pressure transmitters for the fuel range are calibrated with the atmospheric pressure and temperature of 100 °C. If the operation conditions exceed the scope of the assumption, recalibration of the transmitter or correction of the measured signals is required.

6.5.2.3 Situation of the Reactor Water Level Measurement During the Accident

The situation of the reactor water level measurement at Unit 1 during the accident are explained as follows.

(1) **Relations with the IC system of Unit 1**

When DC power was lost at around 15:40 on March 11 after the tsunami struck, the reactor water level instrumentation system in the central control room was disabled, and the operation condition of the IC system, the only reactor heat removal system at that time in Unit 1, became not clear. The station blackout disabled the operators to activate the IC system.

At around 16:42, the reactor water level indicator was restored for unknown reasons and −90 cm was indicated in the wide band. After −150 cm was indicated in the wide band at around 16:56, the reactor water level instrumentation was lost again. Based on this trend, the engineering group of the Nuclear Emergency Response Headquarters estimated the water level would reach the top of active fuel (TAF) at around 18:15.

At around 21:19, the reactor water level indicator was restored by the temporary battery, and the on-site and Tokyo headquarters both received the data of TAF + 200 mm ((A) system in fuel range) for the reactor water level of Unit 1.

Based on this background, the on-site and Tokyo headquarters continued to misunderstand that the IC system was in operation. The report at around 21:19 made the correction of miscomprehension more difficult.

If the reactor water level indicator had functioned normally, deactivation of the IC system after the tsunami could have been detected much earlier, and the handling of Unit 1 could have been different.

The subsequent analysis suggests that core damage began before 19:00, and the reactor water level went below the BAF at around 21:19.

(2) **Incorrect indication of reactor water level indicator**

At 21:19 on March 11, the reactor water level indicator was restored by the temporary battery, and the water level gradually increased and reached the

TAF + 590 mm at 23:24 ((A) system in fuel range). The rise in the water level at this time was unlikely as external water injection into the reactor had not begun yet. The reactor water level indicator ((A) system in fuel range) continued to indicate the TAF + 1,300 mm from 0:00 to 6:30 on March 12, then gradually decreased, reached the TAF – 1,700 mm at 12:35 on March 12, and stopped thereafter.

6.5.2.4 Potential Cause of Incorrect Indication During the Accident

The cause of three events in Sect. 6.5.2.3(2) are estimated according to the principles of measurement of the differential pressure instrumentation as follows.

The water level of the reference water level container is maintained to a certain level during normal operation. When the PCV inside temperature exceeds the saturation temperature during a severe accident, the water in the reference water level container evaporates, and the hydraulic head pressure of container decreases. This raises the pressure of reactor tap side relatively high, and a signal for gradual increases in the water level is generated (events from 21:19 to 23:24, March 11).

The value was fixed to the TAF + 1,300 mm potentially because the PCV inside temperature exceeded the saturation temperature, and the water was lost from the instrument pipeing of reference water level container (events from 0:00 to 6:30, March 12).

When the reactor water level drops below the bottom of active fuel (BAF) while water injection into the reactor is continued, water is not supplied to the reactor tap side piping from the reactor. Water in the instrument piping at the reactor tap side evaporates when the PCV inside temperature exceeds the saturation temperature due to a severe accident. Accordingly, the water in the instrument piping in both the reference water level container side and reactor tap side is lost, and the differential pressure becomes constant (the event after 12:35, March 12). TEPCO has supplied water in the instrument piping twice, once at Unit 1 and once at Unit 2, since the accident up until now. The evaporation of water in the instrument piping was confirmed at these times.

The final report of the Government Accident Investigation Committee described "simulation test is required particularly for the incorrect indication of the water level indicator", but it may be unnecessary because of the above evaluation.

6.5.2.5 Issues and Measures in the Future

Based on the events and issues identified in this accident, the investigation and development to allow the measurement of the reactor water level during a severe accident are required. The reliability of levels 1 to 3 in defence in depth is not necessarily achieved during a severe accident, but reliable instrumentation is required to understand the plant condition. The critical instruments such as the

Table 6.19 Events and issues in reactor water level instrumentation system

Event	Issue	Remark
Unable to measure due to the loss of power	More reliable power system	Necessity of power supply vehicle, battery charger
	Use of portable power supply	
	Use of portable instruments	
Incorrect measurement due to evaporation of water in the instrument piping	Prevention of evaporation of reference water	Setting of severe accident environment conditions
	Use of Water injection equipment	
Multiple measures for common causes as indicated above failed to work	Investigation of multiple measuring methods	Development of a measuring method other than differential pressure system
Other	Use of auxiliary water level indicators	

reactor water level indicator, must be provided with a means to confirm their credibility. It is also important to take measures for the assumption that the reactor water level is unable to be measured. Table 6.19 summarizes the events observed at the Fukushima Daiichi accident and issues, etc.

6.5.3 *Coolant Injection and Heat Removal Systems*

6.5.3.1 Details of Events

In the Fukushima Daiichi accident, the tsunami caused a station blackout at Units 1 and 2, and the emergency core cooling function at level 3 of defence in depth was lost. All DC power was spent after the tsunami struck at Unit 3, resulting in a station blackout, therefore Unit 3 became the same situation as those of Units 1 and 2. The tsunami also disabled the seawater pumps which transfer heat to the ultimate heat sink, the ocean. The core decay heat removal function up to level 3 was, therefore lost. In addition to the safety systems, the isolation condenser (IC) at Unit 1, which had been expected to operate after the tsunami, was also disabled. At Unit 2, the reactor core isolation cooling (RCIC) system operated for about 3 days to supply cooling water to the reactor without control DC power. At Unit 3, DC power was maintained after the tsunami, enabling the RCIC/HPCI systems to operate to supply cooling water to the reactor until DC power was used up.

Consequently, Unit 1, which lost all cooling systems after the tsunami attacked, came to core damage first, presumably before 19:00 on March 11. Core damage began at Units 2 and 3 several hours later the injection of cooling water stopped, probably at around 19:20 on March 14, and at around 10:40 on March 13, respectively.

The implementation of various countermeasures for the accident management had been completed in May 2002 at the Fukushima Daiichi NPS. The cooling water injection and PCV heat removal systems implemented were as follows:

(1) **Cooling water injection**
The fire protection system was connected to the ECCS piping through the makeup water condensate (MUW) system to make them both available for cooling water injection.

(2) **PCV heat removal**
The reinforced vent line between the PCV and exhaust stack was installed to prevent PCV from overpressure (it also operates to remove the heat of PCV).

The water injection line from the fire protection system was used to inject the water to the reactor during the accident, and helped prevent the further escalation of a severe accident. However, these accident management measures ultimately did not work well, and could not prevent core damage.

6.5.3.2 Analysis of Events

(1) **Cooling water injection**
The reasons why the above mentioned accident management measures failed to function are summarized below.

- The cooling water injection systems implemented at the Fukushima Daiichi did not work when the reactor pressure was high, and the main steam safety-relief valve (SRV) had to be operated to depressurize the reactor. DC power and compressed air were required to operate the SRV, but preparations took time at the Fukushima Daiichi resulting in the delay of water injection.
- In reality, fire engines were used for water injection. The fire protection and MUW systems could not be used as initially planned because of malfunction and the lack of power supply.

(2) **PCV heat removal**
Similarly, the accident management measures and facilities could not be used for the following reasons.

- The ultimate opening of air operated valve is required to construct a PCV vent line, which requires DC power and compressed air as the SRV. The preparations to meet these requirements took time, and some valves were installed in locations difficult to operate.
- The PCV vent line was equipped with a rupture disk to ensure seal performance in normal plant operation. The pressure to rupture this disk was designed to exceed the maximum operating pressure of the PCV, and it was impossible to start venting at the pressure under the maximum operating pressure.

6.5.3.3 Countermeasures

(1) **Accident management**

The new regulatory standards present several examples of sequences leading to core damage. Facilities to resolve these sequences are required. The accident at the Fukushima Daiichi is one of these accident sequences. In addition to those sequences which are contained in new regulatory standards, there are sequences to be considered continuously to minimize the risk which may otherwise develop to a significant environmental contamination.

According to the defence in depth philosophy, various means need to be provided in level 4 for coping with severe accidents. When a severe accident is identified, the person responsible for the accident management on site must make use of every available facility to settle the accident (management). Prior discussion and on-site training are crucial to find problems and solutions.

(2) **Cooling water injection facilities to avoid accidents**

The event that took place at the Fukushima Daiichi was as follows: the reactors isolated, and nuclear reaction in the reactor core stopped, but the continuous generation of decay heat continued. If the facilities for levels 1 to 3 of defence in depth were used to control this event, the procedures to settle the event are as follows under high and low reactor pressure conditions (corresponding to Units 2 and 3).

(a) High pressure condition: The SRVs maintain the reactor pressure and the RCIC or HPCI system injects reactor cooling water.

(b) Low pressure condition: The SRVs are kept opened to depressurize the reactor, and the LPCI or CS system injects reactor cooling water.

In the conventional AM, various means of water injection were provided, including the control rod drive water pressure system, seawater pumps, MUW system, and fire pumps. Under the station blackout, however, only diesel driven fire pumps were available. The power loss at the Fukushima Daini was not as serious as that at the Fukushima Daiichi, AM measures worked well and cold shutdown was achieved at all reactor units, unlike the Fukushima Daiichi.

The cooling water injection system required in the sequence of this accident is either a high-pressure injection system or a combination of SRVs for depressurization and a low-pressure injection system, as described above. Even if cooling at a high pressure is stabilized, the high-pressure system must be switched to low-pressure system with depressurization at some point for cold shutdown. This means various facilities are required particularly for depressurizing and low pressure water injection system. For example, it may be effective to install auxiliary, permanent or portable equipment for DC power and compressed air to operate the SRVs for depressurization as level 4 equipment.

(3) **PCV heat removal facilities based on the accident**

The PCV vent line installed was an independent facility as level 4, and if this line was activated before leaks from the PCV, significant environmental contamination could be avoided. Improvements are required for use in a severe accident. The new regulatory standards require the installation of a filtered vent or the equivalent.

The PCV vent line reduces the PCV and reactor only to 100 °C. Additional heat removal equipment is required for the cold shutdown of the reactor. The use of the RHR system is most effective, but the restoration of the RHR system in the sequence of this accident is difficult. An alternative heat removal equipment may be required.

The systems in levels 1 to 3 having the heat removal function are the residual heat removal (RHR) system (the containment cooling system in case of Unit 1) and seawater cooling system.

6.5.4 Importance of Management

6.5.4.1 History of Accident Management

The accident management defines the whole response (response to beyond-design-basis events) in the level 4 of defence in depth. In addition to the deterministic approaches to design basis events in levels 1 to 3, the approach to prevent those events from escalating to severe accidents by conducting the accident management that assumes the possibility of severe accidents has been considered common as the international standard. Discussion of the accident management assuming severe accident started in the 1990s, and measures according to global situations and required hardware have been developed. The Nuclear Safety Commission submitted a report summarizing the conditions and responses in 1992. The approaches in Japan at that time could be understandable, but once the accident happend, and the report was read back, discussions were not necessarily sufficient. For example, either filtered vent or pool scrubbing of suppression pool was designated for the vent of steam full of radioactive materials to prevent causing damage to the PCV. The pool scrubbing is valid only when the PCV vent line is linked, but whether sufficient consideration was given to system configuration is doubtful.

A serious problem is that this report has not been revised for over 20 years. While continuous improvements are the only means to ensure nuclear safety, practically no discussion about improving the accident management has been conducted in Japan. While the world discussed and continuously improved for 20 years, Japan did nothing but largely lagged behind the international standards.

Hence, catching up with international standards for critical matters and continuous improvement can be the lessons learned from our experience. The introduction of new regulatory standards in July 2013 based on the lessons learned from the Fukushima Daiichi accident is a large step forward to disseminate the accident management which used to be at the discretion of licensees. However, some argue that values stricter than the international standard should be used, but this does not always guarantee safety. A more comprehensive perspective may be important including the pros and cons of additional measures.

What to be noted in the new regulatory standards, including comprehensive risk assessment, organizations, systems, education and training is discussed below.

6.5.4.2 Defence in Depth and Design Concept

The concept of design corresponding to level 3 is considered correct even when the accident happened. The concept is to determine a design basis, which ensures safety with high reliability.

Considering Ss as design basis events for earthquake, and under these conditions, safety was ensured with sufficient allowance. The design basis events were set with fewer uncertainties based on the lessons learned from the Great Hanshin-Awaji Earthquake. However, because uncertainties in design basis events are significant, a significant safety margin was placed with simulation uncertainties in mind. Thanks to the effectiveness of this "concept," there was no serious damage to facilities due to the earthquake. No damage could not be declared when the on-site confirmation was impossible, but if there was, it could not be serious. However, we should not stop considering when the planned measures were successful. The insights obtained from this earthquake must lead to improvements as new insights. For example, activities like more active reduction of risks in which B and C class facilities may affect critical facilities, or reevaluation of taking allowance to target more safety are required.

Meanwhile, the design height of the tsunami was overly optimistic. The safety factor was insufficient in providing allowance, and common cause mode system failures were not discussed completely. That is, settings of the tsunami in the design basis were inappropriate in reflecting new knowledge or finfings and the allowance setting was problematic. However, the "concept" of designing events using deterministic approaches based on the design basis with the safety factor in mind is correct. It is an important indication that the design basis was determined based on insufficient knowledge. The design basis must be determined first. Even if the design basis is good, it cannot eliminate risks to zero. The thought, the stricter the design basis, the better, is an unsafe and unscientific mindset.

When, for example, the design basis event is selected so that the initiation probability is 1/10,000, deterministic approaches are required in designing the event to provide sufficient allowance.

The design basis events are not only to target natural disasters. For example, the LOCA (loss of coolant accident) and SGTR (steam generator tube rupture accident) are also design basis events. The deterministic approaches are also used to design these events. Moreover, allowance of 30 min was given as the design basis event for the station blackout (SBO). The problem was excessive focus placed to meet these design basis events, and the allowance for the occasion exceeding the assumption, and smooth link to the subsequent level 4 (accident management) could not be achieved efficiently.

Then, especially important is that the design basis which must take into consideration not only external events but also internal events, is needed to redefine with the initiation probability as an index in a rational way. Furthermore, assuming that the design basis is exceeded, design allowance must be reviewed in terms of accident management.

6.5.4.3 Severe Accident Response

The possibility of exceeding the design basis has been recognized. The accident management in level 4 for the excess of design basis has been considered accordingly, but obviously this was insufficient. The concept of accident management should be reviewed from the basics.

The accident management was provided for the station blackout for more than 30 min (design basis event) in the Fukushima Daiichi NPS, and responses for the SBO for 8 h were determined in the manual. This manual, however, assumed the use of DC power, and when this assumption was exceeded, or the situation worsened without DC power, none of the measures were useful.

Significant tsunami that exceeded the design basis attacked the Fukushima Daini NPS, causing the loss of ultimate heat sink (LUHS), but the predefined accident management worked and the reactors could be shut down safely. Of course, ingenious attempts were made on the site. Beyond design basis tsunamis were also assumed to some extent, and the critical facilities along the coast were equipped with watertightness measures. The heat exchanger building was provided independently for two reactor buildings in every four units. It was built on the site 4 m above sea level, which indicates almost no allowance for the design basis, hence each reactor building was equipped with watertight doors. The height of the tsunami that struck the Fukushima Daini NPS was about seven to eight meters, and almost all watertight doors could not withstand the water pressure, and broke. The watertight door of only the heat exchange building on the south side of Unit 3, at the center of the site, could prevent the building from being submerged. It was unclear whether this was due to weakened waves or the sufficient allowance of the watertight door, but switchgears and motors in this building were available after the tsunami. The decay heat continuously generated in the reactor core of Unit 3 could be removed using the residual heat removal cooling systems, etc. installed in the building and the reactor could be shut down safely. The watertight doors of the other seven buildings were destroyed, and switchgears and motors in these buildings were disabled, making it impossible to remove the decay heat in three units other than Unit 3. The predefined accident management was subsequently applied, but a number of events which were not covered by the manual took place. Flexible responses to these events were successful, and all reactors could be shut down safety. The predefined accident management was not necessarily complete, but the on-site operators covered the shortcomings. For example, temporarily procured power cables were laid out outside of the building to supply power to the heat exchanger buildings, failed motors were placed with motors procured from all over the country. These are good emergency management practices outside the plant and must be evaluated in more detail.

In conclusion, the events at the Fukushima Daiichi and Daini NPSs clearly showed that the accident management for the beyond design basis events had been planned but did not function so well. The shortcomings of the accident management should be improved to prevent the recurrence of such a severe accident. The beyond design basis assumptions cannot be controlled as the manual explains. It is, therefore, important to provide as many data as possible to determine appropriate actions according to the situation.

The critical perspective is that the accident management is not merely an extension of the design in levels 1 to 3. Using the same old method for evaluation would result in the same mistakes. Determining basic scenarios and design basis events and relevant measures is insufficient or wrong. It is unlikely that the things will proceed as assumed in the scenarios. The key perspective of defence in depth is the independent effects at each level (independent effectiveness). The management is discussed from different perspectives from the preceding three levels (safety design). Focus should be placed not only on the hardware but also software, i.e. management, particularly to enhance the management capability of the power plants and regulatory agency. Good management practice in the Fukushima Daini NPS must be evaluated in detail. It should be noted that there were also many good practices even at the Fukushima Daiichi NPS where the incomplete accident management could not prevent the accident, as well as other nuclear power plants. The accident management, having different effects from the preceding three levels, where the focus is placed on the hardware, must be discussed. That is the important perspective.

6.5.4.4 Improvements in Accident Management

Based on the idea described above, actions to enhance the management are discussed here. Although enhanced management capability of the regulatory agency is important, the management capability of the power plants is considered. The hardware is an important tool for supporting the management.

The optimized management and supporting hardware vary depending on power plants, but measures to mitigate risks and minimizing their scale of risks to an allowable level are required. We tend to think the risk is the risk assumed in the design including PRA, but should not forget that severe accident management (SAM) is a form of management. The reduction of comprehensive risks, including those for operation and maintenance is important. Every safety measure may bring new risks. The safety measures must be evaluated for their effectiveness to reduce the risks and the potential to bring new risks. Otherwise, accidents may occur due to the execution of SAM.

In the PRA, discussions are on the scenario basis. In actual severe accidents as shown in the responses at the Fukushima Daini NPS, however, flexible actions with the use of every possible materials and human resources are required. The organization must be familiar with details of the plant including valves and terminal boxes for taking suitable actions promptly. In the accident management, any situations including the beyond assumption, must be dealt with regardless of scenarios.

It is also important in the accident management in level 4 to consider an interaction with level 5 disaster prevention. The effects of levels 5 must be independent from those in level 4.

The important issues in launching the accident management are comprehensive risk prevention measures for the plant including operation and maintenance, actions independent of scenarios, including those for beyond assumption, understanding the plant, and the management using every possible resource. The top priority should never be given to the hardware. The management is centered for discussion with required hardware and software. When developping the accident management of level 4, the smooth shift from the design basis events in level 3 through level 4 to disaster prevention in level 5 should be taken into account.

6.5.4.5 Comprehensive Risk Management

The most suitable or effective measures to reduce the risks depend on the indivisual plant. A comprehensive consideration of the risks is required for taking measures. The requirement of comprehensive risk assessment was largely recognized after the 9.11 terrorist attacks in the U.S. For example, the parent would drive their children to the school when hearing the rumor of suspicious persons around the neighborhood. However, the probability of accidents of pedestrians is lower than the probability of car accidents, and the children in the car may die in a car crash. Of course, the risk in either case is very small. Measures for low risks must be discussed to make risk reduction comprehensive.

The nuclear power plant is a completed system. The system that effectively functions in a severe accident that exceeds design basis once every 10,000 years may largely raise the plant risks during normal operation which accounts for almost all operation in the life of the plant. The system may operate in the direction to escalate the severe accident depending on the situation. For example, the tsunami exceeding the tide wall could make drainage difficult, preventing accident management. The malfunctioning filtered vent may be heated by steam free from radioactive materials and when radioactive materials are actually released, it would lose their filtering capability. There are unlimited number of possible scenarios. When the hardware is centered on the management, the overall risks may be increased. The software centered management with supporting hardware must be implemented. The hardware based and software based concepts seem to be the same, but the content is totally different. In terms of software and software based management, there are unlimited number of systems and management better than the filtered vent.

When new hardware is added to the new plant, the design can be revised. Revision can be indefinite. As far as the new plant is concerned, the hardware-based design may be appropriate, while adding the hardware to the existing plant may incur considerable risk. Sufficient discussions are required before introducing the new hardware.

Software-based management should be considered to enhance management at the existing plant. The supporting hardware should be treated according to the comprehensive risk management.

6.5.4.6 The Ability to Assume Beyond Design Basis Assumptions

The tsunami struck the Fukushima plants was just out of the scope of assumption at that time. The SAM provided for beyond design basis assumptions contained this type of event to some extent. The measures taken at the Fukushima Daiichi and Daini NPSs were successful or failed, but the management capability of the plants was maximized. It is one of the most important lessons to assume beyond design basis assumptions for the similar accident in future. Discussions based on various scenarios are critical but the work is similar to the design in level 3.

Human perceptions on natural disasters have limitations. It is impossible to assume all beyond design basis conditions. We must hone our experiences and enhance support capabilities, including the development of potential scenarios, to overcome risk under any condition.

The severe accident management standard[14] of the AESJ requires consideration of all events, including those with small probability. For example, direct impact of meteorite falls and cyber terrorism are included. Specifically, brainstorming on these matters in the plant is required as part of education and training. When beyond design basis assumptions are identified in advance, real beyond-design-basis events can be suitably handled. Repeated education and training are the only way to respond to beyond design basis assumptions.

Limited human knowledge should be recognized. Repeated simulation of severe events, and repeated discussions of events seemingly with no way out may produce new ideas and improves capabilities to respond to severe accidents.

6.5.4.7 Enhancement of Management Capability

As described above, measures that reflected emergency experience successfully prevented the escalation of the accident at the Fukushima Daini NPS. In the Fukushima Daiichi accident, however, significant releases of radioactive materials could not be prevented. The management at the plant must be enhanced to respond to any situation. Repeated education and training to have trainees experience

[14] AESJ standard "Development, maintaining and improvement of severe accident management in nuclear power plants:2014" (Note: The accident management and severe accident management are used as the same term here.)

various situations, and continuous improvements of required hardware are the only ways to make management more effective. In particular, experience of failures is valuable. Experience of failures would improve capabilities in actual operation. The workshop that does not allow failures can't achieve improvement.

Certain risks may be temporarily increased to reduce overall risks. For example, online maintenance would increase the plant risks temporarily, but once it is complete, the plant reliability is enhanced and the risks are reduced. That is, temporary increases in a small number of acceptable risks may enhance the reliability of the entire plant, and reduce the overall risks. This is not merely the index of plant reliability, but leads to the increased plant management capability. It is a method serving a dual purpose of risk management and increases in reliability and management capability, and there are many such methods except for online maintenance. These methods should be actively used to enhance the management capability.

As mentioned above, beyond design basis assumptions must be assumed. Capacity for imagination has limitations. For the situation out of scope of the assumption, plant organization, human resources courage and experience decide everything. Including a number of failures, accumulation of experience which is the implicit knowledge is crucial. Experience and correct judgment are improved on the premise of out of the scope of the assumption. A plural number of diversifying materials and facilities should be provided. In "Apollo 13", the astronauts got through the risk using their experience and ideas as well as those of the ground staff in spite of limited tools in a limited space of a spaceship. This is an example of management.

As stated above, the clear intention of continuous improvements in management capability in the nuclear plant is important. There is no assessment limit for enhancing the management capability. Continuous improvements should always be pursued. It is also necessary for the third party organization such as international agencies to verity the continuous improvements in the management capability of licensees and regulatory agency.

6.5.5 Multiple Reactors in the Same Site

The Fukushima Daiichi NPS was flooded by the beyond design basis tsunami following the magnitude 9 class Great East Japan Earthquake on March 11, 2011. All area containing plant buildings were affected by the wave. The critical safety facilities were disabled, and the equipment required for accident response was affected. These plants simultaneously underwent "the loss of all AC and DC power for an extended period" and "the loss of emergency seawater heat removal system for an extended period." Almost all auxiliary facilities, essential for smooth accident response, such as monitoring systems, lighting and communication means, were lost. While the state of these plants was deteriorating, obstacles for work largely increased. There was also a shortage of backup members who support accident responses to multiple reactors for days at the beginning of the accident, and the roles and responsibility of the control room, emergency support center at the site, emergency support center in the main office, and the government agencies were unclear. They failed to function as planned in advance.

As a result, stable cooling of reactors failed, and core damage (severe accident) occurred in Units 1 to 3, which led to the hydrogen explosions inside the reactor building at Units 1, 3 and 4, and significant release of radioactive materials. Many residents had to evacuate for an extended period. This status has continued to date.

The earthquake and tsunami also struck the neighboring Fukushima Daini NPS, and generated serious consequences, both human lives and assets, all over the Pacific coast of the Tohoku region. The central government and local municipalities faced unprecedented disasters. Communications were disrupted, and confusion prevailed. The off-site center, which was supposed to have been the center for accident response, did not work as expected due to the disrupted communications and delayed or non-existent responses to the problems. Failures of monitoring equipment hindered radiation measurement. The fundamental infrastructure to respond to nuclear hazards was seriously damaged.

The lessons learned from the Fukushima Daiichi accident for the multiple reactors constructed at the same site are pointed out. Discussions are based not only on new regulatory standards but also on the broader perspective including the hardware and software.

6.5.5.1 Accident Assumptions

- The licensee, TEPCO, did not assume the conditions leading to severe accidents in multiple reactors simultaneously, and did not provide the personnel, facilities, training and procedures required for these accidents. The beyond design basis tsunami caused accidents at multiple units simultaneously, and the progression of the accident at a plant affected the emergency responses of the neighboring plant. Responses to an assumption of serious accidents at multiple units are required to improve safety measures in nuclear power plants.
- The government and municipalities did not assume a compound disaster in which nuclear accident and natural disaster developed at the same time, and did not take disaster prevention measures for these situations. The off-site center was built at the wrong site and provided with wrong facilities according to insufficient countermeasures for damage to roads or disruption of communications due to earthquakes. It was immediately disabled by the compound disaster. The insufficiency of Japan's emergency management to maintain the safety of nuclear power plants, and surrounding communities was revealed. A large-scale compound disaster must be taken into consideration to improve the safety measures of nuclear power plants.
- The emergency response and restoration have features depending on accidents at nuclear power plants and natural disasters. Assumptions on these events and corresponding responses are required first of all. In addition, assuming the potential for the occurrence of nuclear and natural disasters, or multiple natural

disasters at the same time or during the restoration from one disaster, safety measures for reactors and disaster prevention of surrounding areas are required.

6.5.5.2 Facilities and materials

- Assuming severe accidents at multiple units, multiplexing and diversification of locations of facilities such as emergency D/Gs and switchgears, and ensuring their independency are required to be ready for the loss of DC power. Storage of necessary facilities and materials is also required.
  The conditions where support from other unit is not expected on the occasion of accidents at multiple units, and facilities and materials are used simultaneously must also be taken into consideration. For significant natural disasters, damage to the social infrastructure outside the plant, and the difficulties in restoration of external power or supply of materials for an extended period should also be discussed.
  Required facilities must be provided or reinforced based on new regulatory standards and comprehensive risk assessment.
- Communications between the central control room and emergency response center, or emergency response center and off-site center, etc. are important for the stakeholders to understand the plant conditions and take necessary measures. Conventional telephones, cellphones, satellite phones, walkie-talkies and pagers may be used for communications.
  Most of these devices are battery driven. Batteries for long-term operation must be reserved. The use of conventional telephones and cell phones may be limited when the infrastructure outside the site (e.g. switchboard or base stations for cell phones) is damaged or destroyed. Walkie-talkie may not be available when the power of transponder is lost.

6.5.5.3 Procedure Manual

- The procedure manual should include the measures for the loss of all AC power at multiple units after the scram and continuation of this situation for days. Assumptions should include damage to the social infrastructure and delayed external support.
- Radiation evaluation software needs to be capable of predicting the dose of releases from multiple reactors and spent fuel pools to take preventive measures promptly. The criteria for measures must be clarified.

6.5.5.4 Education and training

- Sufficient education and training should be provided on the assumption of severe conditions such as the loss of all AC power at multiple units for an extended period. The following training may be useful to improve the capability of emergency personnel:

- Source of information (SPDS, etc.), equipment, and facilities are not always available. Partial omission of them would be useful. Transfer of incorrect information may test the judging capability and give higher training for emergency personnel.
- In emergency drill, the line of command, decision making, prioritization and disaster countermeasures are provided to understand the framework of disaster control, and practice the measures. For the accidents at, external supports are important. Required off-site equipment and support, and means to multiple units obtain them should be included in the training.
- Specific programs are required for the engineering investigation team of emergency response centers to learn actual and expected plant responses and predict the accident progression.
- For the events involving multiple reactors for an extended period, the strategy and infrastructure for receiving, integrating and sharing an enormous amount of information are required. The procedure manual, method of integrating data and communication protocol must be provided and used for training.

6.5.5.5 Personnel and reinforcement of organization

- For the accidents at multiple units, the persons in charge must confront stressful situations for an extended period. Enough personnel should be procured at the control room, emergency response center at the site, and emergency response center in the main office. The line of command and personnel, and a base for actions for emergency responses are required. The person responsible for the emergency response center should investigate necessary measures according to the plant conditions, and determine priority immediately.
- A system for long-term accident response is required. For seamless responses, the persons in charge, including the supervisor who decides the operation must be available 24 h a day for an extended period. Minimum requirements (closing, food and housing) must be provided for long-term accident response.
- A person responsible for answering external inquiries must be selected in the work team for complicated events to minimize the confusion of the team leader, and concentrate on the supervision of operators as well as ensuring timely and accurate flow of information.

6.6 External Events

One of the key factors that caused the severe accident at the Fukushima Daiichi NPS was insufficient provision in design for natural hazards, particularly, tsunami events. A comprehensive risk assessment, design basis settings conforming to required performance targets must be provided for managing external events (natural hazards), together with a design incorporating safety margin and defence in depth in the management of beyond design basis external events. This section explains the approach in managing external events, and defence in depth, the core element in responding to external events is addressed separately in Sect. 6.3.

6.6.1 Seismic Hazard Management

6.6.1.1 Integrity of Nuclear Power Plant Facilities by Seismic Ground Motion

The reference seismic ground motion used as seismic design basis, and assumptions on maximum tsunami height have been formulated as results of discussions in the academia and professional societies, and established on the accord of parties involved in all regulatory aspects, the academic experts and engineers, and applied in safety assessments. However, the actual seismic ground motion and maximum tsunami triggered by the earthquake observed were far beyond these values.

The magnitude of the seismic ground motion that occurred has been presented through investigations conducted. With the occurrence of the earthquake, all control rods were inserted into the 12 reactors in operation on the Pacific coast without incident and reached a cold shutdown.

Although seismic ground motion at Fukushima Daiichi NPS Units 2 and 3, and Onagawa NPS Units 1 to 3 partially exceeded the design basis, no abnormal conditions in measurement data or visible damage to critical equipment were observed.

With view to the assessment on seismic resistance design and breach of structural integrity at nuclear power plants by seismic ground motion, sufficient margin provided against design basis ground motion for TEPCO's Kashiwazaki Kariwa NPS was verified during the Chuetsu Offshore Earthquake; as well, the plant facilities of Onagawa NPS, located closest to the hypocenter maintained integrity during the Great East Japan Earthquake. There were no plant measurement data that indicated safety-critical "behavior" at the Fukushima Daiichi and Daini NPSs. Tables 6.20, 6.21, and 6.22 shows seismic response values of the Fukushima Daiichi, Fukushima Daini, Tokai Daini and Onagawa NPSs.

The Diet Accident Investigation Committee suggested that while no direct evidences of damage to piping were found, the "possibility cannot be denied". The Government investigation committee negated damage to the piping. The public hearing held by the Nuclear and Industrial Safety Agency concluded that there was no significant piping damage affecting safety functions of the plant based on analysis results, etc.

Magnitude 9 Great East Japan Earthquake induced by the interlock of an area of 450 km in length and 200 km in width largely exceeded the assumptions

Table 6.20 Seismic response at Fukushima Daiichi and Daini NPSs

Monitoring point	Records (Maximum acceleration value)			Maximum acceleration response for design basis seismic ground motion Ss		
	S–N direction	E–W direction	Ver. direction	S–N direction	E–W direction	Ver. direction
Fukushima Daiichi						
Unit 1	460[a]	447[a]	258[a]	487	489	412
Unit 2	348[a]	550[a]	302[a]	441	438	420
Unit 3	322[a]	507[a]	231[a]	449	441	429
Unit 4	281[a]	319[a]	200[a]	447	445	422
Unit 5	311[a]	548[a]	256[a]	452	452	427
Unit 6	298[a]	444[a]	244	445	448	415
Fukushima Daini						
Unit 1	254	230[a]	305	434	434	512
Unit 2	243[a]	196[a]	232[a]	428	429	504
Unit 3	277[a]	216[a]	208[a]	428	430	504
Unit 4	210[a]	205[a]	288[a]	415	415	504

Comparison of records (observed) and response values for design basis seismic ground motion Ss (unit: Gal)
Note: The monitoring point is the base mat of the reactor building
[a]Recording period of approximately 130–150 seconds

Table 6.21 Seismic response at Tokai Daini NPS (Maximum acceleration of reactor building (unit: Gal))

	Records			Design basis seismic ground motion		
	S–N	E–W	Ver.	S–N	E–W	Ver.
6th fl	492	481	358	799	789	575
4th fl.	301	361	259	658	672	528
2nd fl.	225	306	212	544	546	478
B2 fl.	214	225	189	393	400	456

Note: Design basis seismic motion: Maximum response acceleration on each floor of the reactor building based on design basis seismic ground motion Ss (600 Gal) set by (Free surface of the base stratum (altitude (E.L.)—370 m)

of earthquake experts. Magnitude 6 upper was recorded in Okuma town and Futaba town of Fukushima prefecture, where Fukushima Daiichi NPS is situated. Maximum acceleration, presenting intensity of a seismic ground motion, is recorded on the seismometer at the base mat of the reactor building. The seismometers at Units 2, 3 and 5 of the Fukushima Daiichi NPS recorded 550, 507 and 548 Gal (cm/s^2), respectively, all of which exceeded the anticipated maximum response acceleration values of 438, 441 and 452 Gal in design basis seismic ground motion Ss assessed in seismic resistance evaluation.

Table 6.22 Seismic response at Onagawa NPS

Monitoring point	Records: Maximum acceleration (Gal) S–N direction	E–W direction	Ver. direction	Maximum response acceleration (Gal) to design basis seismic ground motion Ss S–N direction	E–W direction	Ver. direction
Unit 1						
Rooftop	2,000[a]	1,636	1,389	2,202	2,200	1,388
Refueling fl (5th fl)	1,303	998	1,183	1,281	1,443	1,061
1st fl.	573	574	510	660	717	527
Base mat	540	587	439	532	529	451
Unit 2						
Rooftop	1,755	1,617	1,093	3,023	2,634	1,091
Refueling fl (3rd fl)	1,270	830	743	1,220	1,110	968
1st fl.	605	569	330	724	658	768
Base mat	607	461	389	594	572	490
Unit 3						
Rooftop	1,868	1,578	1,004	2,258	2,342	1,064
Refueling fl (3rd fl)	956	917	888	1,201	1,200	938
1st fl.	657	692	547	792	872	777
Base mat	573	458	321	512	497	476

Note: Maximum value of each monitoring point if multiple monitoring points are present in the horizontal and vertical directions
[a]Reference value because it exceeds the maximum upper limit (2000 Gal) of the seismometer

In the simulation analysis using seismic ground motion recorded on seismometers, the seismic load impact on critical safety systems, equipment, piping and structure related to reactor shutdown, core cooling and isolation of radioactive materials at Units 1 to 3 (in operation), and Units 4 to 6 (shut down) turned out to be significantly lower than the seismic evaluation standard (allowable stress, etc.) showing a considerable margin. The results of assessment on accident sequence include: operation data of each reactor unit after the earthquake occurrence showed no abnormalities in safety; structural strength assessment on key parts of the SSCs by vibration response observed during the earthquake at Kashiwazaki Kariwa NPS showed sufficient margin in capacity against design; and similar results are assumed for Fukushima Daiichi NPS; and no damages to safety-critical function were found by the plant walk-down at Unit 5, where the maximum acceleration values exceeded the design basis.

Sufficient margin on seismic ground motion had prevented damages with serious consequences on safety functions. However, because identifying minor leaks and damages not evidenced on plant parameters is difficult, onsite survey should be conducted to the extent possible to examine and clarify damages on critical safety components.

6.6.1.2 Significance of Design Basis Seismic Ground Motion

The important perspective is that the observed seismic ground motion exceeded the design basis designated in seismic resistance design. Although Onagawa NPS and Kashiwazaki Kariwa NPS had experienced beyond design basis events a number of times in the past, the SSCs maintained integrity, with no abnormal occurrences as shown in relevant reports.

Design seismic standard was tightened consecutively with the revision of the Regulatory Guide for Reviewing Seismic Design of Nuclear Power Reactor Facilities the year (2006) preceding the Chuetsu Offshore Earthquake, and after, due to damages to the Kashiwazaki Kariwa NPS by the earthquake. The new standard was reflected on the back-check requirements on all NPS in Japan to enhance seismic resistance and ensure integrity.

Around the same period, the adequacy of seismic assessment method based on response acceleration was in dispute, with a controversy over the use of a more appropriate method such as those based on velocity or energy rate in assessing "rupture". The debate had continued with no conclusion drawn until the 3.11 incident. No one had seriously considered the significance of, nor response to seismic ground motion exceeding design basis. This section addresses the issue because seismic ground motion exceeded design basis at both Onagawa NPS and Fukushima Daiichi NPS on 3.11, and beyond design basis conditions falls in the region of accident management. This is linked to important decision making on accident management preparedness and response, and calls for the re-evaluation of seismic design basis criteria.

6.6.1.3 Assessment on Structural Integrity and Seismic Ground Motion

Structural integrity of SSCs by seismic ground motion is assessed by comparing results of assessment on seismic ground motion of Great East Japan Earthquake, response assessment on Kashiwazaki Kariwa NPS by Chuetsu Offshore Earthquake and onsite surveys conducted on the sites. "Measures taken by the Nuclear and Industrial Safety Agency concerning the Kashiwazaki Kariwa NPS during and following the Chuetsu Offshore Earthquake in Niigata Prefecture (3rd interim report)" (Nuclear and Industrial Safety Agency) will be used as a reference source.

Table 6.23 shows seismic response values of the reactor units of Kashwazaki Kariwa NPS and design basis seismic ground motion in parentheses. The response exceeded seismic design values by about 50 % at all units, with the maximum value of more than three times. Table 6.24 compares the acceleration values of seismic ground motion on the free bedrock against the design basis. The design basis on the free bedrock is 450 Gal. The seismic ground motion of 1.5–2.0 times the design basis was estimated. It was reported that reevaluation based on this estimate confirmed seismic safety of key facilities at each reactor unit. Figure 6.22 shows

Table 6.23 Maximum acceleration at Kashiwazaki Kariwa NPS during Chuetsu Offshore Earthquake (Recorded on the base mat of the reactor building) (Unit: Gal)

	S–N direction	E–W direction	U–D direction
Unit 1	311 (274)	680 (273)	408 (235)
Unit 2	304 (167)	606 (167)	282 (235)
Unit 3	308 (192)	384 (193)	311 (235)
Unit 4	310 (193)	492 (194)	337 (235)
Unit 5	277 (249)	442 (254)	205 (235)
Unit 6	271 (263)	322 (263)	488 (235)
Unit 7	267 (263)	356 (263)	355 (235)

Note: Numbers in parentheses indicate the maximum acceleration
"Measures taken by the Nuclear and Industrial Safety Agency concerning the Kashiwazaki Kariwa NPS during and following the Chuetsu Offshore Earthquake in Niigata Prefecture (3rd interim report)"

Table 6.24 Seismic ground motion assessment at Kashiwazaki Kariwa NPS reactor Units (horizontal direction (upper) and vertical direction (lower))

Seismic ground motion (Gal)	Unit 1	Unit 2	Unit 3	Unit 4	Unit 5	Unit 6	Unit 7
Chuetsu Offshore Earthquake (location of seismometer: base mat of reactor unit)	680	606	384	492	442	322	356
	408	282	311	337	205	488	355
Response to design basis seismic ground motion (base mat of reactor unit)[a]	845	809	761	704	606	728	740
	–	–	–	–	–	775	775
Maximum design basis seismic ground motion (on free bedrock surface)[b]	2,300				1,209		
	1,050				650		
Chuetsu Offshore Earthquake (estimated values on free bedrock surface)	1,699	1,011	1,113	1,478	766	539	613
	591	545	618	749	262	422	460

[a]Responses in the vertical direction of Units 1 to 5 were reported by TEPCO in future seismic safety assessment
[b]The design basis seismic ground motion at the time of reactor unit installation was 450 Gal
"Measures taken by the Nuclear and Industrial Safety Agency concerning the Kashiwazaki Kariwa NPS during and following the Chuetsu Offshore Earthquake in Niigata Prefecture (3rd interim report)"

a comparison of the results of the report by TEPCO and the evaluations by the Nuclear and Industrial Safety Agency. The maximum stress values of SSCs by the seismic ground motion was significantly lower than the design upper limit (indicated as design basis values). Figure 6.23 shows the relationship between the actual and design values, with the degree in the margin of each assessment technique to date clearly presented.

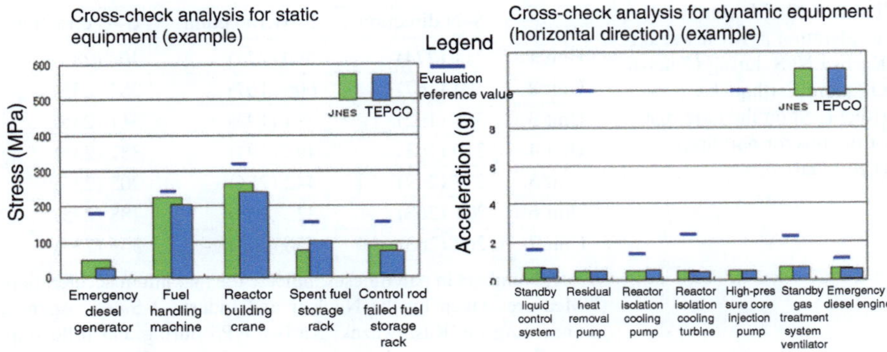

Fig. 6.22 Examples of cross-check analysis (Measures taken by the Nuclear and Industrial Safety Agency concerning the Kashiwazaki Kariwa NPS during and following the Chuetsu Offshore Earthquake in Niigata Prefecture (3rd interim report))

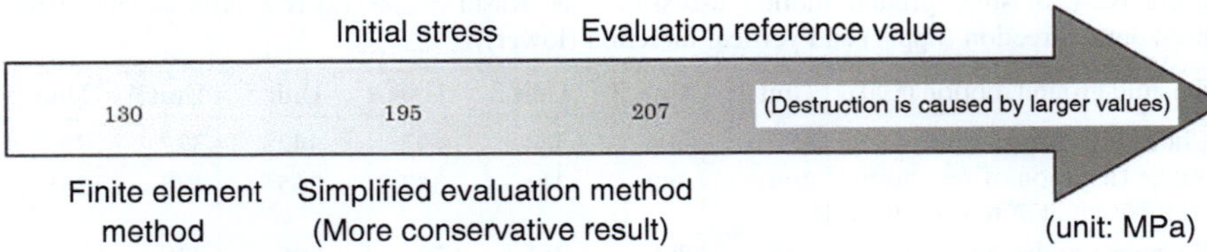

Fig. 6.23 Relationship between evaluation result and evaluation reference value (Measures taken by the Nuclear and Industrial Safety Agency concerning the Kashiwazaki Kariwa NPS during and following the Chuetsu Offshore Earthquake in Niigata Prefecture (3rd interim report))

6.6.1.4 RPV and PCV Integrity

RPV and PCV integrity of units 1 to 3 of the Fukushima Daiichi NPS has been confirmed through analysis on operation data during the accident.

(1) **Plant parameters**

Figure 6.24 shows the timeline of the parameters related to cooling such as water level and pressure, etc., of the reactor at the Fukushima Daiichi NPS Unit 1. According to the trip sequence record, the MSIV isolation signal was issued at 14 h. 47 min. 51 s. 730 ms.

The reactor pressure started increasing due to decay heat after the MSIV was isolated and peaked at 7.2 MPa, but rapidly decreased at 14:53 when the isolation condenser (IC) automatically started by detecting high reactor pressure (7.13 MPa for 15 s). The primary coolant was cooled at a temperature decreasing rate of 150 °C/h. The operators stopped the IC about 16 min after the scram to keep the primary coolant temperature changing rate within the upper limit of 55 °C/h, and continued to control reactor pressure via an IC system, system A. After pressure decrease to 4.2 MPa, it rebounded to increasing, and back and forth as the operators started and stopped the IC system A.

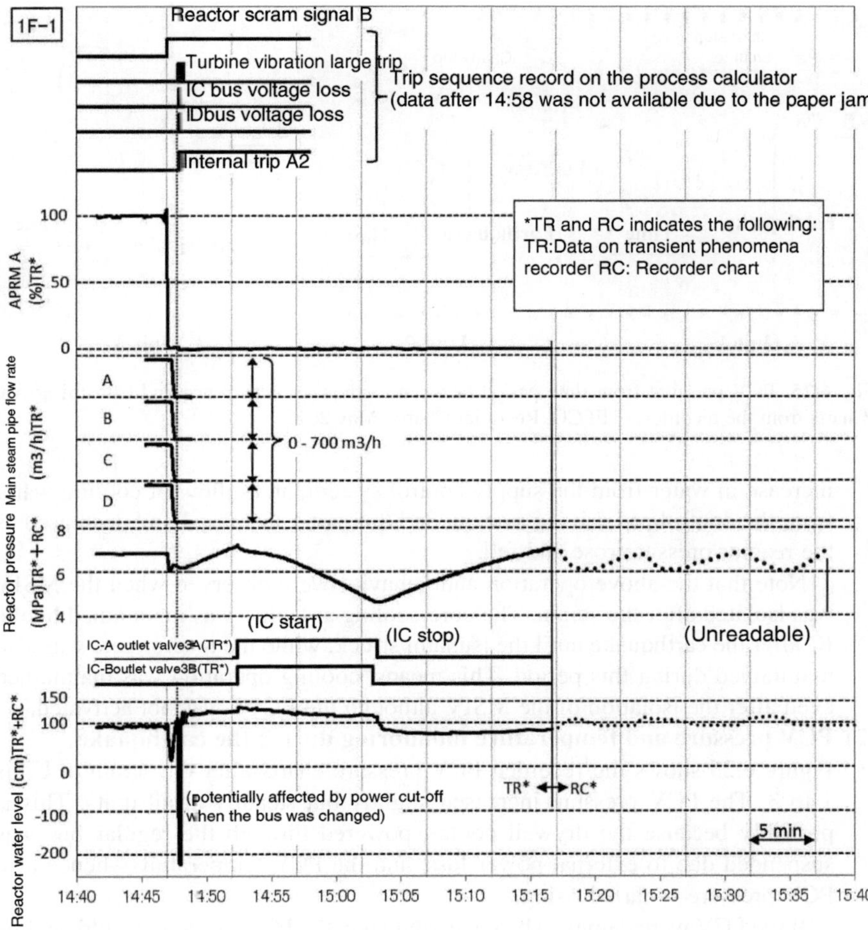

Fig. 6.24 Timeline data on parameters related to cooling—reactor water level, pressure, etc. at Unit 1, Fukushima Daiichi NPS (Kobayashi and Narabayashi [17])

The reactor coolant temperature changing rate of 55 °C/h is a value under ordinary start and stop conditions used in thermal fatigue assessment in RPV design, and a requirement in the safety regulations that operation must be kept within this reactor coolant temperature changing rate. The operators, therefore, stopped IC system B, and turned IC system A on and off according to regulations to prevent a rapid decrease in pressure until immediately before the tsunami struck. The impacts of the tsunami began to emerge at around 15:30, at which point the IC was stopped because the reactor pressure showed an increase immediately before the loss of signals due to the tsunami.

The reactor water level decreased temporarily because of the disappearance voids after the scram, but immediately recovered due to a temporarily rapid

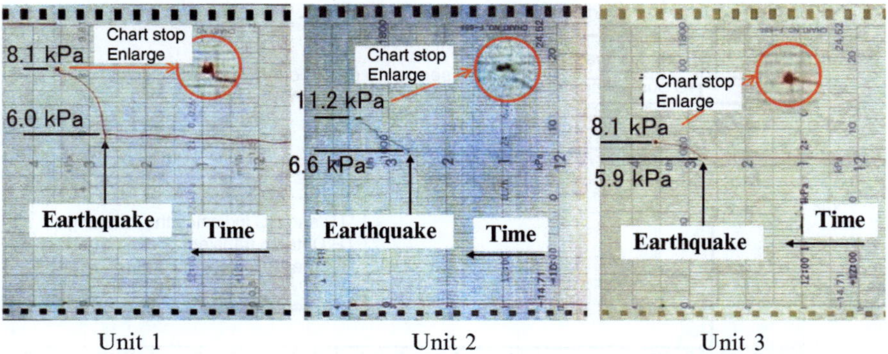

Fig. 6.25 PCV pressure from time period between earthquake occurrence and tsunami arrival (Charts from the recorder) (TEPCO, Recorder Charts (May 2011))

increase in water from the supply control system, and inflow of cooling water from the control rod driving system, and thereafter increased and decreased as the reactor pressure rose and fell.

Note that the above operation and behavior were observed when the MSIV was isolated after the scram. The core cooling at Unit 1 was maintained by the IC after the earthquake until the tsunami struck, while the reactor core was also maintained during this period. This means, cooling operation was maintained even after the isolation of the MSIV although the ECCS was not activated.

(2) **PCV pressure and temperature monitoring during the earthquake**

Figure 6.25 shows the recorded PCV pressure charts after the scram at Units 1 to 3. The PCV pressure increased slightly but steadily at all units. This is probably because the drywell cooler, powered through the regular bus was suspended due to external power loss, and the PCV temperature—hence also PCV pressure—started rising.

If the PCV were damaged by the earthquake, the PCV pressure would decline to atmospheric pressure, and no pressure increase as described above would be observed. The pressure increase indicates that the PCVs in each unit maintained integrity.

The charts output were suspended at all plants at around 15:40 due to the arrival of tsunami. The figure shows the enlarged part of charts at this time period. The chart output was stopped but the swollen signal lines indicates that the PCV pressure signal remained alive for a while, and no significant pressure increase was observed immediately after the tsunami.

The potential rise in pressure by temperature increase was examined.

The equation of state for the ideal gas was:

$$PV = nRT \tag{6.1}$$

Where: P is the pressure of the gas (Pa), V the volume (m^3) of the gas, n the amount of substance (number of moles), R the gas constant of a mole = 8.314472 (75) ($\text{Jmol}^{-1}\ \text{K}^{-1}$), and T is the absolute temperature of the gas (°K). Applying Eq. (6.1) to the PCV, and assuming the initial state (reactor scram) is suffix (0), and the state immediately before the tsunami arrival (chart output stopped (around 15:40)) is suffix Eq. (6.1),

$$\text{Before earthquake } P_0V_0 = nRT_0 \quad (6.2)$$

$$\text{Immediately before tsunami } P_1V_1 = nRT_1 \quad (6.3)$$

To ratios (2) and (3):

$$P_0V_0/P_1V_1 = T_0/T_1 \quad (6.4)$$

Hence:

$$P_1 = P_0(V_0/V_1)(T_1/T_0) \quad (6.5)$$

If the thermal expansion of the PCV volume for minimal temperature change is ignored:

$$P_1 = P_0(T_1/T_0) \quad (6.6)$$

Thus, the changes in PCV pressure can be estimated from the temperature change when the PCV confinement is maintained.

Table 6.25 shows an example of PCV pressure computed from the temperature change at Unit 1 of the Fukushima Daiichi NPS. The mean values of return air temperatures of the drywell cooler are used for the temperature change, and the calculation results are almost the same as the measured values. The same results were obtained from assessments on Units 2 and 3.

The other factors that caused the PCV pressure increase are high-pressure, high-temperature primary coolant leaks, and high-pressure nitrogen leaks, etc. In both cases, the pressure continues to rise, but the record did not show this

Table 6.25 Variances in PCV temperature and pressure of Unit 1, Fukushima Daiichi NPS

	Initial value (°C)	After tsunami (°C)
Return air Temp. DUCTHVH-12A	43.2	53.4
Return air Temp. DUCTHVH-12B	50.1	55.8
Return air Temp. DUCTHVH-12C	47.4	54.4
Return air Temp. DUCT HVH-12D	44.1	54.4
Return air Temp. DUCT HVH-12E	49.5	55.1
Mean value	46.9	54.6
PCV pressure (measured)	6.0 kPa (g)	8.1 kPa (g) (106.2 kPa (a))
PCV pressure (calculated)	–	8.5 kPa (g) (106.6 kPa (a))

Note: Initial value: during the scram, tsunami arrival: chart output stopped (around 15:40)
(g) is the gauge pressure, (a) is the absolute value

even after the chart output was suspended. Hence, it is assumed that these other factors did not give rise to the pressure increase. The findings from the analysis include:

(a) The reactor scram occurred with the seismic acceleration high signal, and all reactor units were shut down.
(b) External power was lost, but onsite power, or EDG was maintained until tsunami arrival.
(c) The MSIV was automatically isolated, which was caused by power failure of MSIV logical circuit, and not by conditions that required closure of the MSIV (e.g., main steam flow high, MS tunnel room temperature high, etc.).
(d) The PCV at each unit maintained integrity until tsunami inundated the site. The PCV pressure rose probably with temperature increase in the PCV after the drywell cooler was suspended.

The issues identified by the Investigation Committee on the Accident at the Fukushima Nuclear Power Stations of Tokyo Electric Power Company (the government), Independent Investigation Commission on the Fukushima Nuclear Accident (civilian), and Fukushima Nuclear Accident Analysis Report of the Tokyo Electric Power Company (Final Report of TEPCO) are discussed as follows.

(i) The government accident investigation committee (Report by the Investigation Committee on the Accident at Fukushima Nuclear Power Stations of Tokyo Electric Power Company) discussed damages to key SSCs including the RPV, PCV, isolation condenser (IC), reactor core isolation cooling system, and high-pressure core injection system. The committee claims that the shutdown function had been normal before the arrival of the tsunami, and there had been no damage to the confinement and cooling functions of key SSCs. Trip sequence output list, graphs on the transient phenomena recorder, and charts from the recorders were used in the analysis, however, they have not been discussed along the time axis.
(ii) The civilian accident investigation committee lists the conditions prior to tsunami arrival, such as the fact that the reactor had automatically shut down due to the earthquake and sub-criticality had been maintained, external power had been lost but power restored with the emergency diesel generator (EDG), and the MSIV had been isolated by the failsafe function due to the power loss, and so forth. There was only a minimal explanation that these conditions have been verified in sources such as the report to the IAEA, interim report of the government accident investigation committee, and material disclosed by TEPCO.
(iii) The final report of TEPCO evaluated Units 1 to 3 in two parts; automatic shutdown after the earthquake, and the operation from automatic shutdown until the tsunami arrival. The former presented the results of normal scram of all units following the earthquake, the loss of external power and restoration of power by EDGs, power failure of the reactor protection

system before the activation of EDGs, and the automatic shutdown of the MSIV, etc. The latter estimated there was no piping rupture on the grounds of a moderate rise in PCV temperature and the constant water level of the floor sump, but without quantitative evaluation. Comparison under the same time axis has also been omitted.

As shown in the reports by the accident investigation committees as well as by other sources, the "shutdown" and "containment" functions had been maintained at all units during the time period from earthquake occurrence until tsunami arrival, and no impacts by the earthquake were observed on the safety functions [17].

6.6.2 Tsunami Hazard Management

6.6.2.1 Awareness on Tsunami Hazards

The tsunami had been evaluated only as an earthquake-derived event in the Regulatory Guide for Reviewing Seismic Design of Nuclear Power Reactor Facilities. There had been little accumulated experiences on tsunami, and efforts to improve assessment methods had been neglected. Assessment methods for developing assumptions on maximum tsunami in nuclear power plant design had been examined mainly by the Japan Society of Civil Engineers, as well as by numerous other academic institutions and societies, and finally was re-established recently by incorporating state-of-the-art technologies such as computer science.

6.6.2.2 Tsunami Height Assumptions

The maximum tsunami that struck nuclear power plants on 3.11 was unprecedented and far beyond expectations. The earthquake that induced the massive tsunami was caused by a colossal crustal deformation exceeding assumptions. The unforeseen and complex crustal deformation had not been considered in tsunami assessments so far, which led to the damages at nuclear power stations and to the nuclear disaster of the Fukushima Daiichi NPS.

The tsunami scale exceeded design assumptions at all nuclear power stations, and even the subsequent revisions in some cases (see Table 6.26). Some of the sites, however, were designed with sufficient margin and maintained safety functions, escaping accident serious consequences. The tsunami scale observed at the Onagawa, Fukushima Daiichi and Fukushima Daini NPSs exceeded the original authorized design values and even the latest revisions. Onagawa NPS missed damage by tsunami height of 13 m by a small margin as its ground height was 13.8 m after ground subsidence by 1 m. At the Tokai Daini NPS, because the cooling facilities were operable in the area in which waterproofing bulkhead construction had just been completed, the reactor successfully reached a cold

Table 6.26 Design by tsunami scenario and observed tsunami heights at NPS

Site	Ground height of major buildings	Construction permit	History of tsunami height assumption after construction permit				
			Method of Japan Society of Civil Engineers	Assumed tsunami in Ibaraki Pref.	Assumed tsunami in Fukushima Pre.	Optimization of sea floor topography and tidal condition	Tsunami height observed in the Great East Japan Earthquake
			2002	2007	2007	2009	2011
Fukushima Daiichi	(Units 1–4) O.P. +10 m (Units 5, 6) O.P. +13 m	O.P. +3,122 m 1966 (Unit 1)	O.P. +5.7 m Max. tsunami originating from Fukushima offshore Elevate seawater pump platform	O.P. +4,7 m No need for countermeasures	O.P. approx. +5 m No need for countermeasures	O.P. +6.1 m Elevate seawater pump platform	Tsunami height O.P. +13.1 m Inundation height O.P. +15.5 m
Fukushima Daini	O.P. +12 m	O.P. +3.122 m 1972 (Unit 1) O.P. +3.705 m 1978 (Units 3, 4)	O.P. +5.2 m Watertight buildings	O.P. +4.7 m No need for countermeasures	O.P. +5.0 m No need for countermeasures	O.P. +5.0 m No need for countermeasures	Tsunami height O.P. +7–8 m Inundation height O.P. +14.5 m
Onagawa	O.P. +14.8 m	O.P. +2–3 m 1970 (Unit 1, literature research) O.P. +9.1 m 1987 (Unit 2, numerical calculation)	O.P. +13.6 m Max. tsunami originating from Sanriku offshore No need for countermeasures	–	–	–	Tsunami height O.P. +13.8 m

Tokai Daini	H.P. +8.9 m Seawater pump height +4.2 m	H.P. +2.35 m 1971	H.P. +5.75 m No need for countermeasures	H.P. +6.61 m Wall height extension around seawater pump (H.P. +7 m)	–	–	Tsunami height H.P. +5.5 m Inundation height H.P. +6.2 m

Note. O.P. + – 0.00 m Indicates 0.74 m under mean sea level of Tokyo Bay in Onagawa (Onagawa Nuclear Power Plant Datum Plane for Construction), or 0.727 m under mean sea level of Tokyo Bay in Fukushima (Onahama Harbor Datum Plane for Construction). H.P. + – 0.00 m stands for 0.89 m under mean sea level of Tokyo Bay at Tokai Daini NPS (Hitachi Harbor Datum Plane for Construction) shutdown. At the Fukushima Daini NPS, although the tsunami height was about 8 m, below the site's ground level of 12 m, the inundation height reached 14.5 m in some areas and damaged many facilities, particularly those in Unit 1. Because accident management had been established, cold shutdown of all reactors were achieved. At the Fukushima Daiichi NPS, the tsunami exceeded the revised design height that reflected the latest insights. Although scenarios on large tsunamis were developed by incorporating latest insights to some extent, the urgency of taking countermeasures was not recognized.

The long-term evaluation subcommittee of the Headquarters for Earthquake Research Promotion of the Ministry of Education, Culture, Sports, Science and Technology assumed M7.5 Off-Miyagi Prefecture Earthquake (slippage of about 16 m) with occurrence probability of 99 % within the next 30 years (M8.0, in the case of an interlock with the Off-Sanriku South Trench Near Area Earthquake, and around M8.5, when the South Sea Earthquake and East-Southeast Earthquake are interlocked) on January 11, 2011. "Evaluation Technology of Tsunami at Nuclear Power Stations," published by the Japan Society of Civil Engineers in February 2002 has been recognized as the standard method of evaluating tsunami in Japan before the accident. Reevaluation of tsunami height induced by the largest earthquake based on this method had been in progress at all power stations in Japan.

The magnitude 9.0 Iwate-Miyagi-Fukushima-Ibaraki Offshore interlocked earthquake on March 11 (spanning over an area of about 450 × 200 km, with maximum slippage 60–70 m), was greater than anticipated, and generated a massive tsunami, equivalent to the large earthquake and tsunami in 869, evaluated as a once-in-a-thousand year incident. These events reveal the limitations of current technologies in predicting the scope of large earthquake and tsunami.

The tsunami waves superimposed off the Fukushima Daiichi NPS coast grew to an enormous height exceeding 15 m. All emergency power supply equipments at Units 1 to 6 of the Fukushima Daiichi NPS, installed at the lowest floor of the turbine building basement were inundated and their functions were lost. The 125 V DC power for Units 3, 5 and 6 were installed on the middle floor of the basement of the turbine building, and the air-cooled emergency diesel generator at Unit 6 installed in the building with the highest ground level of 13 m were the only facilities that were functioning. Inundation became the cliff edge that determined whether the facilities maintained functionality, or not. The fact that the slightest difference in the vertical locations determined the functionality of SSCs was a valuable lesson learned from the accident. Facilities and components required for emergency power such as emergency diesel generators, DC power, and switchgears were installed at the basement of the turbine building with no water tightness. This could be the direct fundamental factor which led to the severe accident, initially, the inundation by the tsunami and failure of the emergency components, and subsequently to SBO, including the loss of DC power, and finally to a severe consequence.

6.6.2.3 Tsunami Simulation Analysis

The tsunami exceeded design assumptions at all nuclear power plants, but some sites evaded accident consequences because of the sufficient margin in design.

In particular, the tsunami that struck Fukushima Daiichi NPS far exceeded the latest design basis values that had just been revised, which underscores the shortfalls in preparedness and management of tsunami events.

The tsunami assessment was not clearly defined as a standard applied to the safety design of nuclear power plants. This seems to have caused inconsistency between safety assessment standards on nuclear power plants and the tsunami assessment standard established by the Japan Society of Civil Engineers. The design of a nuclear power plant has performance goals (draft) of core damage factor (CDF) of 10^{-4}/(core-year), containment failure frequency (CFF) of 10^{-5}/(core-year). However, the performance goals (draft) for external events was not clear. The preconditions for tsunami events was considering a historical tsunami of once every century. Although these matters were discussed, it seemed no agreement was reached to incorporate them into the design basis. Hence, the tsunami resistant design at the time nuclear power plants were constructed was in accordance with the insights at the time, however, updates on the latest insights and countermeasures were not sufficiently incorporated into the design.

While the development of a harmonized and consistent standard was slow, the licensees took a substantive approach in implementing measures on their own. Whether a plant was afflicted by disasters depended on whether measures had been taken, or not. It can be said that the fear of tsunami threat saved Onagawa NPS from serious consequences, and that insight on response and management of natural forces is important.

6.6.2.4 Accident Progression Caused by Electric Equipment Failure

External power was lost immediately after the earthquake. However, emergency diesel generators (EDGs) were quickly started, which activated the reactor core isolation cooling system to resume reactor cooling. A detailed analysis and evaluation of the reactor pressure and PCV pressure confirmed the piping inside the PCV and critical equipment were not damaged. The tsunami that followed disabled the emergency generators and switchgears in the turbine building, causing a station blackout. The loss of DC power led to a fatal event. Torrents of abnormal signals were issued after the power cut off of the logical circuit of the control panel, which included a signal to close the valves of the isolation condenser (IC).

The IC was activated in Unit 1 immediately after the earthquake, which decreased the reactor pressure from 7 to 4 MPa in approximately 15 min. Had the control panel been operative, the indicator on the control panel would have shown that the IC was suspended. Communication cut off due to power loss of onsite PHS significantly delayed emergency notices and instructions.

The remote operation of the hardened vent from the control panel was disabled, causing a fail open of the valve of the emergency gas treatment system (SGTS). The valve open was confirmed in Unit 4 in August 2011, and in Unit 3 in December 2011.

6.6.3 External Events and Natural Hazards Management

6.6.3.1 Management of Compound Events

Compound disasters caused by combined earthquake and tsunami events is one of the key issues on safety that must be addressed. The power board caught fire during the earthquake at the Onagawa NPS, (however, it did not lead to a serious incident) indicating the likelihood of a compound disaster involving earthquake and fire. Compound events and countermeasures, including those related to consequences of fire under disaster conditions should also be examined.

6.6.3.2 Scenario on Multiple Component Failures and Common Cause Failures and Accident Scenarios

The tsunami triggered damage and loss of functions of many components, including those with redundancies almost simultaneously. The primary causes for the accident progression that have gradually become clear are: (1) SBO; (2) loss of cooling systems; and (3) loss of heat sink of the plant facilities. Issues related to accident management (AM) measures include inadequacies in: (1) alternative power sources; (2) alternative pump capability (e.g., fire engine); and (3) preparedness for occurrences of unanticipated events (hydrogen explosions, PCV damage, SBO, etc.).

Underlying these causes is the fact that beyond design basis assumption had not been considered sufficiently, probably due to a fear of upsetting the long established safety myth. Evaluation to quantitatively ensure plant safety so far had been based on strict compliance with the Review Guide for Safety Design, requiring functional maintenance and integrity of the components based on a single failure criterion under internal event accident conditions. As a result, scenarios on simultaneous failures of multiple components having the same function and common cause failures have not been taken into account.

Significance of a broad range of assumptions in developing accident scenarios has been re-acknowledged by the 3.11 accident. Because beyond design basis accident scenario, including how and when the fuel and PCV are damaged, and the subsequent occurrences had not been taken into account, and no measures under these conditions had been established, everything went out of control.

Safety assessment had so far focused on internal events, but not on impacts of external event on multiple-unit sites or offsite. Impact assessment must be extended to multiple-unit sites and to offsite in view of the fact that the severe accident afflicting multiple units simultaneously hampered accident management and offsite response enormously at Fukushima Daiichi NPS.

6.6.3.3 Comprehensive External Event Impact Assessment

For example, IPEEE, or evaluation on the consequences of risk on individual plant arising from external events such as earthquake, fire, gale (storms, twisters), flood, avalanche, volcanoes, freezing, high temperature, low temperature, accidents related to transportation or nearby facilities, aircraft fall, and plant vulnerabilities is conducted in the US. However, a comprehensive assessment on external events had not been conducted and there was no solid understanding on plant vulnerability in Japan. Compared with the established PRA quantitative assessment on internal event risk, comprehensive or quantitative assessment on external event risk had not been conducted partly because the PRA on external events was still in the process of development. Accordingly, the IPEEE must be conducted on not only earthquake and tsunami events, but over a broad range of external events. It is essential that in the evaluation process, plant vulnerabilities are identified by impact assessment on external events containing key dominant risk factors using PRA, etc., to enhance plant safety through continuous improvements. Due account must also be given to the frequency of external events, plant impacts with the design margin given consideration, time margin, distance between the hazards and the plant, including their uncertainties in the evaluation process. In view of the compound disaster caused by the earthquake and tsunami, simultaneous occurrences of multiple natural hazards should be taken into account with consideration given to the coincidental, consequential and correlated factors associated.

6.6.3.4 Cliff Edge Effects

With consideration given to seismic vibration affecting all facilities by earthquake events, sufficient margin for all levels of defence in depth is taken into account in plant seismic design. Together with conservative component and structure design, this provides a barrier against the emersion of cliff edge.

Whereas, because the tsunami impacts propagates from offsite and gradually progresses onsite, emphasis was placed on controlling offsite impacts by the tsunami. Accordingly, defence in depth was not incorporated into tsunami-resistant design, which consequently led to the emersion of cliff edge when tsunami height exceeded a certain level, causing functional failure of many safety systems.

Safety assessment so far concerned only external events within the design basis. Hence, the likelihood of cliff edge effect caused by beyond design basis external events was not given consideration, nor was there understanding on plant behavior in the event of a cliff edge, nor countermeasures developed to this end. In the regulatory requirements, only events within design basis over the scope of defence in depth of up to level 3 was subject to safety assessment and considered in plant safety. These were the fundamental causes for the shortfalls in preparing for cliff edge events.

6.6.3.5 Management of Beyond Design Basis Events

Japan had experienced five beyond design basis seismic ground motions including the Chuetsu Offshore Earthquake. As a result, the approach on "residual risk" was introduced in the revised Regulatory Guides for Reviewing Seismic Design of Nuclear Power Reactor Facilities in 2006. The requirement in the revised guideline called for the licensees to develop measures against the likelihood of beyond design basis seismic ground motion, or to reduce residual risk. Some plants like Hamaoka NPS (Units 1 and 2) made a decision to decommission in view of the cost and benefit in implementing such measures.

What was the approach on tsunami measures and design basis? Unfortunately, tsunami resistant design, design basis tsunami, and measure to this end had not been clearly established. Accordingly, facilities had not been arranged to safeguard against beyond design basis tsunami conditions. Not only in terms of the absence of regulatory requirements related to tsunami, but the concept of tsunami design basis itself had not been established at the time, and hence, there was no tsunami preparedness and response in place.

In external event management, design basis settings with sufficient reliability for hazards identified as containing potential risk is not sufficient. In view of the uncertainty contained and the likelihood of beyond design basis conditions of external events, countermeasures must be established in advance to cope with these conditions. Criteria on both design basis and beyond design basis conditions must be discussed and developed regarding external events.

Core damage frequency (CDF) 10^{-4}/(core-year) and containment failure frequency (CFF) 10^{-5}/(core-year), defined in the report by the former Nuclear Safety Commission were considered and used as performance goals. The safety goals have been set forth limiting death rate of the site boundary under accident conditions to 10^{-6}/(man-year). The Nuclear Safety Commission report defines safety goal as "the mean value of acute fatality risk of the public in the vicinity of the site boundary of the nuclear installation and the mean value of fatality risk by cancer caused by radiation exposure resulting from a nuclear facility accident of individuals of the public residing in the area, but with some distance from the facility, should not exceed the probability of 1/1,000,000 per year (10^{-6}/(man-year))." The risk assessment (PSA/PRA) so far had not considered events with frequency smaller than 10^{-7}/(core-year) which was regarded as insignificant.

In meeting the performance goals, the design basis must be formulated with consideration given to hazard curves, etc., and the maturity of assessment technique on external event impacts. The performance goals (draft) determined by the former Nuclear Safety Commission must be met in accordance with the annual exceedance probability of design basis external events (probability of external events exceeding design basis conditions) and safety measures. It is assumed that because the design basis tsunami was set according to historical tsunami in the order of 100 years, exceedance probability of tsunami design basis was set forth in the order of 10^{-2} to 10^{-3}/year, which was not consistent with the required performance goals, including plant safety measures. Since design basis criteria on external events may be formulated by professional societies in related fields, care must be taken to keep close dialogue with the nuclear safety stakeholder in the formulation.

In addition to controlling beyond design basis risks, countermeasures against occurrences of beyond design basis events must be discussed and developed. Accident management as response to beyond design basis conditions, however, had not been sufficiently organized. The root cause for this may be attributed to the lack of a fundamental approach on accident management and management of severe accidents—a resistance to anticipating the likelihood of risks. Up to present, accident sequences have been developed on the basis of internal events initiated by a single failure of constituting components in accident management, for which measures would quantitatively ensure plant safety. Damage causing simultaneous failures of components having the same functions, or common cause failures were not considered, being very small probability events in previous assessments which led to the poor accident management at Fukushima Daiichi Plant.

Significant uncertainty contained in assessment on external events (as compared with that for internal events) should be dealt with by the application of defence in depth concept and safety margin. Safety margin should be verified by conducting stress tests. In addition, safety design and measures must be founded on defence in depth that extends over the realm of beyond design basis conditions. For example, tsunami resistance design may include measures for preventing inundation onsite, including reactor buildings, buildings with critical safety components, and installation of alternative equipments on higher grounds. Since breach of multiple levels of defence in depth may occur simultaneously depending on the magnitude of external events, not only safety margin, but accident management utilizing both permanent components with functions for ensuring safety of up to defence in depth level 3 and redundant means of alternative and transportable safety components with diversity in design should be provided to ensure the effectiveness and independence of the safety measures.

It is also crucial to have in-depth knowledge and understanding on the progression of accidents exceeding design basis. Accident sequences, as when the fuel is damaged, how the containment is damaged, and what happens after containment damage, etc., had not been thoroughly discussed, nor accident response for these conditions been considered. In Unit 1 for example, it would have been effective had arrangements for activating core cooling system utilizing the isolation condenser (IC) in the event of loss of permanent DC power, and ensuring operability of the PCV vent system under power loss conditions been made in advance. However, in reality, PCV isolation was given priority, which delayed accident response and led to the negative turn of events.

Preparedness and response measures for beyond design basis conditions of not only earthquake and tsunami events, but any other external events including natural hazards and terrorist events must be clearly defined.

6.6.3.6 Conclusion

The approach on external events and measures were discussed in this section. The following concept and measures are important for dealing with external events:

(1) Execution of IPEEE to comprehensively evaluate external hazards and understanding of plant vulnerability as well as countermeasures using PRA, etc. (Continuous enhancement process).
(2) Comprehensive evaluation including multiple-unit sites and the impact of off-site events.
(3) Evaluation of cliff edge for beyond design basis external events.
(4) Development of design basis consistent with the performance goals combined with safety measures.
(5) Measures for beyond design basis external events based on insights on risk and concept of defence in depth.

6.7 Radiation Monitoring and Environment Remediation Activities

6.7.1 Environmental Radiation Monitoring as an Initial Response to the Environmental Remediation

6.7.1.1 Actions in Response to the Accident of Tokyo Electric Power Company, Fukushima Daiichi Nuclear Power Station

The local nuclear emergency response headquarters established in the Off Site Center, which was to be a command system in emergency monitoring, was relocated to Fukushima city due to the breakdown in communication systems caused by the Earthquake and the increased radiation dose in surrounding areas. Meanwhile, the Government encouraged the Ministry of Education, Culture, Sports, Science and Technology (MEXT) to actively lead the implementation of monitoring immediately after the accident. Accordingly, the Government, local governments and relevant organizations jointly conducted the monitoring, the results of which were then officially announced by MEXT. However, the results of monitoring carried out by MEXT and the electric power company were still not sufficiently consolidated and shared. The Government instructed MEXT, the Nuclear and Industrial Safety Agency (NISA) and the Nuclear Safety Commission (NSC) that the results should be coordinated and officially announced by MEXT, assessed by the NSC and measures taken by the Government Nuclear Emergency Response Headquarters (GNERH) based on the assessment. Monitoring of food was also conducted by the Ministry of Health, Labour and Welfare (MHLW) and agricultural and livestock products by the Ministry of Agriculture, Forestry and Fisheries of Japan (MAFF). To implement post-accident monitoring more extensively and for a longer period, a state system assisting local government did not function properly, which caused many problems such as ensuring materials and equipment, allocating manpower and rapid communication between organizations and local populations.

The 1st Monitoring Coordination Meeting (hereinafter referred to as the Coordination Meeting) was held on July 4, 2011 to unfailingly and systematically implement radiation monitoring of the accident of the Fukushima Nuclear Power Station with the purpose of coordinating the radiation monitoring carried out by the Ministries and Agencies concerned, the local government and electric power company. At the Coordination Meeting, to organize the monitoring, it was discussed and agreed to implement "fine-tuned monitoring" to rehabilitate the environment in the areas around the Fukushima Daiichi Nuclear Power Station and respond to requirements of the health of children, public safety and confidence. Moreover, it was also agreed that the Government should be responsible for coordinating with local government and the licensee of nuclear energy-related activity to avoid "deficiencies" in the implementation of radiation monitoring.

On August 2, 2011, the Comprehensive Monitoring Plan was established based on the discussion of the Coordination Meeting. In December the same year, the Comprehensive Monitoring Plan was revised to cope with new problems and the emergency monitoring (high-frequency monitoring at a point, etc.), which was responding to the bulk release of radioactive material and had continued until then, was reviewed due to the decrease in the release of radioactive material and temporal variation. On April 1, 2012 and in April 2013, the Plan was revised from the perspective of boosting the discussion of future measures, assessing the overall impact on the surrounding environment and focusing on changes in the evacuation areas and perceiving the mid- and long-term radiation dose.

In the plan, measurements of air dose, determined cumulative dose, radionuclides in airborne dust, soil and index materials such as pine needles (plants obtainable year-round) have continued as wide-area monitoring covering the whole of Fukushima prefecture. The radioactivity monitoring survey by prefectures, measurement at additional monitoring posts and monitoring with aircraft etc. have all been implemented.

Various surveys and monitoring such as measurements of detailed air dose rates, car-borne surveys to be contributed for decontamination, air dose rate measurement in restricted residential areas (confirmation of doses less than 20 mSv/year) and detailed monitoring for wide-area infrastructure remediation work etc. have been sequentially implemented in the restricted areas and deliberate evacuation areas. Monitoring to assist with remediation and efforts to resume life were also to be implemented in areas where evacuation directives were lifted or expected to be lifted.

Sea area monitoring was to be widely implemented around the Fukushima Daiichi Nuclear Power Station as well as 30 km offshore from the shoreline of the Tohoku and Kanto regions, offshore areas and outer seas, monitoring seawater and radionuclides in sea water, sea-bottom soil and sea products were measured. On March 30, 2012, it was decided to enhance the accuracy of analysis of seawater samples, investigate the variability and nature of sea-bottom soil, and time variation of concentration of radionuclides in sea water, sea-bottom soil and marine life (diffusion, deposition, movement and migration) based on the "Guideline for offshore areas monitoring 2012". Monitoring of flow pathway from rivers to the sea was also enhanced taking the outflow routes into consideration. Monitoring of seawater to gage new leakage of radioactive materials from the Fukushima Daiichi Nuclear Power Station was additionally implemented. In addition, it was shown that operational plans for monitoring the land water environment, natural parks, waste material, agricultural soil, forest, feed crops and food, and a plan to effectively utilize the simulation results to calculate the lower detection limit were all implemented.

6.7.1.2 Environmental Radiation Monitoring to Respond to Future Emergencies and Nuclear Facilities in Normal Operations

The Nuclear Regulatory Commission (NRC) started discussing the future emergency monitoring described in the Nuclear Emergency Response Guideline (NERG) in December 2012. At the meeting, the roles and tasks of agencies concerned with emergency monitoring, an operational plan for the emergency monitoring and operational intervention level (OIL) etc. were discussed between the NRC and Nuclear Facilities Radiologic Investigation Agencies Liaison Council, which was a liaison council for those in charge of environmental radiation monitoring of the municipalities where nuclear power stations were existing. Consequently, it was confirmed that the Government would be responsible for direction and supervision systems, which caused confusion in the emergency monitoring at the time of the accident involving Fukushima Daiichi Nuclear Power Station. It was also confirmed that the Government, local government and licensee of nuclear energy-related activity should share the purpose of monitoring and cooperate in implementing emergency monitoring, designated public agencies should assist with every aspect of emergency monitoring and an emergency monitoring center led by the NRC would be established. On June 5, 2013, the above items and results of the discussion on the distribution and administration of stable iodine from medical treatment for exposure in future were consolidated and these visions were reflected in the NERG.

6.7.1.3 Future Tasks

From the time of the accident to the remediation period, the Government and local government took various measures as described above to respond in a fine-tuned manner and avoid any deficiency in future emergency monitoring.

Regarding the radiation monitoring around the area of the Fukushima Daiichi Nuclear Power Station, the incidental release should be carefully monitored and distribution of the concentration and dosages of radioactive materials, which were already widely diffused, should be continued to assess the impact on the health of the local population in future over the longer term, discuss how to reduce exposure and formulate a protection plan. There is also a need to develop new technology to enhance measurement accuracy and the speed of assessment in wider areas. In addition, relevant research should be continued to accumulate data on the diffusion and migration of radioactive materials in the environment.

As described in the "interim report on the activities of the Task Group on Aspects of the Fukushima Daiichi Nuclear Power Plant Accident of the Nuclear Safety Investigation Committee of Experts", when the radiation dose or amount of radioactive material released are assessed without survey data, immediately after the accident, it may be effective to estimate them by calculation using air- and sea-diffusion models. As this method has already been utilized to reconstruct the internal exposure radiation dosage caused by inhaling iodine at the initial stage of the accident, it can be utilized together with the survey data to enhance accuracy. The calculation method should also be maintained to respond to various needs such as rapid local detailed assessments and wide-ranging assessments.

For future emergency monitoring, the NERG specified matters such as the means of implementing a system of emergency monitoring, consolidating the systems of the emergency monitoring center as a proactive step by Government, formulating an emergency monitoring plan after the accident occurred and centralizing the analysis and assessment. However, monitoring of mid- and remediation terms other than initial monitoring were not discussed. During these periods, monitoring should be implemented to review the evacuation areas, judge whether to lift evacuation areas, manage exposure to radioactivity, decide on dosage reduction methods and estimate current and future exposure to radioactive dosage. These methods should be discussed sufficiently based on the current circumstances of Fukushima. In addition, there is a need to establish a centralized system to collect and store data to utilize the monitoring data effectively and functionally.

After the accident has occurred, monitoring the individual radiation dose as well as environmental monitoring may enable more accurate assessment. In this accident, as well as individual dosimeters installed at the Power Station, whole body counters were also rendered unusable by the tsunami and responses were limited. Moreover, the difference between phantoms used to calibrate instruments sometimes resulted in deviations in measuring values. There is a need to discuss the utilization of effective whole body counters, a unified assessment method and intercomparison, as well as methods to assess children's radiation dose. Information on places where the exposure occurred is also important to perceive the individual radioactive dose. To obtain this information, it would be effective to develop a new method of individual dosimetry.

Rapid and adequate response during emergencies cannot be achieved solely by clearly deciding roles and preparing materials, equipment and systems. It is also important to review plans regularly in normal times, share information on disaster progress when emergencies occur and ensure the source of release between agencies concerned by conducting training in normal operations, and avoiding misunderstandings on decided matters. In addition, it is desirable to conduct training courses for those relevant to disaster prevention to learn basic knowledge of nuclear radiation and measuring technology and develop human resources to respond to emergency monitoring.

The radiation monitoring information thus obtained should be provided in an integrated fashion by the Government, local government and licensee of nuclear energy-related activity to avoid confusion and reduce residents' concerns. Also, supplementary information should be provided for residents and other relevant parties to determine and understand the accident circumstances.

6.7.2 *Effects of Radiation*

6.7.2.1 Workers' and Residents' Dose, Effects of Radiation

As mentioned in Sect. 5.3.3.2, the dose limit for people working in emergencies at the Fukushima Daiichi Nuclear Power Station was increased to 250 mSv immediately after the accident. Some workers engaged in the emergency work were internally exposed over the 250 mSv limit due to the misuse of protectors, while others had their skin contaminated with radioactive material, but there were no radiation hazards clearly and clinically identified in either case.

The estimated thyroid equivalent dose for most children in the area around the Fukushima Daiichi Nuclear Power Station in the initial stages of the accident was identified as less than 30 mSv, as also mentioned in Sect. 5.3.3.2. The value is less than the 50 mSv limit value triggering the administration of stable iodine as stipulated by the IAEA, and also less than the 100 mSv screening criterion of the equivalent dose of the thyroid. For the external exposure, the maximum value was around 25 mSv, the effects of radiation were not clinically identified.

In this accident, the radionuclides important to assess the dose of internal and external exposure for both workers and residents are radioactive cesium (134- and 137-Cs) and radioactive iodine (131-I).

Cesium is an alkali metal and its behavior inside the body resembles that of potassium, also an alkali metal and indispensable for life. Cesium does not accumulate in specific organs, has an effective half-life of 70–100 days and approximately half is excreted.

The effective half-life of the iodine is approximately 7 days inside the body. As iodine is an element necessary to synthesize the thyroid hormone, 30 % of its intake is accumulated in the thyroid. An epidemic of thyroid cancer in children caused by intake of iodine-131 emerged in the areas around the accident at Chernobyl Power Station, due to the delay in restricting the intake of affected food and the fact that children consumed milk contaminated by iodine-131. It was also due to routine shortages of iodine inside the body due to the characteristics of inland areas [18]. In the Fukushima accident, some of the drinking water, agricultural and livestock also contained radioactive iodine and cesium, concentrations of which exceeded the provisional regulation values. However, the amounts of radioactive material released into the air were smaller than those of Chernobyl and restrictions on intake and shipments were rapidly imposed, which meant effects on residents were moderate.

The MHLW estimated the value of internal exposure as 0.003–0.02 mSv/year if a single individual were to consume radioactive material contained in food from Tokyo, Miyagi and Fukushima prefectures in the long term. This exposure dose is less than 1/100 of that of natural radiation sources (2.1 mSv) such as 238-U, 232-Th and 40-K contained in soils and 40-K contained in food.

As described in Sect. 5.3.3.2, Fukushima prefecture has been conducting thyroid examinations as part of its health management program for prefectural residents, which exposed thyroid cancers among children. However, it is considered unlikely that these cancers were attributable to the effects of the accident, because cancer due to radiation exposure is believed to develop several years to decades after the exposure.

6.7.2.2 International Mindset on Protection Against Radiation and Effects of Radiation

A recommendation from the International Commission on Radiological Protection (ICRP) in 2007 noted the relationship between a one-time dose of less than 100 mSv or an accumulated dose of less than 100 mSv/year and the effects on human health as follows [19]: As for the deterministic effects (tissue reaction), "However, in the dose absorbed dose range up to around 100 mGy (low Linear Energy Transfer (LET) or high LET) no tissues are judged to express clinically relevant functional impairment." as the equivalent dose (excerpt from paragraph 60). As for the stochastic effects, "There is, however, general agreement that epidemiological methods used for the estimation of cancer risk do not have the power to directly reveal cancer risks in the dose range up to around 100 mSv" (excerpt from paragraph A86). Therefore, in this accident, the effects of cancer on residents, based on their estimated exposure dose, remain negligible.

The ICRP pointed out that the solid lifetime cancer risk triggered by radiation of a fetus will be 2–3 times larger than the group-wide risk. However, this result may contain significant uncertainty. The United Nations Scientific Committee on the Effects of Atomic Radiation (UNSCEAR) recommended that the radiation sensitivity of children be further researched in future.

Apart from the epidemiological study on survivors of atomic bombings, a study on the relationship between cancer death rate and radiation doses in Karunagapally, Kerala province in India, where high natural radiation is found, was recently reported. There are some places where the outdoor radiation doses reach 70 mGy (about 30 times higher than the global average) in the area and some people have a cumulative dose exceeding 500 mGy. However, there is no meaningful trend showing an increase in the risk of cancer caused by radiation [20].

The UNSCEAR decided to assess the level of radiation exposure and risk to human health caused by the accident of the Fukushima Daiichi Nuclear Power

Station at the 58th Meeting in May 2011. The Comprehensive Nuclear-Test-Ban Treaty Institution (CTBT), Food and Agriculture Organization (FAO), International Atomic Energy Agency (IAEA), World Health Organization (WHO) and World Meteorological Organization (WMO) also participated in the assessment. The preliminary report was submitted at the 59th Regular Meeting in May 2012 and the draft final report was prepared at the 60th Regular Meeting in May 2013. At the Meeting, it was reported that there were no acute disorders (deterministic effects) among workers and residents in the vicinity, and also few possibilities to such effects of radiation emerging among many workers and the general public in future. Grounds for the conclusion are (1) residents in the vicinity were exposed to iodine-131 but their dose equivalents of thyroid were several dozen mGy (Sv); (2) their whole body exposure doses of 134- and 137-Cs received within several weeks were 10 mSv and less; (3) and people living in areas far from Fukushima would be exposed to an additional dose of about 0.2 mSv from the intake of food. In addition, causes to further limit the exposure doses for residents living in the area were (1) the amount released was smaller compared with the Chernobyl accident; and (2) protective measures (evacuation and sheltering) were taken. However, it was reported that health surveys should continue in future due to various factors, including the shortage of measured data, significant uncertainty over the effects of lower doses and dose rates as mentioned above and the effects on human health caused by other factors. As for some workers who were exposed to higher radiation doses, it was reported that there would be few possibilities of developing thyroid cancer excessively but it was recommended that medical examinations on cancers of the thyroid, stomach, lung and large intestine should be continued in future. As described in Sect. 5.3.3.2 and as will be discussed later, the radiation dose assessment and health management survey on workers and residents have already been conducted appropriately.

6.7.2.3 Efforts to Reduce the Effects of Radiation and Health Promotion for Workers and Residents in Japan

As described in Sect. 5.3.3.2, the radiation exposure situations of personnel engaged in radiation work have improved since April 2012 compared with those immediately after the accident, since which time it seems that the dose limit of 50 mSv/year can be maintained. However, as about 5 % of the workers have exceeded 20 mSv/year, measures should be taken to avoid excessive burdening specific people with work to maintain the dose limit of 100 mSv per 5 years. In addition, competent experienced workers capable of coping with emergencies should be secured to prepare for unforeseeable circumstances. It is also important to maintain the radiation protection optimization principle whereby radiation exposure must be minimized as far as reasonably possible in radiation management for various work towards the reactor decommissioning.

The MHLW, considering workers' unease over their physical conditions and the increased risk of mid- and long-term health disorders, decided to offer additional medical examinations for the eyes, thyroid and cancer (stomach, large intestine and lung) etc. to workers engaged in emergency work at the Fukushima Daiichi Nuclear Power Station, who worked under extraordinary circumstances and exposed to high radiation doses in duties far different from ordinary radiation handling work. Thus, the long term health management survey is thought to be planed and conducted adequately.

Hereafter, publicizing and information services will be important to ensure all workers can be covered by this health management survey. Radiation exposure management suitable for various works is also required to accomplish the long and difficult task of reactor decommissioning.

As for the exposure dose assessment, UNSCEAR pointed out that the results of the workers' estimated exposure doses as provided by the Government of Japan, which were estimated from values measured at the thyroid, might be underestimated by approximately 20 %, because short-lived nuclides such as iodine 133-I (half-life 20 h) were not assessed in the initial stage of the accident in October 2013. It may be said that estimation of internal radiation exposure is generally associated with more factors of uncertainty than external radiation exposure due to various effects such as the time of intake, chemical form of the radioactive nuclide and transfer rate in the body. However, subsequently, more accurate estimation can be expected by introducing the contribution of the short-lived nuclide. In addition, to prepare for emergencies and reactor decommissioning in future, the establishment of medical systems capable of coping with accidental radiation exposure under further complicated radiation conditions, including external exposure under mixed situations of beta and neutron radiation in addition to the external exposure of gamma radiation, alpha radiation release nuclides, internal exposure of pure beta radiation, and also the development of human resources for these purposes must be reminded.

As described in Sect. 5.3.3.2, the local Fukushima prefecture government has been conducting programs for residents such as the behavior survey, thyroid medical examination, mental consultation and consideration for pregnant women. These programs will continue on a long-term basis and the data obtained and information management will be unified. However, the collection rate of behavior records, which is indispensable to assess the initial individual exposure doses, remains insufficient and problematic gaps among areas have been highlighted. Although the local government has striven to improve the collection rate, in cooperation with the municipalities, cooperation with medical and educational institutions is also required. Before people's memories have receded, these data and information must be promptly collected, which means the health management survey must be continually implemented for all object people without fail.

6.7.2.4 Conclusion and Future Tasks

The UNSCEAR recognized that no acute disorders (deterministic effects) were found in the Fukushima Daiichi Nuclear Power Station accident and there was little potential for the effects of radiation to emerge among workers and residents in the vicinity in future. This perspective was also shared by the international communities and the Atomic Energy Society of Japan (AESJ). In addition, the health management and medical examinations, including the dose assessment and radiation effects on workers and residents, have been adequately implemented and a database to manage these data and information in an integrated fashion is to be steadily prepared. The medical examination will be continually implemented henceforth and the data and information thus obtained will be managed in an integrated fashion. To ensure the system remains effectively operational, a review will be required to ensure the adequacy of items and periods when needed.

As for exposure dose assessment in the initial stages and dose reconstruction, there is currently significant uncertainty due to the shortage of radiation monitoring data and behavior records, which means more accurate exposure dose assessment is desired in future.

6.7.3 Decontamination Measures: Legal Framework and Guidelines

A large amount of radioactive materials was released following the accident at the Tokyo Electric Power Company Fukushima Daiichi Nuclear Power Station, contaminating the surrounding environment over a wide area, but Japanese law did not assume this situation would occur before the accident. Subsequently, stipulations on radioactive material in laws relevant to the environment have been significantly revised. The Basic Environment Law (Law No. 91 of November 19, 1993), which previously did not cover radioactive material, was revised to do so. Under the Atomic Energy Basic Law (Law No. 186 of December 19, 1955), "Preservation of the Environment" was added to the definition of nuclear safety. Following this change, "Preservation of the Environment" was added to the object of the Act on the Regulation of Nuclear Source Material, Nuclear Fuel Material and Reactor (Act No. 166 of June 10, 1957) (hereinafter referred to as the Reactor Regulation Act).

As contamination of the environment by radioactive material was not juristically assumed and no legislative framework existed, various frameworks were established after the accident; the most basic Act of which was the Act on Special Measures Concerning Response to Environmental Contamination by Radioactive Material Released from the Accident of the Nuclear Power Station Caused by the Great East Japan Earthquake, which occurred on March 11, 2011 (Act No. 110 of August 30, 2011) (hereinafter referred to as the Act on Special Measures). The Act on Special Measures has significant meaning as a broader framework; legally stipulating the segregation of duties for the parties concerned, but preparation of additional laws and regulations is desired when needed.

The Ministry of the Environment formulated the "Guideline relevant to waste material" and "Guideline relevant to decontamination" based on the Act on Special Measures in December 2011 to explain the process of decontamination and handling waste material in a specific and easily understandable way. Guidelines on waste material and decontamination were also revised to reflect cumulative knowledge and experiences in March 2013 and May 2013, respectively. This paragraph discusses the concept of the Act on Special Measures and future tasks for the Guideline, taking the ongoing decontamination situation into consideration.

6.7.3.1 Concept of the Act on Special Measures and Relationship with Existing Laws and Regulations

The Act on Special Measures was enacted to conduct decontamination safely, smoothly and promptly, and treat, store and dispose safely of radioactive waste derived from the decontamination. The basic concept of the Act on Special Measures is to divide objects into waste contaminated with radioactive material derived from the accident and into soil (including vegetation and artificial structures) contaminated in the same way, and ensuring each object is subject to measures such as treatment, decontamination and disposal. Another characteristic of the Act is that it defines the roles of the Government, local governments and licensees of nuclear energy related activity; including their burden of expenses.

In addition, the Act on Special Measures classifies the contaminated areas subject to decontamination into "special decontamination areas" and "intensive contamination survey area". It defined that the Government formulates a decontamination plan and conduct decontamination work for the "special decontamination area". The "special decontamination area" is referred to as the "deliberate evacuation area" where the cumulative dose was expected to exceed 20 mSv/year after 1 year of the accident and the areas located within the 20 km radius zone of the Fukushima Daiichi Nuclear Power Station, which have been designated as "restricted area". The "intensive contamination survey area" is an area designated based on the Act on Special Measures, in which the additional exposure dose exceeds 1 mSv/year (equivalent to 0.23 μSv/h).

Pursuant to laws and regulations relevant to the nuclear and radiation implemented before the Act on Special Measures was enacted, such as the Reactor Regulation Act, and Act concerning Prevention from Radiation Hazards due to Radioisotopes, etc. (Act No. 167 of June 10, 1957) (Radiation Hazards Prevention Act), places where there is scope to generate radioactive waste were restricted as "controlled areas". The Act on Special Measures is not basically applicable to waste contaminated with radioactive material generated from the facilities within the controlled area, which are subject to the restrictions of conventional laws and regulations. Pursuant to Reactor Regulation Act and the Act on Special Measures, waste materials contaminated with radioactive materials are classified, but the methods of treatment, disposal, and regulatory limits may differ, while procedures implemented to conduct the projects differ and waste can be generated from both the nuclear facilities and the accident as contamination sources. Therefore, the additional development of legal systems is desirable to avoid confusion.

The Waste Disposal and Cleaning Act (Act No. 137 of December 25, 1970) (hereinafter referred to as the Waste Cleaning Act) applies to materials contaminated with radioactive materials generated from the accident, but the Waste Cleaning Act applies to a more limited scope of objects, laws and regulations may be required or administrative measures may be required for contaminated materials not covered by the Waste Cleaning Act (e.g. paid-for assets).

6.7.3.2 Guideline Relevant to Decontamination and Its Problems

Targeting the additional exposure dose of less than 1 mSv/year as a long-term objective, implementing smooth and effective decontamination and reducing the effects of radioactive materials generated from the accident, the Ministry of the Environment prepared the "Guideline relevant to decontamination" and the "Guideline relevant to waste materials" in December 2011. Decontamination work has already been started by national and local governments, but there are some problems in the Guidelines and relevant manuals.

The "Guideline relevant to decontamination" has described methods used to decontaminate various objects, but it is pointed out that some methods may not be necessarily effective under certain circumstances. When a method not described in the Guideline is employed, all the expenses required to implement the method should be borne by the local government if not approved in the individual consultation with the Ministry of the Environment. There is a need to reflect effective decontamination methods on a timely basis to the Guideline. Moreover, said Guideline did not include sufficient decontamination methods for forests and agricultural land. As to decontamination of forests, e.g. from the human health protection perspective, it emerges that decontamination may be adequate for fallen tree leaves and branches, which are effective means of reducing the air dose, within 20 m of the fringe of forests, for forests around residential areas. Meanwhile, the Environmental Remediation Review Meeting offers various opinions, including scope whereby the decontamination may not be limited to an area of 20 m depending on circumstances. Taking the problems mentioned above into consideration, the "Guideline relevant to decontamination" was revised in May 2013. The revision also introduced new decontamination methods such as shot blasting and ultra-high pressure washing, but decontamination for forests was not revised.

As the "Guideline relevant to waste materials" is based on the Waste Cleaning Act, it clearly decided on methods for disposal of waste materials with lower level contamination, but disposal methods for specified waste with contamination exceeding 8,000 Bq/kg were not clearly decided. In March 2013, said Guideline was also revised and supplemented with an explanation on the method of landfill disposal for specified waste with contamination exceeding 8,000 Bq/kg. New promulgation and notification were added and specific examples upgraded and expanded.

As for common problems of both Guidelines, there is concern that they have described only the nuclides of 137 and 134-Cs but not strontium 90 and other nuclides. Attention will be required when the decontamination is implemented in areas where significant strontium 90 still exists. Moreover, the decay of 134-Cs should be taken into consideration accordingly.

Improvement to promote reuse of decontaminated waste materials is also required. Although the clearance level of 137-Cs is 100 Bq/kg, the Ministry of the Environment reported that concrete with a contamination level of less than 3,000 Bq/kg could be reused provided certain conditions are met. As differences emerge in terms of environmental backgrounds and living conditions between the decontamination object areas and elsewhere, various methods of reuse according to each condition can be implemented.

Meanwhile, as for exposure caused by the release of radioactive materials accompanying measures within the nuclear power station site where the accident occurred, the additional dose outside the site is deemed to be less than 1 mSv/year. However, given the fact that the background dose in areas affected by additional exposure far exceeds 1 mSv/year, or no residents are currently living there due to it being a restricted area, this value is not necessarily reasonable. The same point was made in the review reported by the IAEA in May 2013. Since the restriction of 1 mSv/year may restrict the potential to implement the reactor decommission roadmap there, measures balancing on- and off-site exposure risks are required.

After the Act on Special Measures is enforced, there are indications of retarding decontamination in some areas. It is presumed that activities of residents and volunteers will slow for reasons such as inability to find temporary storage yards or conduct waste treatment. Flexible management of the Guidelines according to need is required.

6.7.3.3 Conclusion and Future Tasks

The fact that legal and technological bases to promote decontamination, treatment and disposal of waste materials have been established by the enforcement of the Act on Special Measures and the publication of the above mentioned two Guidelines should be prioritized. Although various problems have been highlighted in the Guidelines, revised versions were prepared in 2013, reflecting the current decontamination performance, which is expected to facilitate implementation of future decontamination. However, various problems remain unsolved, including the lack of temporary storage yards in some areas, or cases where decontamination has not been effective and so forth. Taking these situations into consideration, points to be improved among the Act on Special Measures and Guidelines are shown as follows:

- The relationship between the Act on Special Measures and existing laws and regulations such as the Reactor Regulation Act should be consolidated and the superordinate concept of these laws and regulations must be effectively integrated. In addition, it is desirable to prepare additional laws and regulations or administrative measures as required.
- Effective decontamination methods should be continued to reflect the Guidelines and others; nuclides to be taken into consideration and response to the same

should be clarified according to contamination circumstances; and the effects of decay over time should be taken into consideration.

- The Guideline should be improved to facilitate reuse and thus promote decontamination activities.
- Monitoring and behavior survey in forests should be continued, targeting the establishment of reasonable and effective decontamination methods.
- Manuals other than the Guidelines should be improved accordingly.
- Exposure risks on- and off-site should be balanced and dose management should be implemented for the nuclear power station site, in a manner which will not hinder the realization of the reactor decommissioning roadmap.
- Promote efforts to locate temporary storage yards; establish the transparency of project safety and procedures taking the accountability of such safety into advance consideration to encourage public confidence in the decontamination projects.
- To avoid any slowdown of decontamination activities, flexible management should be ensured.

6.7.4 Establishment of Areas Subject to Decontamination

6.7.4.1 Criteria to Establish Areas to be Subject to Decontamination and Areas to Fall Under the Category

The Act on Special Measures [21] was promulgated in August 2011 responding to environmental contamination by radioactive materials and in November the same year, basic policy based on the Act was approved in a Cabinet meeting. Following these events, the "special decontamination area" where significant environmental contamination exists and where the Government must implement decontamination and the "intensive contamination survey area" where full operation of survey on radioactive materials contamination situations caused by the accident is required, are defined and classified. Specifically, the "intensive contamination survey area" was deemed to cover areas with an air dose rate exceeding 0.23 μSv/h. This is because the additional exposure dose per year in the 0.23 μSv/h environment is equivalent to 1 mSv/year.

- Details of the 0.23 μSv/h:
  Natural radiation dose from the Earth: 0.04 μSv/h
  Additional exposure dose caused by the accident: 0.19 μSv/h
- Presuming an ordinary life pattern spending 8 h outdoors and 16 h indoors (wooden house with sheltering effect of 0.4):
  $0.19\ \mu Sv/h \times (8\ h + 0.4 \times 16\ h) \times 365\ days = 1\ mSv/year$

The contamination implementation area is that designated by the decontamination plan within the "intensive contamination survey area". The municipalities, including the designated "special decontamination area" and "intensive contamination survey area" are shown in Tables 6.27 and 6.28.

Table 6.27 Special decontamination areas (as of December 28, 2011)

	Number of municipalities	Designated areas
Fukushima Prefecture	11	Narahara town, Tomioka town, Okuma town, Futaba town, Namie town, all areas of Katsurao village and Iitate village, and Tamura and Minami Soma cities; restricted areas and deliberate evacuation areas within Kawamata town and Kawauchi village

Source: Press release from the Ministry of the Environment, on December 19, 2011

Table 6.28 Intensive contamination survey areas (as of December 27, 2012)

	Number of municipalities	Designated areas
Iwate Prefecture	3	Ichinoseki city, all areas of Oshu city and Hiraizumi town
Miyagi Prefecture	9	Ishinomaki city, Shiroishi city, Kakuda city, Kurihara city, Shichikashuku town, Ogawara town, Marumori town, all areas of Watari town and Yamamoto town
Fukushima Prefecture	40	Fukushima city, Koriyama city, Iwaki city, Shirakawa city, Sukagawa city, Soma city, Nihonmatsu city, Date city, Motomiya city, Kori town, Kunimi town, Otama village, Kagamiishi town, Ten'ei village, Aizubange town, Yukawa village, Yanaizu town, Mishima town, Aizumisato town, Aizumisato town, Nishigo village, Izumizaki village, Nakajima village, Yabuki town, Tanagura town, Yamatsuri town, Hanawa town, Samegawa village, Ishikawa town, Tamakawa village, Hirata village, Asakawa town, Furudono town, Miharu town, Ono town, all areas of Hirono town and Shinchi town, and areas in Tamura city, Minamisoma city, Kawamata town and Kawauchi village except the restricted areas and the deliberate evacuation areas.
Ibaraki Prefecture	20	Hitachi city, Tsuchiura city, Ryugasaki city, Joso city, Hitachiota city, Takahagi city, Kitaibaraki city, Toride city, Ushiku city, Tsukuba city, Hitachinaka city, Kashima city, Moriya city, Inashiki city, Hokota city, Tsukubamirai city, Tokai village, Miura village, and all areas of Ami town and Tone town
Tochigi Prefecture	8	Sano city, Kanuma city, Nikko city, Otawara city, Yaita city, Nasushiobara city, all areas of Shioya town and Nasu town
Gunma Prefecture	10	Kiryu city, Numata city, Shibukawa city, Annaka city, Midori city, Nakanojo town, Takayama village and all areas of Higashiagatsuma town and Kawaba village
Saitama Prefecture	2	All areas of Misato city and Yoshikawa city
Chiba Prefecture	9	Matsudo city, Noda city, Sakura city, Kashiwa city, Nagareyama city, Abiko city, Kamagaya city, and all areas of Inzai city and Shiroi city
Total	101	

Source: press release from the Ministry of the Environment, on December 14, 2012

Relating to the implementation of decontamination, instead of the existing areas to which evacuation orders have been issued (restricted areas and deliberated evacuation areas), three kinds of evacuation-directive area (areas in which evacuation orders are ready to be lifted (<20 mSv/year), areas in which residents are not permitted to live (20–50 mSv/year), areas in which residents will face difficulties in returning for a long time (>50 mSv/year)) was newly established in April 2012 and policies to decontaminate these areas were shown according to the extent of the radiation dose [22]. These new evacuation-directive areas were implemented based on consultation and coordination between the prefectures, municipalities and residents concerned.

6.7.4.2 Problems to Establish Areas Subject to Decontamination

The criterion to establish areas to be subject to decontamination was shown by the Cabinet decision based on the Act on Special Measures as mentioned before. The decontamination areas have been established for areas where the additional exposure dose exceeds 1 mSv/year regardless of other factors, such as extent of contamination in the area, population, land use and evacuation situations of residents, etc.

The International Commission on Radiological Protection (ICRP) has shown an international basic concept on radiation protection for circumstances where radioactive materials generated by the nuclear accident are extensively deposited, contaminating areas for an extended period and where radiation protection management is required (currently existing exposure situation) [23]. Based on existing exposure, it is recommended that a "reference-level" target value for radiation protection measures be established instead of the "dose limit" applied at normal times; taking socioeconomic factors into consideration and ensuring optimization of protection measures based on the value. It is recommended to select a reference level ranging between 1–20 mSv/year.

Under the Act on Special Measures, it might have been very difficult to establish the criterion and decide the decontamination areas in accordance with the basic principle of the radiation protection recommended by the ICRP, considering the post-accident circumstances in Japan, including confusion over the effects of radiation and the protection criterion, residents' strong desire for proper decontamination and the lack of confidence among stakeholders. The decision to adopt 1 mSv/year, the lower limit of the additional exposure dose recommended by the ICRP for the currently existing exposure is understandable from the perspective of obtaining residents' confidence.

Nevertheless, the decision to designate areas with an additional exposure dose over 1 mSv/year subject to decontamination may not concur with the abovementioned optimization principle. It may be useful to discuss means of determining an optimal approach, including factors other than decontamination for areas with lower radiation doses, for example, assessment of decontamination cost-efficiency, review based on the individual exposure dose (individual annual effective residual dose), etc. For areas with relatively higher radiation doses, there is a need to formulate remediation plans to improve local infrastructure based on results of future decontamination model projects and taking residents' wishes to return home into consideration. In addition, an optimized option must be carefully discussed to decontaminate vast areas of forest through ongoing model projects and the results of behavioral observation.

The additional exposure dose of 1 mSv/year, reference to which was made to establish decontamination areas, was adopted from the local air dose rate and averaged behavior pattern. However, past experience has shown that exposure levels are determined by individual behavior (such as place of residence, workplace, occupation, time spent in contaminated areas, work performed in the contaminated area and individual living habits including dietary habits). Therefore, the ICRP recommends that the reference level, which is determined by the individual annual effective residual dose, should be used instead of using "averaged individual" figures to manage exposure in the contaminated areas. Recognizing the facts and the protection concept, an appropriate review is required to implement decontamination by perceiving not only the averaged air dose rate but also the distribution of the individual annual effective residual dose, based on individual dose measurement results. It is thought to be effective not only to implement decontamination but also to strive to reduce the exposure dose, including "protection measures by self-supporting efforts" such as improving the behavior pattern. Consideration of this point is desirable in future. In accomplishing the long-term target of an additional exposure dose of less than 1 mSv/year, it may be adequate to judge the result not only by the air dose rate but also the individual annual effective residual dose.

Local governments are currently formulating an operational plan for the decontamination and determining the decontamination implementation areas. A priority concept is also introduced into some formulation and it is expected that decontamination will be established while maintaining a balance between complete radiation protection and socioeconomic factors, with discussion of optimization among stakeholders involved in the process of formulating plans and reviews in future.

6.7.5 Decontamination Framework of the Central and Local Governments

6.7.5.1 Decontamination by the Government and the Municipality

(1) **Framework of decontamination**
The Government sets a long-term target to reduce the additional exposure dose less than 1 mSv/year. To accomplish this, decontamination is required; not only for housing areas and agricultural land but also wider living areas, including public facilities, roads and a part of forests. In FY 2011, the Cabinet Office conducted a decontamination model project by entrusting the implementation to the Japan Atomic Energy Agency (JAEA) and the technology employed for

decontamination of housing areas and others was assessed. In addition, the Act on Special Measures (refer to Sect. 6.7.3) was established, and systems and criteria for decontamination and management of radioactive waste materials generated from the Fukushima Daiichi Nuclear Power Station accident were established. Accordingly, it was decided that the Ministry of the Environment (MOE) should oversee the overall environmental remediation in areas outside the Fukushima Daiichi Nuclear Power Station site.

Based on the Act on Special Measures, relatively higher contaminated area of 11 municipalities in Fukushima prefecture was designated as "special decontamination area" and relatively lower contaminated area of 101 municipalities in Fukushima and seven other prefectures was also designated as "intensive contamination survey area" (refer to Tables 6.27 and 6.28). It was decided that the decontamination of the former area should be directly handled by the Government and that of the latter area with additional exposure dose exceeding 1 mSv/year should be conducted by the municipalities. Accordingly, the MOE established Fukushima Office for Environmental Restoration on January 1, 2012 and has been performing the decontamination of special decontamination area and supporting the municipalities to make their plans of the decontamination in intensive contamination survey area.

(2) **Formulation of the decontamination plan**

The MOE has shown a decontamination plan to divide special decontamination area into three areas according to the annual additional exposure dose (refer to Sect. 6.7.4). Therefore, the MOE formulates decontamination progress schedules for each area and the full-scale decontamination of housing areas in areas, to which evacuation orders are ready to be lifted, was planned and carried out from the first quarter of FY2012. As for the intensive contamination survey area, where decontamination should be implemented by municipalities, the designated municipalities are formulating their decontamination plans based on the Act on Special Measures taking the operability and current contamination circumstances into consideration. Suitable decontamination methods should be selected according to the guideline relevant to decontamination publicized by the MOE in December 2011 (refer to Sect. 6.7.3). The Guideline describes the survey methods used for the contaminated portions (to determine measuring points and methods) and decontamination methods for houses and buildings; specifically roofs, gutters, side ditches, outer walls, garden trees, fences, walls, benches and play equipment. In addition, methods to decontaminate roads are described for side ditches, pavement surfaces and unpaved roads. As for the decontamination of soil, it describes methods used to decontaminate the soil of schoolyards, gardens, parks and agricultural land. To decontaminate shrubs and trees, methods to decontaminate plants in living areas such as lawns, street trees, and forests are specifically described.

Current progress in decontamination in intensive contamination survey area varies significantly according to municipalities. To accelerate decontamination progress, temporary storage yards of waste such as removal soil is necessary and understanding and consensus among local residents will be indispensable.

6.7.5.2 Conclusion and Future Tasks

(a) In 2011, immediately after the accident, no unified practical approach to decontamination had been effected, but following the establishment of the Act on Special Measures and the Fukushima Office for Environmental Restoration, MOE, unified policy was implemented for decontamination target objects managed by the ministries and agencies. Meanwhile, there are circumstances where individual objects such as housing areas, agricultural land or roads are independently decontaminated and certain areas are not totally decontaminated. Total decontamination in an area by cooperation of ministries will be required to decrease the radiation dose effectively.
(b) It is rational that decontamination and waste management implemented by municipalities are advanced in consideration of the situation of local areas. Therefore, flexible and prompt action to revise decontamination plan is needed. A system to allow the prompt decision-making will be required, including the discretion of the municipalities.
(c) The temporary storage yards are indispensable to advance decontamination. The Government, prefectures and municipalities make best effort for consultation with stakeholders for installation of the temporary storage yards.
(d) The Government should explain transparency and safety of decontamination project to promote installation of temporary storage yards, interim storage facilities and so on
(e) In addition, the Government, prefecture, municipalities, organization and company relevant to nuclear energy must coordinate and cooperate to implement effective and prompt decontamination in the contaminated areas. Steady decontamination must be implemented under cooperation with local residents.

6.7.6 Decontamination Technology

6.7.6.1 Outline of Decontamination Technology

(1) **Definition of decontamination technology**
Decontamination technology is generally defined as that of removing radioactive materials. In this paper, it is also defined as technology of diluting radioactive materials emitted by the Fukushima Daiichi Nuclear Power Station accident, in order to decrease the additional exposure dose of local residents (MEO Guideline relevant to decontamination, December 2011, first edition). Additionally, technologies of preventing the migration of radioactive materials from soil to crops on farmland and of shielding radiation such as Tenchi Kaeshi (upside-down) and soil covering are also included in the wider definition of decontamination technology.
(2) **Decontamination object**
Currently, as the short half-life of radioactive nuclides such as iodine and tellurium has become slightly problematic, the decontamination objects are

134-Cs and 137-Cs generated from the accident. The β-emitting energies of 134-Cs and 137-Cs are 0.658 and 1.176 MeV, with maximum ranges in the air of 1.9 and 4.2 m respectively. The average γ-emitting energies of these nuclides are 700 and 662 keV respectively, while their mean free ranges in the air are approximately 110 m.

Emission of strontium is approximately 1/100 of that of cesium and its cumulative amount over wide areas may not be problematic. It will be subject to decontamination if detected.

(3) **Classification of decontamination technology**

Decontamination technology is classified into three categories as follows:

(a) Removal technology; removing radioactive nuclides from contaminated media, or removing contaminated media with nuclides such as contaminated soil,
(b) Washing technology; washing media such as contaminated asphalt road to remove radioactive nuclides,
(c) Diluting/solidifying technology; diluting or solidifying radioactive nuclides.

In practical, multiple technologies is to be used for achieving effective and reasonable decontamination in terms of total time and costs rather than using a single technology. After the accident, various bodies, including national and local governments, have implemented "Decontamination Model Projects" and "Decontamination Demonstration Tests of Decontamination Technology". The following is a summary of results obtained from the model projects and demonstration tests. The newly obtained knowledge and future issues are also described.

6.7.6.2 Environmental Remediation Model Project

Since the mean free ranges in the air of gamma rays of 134-Cs and 137-Cs are approximately 110 m as noted above, decontamination over a certain wider area is considered to be more effective for the decrease of air dose rather than decontamination of localized narrow area. Accordingly, the environmental remediation model projects are intended to prepare technological data and information to help municipalities plan decontamination activities. Therefore, the projects aimed to verify the effectiveness of various decontamination technologies and area-based decontamination of relatively high radiation dose areas.

In the model project implemented by Cabinet Office and entrusted to the JAEA, the 11 municipalities including restricted areas are classified into three groups according to the variation in air dose and land use, and presented details of the effectiveness and applicability of decontamination methods and safety measures for workers are reported based on actual data obtained from the project.

(a) Fukushima prefecture Area-based Decontamination Model Project (Fukushima prefecture)

Duration: November 2011–February 2012
Place: Takinoiri, Kotakinoiri and Otaki districts, Onami, Fukushima city
Work contents: Area-based decontamination using technologies expected to be effective was implemented in the area with additional exposure dose of 1–20 mSv/year. Technical data and information of decontamination was summarized to help municipalities implement the decontamination activities in the future and "Guidance for area-based decontamination" was provided for officials of municipalities.

(b) The decontamination model demonstration project in restricted and deliberate evacuation areas (Cabinet Office (JAEA)).

Duration: November 2011–March 2012
Place: Eleven municipalities within the restricted area and deliberate evacuation area divided into three groups
Work contents: Area-based decontamination of relatively high radiation doses area was implemented and the following outcomes including applicability of decontamination technology and safety measures to protect workers against radiation were obtained:

- Development, an applicability and effectiveness of the decontamination technology,
- Planning implementation and evaluation of decontamination,
- Planning, implementation and evaluation of monitoring,
- Reduction of air dose rate by decontamination,
- Planning of radiation and safety management, and implementation and evaluation of radiation and safety management,
- Planning, implementation and evaluation of disposal of removed materials generated from decontamination.

6.7.6.3 Decontamination Tests of Decontamination Technology

The purpose of demonstration tests is to find valuable technologies from effective, economical and safety point of views for the future decontamination work by the public. The Cabinet Office (entrusting the work to JAEA) received 305 proposals to improve the efficiency of decontamination work and reduce the volume of removed contaminated materials; 25 of which were selected by a committee, including some external experts, whereupon related demonstration tests were implemented. The MOE, MAFF, Forestry Agency and Fukushima prefecture also implemented demonstration tests on specified tasks such as reducing the dose, reducing the volume and improving work efficiency.

(a) Decontamination demonstration tests, FY 2011 (the Cabinet Office (the JAEA))

Duration: November 2011–February 2012
Place: Fukushima prefecture and others, including restricted area and deliberate evacuation areas
Work contents: Implementing demonstration tests for 25 proposals to improve the efficiency of decontamination work and removed contaminated materials and assessing the effectiveness.

(b) "Development of technologies to remove and decrease radioactive materials from facilities around forests and farmland", consigned research project financed by the third supplementary budget in FY 2011, (MAFF)

Duration: November 2011–FY 2012
Place: Iitate village and Kawamata town in Fukushima prefecture
Work contents:
(1) Establishing methods to safely remove fallen leaves in forests adjacent to farmland and settlements, and developing technologies to reduce the radiation dose and prevent radioactive materials in forests from spreading to surrounding areas
(2) Development of equipment to decontaminate agricultural facilities such as channels and drainages, dikes, farm roads and adjacent areas to prevent the farmland from re-contamination
(3) Development of technologies to prevent contamination of surrounding areas from drifting dust and technologies to make the dust into pellets or chips to reduce its volume and stabilize the removed plants and others.

(c) "Verification and development on technology of forest management", among the technology demonstration and development projects to prevent the diffusion of radioactive materials in FY 2011, (Forestry Agency)

Duration: November 2011–March 2012
Place: three places in Hirono town, Futaba county, Fukushima prefecture
Work contents: Verification and development of technology to prevent radioactive materials generated by the accident from diffusing in the forests, which occupy 70 % of this area and have a function for the public benefit such as recharging water resources. The effects of forest management measures such as planting and logging on preventing diffusion and reduction of radioactive material were verified, while the effects of surface soil erosion prevention work and muddy water effluence prevention work on preventing diffusion of radioactive materials were also assessed.

(d) "Decontamination technology demonstration project, FY 2011 (Ministry of the Environment)

Duration: April 2011–September 2012
Place: prepared by applicants
Work contents: To find technologies useful for future decontamination and confirm their effectiveness, economic efficiency and safety, (1) Technology to make decontamination efficient, (2) Technology to reduce the volume of removed contaminated materials such as soil, (3) Technology to process waste materials contaminated with radioactive materials, (4) Technology to collect and process discharged water, (5) Technology to transport and temporarily store removed materials and (6) Technology relevant to assisting decontamination are all demonstrated.

(e) Decontamination technology demonstration project (FY 2011), Fukushima prefecture

Duration: November 2011–January 2012
Place: Areas in Fukushima prefecture
Work contents: About 20 proposals on improved decontamination technologies including technology to decontaminate building structures (roofs, rooftops, wall surfaces, bottom surfaces etc.), technology to reduce the volume of soil (other than farmland) and other technologies relevant to decontamination were adopted for the demonstration. The results were also publicized to promote effective and efficient decontamination and facilitate future decontamination activities expected to be implemented in the prefecture.

(f) Study on the behavior of radioactive cesium in hydroponic work and decontamination (FY 2011)
Study on the behavior of radioactive cesium in a pilot hydroponic culture farm and decontamination (FY 2012) (Field Test Working Group, Clean-Up Subcommittee, Fukushima Special Project, Atomic Energy Society of Japan).

Duration: August–November 2011
May–October 2012
Place: Hirohata district, Baba, Minamisoma city, Fukushima prefecture
Work contents: Field tests were implemented on a decontamination technology called "Shirokaki", which is applied to hydroponic rice fields and which is little-known overseas, using a rice field in Minamisoma city in FY 2011 and the behavior of radioactive cesium in the hydroponic rice field and effectiveness of decontamination were assessed. In addition, in FY 2012, soil and rice plants collected at various stages of paddy cultivation in the pilot farm had their radioactive cesium concentrations measured, while the effects of zeolite dispersion and potassium fertilizing on the migration behavior of radioactive cesium relative to unpolished rice were observed and assessed.

6.7.6.4 Conclusion and Future Issues

The following knowledge was obtained from the decontamination model projects and the decontamination technology demonstration tests:

(1) **Area-based decontamination**

- It was recognized that the area-based decontamination decreased the air dose rate of the whole area.
- The reduction effects of decontamination were higher in areas with higher air dose rates.
- The results showed that the effectiveness of decontamination was higher on soil and concrete surfaces and lower on grassland and forests.

(2) **Applicability of the individual decontamination technology**
New knowledge was obtained regarding the applicability and effectiveness of three categories of decontamination technology for various objects (building structures, soil, arable land, roads, forests, ponds, organic substances and timber) and water-processing technology as shown below.

(a) Removal (removal decontamination)

- As for the soil, like the Chernobyl accident, 90 % of cesium remains around 5 cm from the soil surface and scraping off the surface soil is recognized as an effective means of decontamination, while a method of spraying fixation agent and scraping off the surface of the soil thinly and evenly is recognized as effective; especially for farmland. However, the volume of contaminated waste materials generated from such decontamination is considerable.
- It was recognized that "plowing" (rough scraping) was effective in decontaminating the paddy fields.
- As for roads, it was recognized that removing deposits from side ditches was effective due to the presence of hot spots with concentrated contamination in ditches rather than on the pavement surface. It also emerged that most cesium remained at a depth of 3 mm from the pavement surface with dense particles and 5 mm from the permeable pavement surface. Therefore, it is recognized that stripping the pavement surface is a highly effective means of decontamination and reduces the volume of removed waste materials.
- It was recognized that stripping such as blasting could also effectively decontaminate the surfaces of building structures. A method to remove cesium together with chemicals, which was done by spraying cohesive chemicals such as paint, waiting some days and breaking away the solidified chemicals together with cesium, was recognized as an effective means of decontamination, but its disadvantages included the several days of curing period required to solidify the chemicals, the reduced decontamination effects in heavily contaminated areas and the relatively

high cost per area. However, the volume of removed waste materials was relatively small.

- In ponds, removing the surface layer of subsoil was recognized as an effective decontamination method.
- As for forests, the higher portion of broad leaf trees tends to have higher dose rate and the great amount of radioactive cesium remains on the leaves and branches. In case of broad leaf trees without their leaves during the accident, radioactive cesium tends to be immobilized on the litter layer (humus topsoil). Removing the litter layer and fallen leaves, cutting branches and pruning evergreen trees are recognized as effective decontamination method of forests adjacent to residential areas. However, it was reported that the air dose rates at boundaries between forests and residential areas were hardly decreased by decontamination in forests over 20 m within the boundary, so decontamination of forests may be limited in areas about 20 m from the boundaries. Decontamination of entire forests is difficult for the abundant waste materials generated from such decontamination, and solving this issue will be a challenge in future. It was also recognized that using zeolite as a filler to prevent surface soil erosion and muddy water was effective for decontamination ("Technical guideline on the removal of radioactive materials and diffusion prevention in forests").
- Phytoremediation is a method to decrease cesium concentration in farmland soil by exploiting the power of plants to ingest nutrition via roots. In case of restoration of farmland in Chernobyl, oilseed rape was reported to be effective. In Japan, the Institute of Environmental Sciences (IES) tested Amaranthaceae (celosia, Achyranthes bidentata var. japonica) and reported their effectiveness. Sunflowers were also tested, but evaluation of their effectiveness varied and the effectiveness of this method is generally lower than that of the removal of soil surfaces.

(b) Washing (washing decontamination)

- High-pressure washing was used in the Chernobyl accident and also adopted during the initial stage of decontamination in Japan, due to the simplicity and convenience of the equipment involved. However, the wiping-off method is recognized as superior to high-pressure washing for decontaminating housing areas and building structures from a feasibility perspective, due to the diffusion of cesium with washing water and the difficulty in collecting all the contaminated water. The collected washing water is processed with flocculant to precipitate and collect the cesium. The washing method is also used to decontaminate surfaces of tree trunks, but contamination of the surface of tree trunks is relatively light compared to that of leaves and branches and significant effectiveness is not expected.
- To disseminate the volume reduction technology, a criterion for reutilization of decontaminated soil must be prepared.

(c) Diluting (dilution)/solidifying (solidification)

- Inversion tillage; "Tenchi Kaeshi" dilutes cesium concentration by stirring or interchanging surface soil and subsoil and contributes the reduction of air dose rate. This method is effective in areas with relatively lower contamination on the surface of farmland soil. Inversion tillage is a method to invert soil to the necessary depth using a tractor with a plow. The "Tenchi Kaeshi" is a method whereby about 5 cm of the surface soil is scraped off, temporarily stored, then interchanged with about 45 cm of subsoil by placing it under the latter. The working progress of inversion tillage exceeds that of "Tenchi Kaeshi". A survey on the distribution of cesium concentration in depth direction and tilling base (the dense hard soil layer formed immediately below the plow layer) is required before implementing these methods.
- Dilution and solidification using chemicals; this is a method to solidify materials using chemicals to prevent the radioactive cesium contained in soil or walls of houses from refloating. The walls of houses are covered by acrylic painting, while soil can be covered by lawn, gravel, asphalt and so on.
- In farmland, to control and prevent the radioactive cesium contained in soil from migrating to food, potassium fertilization (dilution) and spraying bentonite and zeolite (solidification) can be used. The effectiveness of this technique in controlling the migration of cesium to brown rice has also been recognized.

There are some other tasks for the rational and efficient implementation of future decontamination work as shown below.

- Even if applying same technology, the effectiveness of decontamination may vary according to places and objects. Decontamination methods should be individually selected according to the characteristics of places and objects.
- In selecting decontamination methods, reasonable and efficient methods should be combined in consideration of the time, costs and volume of waste materials.
- A system to integrate the obtained data and reflect the knowledge and results into decontamination policies and guidelines timely should be established by cooperation and coordination between national and local governments.
- Development of new decontamination technology should be promoted by industry-government-academia closely collaborating under clarification of the purposes and time schedule for practical application.
- Radioactive cesium, which is not solidified, migrates from densely to thinly concentrated places by the movement of wind, rain, people and vehicles. Therefore, it is necessary to continue radiation monitoring, because the radioactive cesium concentration may increase even in decontamination areas.

6.7.7 *Volume Reduction*

To process and dispose of waste contaminated by radioactive materials or secondary waste generated from the decontamination, reducing the volume is useful to decrease the total volume of waste and indispensable to secure waste storage sites. However, full attention on the behavior of radioactive materials is required during the volume-reduction process and when handling the concentrated radioactive materials. The Special Measures Act (refer to Sect. 6.7.3) stipulates that designated waste should be processed by Government (Article 19), sets out criteria for processing specified waste (Article 20) and the applicability of the Waste Management Law (Article 21). It also stipulates methods to process the designated waste by incineration, shredding, and exhaust, drainage and measures and processing methods for dust (Article 25 of the Ordinance for enforcement of the Special Measures Act).

6.7.7.1 Classification of Volume-Reduction Methods

Various methods expected to effectively reduce volume have had their characteristics and other issues compared. Volume reduction is a technology to reduce the volume of waste by physically or chemically separating specific components from the contaminated waste or reducing the occupied volume by reshaping. The former is divided into two methods, i.e. (a) method to separate volatile portions by heating, (b) other methods without using heating. Method (a) is classified according to the temperature applied; (1) melting, (2) high-temperature incineration, (3) - low-temperature incineration and (4) desiccation. Method (b) is classified as (5) washing, (6) sorting, (7) compression and (8) crushing. The volume reduction may be accompanied by changes in weight and (movement) migration of the radioactive materials. If these changes result in an increase in radiation, classification of radioactive waste and preventive measures are required. If the changes result in a decrease in radiation, preventive measures applied to the secondary waste generated, such as discharged gas and water, are required.

The above methods are classified by objects as shown in Table 6.29.

Table 6.29 Classification of volume-reduction methods by objects

Objects	Volume-reduction method
Soil	Melting, incineration, washing and sorting
Wood	High-temperature incineration, lower temperature incineration, washing and compression
Grass, rice straw	High-temperature incineration, low-temperature incineration, washing and compression
Concrete	Compression and crushing
Polluted mud	Desiccation (Drying) and washing

(1) **Melting method**
This method involves melting solid waste at an ultra-high temperature (over 1,200 °C) to reduce its volume. The method is very effective at decontamination and generates stable solidified molten waste. However, the method also generates abundant secondary waste due to the volatilization of radioactive materials and collection of hydrated silica, making it difficult to apply when processing contaminated waste in bulk, considering workability under ultra-high temperature conditions and heating costs.

(2) **High-temperature incineration method**
This method involves heating contaminated waste over 1,000 °C in air (with heavy oil combustion etc.) and reducing its volume by volatilizing the water it contains as well as other volatile oxides.

(3) **Low-temperature incineration method**
The method involves heating wood and other materials at a temperature of 600–800 °C and carbonating the materials to reduce their volume. The volume-reduction rate is about 90 % due to volatilization and removal of hydro-carbon components, volatilization of cesium is prevented and no secondary waste is generated. However, as the cesium is condensed and the radioactive concentration increases, full attention must be paid to handling and storage methods, particularly when it exceeds 8,000 Bq/kg.

(4) **Desiccation**
This method involves placing materials at an ordinary temperature or heating them to under 100 °C to evaporate water. The method effectively reduces both the volume and quantity of materials with significant water content such as sludge. For grass and trees, desiccation is useful to reduce volume and also eliminates the need for secondary processing caused by decomposition.

(5) **Washing**
This method involves dissolving soluble components in materials by washing with highly pressurized water or sousing in water to reduce the volume. In addition, fine particles are also suspended and filtrated for solid–liquid separation to reduce the volume. Although the method has decontamination effects, the waste collected has high water content and a volume-reduction process with desiccation is needed. Moreover, as the radioactive cesium has been condensed in the collected water, contaminated water processing using absorption and separation is needed.

(6) **Sorting**
This method involves separating and sorting fine particles such as clay to which radioactive cesium adhere by screening and reducing the volume of contaminated soil. The wet sorting method offers superior separation capability, but increased secondary waste, such as contaminated water. The dry sorting method separates materials coarsely, but does not generate any secondary waste. If the effects on radioactive materials decontamination are gained by eliminating particles by sizes, the method may be an effective means of preprocessing.

(7) **Compression**
This method involves compressing waste materials of low bulk density to reduce their volume. Although it is not linked to the migration of radioactive materials, the waste becomes highly dense after processing and the radioactivity per unit weight increases, which means full attention must be paid to handling and storage methods. The method can effectively decontaminate wood and grass waste.

(8) **Crushing**
Storage of rectangular or spherical contaminated materials requires a relatively large cubic capacity. To avoid this, this method involves crushing and compacting these types of contaminated waste to decrease the storage volume of the materials. There is no migration of radioactive materials and no decontamination effects are expected. Precautions must be taken to prevent materials from spattering during the crushing.

6.7.7.2 Model Volume-Reduction Projects

As for volume-reduction projects, the Japan Atomic Energy Agency (JAEA), as a designated public institution, invites public proposals on decontamination model demonstration projects and tests on the processing and disposal of contaminated soil are implemented. The MOE and MAFF also conduct decontamination model projects on forests, grass and trees (refer to Fukushima Technological Headquarters, JAEA website on the results of assessment of decontamination technology demonstration projects in FY 2011; http://www.jaea.go.jp/fukushima/kankyoanzen/d-model_report_/report_3.pdf and http://www.jaea.go.jp/fukushima/techdemo/h23/h23_techdemo_report.himl)

6.7.7.3 Conclusion and Future Task

The methods of reducing the volume of contaminated waste are summarized above. The contaminated waste is kept in temporary storage yard, subsequently moved to interim storage and finally disposed of in the final disposal site. During this process, the fewer transported materials are involved, the easier they can be managed. To process, store and dispose of the primary pollutant and radioactive waste generated from the decontamination, volume-reduction processing and reutilization are crucial and various volume-reduction processing systems are proposed according to the objects. The following are future tasks:

- It is desirable to develop suitable methods to reduce the volume of soil, which can be practically applied from secondary waste generation and economic efficiency perspectives.
- High- and low-temperature incineration methods are supposed to be useful to reduce the volume of woods, grass and rice straw. The former method requires

the secure processing of volatile cesium, and while the latter method may be useful, a demonstration is required.

- The volume-reduction processing and overall disposal should be collectively assessed from the perspectives of the volume of secondary waste generated and economic efficiency.

6.7.8 Temporary Storage Yard, Interim Storage Facilities and Final Disposal Site for Waste Generated from Decontamination

6.7.8.1 Importance of Promptly Establishing a Reasonable Storage System for Waste Generated from Decontamination

Waste generated from decontamination to restore the environment of Fukushima includes removed soil, grass and trees, rubble and incinerated ash of combustible materials and sewage sludge processed with incineration. This waste generated from decontamination must be securely stored to prevent public exposure as far as possible until being disposed of in the final disposal site. The following three methods are determined by Government policy [24]:

- On-site storage: small volume of removed soil etc. is temporarily stored on site.
- Temporary storage yard: waste is accumulated in temporary storage yard established by municipalities and stored there for about 3 years.
- Interim storage facilities: waste is stored in the interim storage facilities expected to be constructed in Fukushima prefecture and stored there for about 30 years.

There is therefore an urgent need to focus on installing and constructing these storage facilities. The Government has designated several candidate sites for interim storage facilities amid ongoing consultation with local people, although no location has yet been determined. Meanwhile, municipalities in Fukushima prefecture are promoting the establishment of temporary storage yard, although some municipalities have fallen behind schedule due to the time taken to obtain local residents' consent. To promptly install and construct such temporary and interim storage facilities, which are crucial to facilitate the decontamination, the key task in future will be understanding and cooperating with residents in the neighboring areas.

6.7.8.2 Amount of Materials in the Process from Waste Generation from the Decontamination to the Final Disposal Site [25]

Specified and other forms of waste such as removed soil will be generated from decontamination in Fukushima prefecture. Specified waste comprises waste in the measures area; estimated at about 500,000 t, and designated waste, which is estimated at about 60,000 t/year. Specified waste is divided into two categories, i.e. one with radiation under 8,000 Bq/kg and another with radiation exceeding 8,000 Bq/kg. The former is processed similarly to waste from outside the measures area, while the latter is processed like designated waste. Combustible materials among the designated waste, such as polluted mud and rice straw, are incinerated and the incinerated ash and incombustibles are divided into two categories, i.e., one with radiation exceeding 100,000 Bq/kg and another with radiation under 100,000 Bq/kg. The former will be disposed of at controlled landfill-type sites while the latter will be sent to interim storage facilities. The amount of removed soil other than specified waste to be generated is estimated at 15–30 million m^3. Combustible materials will be incinerated and processed using the same method applied to incinerated ash from designated waste. Incombustible material is temporarily stored in the temporary storage yard, before being transferred to interim storage facilities. Incinerated ash and incombustible materials with radiation exceeding 100,000 Bq/kg will be stored in the interim storage facilities to be constructed in Fukushima prefecture for 30 years and eventually disposed of at final disposal facilities.

The amount of specified waste generated outside Fukushima prefecture is estimated at 80,000 t/year; all of which constitutes designated waste with radiation exceeding 8,000 Bq/kg. These combustible materials such as polluted mud and rice straw are incinerated, and the incinerated ash and incombustible are divided into two categories, i.e. with radiation under and over 100,000 Bq/kg respectively. The former will be disposed of at controlled landfill-type repository, while the latter will be disposed of at isolated-type disposal sites. The amount of soil and waste other than specified waste is estimated at about 1.4–13 million m^3. Combustible materials are incinerated and the incinerated ash is processed using the same method applied to incinerated ash of designated waste. Incombustible material such as the removed soil is temporarily stored in a temporary storage yard, and will subsequently be disposed of at a controlled landfill-type repository.

6.7.8.3 Ensuring the Safety of Temporary Storage Yards and Installation Conditions

The temporary storage yard is a temporary storage facility for removed soil and waste generated from decontamination. In Fukushima prefecture, when interim storage facilities are available, the material involved will be gradually transported to such facilities, whereupon sites will be restored to their original states.

The storage duration at such temporary storage yard is defined at approximately 3 years and the installation conditions of the temporary storage yard are shown in Fig. 6.26 [26]. Measures to ensure the safety of the removed soil, including preventing scattering and outflow of the stored soil (soil covering or soil containing in container), preventing rainwater ingress (rainwater permeation prevention sheet), measures to prevent subsurface contamination (water shielding sheet), protection

Examples of ground-type temporary storage yards | Example subsurface-type temporary storage yards

Fig. 6.26 Example of temporary storage yard installation conditions. *Source*: Material for reporting meeting on the results of decontamination model demonstration projects (Nuclear Suffers Life Support Team, Cabinet Office; Ministry of the Environment; Japan Atomic Energy Agency (JAEA), March 26, 2012)

measures against radiation (placing areas off-limits) and measures to prevent fire disaster (when storing combustible materials) are all made compulsory.

Venues established as temporary storage yard should basically be prepared by municipalities or local communities. For Fukushima prefecture, the MOE directly prepares such venues as special areas for decontamination (including Futaba town and ten other municipalities) with the cooperation of local governments. The national government is responsible for providing financial and technological support to the decontamination implementation areas, while municipalities will set up temporary storage yards. In view of the crucial need for understanding and cooperation from local residents to establish such temporary storage yards, the Clean-up Subcommittee of the AESJ prepared a Q&A pamphlet to explain the requirements for temporary storage yards and safety measures for residents and local governments, which was posted on their website in May 2012.

6.7.8.4 Transportation of Decontamination Waste Accumulated in Temporary Storage Yards

In association with the progress of decontamination work, considerable decontamination waste will be generated and it is crucial to ensure that it is transported safely and efficiently. Moreover, when transporting bulk removed soil accumulated in on-site and temporary storage yards scattered over a wide area to interim storage facilities, it is particularly important to select suitable transportation methods and routes. The Government is deliberating these problems via the Deliberation Committee on Safety Measures for Interim Storage Facilities, including spatial and temporal separation from residential areas and general traffic and large-scale package transportation. There is also specific consideration of transport routes, time zones, vehicle types, packing types and possible transport amounts for the transportation of removed soil and other materials from temporary storage yards to interim storage facilities based on current circumstances and taking future transfer into consideration [27].

6.7.8.5 Concept of Interim Storage Facilities and Installation Plan

As for bulk waste expected to be generated as the decontamination in Fukushima prefecture progresses, such as removed soil and other than specified waste, no definite final disposal policy has yet been formulated. The current policy thus involves storing them safely and intensively for a certain period; for which a number of interim storage facilities will be constructed in Fukushima prefecture. The Government requested Fukushima prefecture and eight municipalities in Futaba county to deliberate on construction of such facilities within the county in December 2011, and made a proposal to implement a survey on the construction of interim storage facilities in candidate areas in Futaba, Okuma and Naraha town in March 2012. Fukushima prefecture and the municipalities in Futaba county intend to cooperate with the Government for the survey, but have not agreed to the construction. As for prefectures outside Fukushima, disposal will be implemented using existing controlled-type repositories within each prefectural area and no interim storage facilities will be constructed.

Images of the interim storage facilities currently planned [28] are shown in Fig. 6.27. The "high concentration elution responsive-type facility" shown on the right of the figure is an interim storage facility to store decontamination waste such as incinerated and volitant ash, which has highly concentrated radioactive materials, via incineration disposal and with the potential of eluting radioactive cesium. The facility includes a concrete pit (artificial structure with an external bulkhead), which blocks radiation to maintain the air dose rate within the site boundary to within a level stipulated by regulations. The facility also functions to seal off radioactive materials and prevent them from leaking outside during the storage period.

The "low-concentration non-elution responsive-type facility" shown on the left of the figure is an interim storage facility to store removed soil generated from

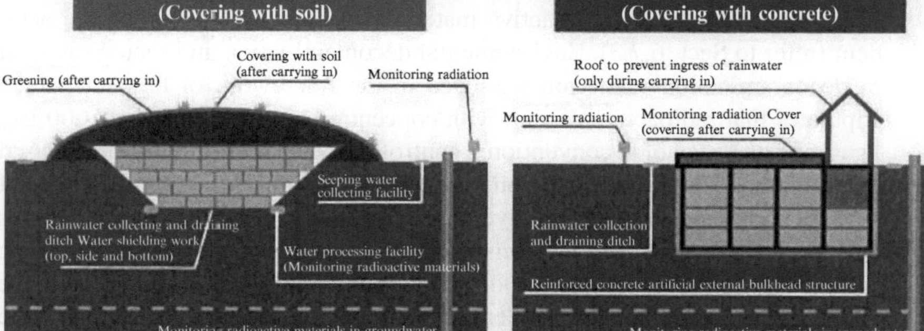

Fig. 6.27 Images of interim storage facilities. *Source*: "Deliberation Committee on Interim Storage Facilities Environment Preservation Countermeasure, the Ministry of the Environment" (the first meeting), Reference 4, "Outline of interim storage facilities", June 28, 2013

decontamination in areas of relatively lower contamination. Part of the facility exposed from the ground surface is covered by soil to shield radioactive materials, while water-shielding work (including water-shielding sheets) is also constructed to prevent the removed soil from coming into contact with rainwater and/or groundwater and thus preventing the leakage of radioactive materials or at least minimizing the same as far as possible.

6.7.8.6 Final Disposal of Decontamination Waste

(1) **Restriction on the amount of decontamination waste generated and disposed of**
According to a recent estimation by the MEO [28], the amount of waste subject to storage in interim storage facilities will be 15–28 million m^3, which will impose a heavy final disposal burden. Limiting the amount of and compacting generated waste will be among the key future tasks to promote environmental remediation. Methods to suppress the amount of generated waste include selecting decontamination methods without generated waste (e.g. Tenchi Kaeshi method) or methods to suppress the generation in the course of environmental remediation. In addition, the volume at interim storage facilities should be reduced, as well as the final disposal volume by volume reduction treatment. With regard to volume reduction, as mentioned in Sect. 6.7.7, various technologies exist such as washing, classifying, crushing, desiccation, compression, melting and some chemical treatment, which can also be applied to reduce the volume. For the technological development of volume reduction, cooperation from experts with experience and knowledge of processing and disposal of radioactive materials, including nuclear operators, Government research institutions, universities and academic societies such as the AESJ is crucial.

(2) **Final disposal methods for decontamination waste**
The Special Measures Act was established to manage and regulate extensive waste contaminated by radioactive materials and released following the accident (refer to Sect. 6.7.3). Environmental decontamination and secure disposal of decontamination waste are regulated in the framework of the Act, which stipulates that radioactive cesium with concentration not exceeding 8,000 Bq/kg can be disposed of at conventional controlled-type landfill repositories based on an exposure dose assessment on residents and workers, assuming the general process and disposal of radioactive waste. In other words, the Act recognizes that radioactive cesium under this concentration meets the regulation value criterion of additional exposure dose for ordinary waste processing and the additional exposure dose of residents in neighboring areas after the end of the controlled period. It also recognizes that the additional exposure dose of workers engaged in the disposal work shall not exceed 1 mSv/year. Decontamination waste exceeding this concentration will be disposed of under the responsibility of the Government as specified waste, pursuant to the Special

Measures Act. In that case, the waste must generally be disposed of at isolated-type landfill depositories and the safety of the final disposal will be maintained by additional measures, including radiation protection for workers engaged in operating the repository. When the character and volume of waste subject to final disposal are confirmed, the safety assessment and design of suitable disposal systems is possible, taking practical conditions into consideration, and disposal can be conducted under optimized conditions.

Meanwhile, it is possible that waste generated from the accident and strongly affected by radioactive materials will emerge in and around the nuclear power station site, particularly the molten fuel debris. The waste in the accident site will be discussed in another section.

6.7.8.7 Conclusion

To facilitate efforts to restore the environment in Fukushima prefecture and neighboring areas, the understanding and cooperation of local residents is crucial. Since a vast quantity of decontamination waste is expected to be generated from environmental remediation, many problems remain to be solved in terms of technology and social acceptance to store the waste and finally dispose of it safely and reasonably. Five primary tasks are shown below as follows:

(1) **Understanding and cooperation of residents in relevant areas**
Sufficient accountability regarding function and roles, the necessity and security of decontamination, temporary storage yard and interim storage facilities is required to obtain the understanding and cooperation of residents. In addition, the residents themselves should actively participate in local decontamination activities to reflect their opinions on environmental remediation activities and widening the scope of such activities is also important.

(2) **Support from relevant nuclear power institutions**
In addition to arranging the system of relevant nuclear power institutions such as the AESJ; supporting decontamination, coordination between local governments and communities is important to facilitate the response to residents, as mentioned in (1).

(3) **Reduction of waste amount**
The estimated amount of removed soil for storage at interim facilities will be 15–28 million m^3 in Fukushima prefecture alone. It is preferable to adopt environmental remediation technology to reduce the amount of soil removed and develop technology to reutilize and reduce the volume of removed soil.

(4) **Safe and efficient transportation of waste**
Transporting the vast amount of decontamination waste safely and efficiently is also important. When transporting waste from a temporary storage yard to interim storage, the routes, time zone, vehicle types, packing type and transport capacity should all be discussed, considering the spatial and temporal separation from residential areas and ordinary traffic.

(5) **Final disposal**
For the final disposal of decontamination waste safely and reasonably after interim storage, a properly designed and operated disposal system should be required.

6.7.9 Environmental Remediation Activities by the Atomic Energy Society of Japan (AESJ)

The AESJ established the Clean-Up Subcommittee under the Nuclear Safety Investigation Experts Committee in April 2011 to analyze decontamination and environmental remediation, offer recommendations and provide information.

Primary environmental remediation activities performed by the Clean-Up Subcommittee are reported below:

6.7.9.1 Recommendation of the Centralization for Monitoring and Environmental Remediation

(1) **Recommendation for establishing a monitoring center**
In a nuclear power station accident like that which occurred in Fukushima Daiichi, sufficient disclosure of information on the circumstances of the accident or the extent of environmental contamination by radioactive materials is crucial. Monitoring information and various data were collected by a number of institutions immediately after the accident, but the data then had to be aggregated and its accuracy assessed. The AESJ recommended establishing the "Environmental radiation monitoring center" as an institution to aggregate the data collected by various institutions and perform integrated analysis, e.g., comparison at measuring points or temporal variation ("Recommendation on environmental recovery caused by Fukushima Daiichi Nuclear Power Station accident", June 8, 2011). This recommendation includes necessity, such as coordination between the local governments concerned, collecting data carefully and disclosing the data and analytical results promptly. It also recommended that an accounting system be set up by radiation protection experts. Subsequently, the MEXT established a monitoring center, which has the functions mentioned above (integrated management by the MEXT).

(2) **Recommendation to integrate the environmental remediation center**
Establishment of an environmental remediation center for environmental recovery in and around the nuclear power station site and prompt verification by decontamination model projects were recommended on July 29, 2011. In this center, the objective is focusing on work outside the nuclear power station site, and its functions include formulating an integrated remediation strategy and programs, demonstration and practice based on strategy and forward-looking programs, specifically, a. investigation for the removal radioactive

materials by applying existing technologies; b. development of new technology; c. demonstration of decontamination technology and d. investigation for the process contaminated materials generated from the decontamination of radioactive materials. The participation of experts from local governments, governmental institutions and institutions for the demonstration experiment was also proposed. The MOE established the Fukushima Office for Environmental Restoration in Fukushima city (January 1, 2012) as an organization undertaking certain of these functions and also opened the Decontamination Information Plaza. In addition, for decontamination technology, the Cabinet Office and other ministries and agencies, including Fukushima prefecture, conducted the decontamination model projects.

6.7.9.2 Presentation of Decontamination Technologies

(1) **Publication of decontamination technologies catalog (referring to EURANOS and so on) and commentary on storage and temporary storage yard**
Investigation of environmental remediation technologies for areas outside the nuclear power station site, environmental remediation strategy, its scenario and remediation technology have been researched; referring to the EURANOS Project and so on. The remediation technologies catalog which include applicable assumptions made for cases in Japan and the AESJ's perspective. Moreover, applicable technologies for 51 objects such as buildings, public facilities, water, agriculture, stock raising districts, forests, water areas, residential areas and rubble etc., are listed, materials for accounting on the formulation of a decontamination plan are prepared and publicized at the AESJ website.

- Presentation of the decontamination catalog version 1.0 [29]

Temporary storage yard, storing removed soil generated from decontamination, is explained on its location conditions, facilities and the management requirement for ensuring safety based on the MOE's "Guideline on storage of removed soil, first edition" (December 2011) and supplementing with items recommended following deliberations of the AESJ Clean-Up Subcommittee.

(2) **Demonstration experiment on rice field decontamination technology**
The decontamination technology catalog mentioned in [1] focused on European cases and knowledge on rice fields was scarce. To perceive points which could not be ascertained in desk study, decontamination technology for rice fields was demonstrated and confirmed. During the first year, demonstration of decontamination methods to reduce radiation exposure of farmers during the farming work was performed, followed in the second year, by the demonstration of a method to suppress migration of radioactive cesium during rice cultivation.

As for decontamination experiments to reduce radiation exposure, the plowing, which farmers could implement and which would generate no waste, was selected from among those recommended for the demonstration.

As for the suppression of radioactive nuclide migration during wet-rice farming, conditional on the presence or absence of zeolite spraying and potassium fertilizing, the effectiveness of cesium migration suppression on raw rice was confirmed by a combination of these conditions. The demonstration work was performed under cooperation of the Agricultural Management and Economy Department, Japan Agricultural Cooperatives (JA) Soma and the owner of the rice field.

(a) Demonstration of radiation exposure reduction effects of the plowing: In 2011, two times of plowing, were performed over a rice field, which had not been plowed since the accident, in Hirohata district, Baba, Minamisoma city, Fukushima prefecture. In both cases, water was drained immediately after the plowing work had been completed.
    As radioactive cesium was present relatively near the ground surface, samples were collected from the surface layer (5 cm from the surface) and plow layer (15 cm depth) after the plowing work and their average radioactive concentrations were compared. It was confirmed that the radioactive concentration had fallen by approximately 50 % after the first plowing (in August), and by a further 50 % after the second plowing (in September).
(b) Confirmation of effectiveness to suppress radioactive cesium migration to raw rice: Factors stimulating the migration of cesium to crops are deemed to be the soil type, irrigation water and fertilizers used. Using zeolite spraying, which is expected to be applicable technology for sanitizing or decontaminating agricultural soil, and potassium fertilizer, which was also used in Chernobyl, the effects to suppress radioactive cesium migration were confirmed by comparing the results with and without applying these technologies.
    In 2012, following the year of the plowing, fertilizing, rice reaping and threshing were performed, and samples of soil, rice and water were collected during each work procedure. The radioactive cesium concentration of all the collected raw rice was far below the standard value (100 Bq/kg) of general food (less than 1/3 of the value), regardless of zeolite spraying and potassium fertilizing, while the migration coefficient of cesium from soil to the harvested raw rice (ratio of concentration between soil and raw rice) was less than 0.01, which means the coefficient is less than 1/10 of that used for radioactive waste disposal.

6.7.9.3 Communication with the Local Communities

The AESJ, as an expert group, held various events to keep the good relationship between national Government and local governments and residents, formulate remediation plans, inform basic selection items and adapt remediation technology. The AESJ also cooperated in operating the decontamination information plaza, which was previously mentioned in the recommendation to integrate the environmental remediation center. The following is a summary of these activities.

(1) **Holding Security Safety Forum**
The Radiation Effects Subcommittee and the Clean-Up Subcommittee, cooperating with Fukushima prefecture, held the "Security Safety Forum", a combination of lecture and panel discussion meetings at Iizaka Spa, Koriyama city, Minamisoma and Iwaki cities in FY 2011. In FY 2012, the Forum developed into "Local Communities Dialogue toward Decontamination Promotion Forum" extending the scope of its discussion by including radiation monitoring, impacts to human health, environmental remediation and temporary storage yard safety. This forum was opened five times.
The main discussion items in the Forums were radiation impacts for children, adequacy of established criteria based on those of Hiroshima, Nagasaki or Chernobyl (Republic of Belarus) and adequacy of the establishment of food regulation criteria. To date, the AESJ has published "Collection of Data", the previously mentioned EURANOS data concerning decontamination translated by the AESJ and "Temporary Storage Yard Q&A", a detailed explanation of the MOE's "Guideline relevant to Waste". The AESJ, as an expert group, must continue providing correct and easily understandable information to local people who may not be experts.

(2) **Cooperation with the Decontamination Information Plaza**
The AESJ decided to support promotional decontamination activities by actively utilizing the Decontamination Information Plaza, which has been jointly operated by the MOE and Fukushima prefecture. Accordingly, the AESJ dispatches experts, provides knowledge on decontamination technologies and radiation effects and supports public relations activities to promote the utilization of the Plaza by local residents. Moreover, the AESJ cooperates to conduct mini-workshops at the Plaza and local venues, while members of the Clean-Up Subcommittee have been working as volunteer advisers for the Plaza since its opening.

6.7.9.4 Conclusion

The AESJ has proposed to the Government that a framework to integrate monitoring and environmental remediation be established in response to environmental remediation, and has presented various decontamination technologies to facilitate selection of methods to decontaminate affected areas. It also conducted literature research into overseas cases as a technology survey and selected subjects considering the applicability of cases in Japan. As for rice cultivation, seldom seen in European case studies, the required data were collected by cultivating rice in disaster-affected Minamisoma city. Forums have been opened and dialogue established with local communities at the Decontamination Information Plaza. These activities should be continued in future, improving methods to convey clear and adequately updated information to be presented to the public.

6.8 Simulation Analysis

6.8.1 Computational Science and Technology Analysis

6.8.1.1 System for Prediction of Environmental Emergency Dose Information (SPEEDI)

(1) **Utilization and criticism during the Fukushima nuclear power station accident**

From 16:00 h on March 11, 2011, SPEEDI started playing a role based on previously determined "Environmental Radiation Monitoring Guideline [30]", i.e. when information on the time trend of atmospheric release of radionuclides (hereafter, source term) could not be obtained from the Emergency Response Supporting System (ERSS) which monitored reactor conditions, or the stack monitor, SPEEDI should provide the predicted results on atmospheric dispersion and radiological doses under the assumption of unit release (1 Bq/h) to assist with the emergency monitoring planning. According to Chap. 2 of the verification report from MEXT, the monitoring in a mountainous area of Namie-machi, at which high radiation doses were recorded on March 15, was instructed by MEXT based on SPEEDI's results, whereupon the results were adequately utilized for the monitoring plan. In addition, after March 16, when the emergency monitoring results were available, the source term was reversely estimated by using the results of emergency monitoring and SPEEDI's unit release results and, then, a diagram of thyroid internal exposure dose was shown on March 23 by using this source term. This diagram was utilized for the screening for children's thyroid radiation exposure.

Despite the utilization of SPEEDI in accordance with the Guideline as mentioned above, the facts that the predictions of the diffusion tendency of radioactive materials were not utilized for reference to the evacuation activities and the disclosure was significantly delayed were exposed to heavy criticism. For such criticism, while the Government report to the IAEA ministerial meeting, the Investigative Committee on the accident at the Fukushima Daiichi Nuclear Power Station of Tokyo Electric Power Company (hereinafter referred to as the Governmental Investigative Committee) and Independent Investigative Committee on Fukushima Nuclear Power Station Accident recognized that predictions on the diffusion tendency of radioactive materials should have been utilized as a benchmark for the evacuation activities and SPEEDI's calculation results should have been disclosed immediately after the accident, the Tokyo Electric Power Company Fukushima Nuclear Power Station Accident Investigative Committee indicated that the SPEEDI results could be utilized to inversely estimate radiation emissions, formulate a monitoring plan and judge the timing of venting or lifesaving activities, but that establishing evacuation areas based on computed predictions was at risk of uncertainty and emergency monitoring should be improved.

To further develop these arguments, the accuracy of predictive information from SPEEDI and the timing of provision to the emergency response authorities must be verified, in which the accuracy should be verified by comparing with subsequent monitoring data. This basic verification is important and should be taken into consideration when utilizing the calculation simulation for protective measures and others in future.

(2) **Verification of SPEEDI's accuracy and timeliness**

Predictive information provided by SPEEDI is divided into two types, i.e. "regular information" and "requested information". Regular information provides for the movement of a radioactive plume in the form of air dose rate distribution, presuming the unit release (1 Bq/h) every hour. The requested information is performed by the request from the Emergency Response Center (ERC) of the former Nuclear and Industrial Safety Agency, Off-site Center (OFC) and the former Nuclear Safety Commission etc. for predicting the environmental effects caused by the events in reactors, planning the emergency environmental monitoring, estimating the source term reversely and/or assessing radiation doses by using estimated source term. There is a report focusing on the verification of information provided by SPEEDI to the authorities concerned in terms of its accuracy and promptness by comparing the prediction results with data from environmental monitoring posts uploaded on the Fukushima Prefecture website on September 21, 2012 [31].

According to the report, although the regular information on the movement of radioactive plume had a maximum error of 2–3 h in several cases, it could temporally and spatially reproduced actual movements from hour to hour. The report from the Governmental Investigative Committee also described the SPEEDI's accuracy was deemed adequate for the utilization to determine the timing of evacuation, showing specific examples such as staying in-house on March 15 and evacuating on March 16 in an area to the northwest of the site. The requested information was via trial calculations for the emission of radioactive materials due to venting, hydrogen explosion and several events on the decrease in pressure inside the reactor. Some of the prediction results obtained before the events were nearly identical to monitoring results in the environment, e.g., predictions for venting and hydrogen explosion of the Unit No. 1 on March 12 and by the probable leakage from Unit No. 2 on March 15. Recently, the accuracy of prediction of atmospheric dispersion of materials and the forecasts of meteorology have been significantly enhanced. Now, stereotypes such as "the weather forecast is not reliable" should be eliminated. However, even with the most advanced meteorological and/or diffusion models, it is difficult to perfectly reproduce complicated natural phenomena. When SPEEDI is used to predict the arrival of plumes, its temporal and spatial information should not be assumed as having pinpoint accuracy but considered as having some uncertainty in time and space.

(3) **To apply calculation simulation for protective measures**

(a) Comparison with measures to natural disasters: The basic policy of disaster prevention involves the prediction of future disaster developments and takes countermeasures for the worst case scenarios. In response to meteorological disasters such as typhoons, heavy snowfall or flooding, computer simulation has been the core methodology for immediate response. Although meteorological calculation forecasts are also subject to uncertainty, experts, who understand this uncertainty well, have a role to predict the damage situation and apply this prediction to measures to evacuate residents by the combination of meteorological observation data, statistical information actually obtained in the past disasters and their experiences. In this nuclear accident, the formal reason why calculation forecasts were not used for the evacuation countermeasure was significant uncertainty due to the lack of source term. However, the actual reasons for non-use may include the lack of knowledge, experience and data collection capability to estimate the uncertainty and enhance reliability by expert's judgement. The gap in experience between nuclear accident and meteorological disasters which occur almost yearly is significant, but it is important to establish a mechanism to integrate calculation forecasts and environmental monitoring data in the Nuclear Regulatory Agency (NRA), which will be a control tower for nuclear disaster prevention measures, and also important to call on experts with experience in developing air diffusion models and verification research by field measurements, etc.

(b) Parallel utilization of calculation simulation and environmental monitoring: Calculation simulations and environmental monitoring are in a complementary relationship, i.e. monitoring undoubtedly shows superiority in terms of accuracy over calculation predictions, but significant time may be required to establish a monitoring system under earthquake or severe weather conditions such as heavy snowfall to obtain accurate data on maximum doses and their distribution over a radius zone of 30 km, which will be a main area taking emergency protective measures. Meanwhile, calculation prediction has a large advantage of forecasting and overall perception characteristics, but also involves uncertainty derived from input data or the use of models. As shown in (3) a., interpretation and correction are required based on meteorological information, environmental monitoring data and experience of experts. In this accident, the intended or unintended emission of radioactive materials into the air occurred following a venting operation at Unit No. 1 and a hydrogen explosion within 1 day from the earthquake while residents in neighboring areas were being evacuated. However, it needed 4 days after the earthquake that the emergency monitoring system was organized on the evening of March 15. At the initial stage of the accident, SPEEDI was the only information source capable of providing data in a timely manner to formulate disaster prevention measures and plan emergency monitoring.

However, it only provided relative distribution of radionuclides with certain reliability and no absolute values could be provided, which meant the information was insufficient to determine the evacuation areas.

Taking these circumstances into consideration, even if the quickness of emergency monitoring or source term estimation are improved in future, it is probably unrealistic to expect that the monitoring or calculation prediction parfectly provides the required information immediately after the accident. Thus, contamination situations should be effectively perceived by utilizing a mutually complementary relationship of both methodologies.

(c) Need for experts: It was already mentioned that, to introduce computer simulation technology which involves characteristics of forecasting capability and uncertainty into the emergency response, the calling on experts having experience on developing atmospheric dispersion models and/or field surveys and the integration of meteorological information and environmental monitoring data are important. In fact, since the Nuclear Safety Commission called on SPEEDI's experts in the evening of March 16, the reverse estimation of source term was started using SPEEDI's prediction results and environmental monitoring data, which successively used for predictions of thyroid internal exposure dose and screening tests to assess children's thyroid exposure on March 23.

The expected potential from the participation of experts, particularly immediately after the accident, is shown using an example of the case of March 15, when a contaminated area was formed in the northwest of the site in the evening following rain and significant emissions.

(i) After an explosion sound at 6.00 a.m., the ERC requested calculations to estimate the environmental effects due to the damage of Unit No. 2's suppression chamber, which was afraided at the time, and received calculation results at 6:51 a.m. from SPEEDI. The comparison of a SPEEDI's forecast including an iodine surface deposition distribution map for the 24-hour period from 9:00 a.m. on March 15 with the surface distribution of cesium 137 subsequently measured by an aerial survey clearly shows that surface contamination in the northwest area, which actually occurred in the evening, had already been predicted by SPEEDI on the morning of March 15, regardless of the significant fluctuation in wind direction that day. If environmental and dose assessment experts exist, they might have requested additional outputs such as temporal changes in plume and rain forecasts and found out the timing of contamination that would be formed from evening. In fact, the regular information provided every hour predicted the movement of the radioactive plume almost exactly, from hour to hour.

(ii) Subsequently, at 8:00 a.m., an unprecedented increase in dose was recorded at the boundary of the site. If some experts had concern about the long-term situation continuing or worsening over time, the rough estimation of contamination level formed in the northwest area during the period from evening to night was possible by simple source estimation and/or fitting of prediction to measured values.

(iii) Subsequently experts aware of the uncertainty might have striven to enhance reliability by correcting and assessing the prediction results, comparing predictions with meteorological observations, weather maps and environmental monitoring data. Moreover, another task will collect knowledge from experts on severe accident analysis and use it to predict contamination for anticipating the worst case of emission scenarios. These contamination prediction methods may not provide high accuracy, given the relative inaccuracy of the source term information, but it is still important to predict the worsened circumstances requiring disaster prevention.

The Nuclear Disaster Countermeasure Guideline [32] published by the NRA stipulates the implementation of protective measures based on emergency monitoring in an area subject to preparatory emergency protective measures. However, in the case when large emissions of radioactive materials is recognized from monitors at the boundary of site before the preparation of emergency monitoring and the areas of high radiation doses can be predicted with a certain reliability using the methods described above, it is questionable whether withholding deliberation on protective measures or countermeasures until the establishment of an emergency monitoring system is appropriate. Predicting worsened circumstances and taking early protective measure are adequate means reflecting a disaster prevention mindset, where a flexible and ad hoc response is desirable. In addition, protective measures such as sheltering and taking stable iodine in relevant areas to avoid exposure should be imposed before the arrival of plume. Obviously, the oncoming plume can only be predicted by calculation simulations such as SPEEDI and in this case, the abovementioned predictions based on expert assessments are also important.

Moreover, SPEEDI prediction results will be disclosed immediately after the accident in the future. However, simple uploading such information online is insufficient. Scientific and easily understood commentary on the significance and accuracy of the prediction results should be added to the disclosure and experts capable of preparing such commentary are also crucial. Although such experts are scarce in the nuclear community, many experts can be found in meteorological and environmental science fields. The cooperation requirement should not be limited to the nuclear community alone.

(4) **Future utilization methods**

As mentioned in (2), SPEEDI is capable of relative distribution prediction with certain reliability, even in the absence of source term information. The provision of prediction results continued over 1 year, showing its robustness as a practical system. The staff also gained valuable experience in organizing and operating SPEEDI. Under these circumstances, SPEEDI should be maintained and developed in future, and as mentioned above, SPEEDI and environmental monitoring data should be integrated to the NRA, as a control tower. Moreover, it is also important to establish a mechanism allowing a group of experienced experts to perform an overall judgment.

Based on past verification, the specific future utilization of SPEEDI will be as follows:

(a) Over the entire emergency response period: Formulation of emergency monitoring plan and assessment of the monitoring results, judgment of evacuation timing, sheltering and taking stable iodine, and judgment to determine the timing of planned emissions such as venting,
(b) When significant radioactive materials are emitted before organizing an emergency monitoring system: Fitting discrete monitoring values with predicted distribution assuming unit emissions, or judging protective measures in the early stages based on contamination prediction of the worst emission cases,
(c) In the period after the emergency monitoring system has been installed: Assessment of the accident scale by inversely estimating emissions and assessing the detailed exposure dose, and selecting food inspection priority areas based on large-scale deposition predictions.

To support these utilizations, it is important to develop technologies, including continuous development and improvement of models for the dispersion of radioactive materials in the atmosphere and surface deposition, systemizing emission inverse estimation, integrating meteorological information, environmental monitoring information and multiplexing systems.

6.8.1.2 Numerical Analysis on Seismic Resistance

According to appendices of the application documents for the license to construct the Fukushima Nuclear Power Plant submitted in July 1966, in the then seismic resistant design (Appendix 8 and others), dynamic analysis by the basic seismic ground motion was carried out. As for the Nuclear Power Plant, a seismic response was obtained from a vibration model considering the vibration characteristics and adequate attenuation. This is specifically referred to experimental formulas provided by Kanai et al. Assessment of structural robustness with seismic response analyses was conducted pursuant to the Building Standards Act; primary and secondary stresses were determined by combinations of allowable stress and load; the obtained stresses were evaluated as safe in the case under 90 % of the yield point; also as long as no excess strain was found in the combined stress of local stress, the allowance was extended up to the yield point. As for equipment and piping systems, dynamic analysis was performed at the installed point considering the response of supports, whereupon the deformation and so on were analyzed by seismic loads introduced from the response acceleration.

The designing for seismic resistance in those days intended to be a rigid structure. Important buildings and structures were primarily constructed on base rock and the seismic resistance was determined in accordance with the importance classification of components (As, A, B and C). The seismic resistance for class C facilities were applied a standard coefficient, 0.24 based on the Building Standards Act (the coefficient for facilities is 20 % larger than buildings). The horizontal load applied to class A on static coefficient is three times larger than for class C. In the dynamic analysis, the seismic ground motion of input maximum acceleration +0.18 Gal was applied. For class As, dynamic analyses and others were executed, premising the maintaining their functions and ensuring shutdown process at 0.27 Gal.

After the construction, the evaluation for seismic resistance was continuously done based on instructions from the Nuclear and Industrial Safety Agency (NISA), METI, when the "Regulatory Guide for Reviewing Seismic Design of Nuclear Power Reactor Facilities" was revised. In here, the evaluation for seismic resistance is referred to when formulating standard seismic ground motion, based on the results of a geological and seismic survey, and evaluating buildings, equipments and pipings by the seismic response analysis with the standard seismic ground motion. In these evaluation processes, the most advanced computational science and engineering technology was introduced in the right moment. In addition to the previous survey, additional geological and seismic survey was also introduced by extending the field. The result of those survey and analysis are reported. The evaluation for seismic resistance had done when "Regulatory Guide for Reviewing Seismic Design of Nuclear Power Reactor Facilities" was revised.

(1) **Issues and lessons on seismic ground motion from the computational science and engineering**

At the time, the past seismic history was learned from the "Science Chronology" Earthquake section, edited by H. Kawasumi, Tokyo Astronomical Observatory, published by Maruzen Co. and all potential earthquakes occurring were estimated. The method was based on empirical knowledge, but it could have also been possible to introduce computational scientific and engineering technology at the time. Even now, although no numerical simulation codes targeting and specializing seismic ground motion propagation is available on the market, it can be found in research in the United States and France. In particular, in the Structural Mechanics in Reactor Technology (SMiRT), the world's largest international conference, the theme has been actively and constructively discussed. Presently, the most advanced code is that used to solve geological scale issues. Not only the methodology used to obtain predictions from experimental such as survey of past seismic history, also it is important to utilize a methodology of estimating more rational seismic ground motion. Efforts to obtain more rational estimation are required such as acquiring the most advanced knowledge as much as possible. In addition, probabilistic safety assessment ought to be introduced and applied to the evaluation for seismic resistance.

(2) **Issues and lessons on numerical analysis for seismic resistant calculations from the computational science and engineering**

Using techniques to compute for the entire and detail structures of nuclear power plant, best efforts were made with the technology in those days. However, constantly advancing in technologies, efforts should be made to acquire technology which is advanced or at least effective on a timely basis, reviewing techniques to simulate both entire and detail structures as well as striving to assess, analyze and confirm safety, robustness, elasticity, reliability and quality as required. In other words, the most advanced technology should be acquired to carry out simulations in a more rational and determining the allowable safety coefficient. Meanwhile, it is important to continue leveraging experience and knowledge and computational science and engineering technology during the design process, and also solve problems of improved efficiency for methods and modeling of data preparation work. Moreover, given progress in computers, computational science and engineering technologies should be utilized always with the most advanced computers and softwares. In analyses for structural and seismic resistant, large-scale computational technology, i.e. three-dimensional analysis by finite element method and time history response analysis should actively be acquired and seismic resistant assessment utilizing the most advanced computers should be carried out. Accordingly, it is deemed crucial to analyze bearing force and confirm the rational safety coefficient.

(3) **Future of numerical analysis for seismic resistance**

After confirming the data in the investigative report of this accident, at least until receiving the tsunami damage, the nuclear power facilities' "cooling" and "stopping" procedures were deemed absolutely functional, which means that the functions of the facilities were maintained and the numerical analysis for seismic resistant and structure at the time were considered sufficient to fulfill their roles. However, given current progress in computers and softwares, it is important to develop forecasting, which evaluates uncertainty of data and scenarios and analyzes risk assessments over and above analysis of standard values of seismic design intensity and methods using computational science and engineering based on scenarios assuming functional maintenance and shutdown process. With a deterministic simulation alone, complete estimation or prediction may be difficult, but it is effective to utilize the deterministic simulation itself, not only computed based on the scenario but also analyzed using data rationally derived from calculation. In addition, the "Investigative Committee on Seismic Safety Issue of Nuclear Power Station" report (Japan Association of Earthquake Engineering), issued in October 2011, prepared an earthquake safety roadmap as a topic for discussion on the "earthquake safety" of nuclear power stations and cited research tasks of computational science and engineering.

6.8.1.3 Tsunami Numerical Calculation

(1) **Tasks and lessons learned on the handling of tsunamis from a computational science perspective**

(a) Establishment of a tsunami wave source: A tsunami is perceived in terms of three phenomena i.e. wave generation, propagation at sea and flooding on land. The key indicator governing tsunami height is the tsunami generation model. For tsunamis caused by earthquakes, seismic magnitude is the primary factor that determines the wave source size.

In Japan, seismic magnitude has been predicted based on records on the largest earthquake in the past. The Headquarters for Earthquake Research Promotion (hereinafter referred to as the Earthquake Headquarters) predicted the largest earthquake as M 8.2 (Sanriku tsunami in the Meiji era) for The Tohoku District—off the Pacific Ocean Earthquake before the Great East Japan Earthquake Disaster. The Tohoku District—off the Pacific Ocean Earthquake, which caused the Great East Japan Earthquake Disaster, significantly exceeded predictions, and was about 16 times larger than the Sanriku tsunami when converted into energy terms. Henceforth, determination of a method to predict the largest earthquake based on past earthquake records will be an important task, which involves not only records on largest historical earthquakes, but also their occurrence frequencies. In addition, the interlocking of earthquake assumed to have been induced by the Great East Japan Earthquake and large crustal movement localized around submarine trenches are the lessons drawn from the Great East Japan Earthquake. Regarding the latter, there is a dominant hypothesis on significant sliding around the trench or spray fault inducing the crustal movement and a survey report that suggests a submarine landslide. These lessons should be taken into consideration for surveys on faults in other marine areas according to their specific features.

(b) Understanding on the behavior of tsunami running up on land: According to the results of tsunami assessment before the Great East Japan Earthquake Disaster, there was little potential for tsunamis to inundate nuclear facility sites and accordingly, assessment on inundation was not prioritized. However, because the tsunami height assumptions in the design (design basis tsunami) was increased after the Great East Japan Earthquake Disaster, and beyond design basis tsunami also had to be considered, corresponding impacts of tsunami flooding and inundations into nuclear sites must also be assessed. For these reasons, understanding on tsunami behavior including (a) on-site flooding behavior, (b) inundation behavior through intake channels and spillways, and (c) tsunami inundations of important structures are all required.

(2) **Tasks and lessons learned on tsunami calculation from a computational science perspective**

(a) Tsunami propagation model: As shown in (1) a., with the increase in the magnitude of earthquake scenario, non-uniformity of the sliding amount for large-scale earthquakes may be considered. Here, the use of a planar two-dimensional numerical model based on conventional nonlinear long wave theory (nonlinear model) may impair accuracy. There may be a need for a planar two-dimensional numerical model based on higher accuracy nonlinear dispersive wave theory (dispersive wave model). Specifically, upheaval and submerging geometry of sea bed, which are directly related to the occurrence of tsunamis, become increasingly complicated and the tsunami's short cycle component increases compared with cases where non-uniformity is not considered. In such case, tsunami height declines due to the dispersibility of the tsunami wave number. This phenomenon cannot be expressed by a nonlinear model, while the computation time on the dispersive wave model exceeds that of the nonlinear model by several to tenfolds. It is advisable to determine the numerical calculation model considering the importance of enhanced accuracy over total assessment compared with the uncertainty of the wave source.

(b) Understanding of tsunami behavior during tsunami running-up on land: Devastation surveys, model experiments and numerical calculations are effective means for understanding on-site tsunami behavior. In numerical calculation, inundation depth and flow velocity may be estimated using a planar two-dimensional (nonlinear) model based on conventional nonlinear long wave theory, as applied to ocean areas. However, there are issues in assessing tsunami behavior for complicated geometric formulation using this method because pressure and hydrodynamic forces affecting structures cannot directly be estimated. This is why the three-dimensional numerical calculation (three-dimensional) model is being more frequently used. Among a number of three-dimensional models available, model using the Volume-of-Fluid method (VOF) is generally preferred due to its superiority in handling water surfaces. The accuracy on simulations using these models needs enhancement. In addition, because significant calculation resources are expended for three-dimensional numerical calculation, resolution is currently limited to within several meters, while the calculation domain may be limited to within several square kilometers radius of a nuclear power station. A further enhancement on efficiency and speed of the calculation model will be required in the future.

Moreover, improvements to the solid and fluid coupled model and simulation codes are anticipated, particularly in response to issues involving drifting debris and sand migration. First of all, there is a need to improve the model used to estimate the generation, migration and colliding force of drifting debris, which include ships and vessels in ocean areas, and objects existing within the site, vehicles and broken trees, etc., on land. Significant changes in terrain features and seawater containing highly concentrated sediments, which are brought up from flooded land and sea bottom from a flow at several meters per second caused by the tsunami may affect the intake system. Although changes to the terrain features and sediment disturbed did not affect the functions of the nuclear power station by the tsunami generated in The Great East Japan Earthquake, evaluation on the event should be made for the development of a more accurate tsunami numerical calculation model and simulation codes.

As for probabilistic assessments, while only water levels on the front side of the power station had been evaluated, no method to evaluate the exceedance probability of flow velocity and force affecting on-site structures has been established yet, which must be addressed in the future.

(3) **Conclusion and course of future action**

The Great East Japan Earthquake was an unprecedented earthquake in Japan, and prediction of tsunami wave source induced by the unanticipated, historical earthquake is a huge challenge. The tsunami consequences on nuclear power station sites has grown with the changes in tsunami predictions, and accordingly, inundation depth, flow velocity, pressure, drifting debris and behavior of sediment onsite, and together with their impacts must be assessed. The numerical calculation model is expected to be instrumental for these assessments, hence the applicability of the three-dimensional numerical calculation model in particular should be improved.

6.8.1.4 Severe Accident Analysis

(1) **Countermeasures for severe accidents**

In response to the 9.11 simultaneous terrorist event, the US had taken measures to mitigate risk on extended loss of entire power, including DC power sources. However, Japan did not develop measures to this end, assuming that all power loss may be swiftly restored. Of course, the Japanese utilities had prepared emergency operation manuals in the event of loss of all AC power as part of accident management. However, they did not predict the extended loss of DC power source, assuming that at least DC power will be available. The Tokyo Electric Power Company believed that DC power sources could be used for around 8h and assumed the power resources would be restored by then. They also prepared an accident management measure to accommodate low-pressure power sources from neighboring plants in case of DC power failure. The Fukushima Daiichi Nuclear Power Station responded to the loss of entire power by collecting car batteries based on the judgment of on-site staff. The batteries did not generate sufficient power for the pumps and were utilized only for valve operations and measurement instrument used to understand plant conditions. In addition, after the loss of core cooling, water injection by fire engines was conducted. These activities were obviously not included in the emergency manual, but was the best judgment at the time. However, the water injection by fire engines could not prevent the core from melting down.

The Fukushima Daiichi Nuclear Power Station experienced simultaneous damages to multiple units (could not accommodate power supply) and extended loss of all power supply that no one had anticipated, which led to the catastrophe. The direct cause of the accident was the tsunami impact. However, considering that measures for loss of entire power including DC power source was taken in the US as terrorist measures, those concerned must feel responsible for the fact no countermeasures for the loss of entire power sources over time had been taken in Japan.

(2) **Severe accident analysis**

In March 2008, the Japan Nuclear Energy Safety Organization (JNES) published analysis on accident sequence behavior under the loss of all AC power sources, using severe accident analysis code, MERCOR in the "Report on analysis of PSA level 2 (BWR)". The results showed that the reactor pressure vessel (RPV) was damaged 15 h after the accident (i.e., 7 h after the DC power source had been lost, assuming AC batteries worked for 8 h after the accident), and the reactor containment (RC) was quickly damaged. From probabilistic assessment perspective, the results did not necessarily provide new information, but it was unclear how the NSC or NISA judged and responded to the analysis showing that the reactor containment was damaged only 7 h after the loss of DC power source. Moreover, the industrial community also did not recognize the need for countermeasures. These attitudes may be influenced by the NSC's decisions in 1990 that "measures for the loss of entire power sources over time need not necessarily be taken into consideration" and "it can be managed by securing alternative power sources at an early stage".

The Tokyo Electric Power Company (TEPCO) submitted the results of analysis on accident sequence behavior using MAAP code together with plant data analysis after the Fukushima Daiichi Nuclear Power Station accident to NISA on May 23, 2011, about two and half months after the accident. The analysis was conducted based on limited plant data after the accident, with the results of assumptions and hypotheses containing uncertainties. In addition, the Agency of Natural Resources and Energy and NISA jointly held the "Technological workshop on reactor core damage situation presumption of Units 1–3 of the Fukushima Daiichi Nuclear Power Station, TEPCO" on November 30, 2011, eight and half months after the accident, where TEPCO's MAAP analysis, the JNES's MELCOR analysis and the Institute of Applied Energy (IAE)'s SAMPSON analysis were publicized. According to the results of these analyses, it was presumed that the reactor core meltdown occurred in all the three units and parts of the reactor core fell to the bottom of the reactor containment penetrating the reactor pressure vessel (RPV). Adjustment of coefficients by user input to reproduce the observed pressure and temperature values, etc., are possible by MAAP and MELCOR analyses. However, there were discrepancies between the results of analyses and actual measurement, showing that the analyses failed to accurately reproduce accident sequences. Portion of the results of SAMPSON analysis, in which coefficients cannot be adjusted by user's input, also did not accurately reproduce accident sequences.

It was observed that the three codes commonly failed to reproduce transient changes after the occurrence of accident shown by plant data. Causes for failure to reproduce accident sequence immediately after the accident can be summarized on the following three points:

(a) Reliability of plant measurement data: Parts of the plant data measured using portable batteries were subject to possible errors in readings and data from failed meters. Signals of the water-level gage were highly likely not to show accurate water levels under drastically changing transient conditions of the plant. The reliability of data could not be confirmed in the early stage after the accident, while a part of measurement data in the early stages did not accurately present plant conditions.
(b) Lack of information on operation and equipment manipulation after the accident: For conducting accident sequence analysis, plant operating conditions and equipment manipulation at the time of the accident must be determined as analysis conditions (constraints). For analysis of early stage in the accident, the conditions were determined on the basis of on-site operating records, or memoranda written on the whiteboard, and for parts of the sequence that were unknown, assumptions and hypotheses were used to determine the conditions.
(c) Various events and phenomena specific to the Fukushima Daiichi Nuclear Power Station accident, which had not been taken into consideration in severe accident analysis code at the time are becoming more clear. It is beginning to be recognized that for accident event reproduction analysis, models that take into consideration these events need to be added and enhanced.

    These phenomena include the following:

    (i) Direct leakage of coolant from RPV to reactor containment dry wells: damage to some reactor instrumentation tubes, direct leakage of high-temperature vapor due to gasket high-temperature degradation at flanges of plumbing connected to RPV (e.g., pipes of safety relief valves)
    (ii) Part-load operation of isolated reactor cooling systems driven by steam turbines and high-pressure water injection systems
    (iii) Leakage from reactor containment to reactor building: leakage associated with high-temperature degradation of the top flange, gaskets of the component hatches, penetrations of electric wiring and instrumentation wiring, etc.
    (iv) Reduction of pressure suppression of the pressure suppression pool (layered temperature or partial steam condensation)
    (v) Existence of branch flow during alternative water injection by fire engine

The latest current analyses using MAAP, MELCOR and SAMPSON were publicized at the NURETH-15 international conference held in May 2013

(NURETH-15-536, -653, -601, -033, -075 and -234). The plant data reproducibility by the three codes have improved compared with the results of analyses immediately after the accident. However, the introduction of new models and improvements on existing models have not completed as the models have not been successful in reproducing all transient changes on the plant data. Currently, a government project, "Perceiving and analyzing conditions inside the reactor" and an international project, "Benchmark Study of the Accident at the Fukushima Daiichi Nuclear Power Station (BSAF project)" by the Nuclear Energy Agency (NEA), OECD are in progress. In these projects, the abovementioned modeling of various phenomena, improvement and analysis of codes are conducted and reproduction and analysis of realistic accident events are anticipated.

6.8.2 Simulations by SPEEDI

6.8.2.1 Circumstances Before the Accident

The System for Prediction of Environmental Emergency Dose Information (SPEEDI) was developed by the Japan Atomic Energy Research Institute (now the Japan Atomic Energy Agency (JAEA)), and the Nuclear Safety Technology Center (NUSTEC) was in charge of technological maintenance and management of the SPEEDI; commissioned by MEXT at the time of the accident. SPEEDI has the capacity to compute a three-dimensional meteorological field using digital meteorological input data provided by the Japan Meteorological Agency (JMA), taking terrain features into consideration, and computes past conditions of concentration of radioactive materials emitted into the air, air dose rate and exposure dose and also predicts their future evolutions for about 1 day. The objective areas are about 25 km^2; centering on the objective facilities (narrow area) and 100 km^2 (wide area). It has been arranged that in the event of an emergency, with instructions given by MEXT to NUSTEC, SPEEDI will shift to emergency mode and frequently collect radiation monitoring data in and around the site; carry out predictions and send the predicted results to the relevant parties including MEXT, the NSC, etc. The arrangement includes predictions on an hourly basis, assuming unit emissions, even in the absence of emission data, and sending the results to the relevant parties. In addition, apart from SPEEDI, which has been operating as a governmental system, the JAEA developed a worldwide version of SPEEDI (WSPEEDI-II) for research purposes. At the time of the accident, although no operation system such as functions to send prediction results or staffing operators had been prepared, researchers in charge of the WSPEEDI-II were capable of conducting the prediction manually. The WSPEEDI-II can compute the concentration of radioactive materials concentration radiation dose for several days in advance over a given area, which is wider than that of SPEEDI, e.g. the entire East Japan.

6.8.2.2 Accident Response

According to the evaluation conducted by MEXT, NUSTEC started operating SPEEDI in emergency mode from the evening of March 11 based on MEXT instructions, while from 17:00 the same day, NUSTEC performed computation on an hourly basis, predicting 2 h ahead within the narrow area (Regular computation) assuming unit emissions, the results of which were sent to MEXT, the NSC, the NISA and others. During the regular computation after 08:00 of March 16, the objective area was expanded to the wide area and prediction was to be 3 h ahead. The emergency response support system (ERSS) was to predict information on emission sources based on information inside the reactor, but the ERSS was not functional for this, so regular prediction computation was done using unit emission amounts. There was insufficient evaluation as to how the result of unit emission computation had been handled by the ministry and institutions received the information. It was said that MEXT deemed the computation to be based on premise and differing from reality, and that decisions were made based on this consideration, which was deemed adequate in the subsequent evaluation. Conversely, MEXT said that the prediction results of the SPEEDI were utilized for emergency monitoring and a discrepancy remained in the evaluation results.

Apart from the regular prediction computation, the NISA, as secretariat of the Nuclear Emergency Response Headquarters (NERH) in the Prime Minister's official residence, performed various computations involving predictions assuming various emission amounts since the evening of March 11. The first computation was performed assuming venting at Unit 2 before dawn of March 12, whereupon the relevant parties received the results immediately after 21:00 on March 11. Subsequently, various computations were performed for various purposes and assuming various events, including Unit 1 reactor containment damage, venting at Unit 1, hydrogen explosion at Unit 1, hydrogen explosion at Unit 3 and confirmation of environmental effects by dry venting at Unit 2 etc. (computation of 45 cases before March 16). Computation of 73 cases was also performed at the Local Nuclear Emergency Response Headquarters from 14 to 31 March to utilize the results to formulate emergency monitoring plans. MEXT computed 38 cases during 12–16 March.

The NSC started computation for specific purposes from March 19, computed 23 cases to estimate emission source information during March, 36 cases to prepare basic materials for advising the emergency monitoring plan and continued the computations, together with a number of additional computations to assess the radiation dose and monitoring results in April and beyond. From 22 to 23 March, the NSC additionally computed the radiation dose accumulated from the initial time of the accident until 00:00 of March 24; using estimated emission information as input data. Among the results, only one sheet showing an isopleth map of the iodine 131 thyroid dose equivalent within the narrow area was publicized on March 23 (Fig. 6.28). Although the map partially contained prediction, it should be noted that the results were mainly retrospectively computed. Until entire ministries

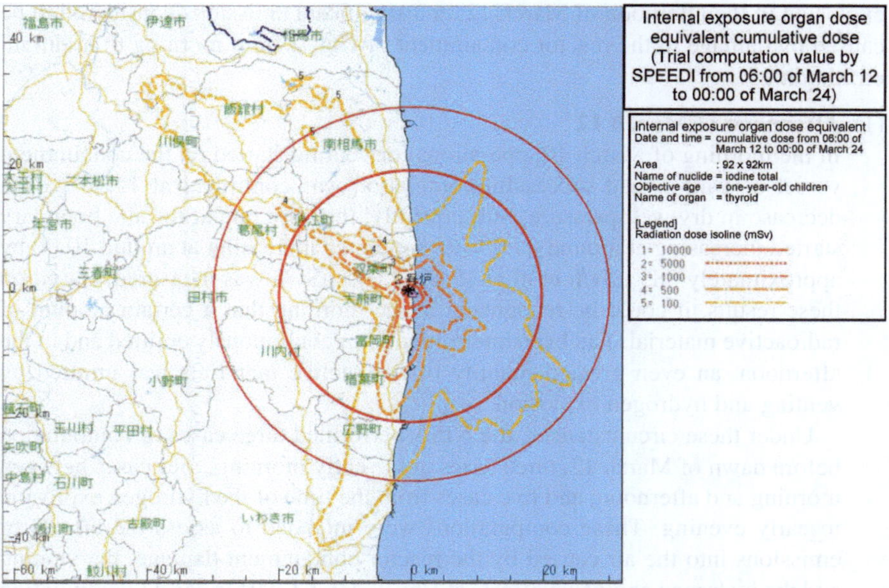

Fig. 6.28 SPEEDI computation results (publicized on March 23, 2011 by the NSC) (Former the NSC website, http://www.nsr.go.jp/archive/nsc/mext_speedi/0312-0324_in.pdf]

and governmental agencies jointly disclosed SPEEDI's computation results in late April, many prediction results obtained immediately after the accident by various governmental institutions remained undisclosed except NSC's disclosure of subsequent computation results similar to the March 23 map at the beginning and end of April.

The WSPEEDI's wide area diffusion prediction computation was performed by the JAEA as instructed by MEXT, whereupon the prediction results were sent to MEXT and the NSC. MEXT instructed the JAEA to perform specific computations; the results of which were received on 15, 24 and 25 March. Moreover, MEXT and the NSC received results computed 3.5 days ahead on a daily basis, on the premise of unit emissions. The objective computation area covered almost half the Honshu island, namely the Chubu and eastern districts, and the geographical distribution of the impacts on the Kanto and Tohoku districts could be perceived from the results.

6.8.2.3 Prediction Assessment

In this section, the prediction results provided by the NSA, MEXT and the LNERH are discussed, based on the perspective of the extent to which SPEEDI's prediction reproduced actual situations and how the prediction could be utilized. Example topics of discussion include two events exerting significant environmental impacts, i.e. the venting from reactor containment of Unit 1 followed by the hydrogen explosion in the afternoon of March 12, and significant emissions considered to be caused by damage to the reactor containment of Unit 2 from morning to midnight on March 15.

(1) **Afternoon of March 12**

In the morning of March 12, operations for venting started on the containment vessel of Unit 1, and successful completion was confirmed at 14:50 with a decrease in dry well pressure. Subsequently, the dose rate at the site boundary started increasing at around 04:00 and continued increasing at around 10:00 by approximately 385 μSv/h until at 15:29, 1,015 μSv/h was measured. Based on these results, it could be recognized in the morning that a certain amount of radioactive materials had been intermittently or continuously emitted and in the afternoon, an even greater quantity of radioactive materials was emitted by venting and hydrogen explosion.

Under these circumstances, the NISA performed three cases of computation before dawn of March 12, three cases in the early morning, four cases between morning and afternoon, and five cases from the time of the hydrogen explosion to early evening. These computations were intended to assess the effects of emissions into the air caused by the reactor containment damage, the venting and the hydrogen explosion effects at Unit 1. MEXT performed 11 computation cases from dawn of March 12 to early evening to predict the effects caused by a series of events of Unit 1. With these computations, it could be deduced in the morning of March 12 that the wind during the period from early evening to night would be blowing in the northwest and north-northwest directions. For example, the results of computation delivered to MEXT at around 09:00 assumed 10 h continuous emissions from 10:00 the same day. Referring to the prediction results on the wind field delivered at the same time, there was reason to believe that, in the morning, radioactive materials would flow south-east toward the sea and the effects on inland areas would be small, but it would flow in a northwest and north direction from afternoon to night, affecting inland areas. The prediction of the diffusion to the northwest nearly coincided with the distribution of the high dose rate due to deposition confirmed by monitoring in March 13. This means the computed prediction almost coincide with the actual conditions.

Based on analysis and evaluation of SPEEDI's computation results, there was reason to believe in the morning of March 12 that emissions would occur in the daytime of March 12, possibly significantly affecting inland areas and emissions during the late afternoon to early evening period would affect the northwest and north-northwest areas. Moreover, in the morning and afternoon of the same day, as significantly high dose rates were measured around the site boundary and venting work was performed, it was deemed that protective measures against the plume, at least instructions to stay in house, should be taken; particularly in areas up to 10 or 20 km from the site to the northwest.

As emission rate information could not be obtained at the time of the accident, absolute values of the abovementioned SPEEDI's radiation dose and concentration predictions are uninformative due to the use of unit or assumed emission rates. From an absolute value perspective, the comment of the organs concerned that "SPEEDI's prediction results do not reproduce actual conditions" applies to a certain extent. However, the concentration and radiation dose distribution and their time of occurrences almost coincided with actual conditions, as described above, and because the radiation dose rate or concentration was proportional to the emission rate, the approximate extent of effects can be estimated with the dose rates obtained by measurements using a proportional relation.

(2) **March 15**

The previous evening, March 14, venting operations were performed, but it was not confirmed whether they were success. However, since an air dose rate of around 3 mSv/h was measured at the main gate in the evening of March 14, it is considered that radioactive materials were emitted in significant amount. At 06:00, March 15, a drop in pressure in the pressure suppression chamber was confirmed, since then dose rates around the main gate fluctuated, soaring to reach about 12 mSv/h at 09:00. It was possible to conclude from these evidences that a considerable and continuous emission was taking place.

The NISA performed five computation cases targeting Unit 2 before dawn of March 15; four of which were computations to predict impact of dry venting, which predicted consequences of short-term emissions for 1–3 h, and which did not correspond to actual emissions. The other case computed the concentration and dose distribution over 24 h; assuming continuous emissions due to failure of the containment function. Under circumstances whereby continuous emissions could be inferred from the temporal variations in the dose rate in the morning, the setting of continuous emission for the prediction was reasonable. The prediction results were distributed at 06:51. According to the prediction results, there was reason to believe emissions which continued from morning would significantly affect the northwest areas within 24 h. In particular, the prediction results showed that areas likely to have high iodine deposition and thyroid dose equivalent by inhalation would extend at least 40 km or more from the site to the northwest.

MEXT also performed five cases of SPEEDI prediction computations in the afternoon of March 15 and obtained prediction results that showed effects would mainly occur in areas from west to north of the site with the heaviest effect in the northwest. The LNERH performed seven cases of computation in March 15 and used the results to formulate monitoring plans.

In these computations, the result with assuming a continuous emission from 01:00 of March 15 is shown in Fig. 6.29. Under situations whereby emissions were supposed to continue from the previous night, this emission setting for the computation was rational. The results showed significant probability of effects on areas to the northwest and south from the site, and the pattern is similar to the deposition distribution revealed by subsequent monitoring. It is a fact that the significant effects on areas from the site to the northwest in the late afternoon were predicted before dawn of March 15. The direction of areas to be affected

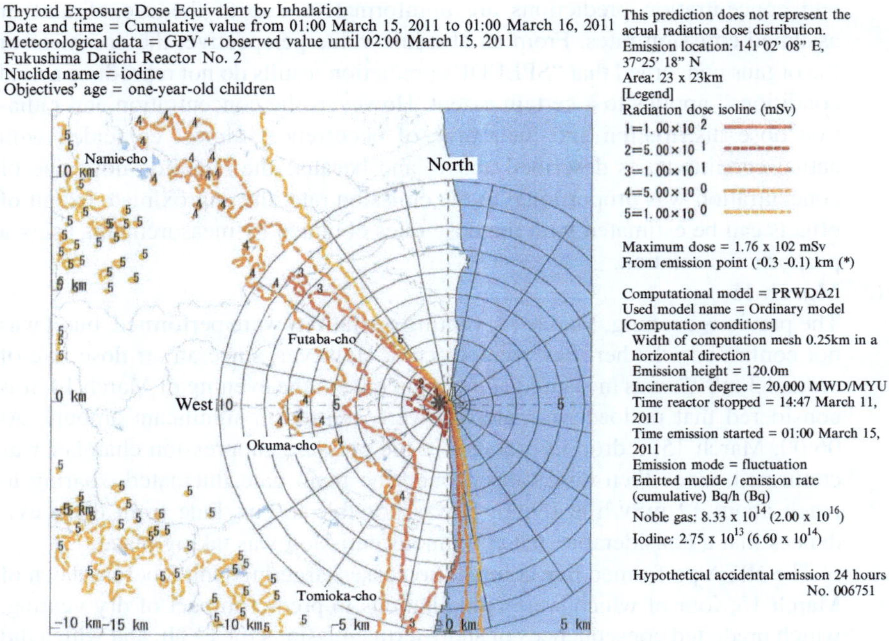

Fig. 6.29 SPEEDI computation results (by the LNERH on March 15, 2011). The NISA website, http://www.nsr.go.jp/archaive/nisa/earthquake/speedi/ofc/003-1103150100-006751.pdf

could be predicted successfully because SPEEDI adequately predicted the wind blowing in the relevant direction for a prolonged period. Although the absolute value of the predicted concentration and radiation dose were uninformative, it is deemed that a rough estimate of the effects on downwind areas could be made with a simple proportional calculation when a dose rate of about 12 mSv/h was perceived at the site.

6.8.3 *Event Sequence Analysis and Source Term Assessment*

6.8.3.1 Background

To reproduce the severe accident which occurred at the Fukushima Daiichi Nuclear Power Station (NPS), the event sequence simulation based on severe accident analysis codes are required. Currently, analysis codes for reproducing the event sequence of severe accident, e.g., MAAP, MELCOR, SAMPSON and THALES, which can simulate various phenomena associated with severe accidents, such as when the reactor core or reactor pressure vessel (RPV) has been damaged or emission amounts of hydrogen or radioactive fission products (FPs). The dominant physical models among complex phenomena during the accident are numerically analyzed based on the analysis codes. The validation of the analyzed results can be evaluated with plant information such as the reactor water level, pressure, containment pressure and radiation dose rate. However, it should be noted that reproduction of the accidents by analysis codes include a certain level of uncertainties; hence research projects should be promoted to solve analytical problems for achieving sufficient capability to reproduce the event sequence, as described in the "Government Accident Investigation Report" [33] II. 1 (3).

Currently, to solve the event sequence of the Fukushima Daiichi NPP accident, many institutes carry out developing and improving simulation codes, while the limited number of the analysis codes have also been acknowledged and efforts to improve the codes are required. To prioritize these efforts, the AESJ established a committee of research experts on severe accident assessment to point out the primary tasks involving development of event sequence analysis codes and source term assessment.

6.8.3.2 Latest Situations of Research on Severe Accidents

Over recent decades, there has been considerable number of research projects on severe accidents and the research results have been accumulated [34]. Such severe accident phenomena often include thermal hydraulic related phenomena, such as zirconium-water reactions, re-allocation of molten cores, vapor explosions, fuel-coolant interactions, direct heating of PCV, molten core-concrete interactions, hydrogen explosions, and behaviors of FP and aerosol. Severe accident management, including in-vessel retention (IVR), is thus considered to prevent RPV break. Experiments to reveal such phenomena have mainly been conducted in Japan, the United States and European countries, based on which analysis codes have been developed. The analysis codes usually used in Japan are MAAP and MELCOR developed in the US, while Japanese-made THALES and SAMPSON have been developed by the JAEA and the Institute of Applied Energy (IAE), respectively.

Based on the experiments and analytical research, the AESJ formulated the level 2 probability safety assessment (PSA) standard stochastically to assess the frequency of event sequences from core melting to the emission of FPs into the environment and source term [35]. The source term is defined as the species, natures, emission amounts, timing and duration of emission and emitted energy of FPs emitted into the environment; i.e. if the source term can be adequately assessed, the amount of radioactive materials emitted outside the site of Fukushima Daiichi NPS during the accident can be evaluated. However, some uncertainties remain in the late stage of severe accidents as well as a number of identified phenomena.

6.8.3.3 Tasks for the Event Sequence Analysis

To understand the event sequence of the Fukushima Daiichi NPS accident, the meltdown of the reactor core, its migration behavior, and the source term must be evaluated by the severe accident analysis codes. Moreover, to address mid and long term decommission measures, including fuel debris taking off, estimation of location and distribution of molten core fragments are essential. Therefore, with efforts to improve the prediction accuracy of the analysis codes to simulate event sequence and determine the circumstances inside the RPV, major tasks of event sequence analysis are pointed out. Firstly, Unit 3 is evaluated, while the scenarios for Units 1 and 2 are accordingly to be supplemented in the framework of the Unit 3 scenario.

A method called a Phenomena Identification Ranking Table (PIRT) was used to select the major tasks. The PIRT combines design evaluation criteria and phenomena emerging in each basic unit related to flow dynamics events as tabular forms, which are then ranked from the perspective of the magnitude of effects on evaluation results. This method can prepare for tables, which can select primary tasks taking the degree of importance of emerging phenomena into consideration, and the results of prepared tables [36]. In Europe, numerous severe accident researchers implemented PIRT targeting the Pressurized Water Reactors (PWR), identified about 1,000 phenomena and pointed out 106 important and unknown phenomena (EURSAFE) [37]. In this section, phenomena unique to the Fukushima Daiichi NPS accident should be acquired targeting Boiling Water Reactors (BWRs).

For Unit 3, the following scenario may be assumed: earthquake occurrence, reactor scram, reactor core isolation cooling system (RCIC) activation, tsunami attack, loss of all power sources, RCIC shutdown (alternative PCV spray), high-pressure coolant injection system (HPCI) activation, HPCI stop, reactor decompression operation, injection from outer sources, core exposure (venting reactor containment), cave-in by core damage, core meltdown materials migration to lower heads, reactor vessel damage, core meltdown material migration to reactor containment and hydrogen explosions. The PIRT is divided into five temporal phases, i.e. the first phase goes from the reactor scram to the beginning of fuel meltdown, the second beginning of migration from the core area, the third until the reactor vessel is broken, the fourth reactor containment break and the fifth hydrogen explosion.

The plant system is roughly divided into three, i.e. inside RPV, inside the PCV and inside the reactor building (R/B). In addition, inside the RPV is divided into ten physical areas (subsystem/equipment), i.e. the core, shroud head, stand pipe and separator, dryer, top head, main steam line, top down comer, lower head and recirculation loop. Inside the PCV is divided into five, i.e. pedestal cavity, drywell, drywell head, venting line/wet-well down comer and wet-well. Inside, the R/B is divided into seven, i.e. emergency condenser, rooms in the R/B, emergency gas processing systems, operation floor, blowout panel, spent fuel pool and equipment pool.

To determine the level of importance of the identified phenomena, key indicators are required. In the PIRT, this is known as the Figure of Merit (FoM) and should be selected at each temporal phase. The FoM is determined at the first phase as the highest temperature of the cladding tube and the fuel highest enthalpy, the second is the average core temperature, the third is the highest temperature of the reactor vessel wall and the highest temperature of the corium in the lower head, the fourth is the highest pressure and temperature in reactor containment and the fifth is the concentration of gases (hydrogen, oxygen and vapor)

To identify the relevant phenomena, brainstorming sessions were to be held, based on currently available information and knowledge. The research expert committee members are experts in fields of thermal hydraulics and severe accident analysis and occasional cooperation from the nuclear fuel subcommittee members was required, whereupon extensive sessions were held twice a week. The discussions focused not necessarily on details but on every conceivable event which could affect the FoM. Consequently, 677 RPV events, 358 PCV events and 124 R/B events, a total of 1,159 events were identified.

Secondly, the degree of impact in terms of effect on the FoM at each temporal phase was defined to rank the identified events. It is coded as follows: High (H) shows the largest effects, Medium (M) shows medium effects on the FoM and Low (L) shows little effects on the FoM. Not Applicable (N/A) shows it does not affect the FoM.

In addition, the current knowledge level (State of Knowledge (SoK)) is also classified into three areas. Known (K) means the event is well understood and there is little uncertainty in the experimental data and analytical model. Partially known (P) means the event is generally understood but experimental data are limited, with a medium level of uncertainty in the analytical model and need for further research. Unknown (U) means the event is not well understood, there is a lack of experimental data, the analytical model has significant uncertainty, the analysis is largely dependent on assumptions and research is required.

The ranking is performed through discussion among experts. Consequently, the degree of importance and level of knowledge are ranked as shown in Table 6.30. The 208 events of importance H and knowledge level P or U among the 1,159 identified events were selected. These 208 events were then rearranged and consolidated into 88 key item events as shown in Table 6.31. As a general trend, the greater the distance from the core, the higher the ratio of P and U knowledge level to K and the more insufficient the knowledge becomes. As for the importance of events, in addition to the phenomenon of thermal hydraulics unique to severe accidents, which may significantly affect the event development, it was revealed that the behavior of portions, which become migration routes for fuel or gases caused by damage to instrument piping, is important. Recriticality may be unlikely, but this is still considered crucial due to its key influence. The nature and physicality value of mixed materials contained in corium are also highlighted, while the effect of seawater, unique to the Fukushima Daiichi NPS accident, is also considered.

Based on this result of the PIRT, a research program for upgrading analysis codes should be specified.

Table 6.30 Phenomena, levels of importance and knowledge identified by the event sequence analysis PIRT

System	Physical area (Subsystem/ equipment)	Identified phenomena	Importance level															Knowledge level			H & P or U	Items arranged
			1st phase			2nd phase			3rd phase			4th phase			5th phase							
			H	M	L	H	M	L	H	M	L	H	M	L	H	M	L	K	P	U		
Inside the RPV	Core	178	16	39	36	52	69	48	47	42	87	5	12	161	4	38	132	67	102	9	54	12
	Shroud head	32	0	1	26	0	1	31	0	1	31	0	3	29	0	6	26	17	12	3	0	0
	Stand pipe/ separator	32	0	0	29	0	4	28	0	4	28	4	3	25	7	2	23	16	14	2	2	1
	Dryer	24	0	0	24	0	4	20	0	4	20	4	1	19	6	3	15	11	6	7	2	1
	Top head	24	0	2	22	1	6	17	1	6	17	4	6	14	3	7	14	11	9	4	1	1
	Main steam line	32	0	7	22	5	8	18	5	8	18	3	7	21	5	7	0	0	5	2	1	1
	Top down comer	31	1	3	26	0	5	26	0	5	26	2	12	17	1	12	18	20	6	5	0	0
	Lower down comer	123	2	6	38	2	7	37	42	49	31	0	23	100	1	9	113	28	82	13	0	9
	Lower head	164	0	3	26	1	1	32	78	41	19	21	72	70	14	52	97	25	123	15	0	18
	Recirculation loop	37	0	0	29	0	2	29	2	4	31	0	3	34	2	8	27	17	13	7	0	0
	Subtotal	677	19	61	278	61	107	286	175	164	308	43	142	490	43	144	465	212	372	67	60	43
Inside PCV	Pedestal cavity	140	0	0	40	0	0	40	0	0	40	69	35	36	54	37	49	24	97	19	67	13
	Drywell	105	0	0	50	0	0	50	0	0	50	46	31	28	39	30	36	16	74	15	45	11
	Drywell head	33	0	1	28	1	1	30	1	1	30	14	5	14	17	2	14	14	17	2	4	4
	Venting line/ wet-well down comer	40	0	0	36	0	0	36	0	0	36	7	7	26	5	5	30	10	23	7	6	5
	Wet-well	40	0	0	33	0	0	34	0	0	34	9	9	22	12	7	21	76	22	6	3	2
	Subtotal	358	0	1	187	1	1	190	1	1	190	145	87	126	127	81	150	73	233	49	125	35

Inside R/B	Emergency condenser	16	0	9	4	3	2	8	2	2	12	0	2	9	0	4	9	7	9	0	2	1
	Rooms in the reactor building	65	0	0	10	0	0	10	0	0	10	0	0	11	17	35	13	5	60	0	17	7
	Emergency gas processing system	2	0	0	0	0	0	0	0	0	0	0	0	0	0	2	0	1	1	0	0	0
	Operation floor	30	0	0	9	0	0	9	0	0	9	0	0	9	4	11	15	0	30	0	3	2
	Blow out panel	4	0	0	0	0	0	0	0	0	0	0	0	0	2	2	0	3	1	0	1	1
	Spent fuel pool	6	0	0	4	0	0	4	0	0	4	0	0	4	0	2	4	0	6	0	0	0
	Equipment pool	1	0	0	0	0	0	0	0	0	0	0	0	0	0	0	1	0	1	0	0	0
	Subtotal	124	0	9	27	3	2	31	2	2	35	0	2	33	23	56	42	16	108	0	23	10
Total		1,159	19	71	492	65	110	507	178	167	533	188	231	649	193	281	657	304	713	116	208	88

Note: Importance level, *H* high, *M* medium, *L* low. Knowledge level, *K* known, *P* partially known, *U* unknown

Table 6.31 Arrangement of phenomena with high importance level and insufficient knowledge level in the event sequence analysis PIRT

No.	Physical area	Phenomena are arranged for large items	No.	Physical area	Phenomena are arranged for large items
RPV	Core	Zirconium-water reaction	PCV	Pedestal cavity	Migration of corium
		Physicality of core materials			Melting corium-concrete inter-reaction (MCCI)
		Heat transfer between core materials			Recriticality
		Radiant heat transfer between corium and structural materials			Oxidation reaction of pedestal wall
		Migration of corium			Fuel-coolant inter-reaction (FCI)
		Melt pool characteristics			Direct containment heating (DCH)
		Corium-water oxidation reaction			Fluctuation of water level in pedestal
		Debris bed characteristics, behavior			Melting structure in pedestal
		Solidification of corium			Corium migration to thumping room
		Melting corium			Deposition conditions of corium on pedestal floor
		Recriticality			Leakage of corium to connecting pipe in thumping room
		Behavior relevant to instrument piping		Drywell	Leakage from opening of reactor containment
	Stand pip and separator	Depositing FP, deposition			Heat transfer between core materials and structure
	Dryer	Depositing FP, deposition			Fluctuation of water level in drywell
	Top head	Depositing FP, deposition			Melting corium-concrete inter-reaction (MCCI)
	Main vapor line	Depositing FP, deposition			Direct containment heating (DCH)
	Lower down comer	Depositing FP, deposition at leakage pass			Formation and physicality of crust/remelting behavior
		Heat transfer between core materials			Recriticality
		Radiant heat transfer between corium and structural materials			Drywell structure oxidation reaction
		Migration of corium			Melting structure inside drywell
		Fuel-coolant inter-reaction			Heat radiation from drywell wall
		Physicality of core materials		Drywell head	Mechanical damage of drywell head
		Solidification of corium			Loss of pressure of plate head

		Melting corium			FP deposition in leakage pass
		Debris bed characteristics, behavior			Direct containment heating (DCH)
	Lower head	Heat transfer between core materials and structure		Venting line/wet-well down comer	Mechanical damage to plumbing
		Debris bed characteristics, behavior			Direct containment heating (DCH)
		Oxidation reaction between structure and molten metal			Change in crack area of bellows
		Migration of corium			Leakage from bellows
		Melt pool characteristics			Fluctuation of water level at exhaust line of dry/wet-wells
		Solidification of corium		Wet-well	Mechanical damage to wet-well
		Fuel-coolant inter-reaction (FCI)			Leakage from wet-well
		Physicality of core materials		Isolation condenser	Leakage from wet-well
		Melting corium (particles and crust)	R/B	Rooms in reactor building	Leakage of gas and water at equipment and personnel hatch
		Rec0072iticality			Leakage of gas and water at instrument pipe
		Radiation heat transfer between corium and structural materials			Leakage of gas and water at cable perforation
		Forming gap between corium and lower head			Deterioration of sheet for reactor containment exhaust pipe at exhaust line
		Forming crack at lower head			Reactor containment exhaust pipe and flowing of gas reflux of stack
		Mechanical and chemical damage at lower head			Flow of gas reflux in containment
		Migration of materials from control-rod guide tube			Mixture and deposition of equipment in reactor building
		Migration of materials from SRM/IRM/ICM		Operation floor	Gases mixture and change in composition ratio
		Radiant heat transfer between corium and structural materials			Hydrogen fire and explosion
		Melting materials at lower head		Blow out panel	Open of blow out panel
		Effects from sea water			
PCV	Pedestal cavity	Heat transfer between core materials and structure			
		Debris bed characteristics, behavior			

6.8.3.4 Tasks for Source Term Assessment

There are two different methods used for the source term assessment of the Fukushima Daiichi NPS accident. One is a source term estimation method from atmosphere diffusion (inverse estimation using the SPEEDI code) and another is a source term estimation method from event sequence analysis (forward estimation using the MELCOR code). Currently, efforts are underway to improve the accuracy of source term assessment by combining both methods. In this section, source term assessment tasks are highlighted to improve the prediction accuracy of the event development analysis codes mentioned above.

As in the previous section, the PIRT method is used to highlight the primary tasks. However, unlike event sequence analysis, the FoM in the source term PIRT is the degree of source terms, e.g., emitted amounts into the environment, hence the identified phenomena and ranking of importance and knowledge levels should be consolidated.

As the later phase of event is important for source term assessment from emissions to environment perspectives, phases 1–3 described in the previous section were packed into one. This also meant the temporal phase was divided into three, i.e. the initial stage from the reactor scram to RPV break, the middle stage from the RPV break to PCV break and the later stage, following the PCV break to a time about 1 week after the earthquake. The plant system was basically divided similarly to the description in the previous section, but the scope was limited to items required for the source term assessment. With cooperation of experts, brainstorming sessions were held and 70 phenomena were identified. To rank the importance and knowledge levels of phenomena like EURSAFE [37], experts engaged in voting for the selection. Following ranking using a threshold and discussion to evaluate the same, it was determined as shown in Table 6.32, while phenomena with importance level H and knowledge level P or U were narrowed down to 29 cases. In the early stages of events, emissions from melted fuel at the reactor core were acknowledged as critical. During the middle and later stages of the event, the gas/aerosol behavior, leakage portions and migration routes, melting erosion of concrete, venting operation and iodine chemicals were all recognized as important. Research plans to enhance the capability of the analysis code should specifically be formulated based on the results of this PIRT.

6.8.3.5 Conclusion

Simulation with analysis codes is crucial to assess the source term, e.g., FP amounts emitted following the severe accident which occurred at the Fukushima Daiichi NPS. The PIRT was performed to highlight tasks and enhance the accuracy of the simulation assessment. Henceforth, research plans including tests and experiments to improve the analysis codes should specifically be formulated based on this result.

Table 6.32 Phenomena, importance level and knowledge level identified by the source term assessment with PIRT

System	Physical area (Source term emission) or characteristics	Identified phenomena	Importance level									Knowledge level			H & P or U
			Initial stage			Middle stage			Late stage						
			H	M	L	H	M	L	H	M	L	K	P	U	
Inside Reactor	Reactor core (emission in reactor vessel)	16	4	11	1	1	7	8	0	3	13	3	5	8	4
	Reactor vessel, plumbing (migration in reactor vessel)	10	1	9	0	1	9	0	0	6	4	9	1	0	0
	Instrument piping etc. (migration from vessel to containment)	3	2	0	1	3	0	0	3	0	0	0	3	0	3
	Subtotal	29	7	20	2	5	16	8	3	9	17	12	9	8	7
Inside Containment	Pedestal cavity (emission in containment)	2	0	0	2	1	1	0	1	1	0	1	1	0	1
	Drywell/wet-well (aerosol behavior in containment)	16	1	6	9	2	14	0	3	11	1	12	4	0	2
	Top head flange etc. (migration from containment to building)	5	0	0	5	3	0	2	4	0	1	0	4	1	4
	Subtotal	23	1	6	16	6	15	2	8	12	2	13	9	1	7
Inside Reactor Building	Reactor building (aerosol behavior in building)	2	0	0	2	0	0	2	1	1	0	1	0	1	0
*	Iodine chemical reaction	13	0	5	7	1	11	1	13	0	0	0	11	2	13
*	Iodine chemical form	3	0	1	2	0	2	1	3	0	0	1	1	1	2
Total		70	8	32	29	12	44	14	28	22	19	27	30	13	29

Note: Importance Level, *H* high, *M* medium, *L* low. Knowledge Level, *K* known, *P* partially known, *U* unknown
Iodine chemicals are marked with * because they are found in reactor containment and reactor building

6.9 Emergency Preparedness and Response

The fifth level of defence in depth describes the need for off-site emergency response arrangements. In addition to the prevention and mitigation of nuclear or radiation accidents (Principle 8), the IAEA's Basic Safety Principles (SF-1) establishes emergency preparedness and response (Principle 9), as the last barrier of protecting people and the environment.

The primary goals of emergency preparedness and response are as follows:

(1) To ensure arrangements are in place for an effective response at the scene and, as appropriate, at the local, regional, national and international levels;
(2) To prevent the occurrence of deterministic effects and to ensure that all reasonable steps are taken to reduce the occurrence of stochastic effects in the population;
(3) For any incidents that do occur, to take practical measures to mitigate any consequences for human life and health and the environment.

The legal basis of the nuclear emergency response system in Japan is based on the Disaster Countermeasures Basic Act ("Basic Act") and the Act on Special Measures Concerning Nuclear Emergency Preparedness ("Nuclear Emergency Act"). The Basic Disaster Management Plan ("Basic Plan") is a planning basis document that describes the roles and responsibilities of the relevant organizations. The Basic Plan specified that the "Regulatory Guide on Emergency Preparedness for Nuclear Installations" ("Nuclear Emergency Preparedness Guide": old guide "emergency preparedness guide" was revised after JCO accident, and the name of guide was changed) issued by the Nuclear Safety Commission (NSC) should be considered an important technical document by national and local governments and utilities for use in establishing an emergency plan and for implementing protective actions. Arrangements are then made at the local level (in the area where the nuclear power stations are located) based on the Basic Plan and the Nuclear Emergency Preparedness Guide. However, there was an absence of a clear concept of operations for the implementation of actions to protect the public (e.g. evacuation, sheltering, or restriction of consumption of food, milk and water) in these relevant emergency preparedness and response (EPR) documents.

The Nuclear Emergency Preparedness Guide, included technical guidelines on: an emergency planning zone (EPZ), where arrangements needed to be made for notifying and communicating with the public, a monitoring system, and the implementation of protective actions; notification criteria and declaration criteria of a nuclear emergency (Article 10 and Article 15 respectively specified in the Nuclear Emergency Act); and dose criteria for protective actions. The concept of implementing protective actions was only described briefly in the Nuclear Emergency Preparedness Guide: "when actually implementing protective actions such as sheltering in a house or a concrete building or evacuation, it is necessary to establish a certain distance in accordance with the abovementioned criteria, taking into account the severity of the emergency and weather conditions, and implement the protective actions gradually".

Moreover, when the Nuclear Emergency Preparedness Guide was revised based on the IAEA safety requirements in May 2007, only one additional provision was added: "Before or promptly after the release of radioactive materials, depending on the actual local conditions and any abnormal situation and future prospects, it is effective to conduct protective actions such as sheltering indoors or evacuation preventively". Thus, the concept of operations was not elucidated, such as the prioritization and implementation of the specific protective actions.

Furthermore, there was an implicit assumption that a situation such as the accident at the Chernobyl nuclear power plant in 1986 would not occur (that resulted in failure of the containment), while an accident similar to that which took place at the Three Mile Island (TMI) nuclear power plant in 1979 (containment did not fail) could happen. Arrangements, therefore, had not been made for preparedness and response to an emergency such as severe damage to the fuel in the reactor that would result in failure of the containment and a release of radioactive material to the environment. Consideration had also not been given to long-term[15] protective actions such as temporary relocation. This was implemented in this accident as the 'deliberate evacuation area'. Thus, the following lesson is learned:

(**Lesson 1**) The principal reason for the weaknesses in the emergency response, such as was observed in previous accidents (TMI, Chernobyl, Goiania,[16] and JCO[17]) is that there was an implicit assumption of both the operator and the regulatory authorities that such severe accidents could not happen and thus enough attention had not been paid to preparedness for such accidents.

In Sect. 6.9.1 the concept for the implementation of protective actions and issues related to radiation protection will be discussed. Following the timeline of emergency response, urgent[18] protective actions (such as evacuation, restrictions on food and drink), termination of urgent protective actions and long-term protective actions will be analyzed. In Sect. 6.9.2, emergency management and operations will be discussed, including the allocation of responsibilities and roles of the operator, national and local governments. The lessons learned will be stated for each issue. Finally, in Sect. 6.9.3, the issues concerning emergency response other than off-site protective actions will be summarized.

6.9.1 Urgent Protective Actions

In the response to the emergency, it is important to develop a common decision-making structure with the relevant organizations, which will facilitate a consistent

[15] Protective actions which can be implemented within days to weeks and still be effective.

[16] Radiation source accident in Brazil, 1987.

[17] Criticality accident in Japan, 1999 at the JCO uranium-conversion plant in Tokaimura.

[18] Protective actions which must be taken promptly (normally within hours) in order to be effective, and the effectiveness of which will be markedly reduced if they are delayed.

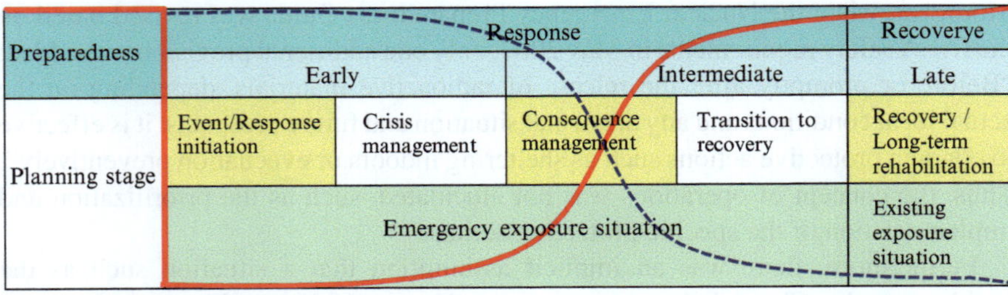

Fig. 6.30 Timeline of emergency management and emergency phase. The figure is quoted from "OECD/NEA, Strategic Aspects of Nuclear and Radiological Emergency Management, OECD, Paris France (2010) and the amount of information or involvement of stakeholders (*broad line*) and uncertainty (*dashed line*) are added

approach during the emergency. Figure 6.30 shows a concept of emergency management that is divided into three parts: preparedness, response and recovery. The response can be subdivided into early and intermediate phases. The initial response and crisis management is undertaken in the early phase, and the consequence management and transition to recovery is undertaken in the intermediate phase. Owing to the limited amount of information available and large uncertainty, a predetermined response is required during the early phase. This should be established in advance. There is also a need to formalize agreements on the adjustment of protective actions with stakeholders in the preparedness phase. The amount of information that becomes available to inform decision making increases over time, with the negotiations with stakeholders on the adjustment of protective actions increasing in importance during the early phase. The concept of an emergency exposure situation, as specified in the 2007 recommendations of the International Commission on Radiological Protection (ICRP), applies to the response during the early and intermediate phases, and the concept of an existing exposure situation applies to recovery in the late phase. In the following section, the issues related to crisis management will be analyzed, in particular its impact on the response to the accident at the Fukushima Daiichi nuclear power plant.

6.9.1.1 Protection Strategy of Precautionary Urgent Protective Actions

In emergency response drills conducted frequently after the JCO accident, arrangements to determine urgent protective actions were established based on the results of dose projections provided by—ERSS (Emergency Response Support System-source term predictions for accident progress and released amount etc.) and SPEEDI (System for Prediction of Environmental Emergency Dose Information). This informed decision making on protective actions by comparing the results obtained using ERSS and SPEEDI with dose criteria specified in the NSC Guide. For example, determining the distance out to which a protective action (e.g. evacuation, sheltering) is effective (i.e. kept below the dose criteria).

Under this system decision making on protective actions is heavily dependent on the dose projections. However, the Basic Plan and Nuclear Emergency Response Manual (not open to the public at the time of the accident) assigns the regulatory body Nuclear and Industrial Safety Agency jurisdiction over ERSS, with the Ministry of Education, Culture, Sports, Science and Technology responsible for SPEEDI. The NSC Guide only mentions "The criteria to take protective measures ......is expressed as a projected dose. The projected dose should be estimated from data such as abnormal circumstances, the expected or actual release of radioactive, emergency monitoring information, weather information, SPEEDI network systems etc.", and does not specifically describe how to use them, including during an emergency response drill. In the Fukushima Daiichi accident, each system in Tokyo started independently, but due to the disruption in power supply and other aspects affecting the communication system caused by the earthquake, information on the condition of the nuclear reactor could not be obtained, and so the source term information from ERSS was not provided to SPEEDI. After the accident, there was a lot of focus on the problems related to SPEEDI. For example, it was reported in the government accident investigation report that SPEEDI did not fulfil its required function during the emergency. However, the technical problem of the role of dose projections in decision-making on urgent protective actions, and the problem of making the results available to the public were misunderstood. These were actually two separate issues. Next, the strategy for the implementation of urgent protective actions will be examined.

On 11th March at 20:50, approximately 6 h after the earthquake occurred, Fukushima prefecture initially issued an evacuation order for the area within a 2 km radius, and, shortly after at 21:23 the national government instructed evacuation out to a 3 km radius and sheltering indoors out to 10 km. Subsequently, at 05:44 on the following day (12th March), the national government extended the evacuation boundary out to a 10 km radius. This meant that urgent protective actions were taken relatively promptly. These protective actions were issued after assessing the situation based on the loss of cooling function of Unit 1 and the increased pressure in the containment vessel. Moreover, on the 12th March at 18:25, after the hydrogen explosion in the Unit 1 reactor building and in preparation for the risk of a simultaneous disaster involving multiple reactors, the evacuation area was increased out to a 20 km radius, and on the morning of 15th March at 11:00, after events in Units 2 and 4 (a loud noise was heard and vibration were sensed around Unit 2 after 06:00 and the roof of the fifth floor on the containment building at Unit 4 exploded with a loud noise and vibrations at 06:12), instructions to shelter indoors within 20–30 km were issued. Sheltering was instructed without any prior planning and no consideration given to the period of implementation. As already described in Sect. 5.2.3, these urgent protective actions contributed to preventing the development of severe deterministic effects from very high levels of dose among the exposed members of the public.

In the absence of information on the source term, an analysis assuming a unit release amount of radionuclides or a certain accident scenario were performed using SPEEDI. The result was distributed to each emergency response headquarter and the NSC. The fact that the SPEEDI results were not made publicly available resulted in a social problem. Some of the inhabitants within a 20 km radius that had been instructed to evacuate on 12th March moved to a shelter that was located in area northwest from the Fukushima Daiichi site. It was later determined that this area was heavily contaminated in the afternoon of 15th March. Consequently, it was claimed that non-disclosure of the SPEEDI results (first announced on 23rd March) caused exposures that could have been preventable. However, the SPEEDI results announced on 23rd March were reverse estimation calculations of thyroid equivalent doses for those located in affected areas, but were based on environmental monitoring results that had been obtained from 15th March onwards. The misunderstanding was that the results were available at the time when the instruction to evacuate was made (i.e. on 12th March, when actually the estimations started to be performed from 16th March).

As described in Sect. 5.2.2 the weather conditions in Fukushima prefecture on 15th March were complex. For example, from early dawn to morning on the site in "Hamadori" area a northerly wind blew, which toward noon turned clockwise to an easterly wind direction, and in the afternoon the wind was mainly in a southeasterly direction. In the evening it began to rain in the north and then a mix of rain, sleet and snow late at night throughout the whole of the prefecture. The SPEEDI results made publicly available that assumed a unit release amount from the early morning of the 15th, as indicated in Fig. 6.29, estimated the transfer direction of the plume and the deposition relatively accurately. However, in the event where the timing and rate of change of the release are unknown, it was impossible to determine the protective actions within a specific area around the site based solely on the predicted passage of the plume and assuming a unit volume of release. In any case, most of those located in the affected area had already evacuated beyond 20 km.

Based on source term analysis information obtained using the severe accident analysis code MELCOR (described in the report submitted to the IAEA Ministerial Conference of June 2011 by the Japanese government), the distribution of Cs-137 contamination was estimated, using probabilistic safety assessment (level 3 PSA) code OSCAAR developed by the Japan Atomic Energy Agency. The result of the above and similar results reported in the side events of the "Fukushima Ministerial Conference on Nuclear Safety" held in December 2012, are indicated in Fig. 6.31. The severe accident analysis codes such as MELCOR, were unable to adequately explain the duration and rate of change of the large release that took place on 15th (believe to be from Unit 2). Therefore, it was not possible to reproduce the distribution of contamination in the northwesterly direction. Despite various information obtained after the accident being analyzed, it was very difficult to reproduce the precise source term to include the timing, rate of change, radionuclide composition, duration and height of the release. Furthermore, a release that warrants protective actions takes place over a prolonged time period, and due to changes in the weather conditions, such as wind direction or rain, results in a complex pattern of deposition. This indicates the significant uncertainty associated with dose projections. In addition, the complex pattern of deposition was also found to be the

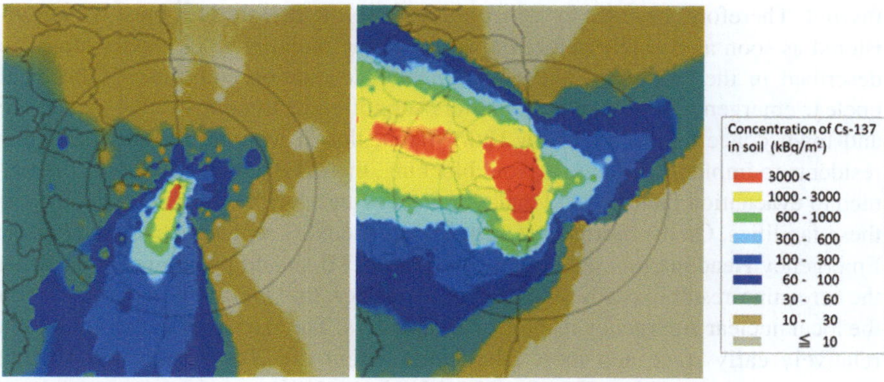

Fig. 6.31 Cs-137 reproduction of contamination distribution computing using OSCAAR code. The *left figure* is reproduced by computing based on the source term analysis information, which was included in the Government report submitted to the IAEA Ministers' Meeting in June 2011. The *right figure* is reproduced by computing based on the source term analysis information, which was reported at the "Fukushima Ministerial Conference on Nuclear Safety" held in December 2012

case in the Chernobyl accident. This means that prompt evacuation before the release of radioactive material is the most effective protective action.

In light of the uncertainty in source term evaluation and environmental impact, at the crisis management stage when there is likely to be insufficient information in the urgent phase of the accident, a framework for decision-making based on prediction systems will not provide results sufficiently quickly or accurately enough that could provide the sole basis for deciding on initial urgent protective actions. Such a decision-making framework also deviates from international standards, such as that of the IAEA. Thus, the following lesson is learned:

(Lesson 2) In implementing urgent protective actions, arrangements must be established based on an assessment performed on the facilities beforehand, so that urgent protective actions within a predetermined zone can be implemented promptly before a release of radioactive material to the environment.

After a release (during the consequence management phase), the effective concept is a framework to identify where additional urgent protective actions may be warranted (beyond those areas already evacuated), using Operational Intervention Levels (OIL). An OIL is a type of action level that is used immediately and directly (without further assessment) to determine the appropriate protective actions or other response actions on the basis of an environmental measurement or laboratory analysis. In this case, the estimate with the computation code may also inform decision making on the protective actions to be taken, for example, utilization of monitoring results to perform reverse calculations of the release source information.

As demonstrated by the Chernobyl accident, one of the urgent protective actions that is important in the crisis management phase is the administration of iodine thyroid blocking agents (ITB). ITB prevents the uptake of radioiodine to the thyroid. Therefore, in order to reduce the dose to the thyroid ITB must be administered as soon as possible before the intake of the radioiodine or shortly after. As described in the "report by the government investigation committee", the local nuclear emergency response headquarters submitted instructions to the prefecture and town concerned at 13:15 on 12th March, that if instructions are issued for residents to implement iodine thyroid blocking, then ITB agents need to be distributed to evacuation facilities and that pharmacists and doctors should be stationed at these facilities. On the morning of 13th March, the NSC sent advice to the Nuclear Emergency Headquarters secretariat (ERC) that ITB should be administered when the screening result exceeds 10,000 cpm. However, this information did not reach the local nuclear emergency response headquarters. The start of evacuation was at a relatively early stage and most inhabitants within the 20 km area had already evacuated without taking ITB. On 15th March the NSC issued advice to patients located within 20 km, and also on 16th March to those that remained in the evacuated area, that ITB should be taken when evacuating, but the prefectural government did not instruct the local nuclear emergency response headquarters, as it had been confirmed that no-one remained in the evacuated area. Conversely, Miharu town and some cities, towns and villages, decided to distribute and instruct intake of ITB (it is reported that Miharu town distributed to 95 % of target group with a pharmacist present on 15th March). The confusion in the administration of ITB, can be attributed to the absence of a Concept of Operations that described how to conduct this urgent protective action. In the NSC guide, the intervention level for the infant thyroid equivalent dose 100 mSv is specified as the criteria for ITB administration, in addition to sheltering indoors and evacuation. A specific suggestion to make the prophylactic administration effective was provided in the "Policy for prophylactic administration of stable iodine in a nuclear emergency" (April 2002, Special Committee on Disaster Countermeasures on Nuclear Facilities, etc. of the NSC), but it was not adequately reflected in the regional emergency response plan. Especially, where promptness is crucial, the channel for instructions went from the off-site center to the nuclear emergency headquarters, from the nuclear emergency division director to the local nuclear emergency response headquarters, and then to the prefectural governors and inhabitants. The problem is the precedence given to decisions being made at the national level government rather than at the location of the accident (or close proximity to it).

Sheltering within 20–30 km that was implemented on 15th March caused significant confusion amongst the public. While the status of the plant was not sufficiently understood, the difficulty of long-term sheltering was observed, owing to the loss of local infrastructures (e.g. closing of shops), and the fact that the government requested the public to voluntarily evacuate on 25th March. Voluntary evacuation meant that those located in the 20–30 km zone could choose whether they will evacuate. However, this proved to be problematic as the public were uncertain as how to decide if they should evacuate. In the previous emergency plan, instructions on evacuation were to be issued to districts with areas that were expected to have very high levels of radiation, and beyond this area sheltering is instructed within a certain range. During the Fukushima accident, it was found that sheltering indoors was difficult to implement long-term. Sheltering in one's own house can be done quickly, and is also advantageous as it enables rapid and easy access to information such as new instructions on protective actions. Conversely, depending on the structure of the house, an effective method to reduce the dose cannot be expected, and considering efforts to secure food etc. sheltering indoors for longer periods is not practical. Instructions for sheltering indoors are temporary measures to reduce exposure from a release of radioactive material. It requires prompt termination of the instruction or changing to an instruction to obligatory evacuation (if it can be done safely), depending on the situation. There is a need to examine the implementation procedures for each of the protective actions in advance i.e. sheltering indoors, evacuation, temporary relocation, and ITB administration, as part of the whole preparedness process.

As described above, in the case of the Fukushima Daiichi accident, urgent protective actions were implemented differently to the framework established and exercised in the emergency response drill at the preparedness stage. Significant confusion arose amongst the inhabitants due to the communication methods used to inform them of the protective actions, and the fact that the evacuation was out to a distance beyond the pre-established emergency zone (EPZ), amongst others. However, to date and as described in Sect. 5.2.3, the development of severe deterministic effects are not expected among the exposed members of the public. Conversely, many people requiring medical support were left behind at the time of the accident in a hospital or a nursing facility within 20 km, resulting in deaths due to inappropriate transport being provided and priority given to monitoring for skin contamination. In the emergency plan, it is necessary that additional arrangements are in place for the evacuation of special facilities such as hospitals or prisons. In the report of the "Investigation Committee for Criticality Accident at Uranium Processing Plant", it was proposed that a study should be conducted on the protective measures, including a response for vulnerable people, but again, there was no adequate arrangements established in advance.

(**Lesson 3**) Advance preparation for safe evacuation of Persons Requiring Support in hospitals is necessary. It is not appropriate to delay treatment or transportation of patients due to performing monitoring and/or decontamination. Sheltering indoors should be conducted only for a short period until safe evacuation and relocation is possible.

6.9.1.2 Protection Strategy for Food and Drink

Contamination of a wide range of food and drink occurred due to the large release of radioactive materials into the atmosphere from midnight on 14th March until the early morning of 16th March. The possibility of consuming contaminated food and drink resulted in concerns raised amongst the public over internal exposure and also economic consequences as a result of rumors. For distribution or consumption restrictions of contaminated food and drink, there are two main problems.

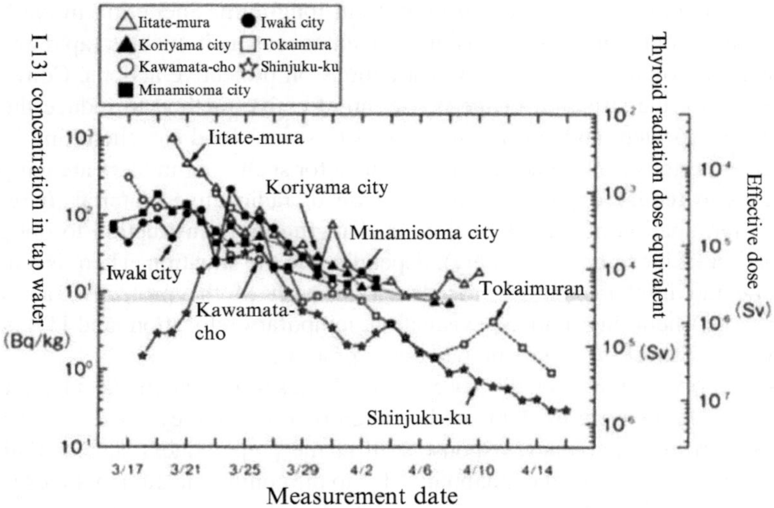

Fig. 6.32 Temporal changes of I-131 concentration in tap water (Kinase et al. [38])

One problem is the need to impose prompt restrictions on distribution and consumption in the urgent phase following passage of the plume. In particular, short half-life radionuclides such as radioiodine and tellurium are principal sources of dose in this urgent phase. Beyond 20 km, high levels of I-131 were detected on 16th March in samples that exceeded the food and drink intake restrictions criteria recommended by the NSC on tap water, milk and leafy vegetables. However, there were no specific regulatory limits that applied to radionuclides in foods produced in Japan established prior to the accident. On 17th March Ministry of Health, Labor and Welfare (the Ministry responsible for food safety) established Provisional Regulation Values for radionuclide levels in food and drink based on the NSC's criteria.

Figure 6.32 shows the time change of the I-131 concentration in tap water detected in Fukushima, Ibaraki and Tokyo prefectures after 16th March. The radioactive material included I-131 and was released into the atmosphere in large quantities on 15th March. It contaminated the ground through dry deposition and wet deposition following interception of the plume by rain. In particular, I-131 flowed into the river with rainwater in the short term and is believed to have flowed into unprocessed water that was used as a water supply. But it tended to decrease in concentration relatively early (approximately 2.8 days), owing to the short half-life of the radionuclides, such as I-131 [38]. MHLW issued a notice on 19th March "Measures to be taken against water supply associated with the accident in the Fukushima No. 1 and No. 2 nuclear power plants" and on 21st March "Measures for infants' ingestion of tap water", and took actions, i.e. in the event that the values exceeded the provisional regulation values, public water supply providers should submit a public information notice to the effect that people should refrain from

Table 6.33 Restriction of drinking water and the averted equivalent dose to the thyroid

Jurisdiction	Date started	Date ended	Averted equivalent dose to the thyroid (mSv)
Iitate-mura, Fukushima prefecture	21-Mar	10-May	8.3
Koriyama city, Fukushima prefecture	22-Mar	25-Mar	0.51
Kawamata-cho, Fukushima prefecture	22-Mar	25-Mar	1.1
Minamisoma city, Fukushima prefecture	22-Mar	30-Mar	1.7
Iwaki city, Fukushima prefecture	23-Mar	31-Mar	2.9
Tokaimura, Ibaraki prefecture	23-Mar	26-Mar	2.1
Shinjuku-ku, Tokyo	23-Mar	24-Mar	0.13

Kinase et al. [38]

drinking tap water. As shown in Table 6.33, due to tap water intake restrictions imposed by each local government, it is estimated that an equivalent dose to the thyroid up to approximately 8 mSv was avoided in the Iitate-mura [38]. As shown in Fig. 6.32, the rate of change of I-131 concentrations in tap water were not uniform across the regions. This is due to variations in the status of the contamination, weather conditions, and water catchment environment. In addition, following detection on 16th March 5 days were needed to impose intake restrictions, which means there can be possibility of uptake of I-131 via tap water during these days. In areas where an individual can directly consume tap water at any time, it is important to avoid unnecessary exposure by imposing restrictions more rapidly. This means that restrictions should be based on such criteria as an increase in the air dose rate measured above contaminated ground due to the passage of the radioactive plume. This is because assessing intake restrictions using the criteria of radionuclide concentrations is relatively slow as it requires time to obtain the results from laboratory analysis. In the case of this accident, the provisional regulation values of I-131 (infant: 100 Bq/kg, adult: 300 Bq/kg) were exceeded, with 20 water supply providers in Fukushima, Ibaraki, Chiba, Tokyo, and Tochigi prefectures imposing limits on the consumption by infants. In Iitate-mura and Fukushima, small water suppliers instructed intake restrictions for the general population. In Iitate-mura a small water supplier continued intake restrictions for infants from 21st March to 10th May and for the general population until 1st April. Other water suppliers continued for a maximum of 9 days. For most suppliers, restriction was terminated in 2–3 days. Conversely, the radiocesium did not exceed provisional regulation values (200 Bq/kg). This is believed to be due to cesium tending to remain in soil at the ground surface. Even if some of the cesium flowed into the river by heavy rainfall [39] it is removed by water treatment processes performed by the water utility, such as coagulation process.

The main cause of the increase in the number of thyroid cancers observed among infants after the Chernobyl accident was due to the delay in imposing food and drink intake restrictions by the former Soviet Government. In particular, uptake of I-131 after consumption of contaminated milk was the main contributor of the dose. In the Fukushima Daiichi accident, very high I-131 concentrations were detected, i.e. up to 151 kBq/kg in soil and 1,440 kBq/kg in weeds on 16th March, and also exceeding the provisional regulation values in raw milk and leafy vegetables (spinach) on the same day. The NSC advised the nuclear emergency response headquarters to impose restrictions on intake for products from the district on 17th March, whereas distribution restrictions were only issued on 21st March by MHLW. Also in this case, there was a delay in determining the protective actions to be taken in the response to contaminated foodstuffs, e.g. for tap water, but it is unclear what the cause of the delay was. Restricting distribution and consumption of food and drink are effective means of reducing internal exposure, but conversely, are regarded as a cause of social concern and contributing to economic consequences as a result of rumors. This also created a problem in determining the scope of restrictions. Initially, the local government established a wide range of restrictions across its administrative boundary. Then later, to prevent the effects of rumors, "Concepts of inspection planning and the establishment and cancellation of items and areas to which restriction of distribution and/or consumption of foods concerned applies" was issued on April 4 by the Government Nuclear Emergency Response Headquarters, which required restrictions to be established more apportioned across the administrative territory.

The concentrations criteria for food and drink intake restrictions specified in the Nuclear Emergency Preparedness Guide (NSC guide) were believed to be appropriate for determining whether restrictions on the consumption of food and drink were warranted. However, in order to ensure the prompt implementation of restrictions on the consumption of food and drink, two phases of implementation should be prepared, i.e. for the first stage, determine where restrictions on the consumption of food and drink are warranted based on the air dose rate OIL; then in the second phase, determine where restrictions on the consumption of food and drink are warranted based on a food, milk and water concentrations OIL, as recommended by the IAEA in its safety guide (GSG-2, 2011).

Another problem of distribution and consumption restrictions of food and drink is the concept of establishing restriction levels on food and drink to protect against the impact of long-term internal exposure by radiocesium as a marker radioisotope. In the Chernobyl accident, the former Soviet Ministry of Health introduced Temporary Permissible Levels (TPL, Bq/kg) of radionuclide content in food and drinking water at that period. Annual consumption by rural inhabitants of the usual food ration, if all components contained radiocesium at the level of TPL-86, would cause an internal dose of less than 50 mSv (at TPL-88 it would be less than 8 mSv and at TPL-91 it would be less than 5 mSv). The fundamental policy of the former Soviet Union and three other countries, the Ukraine, Belarus and Russia was to lower the numerical values of the dose and TPL, in accordance with the improvement in the radiological conditions of the environment following decay of the radionuclides, their permeation into the soil and stabilization. The levels of TPL were established by experts, taking into account a balance between the request of inhabitants to restrict internal exposure and requirements to maintain farming production in the restricted zone.

In the Fukushima Daiichi accident, the MHLW initially established a provisional regulation value, and based on the food health impact assessment report (October 27, 2011) of the Food Safety Commission, engaged in a discussion to establish measures against radioactive material during a sectional meeting with the Pharmaceutical Affairs and Food Sanitation Council, whereupon a new standard value was established, effective on 1st April, 2012 for the longer-term.

The criteria to determine the restriction of food and drink is calculated using the following formula (known as the Derived Intervention Level):

$$DIL = \frac{RL}{f \times I \times DF}$$

Where:
DIL is the Derived Intervention Level (Bq/kg)
RL is the reference level of dose (mSv)
f is the contamination ratio
I is the Intake (kg) and
DF is the dose conversion factor for ingestion (mSv/Bq)

The provisional regulation value of radiocesium for vegetables, grains and meat, egg, fish and others was decreased from 500 to 100 Bq/kg because the RL was reduced from 5 to 1 mSv. According to the concept whereby the MHLW promotes re-examination in the new standard setting, safety is secured by a provisional regulation value, but from the perspective of securing safety and reliability for food furthermore, the reference standard value was reduced to 1 from 5 mSv a year. This was based on: (1) Joint FAO/Codex Alimentarius Commission criterion for food is 1 mSv a year (effective dose), (2) following the most recent sampling results, the concentration detected in food shows a considerable reduction over time. The concept of the food health impact assessment by the Food Safety Commission (which the new standard was based on), was developed for application under normal conditions (i.e. non-emergency). Conversely, the standard to which the Joint FAO/Codex Alimentarius Commission referred to in (1) is an international standard related to the import and export of food after a nuclear power plant accident. Also, the reason for (2) can be similar to the case of the Chernobyl accident, in that it facilitates the promotion of activities by producers to further reduce contamination.

In the radiation protection concept of the ICRP (2007) recommendation, the implementation of protective actions such as food and drink intake restrictions is one of various protective actions that should be considered as part of the process to optimize the overall protection strategy. The following should be considered when deciding on restrictions for the intake of food and drink:

- The Impact on the diet the restrictions will have (e.g. the availability of alternative products)
- Contribution of the dose from ingestion to the total dose

- Realistic evaluation of dietary habits (intake I) and contamination ratio f
- Food safety for consumers and the status of producers located in the affected area

In the international standard of the Joint FAO/Codex Alimentarius Commission relating to the import and export, the contamination ratio is 0.1, whereas the NSC adopts a contamination ratio of 0.5 for establishing the food and drink intake restrictions criteria after an accident. Even though the new standard in Japan applies for normal food safety (i.e. non-emergency), it adopted a contamination ratio of 0.5. This resulted in a Derived Intervention Level that is extremely conservative compared with international standards. However, food safety was a problem that caused significant concern amongst the public.

(**Lesson 4**) For restrictions on food and drink during the crisis management of the early phase, the air dose rate OIL should be used to obtain data promptly that can inform decisions on restrictions of intake for food and water.

(**Lesson 5**) Long-term restrictions on food and drink should be discussed in the justification and optimization process of the overall protective action strategy, taking into account the actual radiological conditions of the affected area when they are better understood.

6.9.1.3 Termination of Urgent Protective Actions and Implementation of Long-term Protective Actions

The establishment of a deliberate evacuation area and the evacuation prepared area in case of emergency on 22nd April is explained in Sect. 5.2.3, and termination of urgent protective actions and the transition to long-term recovery operations is explained in Sect. 5.2.4. At the time of the accident, the NSC guide did not include criteria for the implementation of long-term protective actions such as temporary relocation. Before the Fukushima Daiichi accident, some aspects of this issue were discussed in the NSC guide working group in 2007. However, it was only identified as a future study issue to be addressed, "As can be seen from the example of the Chernobyl accident, particular attention should be paid to the fact that implementation of long-term protective actions becomes more complex compared to that of short-term protective actions. This is due to the need to increase involvement of stakeholders in the decision making process, and the need to take into account all aspects of daily life that are affected".

At the time of the accident in 2011, international recommendations had already been specified in ICRP (2007), and radiation protection had evolved from the previous process-based approach of practices and interventions to an approach based on the characteristics of radiation exposure situations: planned, emergency and existing situations. In 2009, the ICRP recommendation Pub. 109 "Application of the commission's recommendations for the protection of people in emergency exposure situations" and Pub. 111 "Application of the commission's recommendations to the protection of people living in long-term contaminated areas after a nuclear accident or radiation emergency" were published. In addition, the IAEA completed its revision to the basic safety standards (BSS) on 21st March 2011, which included radiation protection safety requirements. It was approved by the Safety Standards Committee (CSS) in May 2011.

The basic concept in emergency and existing exposure situations is to determine the most suitable set of protective actions based on the residual dose, while considering the overall protection strategy. This is instead of evaluating the effectiveness of individual protective actions to determine the averted dose, taking into account all exposure pathways and protective actions. This approach is intended to ensure that overall the most effective protective actions are implemented, even when it is difficult to ensure sufficient protection with a single protective action. However, at the time of the accident, the NSC guide was based on the predecessor of Pub.109 and Pub.111—the ICRP 1990 recommendations and the predecessor of the IAEA's BSS. Discussions were held on whether or not additional protective actions were necessary, prior to the establishment of the deliberate evacuation area, with consideration of the evacuation criterion of 50 mSv, which was based on the concept of averted dose.

Finally the NSC, taking into account a letter from ICRP dated 21st March, advised the government to establish a deliberate evacuation area (temporary relocation) and termination of the recommendation to shelter indoors (establishment of evacuation prepared area in case of emergency) on 10th April. For the deliberate evacuation area, the value of 20 mSv a year was selected, from the reference level 20–100 mSv (residual effective dose) based on the concept of the emergency exposure situation of ICRP 2007 recommendations, and temporary relocation was advised to inhabitants in the area where values were expected to exceed 20 mSv. Thus, the implementation of these additional protective actions were delayed, which was due to several reasons that included: the absence of criteria for the initiation and termination of long-term protective actions, the new concepts of international radiation protection had not been incorporated into national arrangements, and it also took time to reach consensus from those involved in the decision making process, amongst others. Furthermore, in order to implement long-term protective actions, discussions with those located in the affected area and with the local government concerned was necessary. The relatively high radiation dose determined by the Heath Management Survey for the Residents in Fukushima Prefecture (as described in Sect. 5.2.3) were observed mainly in inhabitants from the deliberate evacuation area. Based on this finding, there was a delay in decision-making. On 30 March, since it appeared that the OIL suggested by the IAEA for evacuation was exceeded in Iitate-mura, the IAEA announced that the Japanese Government had been advised to carefully assess the situation. This demonstrates the importance of a framework to facilitate the implementation of protective actions after the release of radioactive materials into the environment. Prompt decision making can be performed quickly using an OIL to assess environmental monitoring results.

(**Lesson 6**) Criteria for the initiation and termination of urgent protective actions and long-term protective actions, including criteria to facilitate resumption of normal life, must be established at the preparedness phase. This should include guidance on the application of the principles of radiation protection.

(Lesson 7) OILs are a crucial concept to inform decision making on protective actions. More detailed international guidance on this concept is necessary.

6.9.2 Emergency Management and Operations

In the general requirements of the IAEA GS-R-2 (Sect. 3.15), at the preparedness phase, "The full range of postulated events shall be considered in the threat assessment. In the threat assessment, emergencies involving a combination of a nuclear or radiological emergency and a conventional emergency such as an earthquake shall be considered". The lack of preparations not only for an emergency leading to damage to the fuel in the reactor core (i.e. a severe release that could result in health effects off site is possible), but also for a combined emergency involving an earthquake (i.e. without due consideration of the experience from the Niigata-ken Chuetsu-oki Earthquake) resulted in inefficiencies in the response of all the organizations involved (operator, national and local governments).

In the NSC there had been discussions on the relationship between the siting of facilities and accident management, or emergency preparedness several times when the Regulatory Guide for Reactor Siting had been reviewed. The document "Reviewing the structure of the NSC Regulatory Guides" (2003), stated that the "Nuclear emergency response plans are established ... based on the Disaster Countermeasures Basic Act, to ensure that the national and local governments can take the most effective and appropriate actions to prevent a disaster, or to reduce the radiological consequences as low as practicable. The protective actions are established beyond the framework of technical aspects of defence in depth and isolating facilities from the public ("No impediments to the prevention of disasters" had been secured in the previous provisions), which are taken to ensure the safety of nuclear reactor facilities, and it should be considered part of defence in depth in a broad sense. Therefore, emergency response planning is a kind of administrative measure prepared independently from the safety regulations under the Nuclear Reactor Regulation Act, and should not be considered as requirements for site evaluation for the construction permit of nuclear facilities." The document stated that accident management was not one of the licensing conditions under the Nuclear Reactor Regulation Act, but a kind of self-controlled measure by the operators related to "operation safety". Accordingly in Japan (off-site) emergency response planning is developed under the Disaster Countermeasures Basic Act, not under the Nuclear Reactor Regulation Act, and the Nuclear Emergency Act clarified the responsibility of operators to prepare the nuclear operator emergency action plan. In this Act, operations are clearly classified into off-site plans for the national and local governments, and on-site plans for the operators only. Conversely, in various foreign countries, the emergency response planning of the on-site is, at a minimum, a licensing requirement for operators. Moreover, in the United States, a system is established where the Federal Emergency Management Agency (FEMA) reviews the off-site regional emergency response plan and provides suggestions to the Nuclear Regulatory Commission (NRC), before the licensing of facilities by the NRC. It will be necessary in Japan to further ensure the checking and review in the preliminary stage of such emergency response planning.

Furthermore, the lack of the initial response by the national government in the JCO accident resulted in the Nuclear Emergency Act increasing the role of the national government in arrangements. Such allocation of responsibility and increased role is retrogressive in light of the timings of the emergency management, as indicated in Fig. 6.30. In the case of this accident, it took 2 h 18 min after the Nuclear Emergency Act Article 15 Report by the operator until the declaration of the nuclear emergency situation (i.e. launch of the response) and it took a further 2 h 20 min until the first instructions on evacuation were issued. At the crisis management stage when there is little information and great uncertainty, it is necessary to establish the framework in which local authorities cooperate with the operator and coordinate at a location close to an emergency, including the prompt implementation of urgent protective actions in accordance with predetermined procedures that are initiated based on conditions at the facility. To that end, as clarified by the new nuclear emergency response guidelines developed by the Nuclear Regulation Authority, it will be necessary to consider in the future that the operator, in addition to providing the criteria for the establishment of the emergency action level (EAL),[19] also has a role in the regional emergency response plan (i.e. modifications made by the local authority), such as providing advice on the urgent protective actions to be instructed to the public located within the vicinity of the facility.

Nuclear emergencies are often emphasized to be a 'special' type of emergency. However, protective actions such as evacuation and sheltering, are typical actions to be taken in the response to a conventional emergency, such as a natural disaster (even if selection of the area and length of time are different). It is the local authorities that should coordinate the response, and it is the police, firefighting and the Self-Defense Forces that should take the lead role in the protection of inhabitants. From that perspective, even if it is not a complex disaster, the management of implementation of the urgent protective actions should be integrated as far as feasible, by making use of the common framework of emergency measures for the response to other types of disasters. In the Fukushima Daiichi accident, the off-site center located approximately 5 km away from the scene of the accident lacked robustness (e.g. lack of air filtering system), and was reported to have be unable to function properly due to the lack of emergency power supply and paralyzed communications infrastructure. Despite these issues, it is still doubtful

[19] Predetermined conditions and instrument readings in the nuclear power plant, if exceeded the staff will immediately notify off-site officials to issue a coordinated response order.

whether the off-site center would have functioned as planned. In order to facilitate integration of the response to a nuclear emergency with the response to a conventional emergency, the equipment and personnel of an existing emergency center in each prefecture should be utilized.

Weaknesses in arrangements (plans, procedures, criteria) are a consequence of responding to emergencies that are beyond the scope of the assumptions made prior to the accident, when these arrangements are developed. This means that, at the same time, it is important to consider how to ensure reasonable preparedness for foreseeable events depending on the assessment of the hazard (threat) and ensuring the flexibility to respond to an emergency beyond what had originally been assumed. In the preparedness phase, effort is required to routinely broaden the scope to manage a situation to be within the scope of the assumption even if emergency occurs. During the crisis management of the response phase, the response should be undertaken in accordance with predetermined criteria, then subsequently responding more flexibly when deviation from the pre-established arrangements is required. There is also a need to develop the capability to resume normal life. To that end, it is necessary to review the responsibility and role of the organizations concerned at each level—operator, regional, national, and international. Moreover, it is necessary to make an agreement between organizations, including arrangements to coordinate a unified response, and re-examination following training to ensure that it functions effectively.

(**Lesson 8**) Arrangements should be made for emergency preparedness and response to take into account the full range of postulated events including events with very low probability of occurrence, and emergencies involving a combination of a nuclear emergency and a conventional emergency, such as an earthquake.

6.9.3 Off-site Emergency Response Other Than Disaster Prevention Measures

As disaster prevention measures are described in the previous section, this section discusses off-site emergency responses other than disaster prevention measures.

6.9.3.1 Full Spectrum of Off-site Emergency Measures

The principal role of emergency measures in an accident is to control on-site accident control activities. Where on-site control activities alone are insufficient, off-site emergency responses play an important role. The overall outline is shown in Fig. 5.1 (refer to Sect. 5).

6.9.3.2 Problems for Immediate Responses

As described in Sect. 5.2, TEPCO reported an emergency notification to the Government based on Article 15 of The Act on Special Measures Concerning Nuclear Emergency Preparedness (ASMCNE) at 16:45 on March 11, 2011. But it took 2 h and 15 min after this report until declaring a nuclear emergency and starting to prepare for the emergency situation by the Government. According to a post-accident analysis, the core melting of Unit 1 was estimated to have started a little after 18:00 and the response is deemed as already too late as an accident prevention measure for Unit 1. Obviously, the top priority is to prevent a severe accident by strict nuclear accident prevention measures, and reinforce measures to minimize the occurrence of severe accidents. However, one of the future tasks will involve reducing the time required to establish an emergency system for nuclear accidents.

6.9.3.3 Allocation of Responsibilities

Where an accident exceeding the design basis like this one occurs, as noted in Sect. 6.5.4, the leader's judgment, i.e. severe accident management, plays the key role. This role involves instructing the parties concerned, perceiving circumstances and promptly judging the most adequate countermeasure to mitigate the impact of an accident. If the allocation of responsibilities is not definitely clarified between the on-site leader on the nuclear operator side, the off-site leader of the Government Nuclear Emergency Response Headquarters (GNERH) and the leader of an off-site center of the Local Nuclear Emergency Response Headquarters (LNERH), things become confused. During this accident, people were well aware of confusion over filling of seawater or implementation of venting via media reports, which was due to the indefinite allocation of responsibilities among the parties concerned. Establishing such firm allocation of responsibilities is also stipulated in the IAEA safety standard of "Preparation and responses to nuclear radiation emergency situations" (GS-R-2). A valuable lesson learned was this insufficiency of readiness, hence the task of promptly solving the same in future. After the accident, the Disaster Prevention Guideline was substantially revised, but "overall unification" and "impact mitigation after the accident" were deemed outside its scope.

6.9.3.4 Problems on Infrastructure

One serious problem in the off-site emergency response was the breaking-off of communication system among the parties concerned due to the complex disaster. As shown in Fig. 6.33, immediately after the accident, communications among the related parties concerned were limited to the TEPCO main office and the site office. No communication was available between GNERH and the TEPCO site office, nor even GNERH and LNERH. They had no other means except relying on

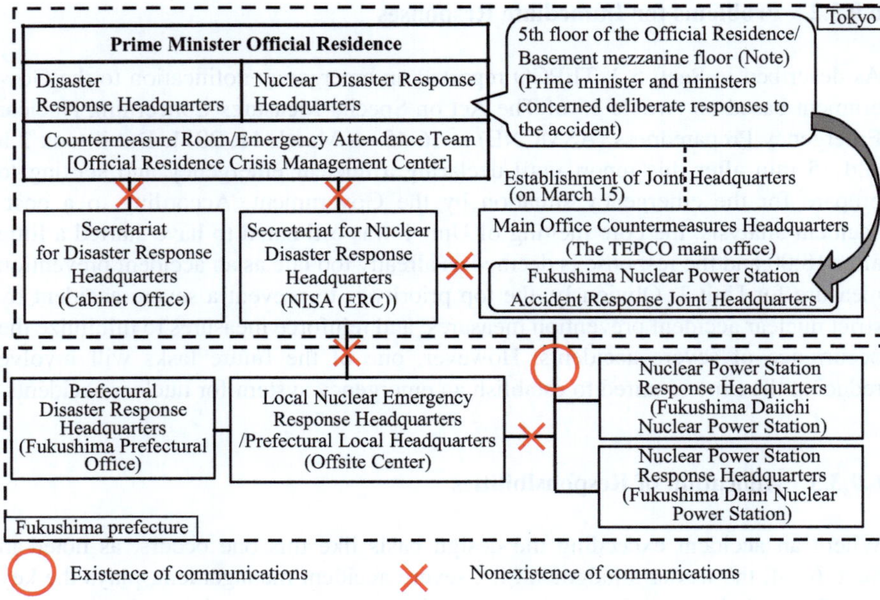

Fig. 6.33 Existence of nonexistence of communications between parties concerned immediately after the accident. *Original source*: Government Accident Investigation Report, Figure III-1 (p. 198)

occasionally connected cellphones. The GNERH was only able to determine accurate site information after the establishment of the GNERH-TEPCO joint headquarters in the TEPCO main office on March 15, 4 days after the accident. One important future task must be to prepare a robust communication infrastructure to predict complex disasters. Another issue related to the infrastructure was the incapability of the off-site center due to the complex disaster. The causes were the location being too close to the site, the fragile earthquake protection and the lack of preparation for high environmental radiation. As for future improvement measures, one option may be to locate the center close to the local government office.

6.9.3.5 Support for Supplies and Equipment

To control the emergency situation, abundant supply and equipment had to be transported from off-site sources. However, as shown in Sect. 5.6, the support operations were not necessarily conducted successfully, primarily due to the lack of a support system at the GNRH. Nonetheless, support would still be difficult, even if a support system had been established, due to the communicational difficulty shown in the previous section. Another cause was the police security system, which prohibited anyone or any transport from entering the evacuation districts from outside. The existing disaster prevention plan only predicted the evacuation of residents from the accident site, with no preparation made to transport supplies and equipment in the opposite direction. As for future tasks, the disaster prevention plan must include establishment of a supply and equipment support system in the GNRH and preparation to transport supplies and equipment from outside in the opposite direction to evacuating residents.

6.10 Nuclear Security, Physical Protection, and Safeguards

6.10.1 Nuclear Security and Physical Protection of Nuclear Material

6.10.1.1 Importance of Nuclear Security

The accident of the Fukushima Daiichi Nuclear Power Station of Tokyo Electric was caused by The Great East Japan Earthquake and resulting tsunamis of March 11, 2011.

Through this earthquake and tsunamis, damages to the external power supply, emergency diesel generators located on the coast side and seawater pumps, which were heat release devices as the last resort, occurred. The following three functions were the major causes leading to the accident: (1) total loss of AC power supply-Station Blackout occurred due to loss of AC power supply for an extended period and loss of DC power supply, (2) the loss of the cooling functions of nuclear reactor facilities, (3) the loss of the cooling function in the spent fuel storage pool. Although the accident of this Fukushima Daiichi was caused by natural disasters, it shows how a similar event could be generated by sabotage, and the importance of this point was recognized as well as safety measures from the nuclear security side in the occasion of this accident.

Following the synchronized terrorist attacks of September 11, 2001 in the United States (hereinafter referred to 9.11), INFCIRC-225 Rev. 4, as guidelines on physical protection of the International Atomic Energy Agency (IAEA) were integrated into a Japanese laws and regulations. Accordingly the Act on the Regulation of Nuclear Source Material, Nuclear Fuel Material, and Reactors ("Nuclear Reactor Regulation Act") was revised in 2007. Following the deployment of special police corps against firearms to Nuclear Facilities in response to a request for ant-terrorism, as well as the reinforcement of physical protection equipment and security precaution by operators, circumstances criticized by the United States such as "Japan lacks armed guards for nuclear facilities" have been resolved.

It has long been thought that nuclear safety/security were assured for equipment in strong buildings (protected areas). However, the accident that occurred in Fukushima Daiichi showed that the loss of all AC power supplies and the cooling function of nuclear reactor facilities and the spent fuel storage pool could be caused intentionally, by simultaneous damages to multiple equipment outside the protected area, thus it clearly resulted in exposing the vulnerability of the nuclear power plant to potential terrorism.

In response to the 9.11 synchronized terrorist attack, B5b (Guidance for core cooling, containment vessel confinement function, and maintenance/recovery of spent fuel pool cooling capability under circumstances where the plant is significantly damaged by an explosion and/or a fire disaster) is established by the U.S. Nuclear Regulatory Commission (NRC). Conversely the lack of response to B5b in Japan was considered a problem neglecting nuclear security. It is said that the U.S. NRC planned to share information on B5b with Japan [40]. If the Japanese Government had recognized the importance of the contents of B5b sufficiently, and the government together with operators had well collaborated, the response at the time of the accident would have been different. As this has already been mentioned in Sect. 6.2.5, there are demands for the government to respond sincerely and positively to the security events in future, by continually and sensitively monitoring changes in overseas movements and responses in confidential and delicate fields such as nuclear security.

To reflect lessons learned from the accident of the Fukushima Daiichi concerning the implementation of various measures in future, there is a need to study reinforcement measures; not only for nuclear safety but also nuclear security simultaneously.

Following the accident at Fukushima Daiichi, an approach to completely ensure safety has been demanded on both sides of the regulator and the nuclear operator. Conversely, the nuclear security of our country, the implementation experience of which lags behind Western countries, and the national situation are considerably different. Henceforth, confirmation of each role and organized cooperation are demanded between operators, regulatory authorities, research and development organizations and concerned scientific society.

6.10.1.2 Postulated Scenario at the Time of Nuclear Terrorism Judging from Damage by Earthquake and Tsunamis

(1) **The situation of the nuclear power plant damaged by earthquake and tsunamis**

The situation of the cooling pump (seawater pumps) to power supply equipment and the final heat radiation source of each nuclear power plant damaged by this earthquake and tsunamis are shown in Table 6.34 in the form of a summary comparison.

As shown in the table, at both the Fukushima Daiichi and Fukushima Daini plant, although the Emergency Diesel Generator (D/G) itself was not submerged in sea water, 12 of 25 D/Gs lost their function due to damage to adjacent power distribution panels and seawater pumps following immersion in water. As seen from the above, it is recognized that securing power supplies is hindered following damage to peripheral devices as well as major equipment, and we must consider such vulnerable points on both the safety and security perspectives.

Table 6.34 The situation of the power supply and seawater systems after the tsunami struck Fukushima Daiichi, Fukushima Daini, Onagawa, Tokai Daini

Site	Unit No.	External power supply[a]	Emergency D/G A	Emergency D/G B	Emergency D/G H	DC power supply	M/C Emergency	M/C Regular	P/C Emergency	P/C Regular	Power supply car	Seawater pump[a]
Fukushima Daiichi	1	All Lost	×	×	—	×	×	×	×	×	Partial utilization (power distribution panel submerged, access difficulty due to another unit, cable laying difficulty and other reasons)	×
	2		×	Δ+	—	×	×	×	2/3	2/4		×
	3		×	×	—	○ → Run out	×	×	×	Δ		×
	4		×	Δ+	—	×	×	×	1/2 (1)	1/1 (1)		×
	5		Δ	Δ	—	○	×	×	×	2/7		×
	6		Δ	○+	Δ	○	○	×	○	×		×
Fukushima Daini	1	500 kV: 1/4	×	×	×	3/4	1/3	○	1/4	○	Partial utilization	×
	2	66 kV: ×	Δ	Δ	Δ	○	○	○	2/4	○		×
	3		Δ	○	○	○	○	○	3/4	○		1/2
	4		Δ	Δ	○	○	○	○	2/4	○		×
Onagawa	1	275 kV: 1/4	○	○	—	○	○	1/2	○	○	External power in good condition, D/G in good condition	○
	2	66 kV: ×	○	Δ[b]	Δ[b]	○	○	○	○	○		2/4
	3		○	○	○	○	○	○	○	○		○
Tokai Daini		275 kV: 1/4	2/3[c]			○	○	○	○	○	D/G in good condition (Backup supply)	For D/G:2/3
		154 kV: ×										For RHR:4/4

Note: Numbers shown in parentheses are under repair. ○: usable (or Available), ×: unavailable, Δ: Unavailable due to submersion of MC and others +: air-cooling, —: not existing, M/C: high voltage switchboard (metal-clad switchgear: 6.9 kV), P/C: low-voltage switchboard (power center: 480 V), *RHR* residual heat removal system

[a]Outside Protected area

[b]B/H system lost due to loss of nuclear reactor supplementary machine cooling pump because of submersion in seawater pumps room opening tide gauge

[c]One seawater pumps for D/G shut off automatically due to submersion of the seawater pump room. Manually stopping D/G

As shown in Sect. 6.10.1.5, "The Reactor Regulation Act" relating to physical protection was revised in March 2012, and countermeasures have been established.

(2) **Assumption of terrorism damage to a level equal to this earthquake, tsunami damage**
In the case of terrorism, it is supposed that the loss of an external power supply, emergency generator, DC power supply, seawater pumps, and the inability of power supply cars may occurred as a single, combined or multiple events. In other words, this accident suggests that destruction of peripheral devices (some important equipment outside the protected area) by terrorism may potentially trigger a severe accident. It becomes important to discuss the scenario assuming these circumstances, namely, measures in response to a severe event caused by terrorism. As for the part related to facilities, the discussion of severe accident assumption in the "safety" in Sects. 6.2–6.5 is thought to be approximately applicable, but for the case of nuclear security, a far more severe scenario, including internal threats (insider) should be discussed. In addition to the above, the response in conjunction with physical protection before and after the occurrence of the accident will be discussed below.

6.10.1.3 Security Systems and Measures at the Fukushima Daiichi NPS

Concerning the security precaution of the Fukushima Daiichi, according to data from the Nuclear Emergency Preparedness Subcommittee crisis management WG (Crisis Management WG) of the Advisory Committee for Natural Resources and Energy Nuclear and Industrial Safety Subcommittee, it is reported that due to the tsunamis resulting from the earthquake which occurred on March 11, 2011, fences, cameras, sensors on the ocean side (also part of landside) were destroyed, while Protection Headquarters were also damaged and submerged in water and their functions were lost.

In addition, security arrangements were reduced as a result of tsunamis and the release of radioactive materials, and furthermore, due to withdrawal of the security contractor, guarding was performed by the employees of TEPCO instead of hired security officers from security contractors.

According to the same data, cited examples of deterioration of access management capability following the disaster included insufficiency in identification such as no collation with the original indenture with photographs for personnel engaging in emergency work, and direct hand delivery to actual persons of entry certificates was not performed. An administrative penalty of written reprimand was issued to the relevant office.

According to the actual site investigation of the Fukushima Daiichi conducted by the AESJ (Atomic Energy Society of Japan) Accidents Investigation Commission of January 9, 2013, part of the fences previously installed in the peripheral protected area were damaged and shredded. We heard that they were planning to establish new fences as installing the fence in the existing position may be difficult when considering the operation of large-sized machinery and materials for restorative construction work. Conversely, upon confirmation from the bus, we got the impression that sufficient work in terms of access management at the Fukushima Daiichi site entrance was being performed to ensure proper identification.

6.10.1.4 The Shape of Support Activity Toward Security (Guarding) During and After the Accident

When observing the situation of the abovementioned Fukushima Daiichi from the perspective of guarding, the security officers had to take refuge themselves while conducting evacuation guidance, since the Protective Headquarters were subject to terrible damage caused by the tsunamis after the earthquake. Here, a report of the non-government accident investigation mentions that "Security officers were seen to have stopped guidance and taken refuge" [41].

As is known from the abovementioned on-site investigation of the Fukushima Daiichi, the dose rate exceeded 1,000 μSv/h when passing through the coast side of Unit 3. Moreover, taking this fact into consideration, apart from the situation just after the accident (excepting this situation), as the later security support, to ensure on-site safety (including reducing radiation exposure at the time of the spot patrol), we consider it advisable to utilize strengthened remote watches effectively.

Taking into account the report that in the Kashiwazaki Kariwa Nuclear Power Station, fences, sensors and cameras were damaged and did not work at the time of the Niigataken Chuetsu-oki Earthquake, functional decline in machines and therefore monitoring function over a certain period may be inevitable. However, if the function of the central monitoring room is secured, we consider it possible to monitor using signals from temporarily installed cameras and sensors (regulatory requirements imposed under the Revised Nuclear Reactor Regulation Act includes redundancy of the Protection Headquarters (monitoring room) for monitoring).

Moreover, under circumstances where permanent sensors and cameras are unavailable, there is a need to develop and stand by temporary sensors and cameras of power supply-free, as alternate means, which can promptly be deployed. One alternative idea would also be to develop and prepare a sophisticated radio control helicopter equipped with a camera or similar device, so that utilizing the same for round inspection of restricted access areas by remote control may be possible. Studying such measures may be useful in efforts to reduce radiation exposure. The security company SECOM announced the development of "a small independent-type flight monitoring robot" reflecting this idea (http://www.secom.co.jp/corporate/release/2012/v_121226_long.html), but this was only intended for use in small spaces, such as rooms, so the development of a "small flight monitoring robot" usable for patrols outside is necessary in future.

Furthermore, as is known from Table 6.35, which shows that a very high maximum dose of radioactivity in the vicinity of the front gate during the period March 11 to April 3, an access restriction fence is installed and access management at the entrance is required by law. Accordingly, when the radiation dose at the

Table 6.35 Security precaution after the March 11 disaster (Extracted from Document prepared by the Nuclear Emergency Preparedness Division dated October 14, 2011 (Crisis Management WG material))

Period	March 11–14	March 15–17	March 18–April 3	April 4–August 22	August 23–
Area for guarding	Guard area	Guard area	Guard area	Guard area	Guard area
Arrangement	Contracted Guard 41 persons and company staff 20 persons	Company staff guard 5 persons	Daytime 2–3 persons Night 2–3 persons	Daytime 3–4 persons Night 2 persons (6 persons × 4 groups)	Daytime 4–8 persons Night 4–6 persons (8 persons × 4 groups, Day shift 2 persons)
	Contract Guard downsize	Contract Guard Withdraw	Company Staff Guard	Company Staff Guard	Company Staff Guard
Patrol	Solely Access Control	Solely Access Control	Case-by-case	Twice/day (May 5–3 times/day)	12 times/day
Basic point	Seismic important building	Seismic important building	Seismic important building	Seismic important building	Front gate station
Maximum dose rate around the front gate	3,130 μSv/h (Mar. 14 21:37)	11,930 μSv/h (Mar. 15 09:00)	1,932 μSv/h (Mar. 21 18:30)	123 μSv/h (Apr.4 04:00)	32 μSv/h (Aug. 24–31)
Remarks	• Contracted Guarding downsized day by day, 20company staff come to the site and respond to guarding	• Instruct company staff to stand by at the evacuation area, due to explosion of building	• For some, impossible to come to work due to the effect of the disaster • Some did not come to work for fear of radiation	• For some, impossible to come to work due to the effect of the disaster • Some did not come to work for fear of radiation	• Radiation dosage decreased. Increasing the number of company staff guarding the station at the front gate

[Extracted from Document prepared by the Nuclear Emergency Preparedness Division dated October 14, 2011 (Crisis Management WG material)]

Situation at the time of the March 11 disaster

- Due to the tsunamis, fences, cameras and sensors on the ocean side (also part of the land-facing side) were destroyed
- Due to the tsunami, the Protection Headquarters were damaged and submerged and their functions were lost.
- Managing protection equipment in the guard district (land-facing side) at the front gate guardhouse case-by-case (-Aug. 22)

entrance to the restricted access area is high (which was installed under normal circumstances), the application of a different substitute plan should be considered on a case-by-case basis.

Situation at the time of the March 11 disaster

- Due to the tsunamis, fences, cameras and sensors on the ocean side (also part of the land-facing side) were destroyed
- Due to the tsunami, the Protection Headquarters were damaged and submerged and their functions were lost.
- Managing protection equipment in the guard district (land-facing side) at the front gate guardhouse case-by-case (–Aug. 22)

In summary, for an important point relating to nuclear security learned through the accident, as mentioned above, with risk management of the Nuclear Facilities in mind, namely the nature of support activity toward security at the time of a severe accident (when it becomes difficult to access the site) the operator should study immediately applicable substitute plans. For this purpose, immediate restoration of the monitoring function as stated above, and study of the introduction of new technologies is desirable, for example an external power supply-free camera sensor or remote controllable camera leveraging new technology.

6.10.1.5 Movement of the Regulator Side Relating to Nuclear Security

(1) **Revision of the Nuclear Reactor Regulation Act in 2012**
The government, which has been studying the reflections of INFCIRC-225 Rev. 5 (formally effective in January, 2011) of the IAEA after the revision of the Nuclear Reactor Regulation Act in 2007, based on lessons learned from the accident at the Fukushima Daiichi which occurred in March, pushed forward a study on problems for each of the boiling water reactors and the pressurized water cooled nuclear power reactor in the Crisis Management WG. The study contents were reported to the Atomic Energy Commission Advisory Committee on Nuclear Security, upon receiving the report (http://www.aec.go.jp/jicst/NC/senmon/bougo/siryo/bougo24/siryo1.pdf) "The Fundamental concept for ensuring nuclear security" based on the study in the Advisory Committee on Nuclear Security to the Atomic Energy Commission, Atomic Energy Commission decision "concerning the fundamental concept for ensuring nuclear security" was released in September 2011 (http://www.aec.go.jp/jicst/NC/about/kettei/kettei110913.pdf).

The specific content of the Revision of Nuclear Reactor Regulation Act related to physical protection/nuclear security, part of which reflected lessons learned from the accident of the Fukushima Daiichi, was prescribed as follows in the Regulations concerning the Installation, Operation, Etc. of commercial nuclear power reactors (Commercial reactor regulations). (The government ordinance for other Nuclear Facilities was also revised).

In the Commercial reactor regulations, of all equipment for supplying AC power, all equipment for cooling the nuclear power reactor facility and all equipment for spent fuel storage tank which might leak specified nuclear fuel material when each function was lost, equipment located in the Protected area is categorized as "Important equipment in the Protected area for physical protection", while equipment outside the Protected area is categorized as "Important equipment outside the Protected area for physical protection". It is demanded that protective measures to be taken respectively.

Accordingly, by imposing laws and regulations compelling operators to ensure protection of the equipment at risk of triggering the loss of the AC power supply in the nuclear power reactor facility, reflecting the lesson learned from the accident at the Fukushima Daiichi, the loss of the function to cool the nuclear power reactor facility, and the loss of the cooling function of the spent fuel storage tank (pool), we consider that this would help solve vulnerable points, and that meaningful nuclear security improvement will be accomplished in nuclear power reactor facilities.

(2) **New safety standards (design-basis) outline (plan), from "January 31, 2013 revised edition"** (http://www.nsr.go.jp/public_comment/bosyu130206/kossi_sekkei.pdf)
In 2013, a Public Comment on the new safety standards of nuclear power reactor facilities was issued by the Secretariat of NRA (S/NRA). In that public comment, safety standards for human-induced external events were shown, and significant progress in terms of safety improvement was seen.

6.10.1.6 The Shape of Support Activity Toward Other Nuclear Security Enhancement; as Learned from the Accident of the Fukushima Daiichi

(1) **Preparation to respond at the time of occurrence of the case (training)**
The response relating to nuclear reactor operation at the time of SBO etc. caused by terrorism is just what was discussed in terms of severe event response in safety terms, but during the response in terms of physical protection, the training, as required by laws and regulations, becomes a particularly key element. It is individuals which actually respond to circumstances, and although improvements in regulations and procedural manuals are expected in future, that is not necessarily enough for individuals to improve their response. The report of the non-government accident investigation states a point highlighted by the United States: "In Japan a script for training is prepared beforehand, which means such training does not reflect the actual circumstances of a terrorist attack" [41]. Script-based training is not always ineffective, but on-site situations are ever-changing and people may not cope with real cases without applicative ability. To cope with real cases, it is important to adopt training methods such as those in which only the starting time is set (so-called Scenario-less Training or Blind Training) or the FOF (Force on Force) training, reflecting training recommended by the U.S., where no scenario is informed to the guarding person side. Guidelines on the IAEA and FOF

training cannot be applicable in the case of our country, because private security officers are not allowed to bear weapons, and these guidelines and training are beyond the range of defensive action for private guard staff. Therefore, integrated exercises in collaboration with special police corps against firearms, which are permanently stationed in Nuclear Facilities, will be crucial. In this case a scenario-less base as abovementioned should be applied.

Furthermore, according to our hearing from the security police authorities, control by weapon firing may not be possible for infiltrators depending on the kinds of weapon used. Different methods must be studied other than the present example, which is intended to control infiltrators. Moreover, because early control is an important aspect of illegal infiltrators, we see the need to establish and reinforce laws and regulations under which interception for control by firing weapon can be performed immediately.

(2) **Information management when the case happens**
Generally speaking, during an accident, information disclosure is demanded as promptly and as far as possible, but from a nuclear security perspective, stricter information control is important, since nuclear security information disclosure could make it easier for intruders to invade the site from the outside, or steal the nuclear material, and furthermore commit sabotage. Information in the nuclear facilities is sometimes thrown open to the public carelessly to explain the accident. On future occasions when presenting information related to facility design, careful consideration is required from the nuclear security perspective. It is important for the government to strive and request the understanding of media and public people concerning considerations from a nuclear security perspective.

(3) **Cooperation of security police authorities and operators**
Actions such as detection at the time nuclear security cases occur, notice and delays of acts come within the scope of the operators. There are matters that private security officers without knowledge of firearms cannot judge solely by watching camera pictures, for example, namely the kind of firearms illegal infiltrators possess, and hence the nature of support required by experts such as security police authorities. Moreover, because the suppression of the trespasser is solely the duty of the security police authorities, clarifying each role and constructing a system of cooperation is important. In addition to the above, to facilitate the allocation of roles, it is crucial to promote a relationship of mutual trust and mutually strive for everyday cooperation.

6.10.1.7 The General Approach Towards Nuclear Security Enhancement and Its Improvement

(1) **Understanding of the nuclear security**
With regard to nuclear safety, after the Chernobyl disaster of 1986, the importance of safety culture has been highlighted, and various kinds of approaches have been performed in our country [42]. Conversely, with respect to nuclear security, it is only very recently that the security culture has been discussed

from an international perspective. Whereas INSAG-4 [43] which is an important report on safety culture, was published in 1991, the implementation guide of the IAEA on security culture [44] was only published in 2008. Judging from the above, the global approach to nuclear security has obviously been subject to significant delay.

The starting point for fostering safety culture is to ensure awareness that safety should be the top priority, throughout the entire organization. However, in the context of nuclear security, there has been a strong belief that protection administrative tasks regarding the prevention of information diffusion should be limited to the department in charge. An approach to reconfigure this culture is necessary, with safety and nuclear security in mind. In the revised Nuclear Reactor Regulation Act (March, 2012), ordinances for managerial responsibility are clearly provided, and henceforth, it is important to push forward sharing of recognition and information on nuclear security, both in operators' and regulators' organization, from top to bottom.

(2) **Preparedness and response related to nuclear security**

A wide range of actions are required, for example, prior preventive measures to prevent cases of nuclear security such as stealing of nuclear materials or sabotage of nuclear facilities, detection and notice of nuclear security cases, actions to delay the act, finding and recapturing stolen nuclear material, and ex-post response action, subsequent support action such as mitigation and minimization of the radiation influence. Such wide-ranging actions cannot be arranged with operators alone (including transport operators in case of transportation) and regulators. In recent years, operators have had to take many measures to enhance nuclear security, but under the current legal framework, the scope within which operators can operate is limited. There is therefore a need to clarify role allotment and responsibility for each division, including security authorities such as police or coastguard, which are authorized to bear firearms. Moreover, each organization concerned should have specific response guidelines, and establish mutual close communication systems. In such cases, with increasing international crimes in mind, guidelines on internationally viable responses are required. Moreover, to respond to circumstances beyond prior assumptions flexibly, it is important to accumulate complimentary data enabling responses depending on the situation.

One of the immature fields is to establish countermeasure to internal threats (insider problems) in response to nuclear security. These internal threats (insiders) mean "threats arising due to illegal acts by employees working inside the Nuclear Facilities" [45]. The internal (insider) threats include three categories: (i) physical protection to prevent insiders from performing illegal acts physically and deterrence through nuclear material measurement and management means, (ii) access control to prevent trespass into vital areas by insiders and preventing efforts to bring in tools to use for destruction work and illegal carrying out of nuclear material, (iii) personnel management targeting the exclusion of potential insiders from certain organizations and districts, and the deterrence of illegal acts by observing behavior [45].

One personnel management strategy is to confirm trustworthiness, which involves "collection and analysis of information to determine in advance those persons (marked men) likely to perform the illegal acts" [45], the establishment, reinforcement, and application of which has already been made in the USA and European countries. In INFCIRC-225 Rev. 5, implementation of trustworthiness confirmation is recommended for persons authorized to handle sensitive information related to nuclear materials or facilities and those with authorized access to important facilities and equipment. In Japan, although trustworthiness confirmation has been discussed, its introduction remained pending as of May 2012 due to concerns over privacy protection, difficulty in securing the system in an effective form, and the specifically Japanese mindset. Conversely, a proposal to start discussion of a specific system was made, targeting the introduction of trustworthiness confirmation in fields relating to nuclear material and facilities as targeted by INFCIRC-225 Rev.5 [46]. Fundamental human rights are constitutionally guaranteed, and the introduction of trustworthiness confirmation is not considered easy in Japan, but with the importance of nuclear security in mind, progress of this approach targeting implementation as early as possible is expected, under the Act on Protection of Specified Secrets. In addition, an incentive measure to secure security, including treatment to a person assigned to important duties in security, is necessary.

(3) **Nuclear security in national security**

For nuclear security for specific nuclear facilities, a response to design-basis threats (DBT), including a detailed description of potential insiders, is required, but the response to nuclear-power disasters (armed attack nuclear-power disasters) in case of an emergency is to be handled under the Act Concerning Measures to Protect Japanese Citizens During Armed Attacks and Others (Civil Protection Act). However, it is hard to say that the issue of securing continuity between both and the issue of facilitating response preparations has been sufficiently discussed, and common understanding is obtained. As a precondition, there is a need to discuss how to evaluate nuclear security in national security and share such recognition between those concerned. In addition, there is a need to review the system of cooperation among organizations, including ministries and government offices in case of any emergency in nuclear security concerned, allotment, the leadership, mainly on Secretariat of NRA (S/NRA) immediately.

(4) **Enhancement of laws and regulations**

To secure these measures, the laws and regulations must be established as a basic grounding.

Revision of the Reactor Regulation Act to reinforce physical protection, including nuclear security to reflect INFCIRC-225 Rev.5 and lessons learned from the accident of the Fukushima Daiichi was made in March 2012. In addition, to establish the undeveloped portion of the trustworthiness confirmation system, ensuring a well-balanced nationwide system is desirable as well as in the field of nuclear energy.

(5) **Adjustment of the interface between safety and nuclear security, and synergy**

Both nuclear safety and security are accomplished by preventing the massive release of radioactive materials. Although these may be implemented differently, many of the protective principles involved are common. Furthermore, many factors or acts have roles of strengthening both safety and security simultaneously. For example, the containment structure of the nuclear power plant provides a strong structure protecting a nuclear reactor from attacks of terrorists, and simultaneously prevents the massive release of radioactive materials into the environment during an accident. Similarly, management to control access to vital areas not only ensures safety functions by preventing or reducing the radiation exposure of workers, but also controls access of personnel with qualifications and serves a purpose in a security context to prevent unauthorized access by intruders [47].

Accordingly, safety and security have many common factors, although some problems remain related to differences in technique and culture between the two fields. For example, the introduction of "delay barriers" for security reasons possibly limits the "quick access" required to respond to safety critical events (limiting emergency exit). From such perspective, the International Nuclear Safety Group (INSAG) of IAEA discusses the importance of adjusting the interface of both [47].

Efforts are required; both for those in charge of safety and security to promote understanding of mutual requirements, and determine optimal policy. In other words, the concept of considering both security and safety requirements when designing the facilities (Security by Design) is important, and Japan must expand the scope of such discussion while monitoring the movements of various foreign countries.

We should focus on discussing synergy around the nuclear security measures relating to accidents and measures against nuclear safety. In the case of nuclear security events and following the emission of radioactive materials, we see the need for disaster prevention measures to be exercised immediately, instead of considering crisis control planning for nuclear security and disaster prevention planning side by side separately.

To achieve nuclear safety and security, Japan should study quickly and precisely from the abovementioned perspective, how best to adjust the interface between both and the synergy of both for assumed accidents.

(6) **Human resources development in the nuclear security field**

To conduct measures against nuclear security surely, those concerned must have sufficient knowledge and experience of nuclear security itself. However, devising educational programs to acquire such experience in assuming real nuclear security cases is difficult. Therefore, we should study personnel training methods which adopt findings from many fields related to nuclear security. Those concerned with nuclear energy lack basic knowledge in the security field, while those concerned in the security police authorities lack basic knowledge in the nuclear energy field. Educational training allowing individuals to acquire such knowledge is important. In such cases, we should adopt not only learning on the job but also a program involving actual maneuvers.

Moreover, solely from the personnel training perspective, tasks related to nuclear security should appeal to young people who are responsible for the next generation. Measures to evaluate achievements fairly are also important; for example, providing an environment in which the outcome of research papers can be presented, establishing new national qualifications for nuclear security and ensuring those with qualifications receive favorable treatment etc.

In Japan, only physical protection has been discussed, and the importance of nuclear security has been downplayed in comparison to nuclear safety. This is clearly reflected by the following facts: (i) slow response to previously explained B5b, (ii) no system having initiative substantial for nuclear security in our nuclear energy administration, (iii) until very recently almost no university offered lessons connected with nuclear security, even in universities responsible for educating on nuclear energy, (iv) absence of a national examination to identify specialties of physical protection managers. Recently, the approach of universities to target personnel training focusing on nuclear security is gradually progressing, and further advances are expected in future.

6.10.1.8 Summary

As already mentioned in Sect. 6.10.1.5 about physical protection, to strengthen physical protection, including INFCIRC-225 Rev. 5 and nuclear security, reflecting the lessons learned from the accident of the Fukushima Daiichi, the Nuclear Reactor Regulation Act was revised in March 2012.

The lessons learned from the accident of Fukushima Daiichi are listed in this chapter, but they may change depending on the site situation amid preparation toward decommissioning of the reactor, and the response to the changing threats on the global situation. Therefore, it is important for us to review the matters continuously.

Japan has less experience of implementing nuclear security than Western countries. Therefore from now, the operators, regulatory organizations, research and development organizations and concerned academic societies must confirm each role respectively, and should seek organic coordination between them simultaneously.

6.10.2 Safeguards and Nuclear Material Management and Accountability

6.10.2.1 Preface

It is important to clearly demonstrate the fact that there is no diversion of nuclear materials in civilian facilities by the nation as well as the nuclear security against non-state terrorists to the global community. Japan has long pursued the peaceful use of nuclear energy, and the people could not envisage any other countries considering the potential diversion of nuclear materials from peaceful purposes to military ends given the nuclear power plant accident. However, erosion of international trust in a country where the myth of nuclear safety collapsed might lead to distrust, even in fields relating to nuclear nonproliferation and also safeguards. Particularly, in relation to the discussion concerning the continuous utilization of nuclear energy, some politicians mention aspects of national security. Under such circumstances, and as a way of emphasizing peaceful use of nuclear energy for the global community, it is crucial and meaningful to demonstrate how comprehensively precise safeguards and nuclear material controls are performed, even during accidents.

6.10.2.2 Response to Nuclear Material Management, Safeguards at the Time of the Accident

To ensure the uses of nuclear energy only for peaceful purposes, Japan concluded a safeguard agreement with the IAEA under the "Nuclear Nonproliferation Treaty" (NPT) signed in 1976. According to this, it is important to detect that there is no diversion of nuclear material of significant quantity within a given time. The specific technical means for this include "nuclear material accountancy" as the basis, and "confinement/monitoring" as an assisting means. Moreover, as a means to confirm the lack of nuclear material and nuclear energy activity not reported, under the Additional Protocol and beyond the scope of the agreement mentioned above, there is also voluntary reporting on the nuclear energy activity of the member states; so-called "extended declaration", and "complementary access" to identify accuracy as the integrity. Moreover, there is also the concept of "integrated safeguards" designed to strengthen the effectiveness and improve the efficiency of the safeguard system. Under this system, the IAEA provide "extended conclusion" for the State concerned so that the IAEA can conclude that the nation as a whole, is not engaged in "conversion of the nuclear material under safeguards" and there is "absence of undeclared nuclear material or nuclear activities.

The "detection time" is the key requirement in terms of safeguards equivalent to the abovementioned "given time" necessary for diversion. The following is determined as the detection time (target) (http://www.rist.or.jp/atomica/data/dat_detail.php?Title_No=13-05-02-04).

(a) within 1 month for non-irradiated direct use of nuclear material
(b) within 3 months for the direct use of irradiated nuclear material
(c) within 12 months for the indirect use of nuclear material

In nuclear power generation, spent fuel corresponds to (b), while new fuel in MOX corresponds to (a) and uranium fuel corresponds to (c). However, in Japan, with the application of Integrated Safeguards, the target detection time for this irradiated direct use nuclear material is eased (extended) to 12 months from 3, and for MOX to 3 months from 1 (http://www.aec.go.jp/jicst/NC/senmon/seisaku/siryo/seisaku07/siryo3.pdf). When a nuclear power plant is operating under normal circumstances in

Japan, such detection targets should be applied. Where the condition is to be applied to a nuclear reactor after the accident, the most significant problem is new MOX fuel. Fortunately, Fukushima Daiichi does not have new MOX fuel, which means the fuel assembly loaded into the reactor and the spent fuel in the storage pool in the reactor are subject to the detection target. Namely, the timeliness target is less than 1 year (including the loading of MOX fuel into the reactor). The timing of any response related to operators at the time of the accident of the Fukushima Daiichi and nuclear material management by the nation has not yet been announced, but if "containment and surveillance" are effectively valid until just before the accident, the nuclear materials should be verified within approximately 1 year. Although the fact that safeguards could not be implemented due to access difficulties at the time of the accident is undeniable, judging from the abovementioned concept, there is reason to believe there were some time allowance for the safeguards verification.

Several new fuel assemblies transferred to the in-service pool from Unit 4, spent fuel in the spent fuel common pool, Units 5, 6 and dry casks are already under IAEA safeguards, while the Physical Inventory Verification (PIV) was also implemented.

After the Fukushima Daiichi accident, verification work on safeguards of Units 1–4 was not conducted due to access difficulty, but if it can be shown that there are no illegal transfers from facilities, judging from the abovementioned concept of detection time, the matter should not cause a big problem.

As for safeguards viewpoint after the abovementioned accident, it is important to verify that all the nuclear fuel has remained in the reactor building. In the implementation reports of 2011 and 2012, the IAEA wrote conclusion of "no diversion of nuclear material" and "absence of undeclared nuclear material or nuclear activities," was detected under Integrated Safeguards (http://www.iaea.org/OurWork/SV/Safeguards/documents/es2011.pdf). This reflects the fact that the IAEA judged that there was no diversion nationwide, including Fukushima Daiichi, as "extended conclusion" mentioned earlier.

6.10.2.3 Response to the Nuclear Material Control in Past Large-Scale Accidents

In addition to the abovementioned concept based on normal time safeguards, consideration of how safeguards are treated at the time of the accident also becomes important. In other words, the abovementioned is only a target related to peaceful use under normal circumstances, so that different concepts are expected to be applied in the event of an emergency.

In connection with safeguards at the time of the accident involving the nuclear power plant, repeated discussion took place in the IAEA, regarding the example of the Chernobyl nuclear power plant (from the Soviet Union which is a nuclear weapon state, the power plant was transferred to the Ukraine, and after the point (1994) when Ukraine participated in NPT, safeguards applied) but even now, over 20 years since the accident, no precise technique has yet been decided. It is assumed that timely establishment of safeguards was delayed due to the fact that the State was a nuclear weapon state at the time of the accident, reports of nuclear materials in the reactor concerned were delayed (an initial inventory report was only made by the Ukraine in 1998) and the nuclear material has not yet been transferred from the nuclear reactor after the accident. As for the TMI accident, although the USA is a nuclear weapon state, report of strict management is required for the USA just as the same as non-nuclear-weapon states [48] and measurement control was performed. The measurement control technique to which the NRC finally agreed involves providing measurement reports, basically after fuel removal work, from the initial inventory at the time of the accident and residual volume measurement after the transfer of almost all the nuclear material. In any case, the IAEA is currently studying safeguards at the time of the serious accident in the nuclear power plant, and it seems there is still no clear idea, but in non-nuclear-weapon states, maintaining safeguards is important, even in the case of an accident reactor, and particularly when the nuclear material is moved, with the application of safeguards based on reported specific accountancy measurement is not avoided.

6.10.2.4 The Guarantee on Nuclear Nonproliferation and Countermeasures in Future

As mentioned earlier, to confirm to the global community that the use of nuclear energy in Japan is continuously limited to peaceful purposes, it is important to clearly determine the fundamental concept and act to ensure nuclear nonproliferation during large-scale accident, as well as to demonstrate Japan's performance in complying with safeguards to foreign countries. To that end, besides cooperating sincerely with the IAEA when accidents in nuclear power plants occur, the government and operators must promptly suggest a clear way forward for accountancy techniques of nuclear material and safeguards supported by the relevant research organizations. With regards to Fukushima Daiichi, the report on the volume of nuclear material based on nuclear material accountancy by the facilities before the accident exists apparently. Therefore, it is important to indicate distribution of the volume of nuclear material after the accident, and rebuild functions of confinement, monitoring immediately including transfer of the nuclear material (including volume of slightly released nuclear material and volume of transition to the cooling system) during and after any accident.

The government is presently studying temporary safeguards with the IAEA. However, to avoid suspicion from the international community that an immediate action such as "monitoring" for containment/surveillance should be taken. Henceforth, to ensure confinement of nuclear materials, it may be necessary to consider the installation of a monitoring camera and radiation measuring equipment for safeguards under high dose of radioactivity, including used fuel and new fuel where there are no plans to transfer from the reactor for the time being. Also, it is important to implement verification in the absence of significant nuclear material released to an outside facility, and precise accountancy measurement when transferring spent and new fuel to a common pool.

Also for the long term, concerning accountancy of nuclear material removed when molten fuel (debris) is being transferred, a key problem involves preparing a concept of the measurement technique and material management. With respect to the safeguards of such debris, as a matter of the fact the nuclear fuel substantially melts and turns into nuclear material to be handled as "bulk" (inspection of the volume of all nuclear materials by concentration) from nuclear material which is targeted for "item count" (inspection of the number of used fuel assemblies), which means the forms of inspection will differ. However, as above, the concept of safeguards for nuclear material is not one which takes the accident into consideration, therefore the nuclear material to handle can also be considered as equivalent to loading fuel in the reactor and spent fuel as previously. In any case, with respect to the nuclear material measurement methods for the debris, it is important to secure transparency for the procedures of removal and storage of the fuel in the reactor, and to establish the technique by which nuclear material accountancy is reasonably performed. At present, the development of application technology to remove debris from the reactor and its storage is mainly initiated by the government, IAEA, and Tokyo Electric, with the cooperation of the Japan Atomic Energy Agency (http://www.meti.go.jp/earthquake/nuclear/pdf/121022/121022_02f.pdf).

6.10.2.5 Summary

Summing up the above, concerning the accountancy and management of nuclear material after the accident, although there is no big problem concerning the response by government and operators at present, to avoid suspicion from other countries, it is important for Japan to us to indicate the fundamental concept and countermeasures for future nuclear nonproliferation for this accident definitely for the global community. Also, to facilitate the above, Japan must ensure security action for nuclear nonproliferation by monitoring as soon as possible, plan and conduct precise accountancy and management to transfer problem-free spent fuel, and to establish a technique for accountancy measurement, including nondestructive measurement of loading molten fuel in the reactor, targeting transfer in future. As the actual reactor situation becomes clear, the applicable means of accountancy and management of nuclear material may change, hence the need for the government and operator (TEPCO) to respond flexibly with the cooperation of research organizations such as the Japan Atomic Energy Agency.

Moreover, to indicate Japan's action for nuclear safely, security and safeguards, namely 3S, to the global community, it is important to proceed with work to maintain transparency continuously in information exchange and close cooperation with the IAEA or the United States.

6.11 Human Resources and Human Factors

In this chapter, the Accident in Fukushima Daiichi Nuclear Power Station from the viewpoint of human factors, as well as human resource staffing and development for operation and management of nuclear power plants are described. A discussion is also made on the position of the chief engineer of reactors to gain an insight for the futures.

From the viewpoint of human factors, first of all, how the operators had recognized the plant conditions until the hydrogen explosion of Unit 1 from the viewpoint of Crew Resources Management(CRM) are as follows. Due to loss of all power, the Main Control Room, etc., had been left without any lights for a long time. In the total darkness, it was confirmed that they had exercised a relatively high CRM skill on the site shaking from frequent aftershocks with the major tsunami alarms blaring out. The operating condition of the emergency condensation system of Unit 1 was hard to be recognized because of the loss of functions such as control panels of the Main Control Room, Safety Parameter Display System (SPDS), etc., which were indispensable of monitoring the situation. Also, the operation manual was no longer applicable to water injection operation into the Unit 3. These severe situations resulted in unsuccessful operation. However, the staffs on the site seem to have taken flexible approaches based on their knowledge and experience. In regards to education and training, although various operation simulator training programs had been provided to respond to different accident scenarios, no effective training programs, assuming severe accidents including station black out (SBO) of electricity during a long time, meltdown, etc., had been provided, which constitutes a major cause of a delay in the restoration from the failure. Since the accident, each electric power company has enhanced and reinforced the emergency preparedness system based on what they learned from the accident, providing improved education and training programs for the operators and disaster prevention staffs. In the future, proposals and recommendations from various institutions should be used to help make more improvements. As for the fields of communication and information sharing, some problems were identified among two groups, operation groups, or order-givers and takers. It is important not to prevent site/task operations in applying plausible measures for the problems. On the other hand, in the analysis of emergency response capability to the accident, many good cases were found in individual and organizational levels, but there were bad crisis responses found in managerial or national levels. Based upon the review results on the operation and obstructive factors of the site, it will be effective to have measures to keep the power source and system function for a long duration. A system design that enables manual operation without an excessive dependence on remote control.

The developing program of nuclear human resources had some challenges before the occurrence of the accident. There had been a lack of understanding of potential risks, excessive confidence in the technology and safety systems. There had also been a lack of understanding of the particular features of nuclear power, shortage of individual facilities and human skills, and ambiguous roles that a chief engineer of reactors should play once a severe accident relating to a nuclear reactor occurs. From the aspect of regulations, too, there were problems in terms of expertness of human resources and advice from the Emergency Technical Advisory Body. The improvements required for the future challenges should include, beyond the conventional assumption and the sphere of response, safety prioritized and committed by the top managers; attitude to learn and keep asking; imagination and expertness in plant design; knowledge and skills through visualization initiative; and expertness, internationalism and judgment ability of regulatory human resources. To this end, it is necessary to enhance basic literacy in science, to cultivate the understanding of nuclear energy systems to young generation, experienced operators, and staff human resources in the fields of education and academics.

The Chief Engineer of Reactors with a national qualification is, considering overseas cases, exempted from providing instantaneous response in the occurrence of an unexpected accident, and takes appropriate measures for himself/herself based on an understanding of the principle and significance of an event. To this end, the Chief Engineer is expected to take substantial responsibility for safety measures on the site, providing advice to the Nuclear Emergency Preparedness Manager who commands the entire organization. Also, the Chief Engineer is expected to act as a manager to promote power electric companies to continuously improve the safety of nuclear power plants at ordinary times.

6.11.1 Human Factors

6.11.1.1 Purpose and Method to Examine

Many problems were posed in the Fukushima Daiichi NPS accident, including recognizing the situation in the plant, information sharing in/out of the power station, decision making, emergency response, education and training on daily basis, instrumentation/control facilities and work environment of the plant, etc. The problems suggested in various reports from the viewpoint of human factors (HF: human factors to ensure safety) will be discussed.

By referring the documents, reports and data published, the following 6 items that are important from the viewpoint of HF are reviewed: (1) assessment of the plant's conditions by operators at Units 1 and 2, and a review on accident response from the viewpoint of CRM (Crew Resource Management) (until the hydrogen explosion of Unit 1 occurred); (2) actions taken by the power station staff (recognition of operating status of the isolation condensation (IC), alternative water injection into Unit 3); (3) challenges in terms of education and training; (4) problems and actions to address these problems in the field of communication and information sharing; (5) emergency response capability of the organization; and (6) factors that inhibited from responding smoothly to the accident and a plan on how to improve from the aspects of the operation of these reactors as well as from the field operation on the site.

Note that the review results in this section may be different from those in other sections in this chapter because the review in this section is made from the viewpoint of human factors by using characteristic analysis techniques in this field such as CRM, etc.

6.11.1.2 Assessment of the Plant's Conditions by Operators at Units 1 and 2, and a Review on Accident Response from the Viewpoint of CRM (Crew Resource Management)

(1) **How to conduct investigation and examination**
Based upon reference [49, 50] and information [51] published by the defunct Nuclear and Industrial Safety Agency, how the operators grasped the picture of the plant's conditions and their accident response from the viewpoint of CRM have been discussed.

(2) **Assessment of Units 1 and 2 conditions until Unit 1 was damaged by the hydrogen explosion**

(a) Assessments from the occurrence of the earthquake to the onslaught of the second wave of tsunami: In the Main Control Room (MCR), operators precisely monitor the automatic operation and plant status through the control panel and take operation steps to shutdown the reactor in accordance with the operation manual. However, they are supposed to have had a sense of uneasiness in the wake of frequent and big aftershock jolts. The major tsunami warning issued at 14:58, with which they may not have assumed such a major tsunami enough to flood the reactor building. If the plant components had not been damaged by the ground motion, the operators were supposed that they would be able to achieve a cold shutdown in accordance with procedures specified in the operation manual. Field confirmation of the damage of plant components would continue for a long time due to frequent aftershock jolts with the major tsunami warning, However, judging from the plant parameter data over time, the operators seemed to have concluded that the main equipment and apparatus functioned well.

(b) Assessments from the onslaught of the second wave of tsunami to temporary lighting-up of the Main Control Room: The AC power supply was totally lost (station blackout (SBO)) due to the damage by the second tsunami wave at 15:32 resulting in the turning off the lighting of MCR and main control panel. At 15:50, power supply for instruments was lost, which made the water level of the Units 1 and 2 undetectable. In the darkness, a review was made on the cause of SBO, how to restore the power source (especially, lighting of MCR, as well as power supply of the monitoring instruments), and how to confirm the operations of Isolation

Condenser (IC) (Unit 1) and Reactor Core Isolation Cooling System (RCIC) (Unit 2). Later, operators seemed to have begun studying how to inject alternative water prepared for unexpected problems.

Judging from the frequent aftershock jolts, the Emergency Preparedness Headquarters seemed to have concluded that it would take time to restore the power source, where General Manager directed to study the alternative water injection at 17:12 based on the assumption of the importance of water injection. Also, at almost the same time, the reactor water level (which indirectly indicates the RCIC operation) was found to be stable. Around the evening, the extent of tsunami damage was identified, when the discussion begun on how to restore the power source by using part of Unit 2 power center with a power source car. Taking the above into account, the focus of operators' attention seemed to have shifted to how to secure water injection line for the alternative water injection and to check IC operation.

They tried to identify the IC operation in vain. Then a review of procedures followed due to their recognition that the containment venting would be needed depending on the future situation. Also, they worked to secure the water injection lines in the darkness in the order of Units 1 and 2 on March 11. In parallel they inspected the location, etc. of field instruments based upon the drawings and entered the reactor building (R/B) to see the reactor pressure and functioning status of the main equipment.

(c) Situations from the temporary lighting of the Main Control Room to the access prohibition to Unit 1 reactor building due to unusual increase of radiation dose: A small generator was installed at 20:49 and temporary lighting was turned on in the MCR of Units 1 and 2. Although the temporary lighting did not serve enough illumination for smooth actions of operators, MCR was no longer in the darkness. Temporary batteries were also connected to the monitoring instruments. They must have been relieved by obtaining the data that showed that the reactor water levels of Units 1 and 2 were above the fuel rod level meaning that the fuel rods were not exposed. As for the unknown status of IC valves, they said operators were dubious if IC did function based upon the result of "opening" operation of MO-3A valve at 21:30.

(d) Situations from the access prohibition to Unit 1 reactor building to hydrogen explosion of Unit 1: Probably, the cause of why the radiation levels rapidly rose was discussed. According to the data indicating the reactor water level of Unit 1 on 22 o'clock, operators may have concluded that fuel melting, if any, was only partial. Power source for control operation was expected to be restored in MCR. However, laying temporary power source lines took long time due to the evacuation by frequent aftershock jolts under the major tsunami warning. The operators focused on the confirmation of the operation status of RCIC in Unit 2. Before dawn of March 12, they obtained a proof of its functioning, and shifted their attention to how to restore the Unit 1's power source while wondering the water source.

Meanwhile, the diesel power generator FP for water injection at Unit 1 was found to be shutdown at 1:48, and it was unable to be restarted. Facing difficulties, they seemed to have recognized that they made a step forward in the operation as they successfully started freshwater injection from fire cisterns at 5:46. They repeatedly studied venting operation procedures for pressure containment vessel (PCV), trying to collect the equipments necessary for the venting. Around 5 o'clock, they were ordered equip themselves with the full mask, charcoal filter, and B apparatus. And the operators took shelter of the Unit 2 side due to an increase in radiation dose from the Unit 1 side. Thus, efficiency in the operations at MCR aggravated further. With the situation worsening, the operators must have believed that some of the fuel rods had exposed. Also, the group on duty from the morning of March 11 had worked 24 h straight.

In an attempt of PCV vent at a high radiation level, operators manually opened motor operated valves (MO valve) in the field. They also handled the air operated small valve (AO small valve) from MCR and tried opening operation of AO large valves by setting up a temporary air compressor. With a decrease in the pressure of Drywell (D/W), the Emergency Preparedness Headquarters concluded that they succeeded in PCV vent. Because the operation of freshwater injection had continued during this time, emergency core cooling system, if not sufficient, may have worked to some extent. Because freshwater from the fire cistern dried up, at 14:54 General Manager directed to start seawater injection to the Unit 1 reactor.

They managed to complete preparation for restoring the power source at around 15:30 and a hydrogen explosion occurred in the reactor building at 15:36. This damaged cables, etc., made all of the on-site staffs to take shelter in the important anti-seismic building.

(3) **Discussion of the accident response from the CRM viewpoint**

(a) Outline of the CRM training: CRM (Crew Resource Management) training program was studied and developed in the late 1970s mainly in the United States, which has been applied in many countries in the world. Its basic concept lies in taking advantage of every available resource to make a best decision, and achieving high team performance as much as possible [52]. This is a non-technical training method to develop appropriate skills for addressing problems under abnormal situation [52]. CRM training generally consists of the following five skills as shown in Fig. 6.34. The training program provides reviews on how their skills were used. Trainees learn how to effectively use their CRM skills found in the course of reflection from not only their failures and shortcomings but also positive and concrete insights.

(b) Discussion from CRM Viewpoint: Despite the strong and frequent aftershock quakes, the operators reported the chief on duty by reading out the indicators of control panels they were in charge of until the onslaught of tsunami waves. Reviewing whether or not the chief on duty, or recipient of

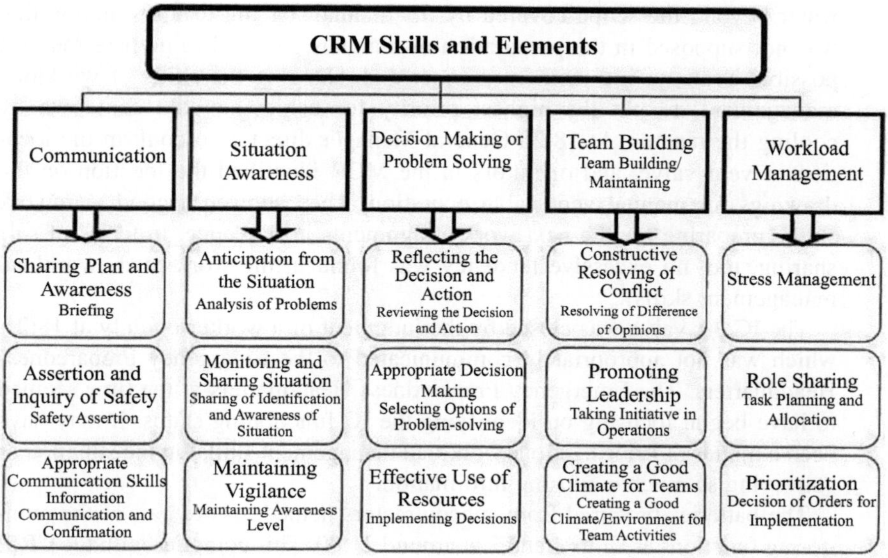

Fig. 6.34 CRM training map

the report, easily understand it, is the debriefing (reporting the situation experienced) that is an important communication skill for CRM. This provides a useful information for responding to similar situations. From the viewpoint of emergent human behavior, warnings, etc., should be visualized and illustrated to display danger points in the whole picture.

The fire alarm set off at 14:52, which made the chief on duty recall his experience that the sensor reacted to the dust generated by a ground motion. And he is reported to have had it reset for alarm stopping. Under the environment suffering from major jolts, he remembered the lesson he learned in the past and examined whether or not it was a false warning. Also, he manipulated one of the two ICs that automatically started at Unit 1 in accordance with the procedures, taking account of the rate of temperature dropping at the reactor. They showed good situation awareness skills and decision making as expected.

The attack of major tsunami waves resulted in SBO, causing black out in the MCR and unable situation of grasping reactor parameters. Because the operators did not identify the cause of SBO, it is important to install monitoring camera systems for the facility to support situation awareness. However, the chief on duty who faced SBO in the MCR reported the summary on what were unidentified factors and what prevented cold shutdown to General Manager at 15:42. As emphasized for CRM, he did report the right information in the right timing.

General Manager directed to examine alternative water injection at 17:12, having the staff review whether or not it was possible to inject alternative water beyond the scope covered by the manual. Facing to the situation that was not supposed in the preparation of manuals, he tried to prepare the best possible strategy in a resilient manner and, showing the skills of workload management, to put the highest priority for achieving cold shutdown by cooling the reactor. At 18:20 General Manager directed to confirm the location of vent valve, and operators in the MCR identified the location on the drawings for manual vent valve operation. They showed a good teamwork skills, preparing for the next work assignments in sequence. In terms of role sharing, too, no excessive tendency was found in the work road (workload management skills).

The IC 3A valve was closed by the judgment of a worker on duty at 18:25, which was not appropriately communicated to the Emergency Preparedness Headquarters. The Emergency Preparedness Headquarters at the time seemed to have begun focusing on identifying the IC functioning status. It may have been a matter of Prioritization (workload management skills) in the situation of conveying so much confusing information.

The batteries procured from subcontractors helped restore part of the monitoring functions at Units 1 and 2 at around 21:00. This complies with the CRM concept where all resources available should be effectively used.

General Manager had understood the need to get the picture of the reactor, to start venting and to inject water, and concluded that entering the reactor building was the only option to grasp the reactor pressure. To this end, he directed to organize a "suicide corps". He showed what is called the workload management skills and decision making skills. Also, this indicates a need to construct mechanisms to measure the reactor pressure without a power source. General Manager directed to be ready to vent PCV at 0:06. It must have been a difficult task to select members who would access to the site amid the increasing radiation level. The MCR began choosing the staffs and reviewing concrete procedures for venting PCV. It can be said that the MCR team had showed good workload management and teamwork skills that would respond to urgent problems. They began studying a possibility of water injection using fire engines at 1:48, and found that the fire cistern was available for use. Their flexible review and judgment based upon decision making skills comply with the CRM concepts which make use of every possible resource.

However, the Prime Minister arrived at the Fukushima first at 7:11. Although the site workers were very busy in preparation for injecting water and venting, they supposedly had to focus on the PM's movements. This was not a desirable event from the viewpoint of CRM, whose lessons must be learned for constructing future crisis control system.

A good leadership and good followerships to support this as CRM team skills were shown for the preparation of seawater injection in face of the great damage by tsunami.

6.11.1.3 Examination from the Viewpoint of Human Factors: Actions Taken by the Plant Operators

The government investigation committee pointed out [49] "Misjudgment of operational situation of the IC (Isolation Condenser) at Unit 1" and "Poor handling of alternative water injection at Unit 3" as some of the causes of the Fukushima Daiichi Disaster. In this section we will again discuss on them from the viewpoint of HFs with using the information written in the published reports including TEPCO's [50].

(1) **Discussion on "Misjudgment of operational situation of the IC at Unit 1"**

(a) Recognition at the Main Control Room (MCR): Warning to inform the IC operating status including isolation signal turning on was installed on the control panel in the MCR. However, these warnings did not work due to a loss of power in the wake of tsunami. They had no means to find the IC operating status as the IC isolation valve indicator did not flash or light up. Therefore, ≪IC operating status unknown≫ (Reference [50] p. 149 in Japanese ver.) was their initial common recognition in the MCR. At 16:44 on March 11, steaming from the IC venting pipe ("we can see a smoke at your right hand") was reported to the MCR. However, due to a small amount of the steam, they ≪doubted IC operating status≫ (Reference [50] Appendix 8–10 ditto) which means the IC might not had been working. At about 18:00, DC power was restored temporally and they found that the IC isolation valve 2A was identified closed. It means the IC isolation signal went on and the signal also closed the IC isolation valves 1A and 4A along with 2A. Expecting the valves 1A and 4A were open by a miracle, they opened the valves 2A and 3A which was manually closed by an operator on duty before the tsunami attack. However, the steam from the IC venting pipe or the sound of the steam disappeared; they recognized that ≪IC does not function properly due to the closure of the 1A/4A valves or water shortage at the condenser of the IC≫ (Reference [50] p. 126 ditto). After 21:00, they opened the Isolation valve 3A again because the concern over the water shortage at the condenser of the IC disappeared but they found no sound of steaming. This led to their conclusion as ≪IC does not work normally≫ (Reference [49] p. 107, 108 ditto).

As shown above, the operators on duty in the MCR do not seem to "have responded to the situation based on their assumption that IC functioned normally". In parallel to identify the IC operating status, they prepared the alternative injection with D/DFP (Diesel Driven Fire Pump) from about 16:30 in case cooling by IC was not available (Reference [50] p. 124 ditto).

The IC isolation valve 3A was manually closed before the tsunami attack in order to maintain the rate of RPV temperature decline not to exceed 55 °C/h. This operation was not shared among the operators in the MCR (Reference [50] p. 143 ditto). Amid conflicting reports and many warnings, operation and shutdown of equipment, it must have been quite difficult to, as instructed, pay a special attention to the report on the valve 3A successfully operated as planned.

(b) Recognition at the ERC (Emergency Response Center) in the power station: Immediately after the tsunami, the ERC was reported a loss of AC and DC power sources from the MCR (where the occurrence of the IC isolation signal was not identified) (Reference [49] p. 91 ditto). Judging from this single report, it is reasonable to recognize that the valves (that lost driving power) had remained the same as what immediately before loss of power. However, this report does not provide a full information about the valve status which had been opened/closed manually before the tsunami. Therefore, there is no suggestion about if the IC was in service or not. On this matter, General Manager of the Fukushima Daiichi power station said in the interview later "we expected IC was working normally, but reports afterwards from the MCR suggested that the IC was not working" (Reference [49] p. 96 ditto). Immediately he directed to the alternative injection by the FP (Fire Protection) System and also made emergency declaration based on the Article 15 of Act on Special Measures Concerning Nuclear Emergency Preparedness and notified the government and associated authorities. The ERC had the recognition of ≪IC might not be functioning≫. The following reports, including those "1 h left before the reactor water reaches down to the TAF (Top of Active Fuel) " at 17:15, "high radiation dose detected at the entrance of the Reactor Building" at 17:50 (Reference [49] p. 108 ditto), did not contradict their recognition that the IC was not in service. The ERC is supposed to have had a recognition that ≪IC would not be operated≫ until around 18:00.

However, after the report "the IC lineup was complete and water injection started" to ERC at 18:21, they received some reports as if the IC had be back in service, including "the IC confirmed operated" at 18:24, and " the reactor water level is TAF + 20 cm" at 21:19. (Reference [50] Appendix 8–10, p. 126 ditto).

Here we will focus on the report "the IC lineup was complete and water injection started" at 18:21. The IC isolation valves 2A and 3A was opened by the operators on duty in the MCR and it was reported to the operation team at the ERC at 18:20 (Reference [50] Appendix 8–10 ditto). However, the report to the ERC from the operation team was only described that "the IC lineup was complete and the water injection restarted", which does not include any information that the operators in the MCR opened the valve 2A (which usually keep open) or the IC isolation signal occurred. The SPDS (Safety Parameter Display System) did not work, which forced the operation team at the ERC to convey all of the information on the plants orally. Amid various conflicting information, they reported only results of operations in an effort to effectively provide the necessary information to the executives of the ERC. However, this report deprived the ERC of the opportunity to notice the occurrence of the IC isolation signal. As the results, they seems to be led to misbelieve that ≪ IC was back in service ≫.

Given the IC resumed its function, their big focus should have been on how much the reactor water level decreased by the time. It was estimated by technical team at the ERC that the reactor water level would reach down to the TAF around 18:15 (Reference [49] p. 108 ditto). But its estimation was based on the read from a malfunctioned indicator of the reactor water level. Furthermore, after the report of "the IC lineup was complete and the water injection restarted" at 18:21, the ERC had some reports indicating as if the reactor water level was higher than TAF. These might have misled them to assume that the IC restored its function and they managed to avoid the core damage.

As described above, the ERC assumed that the IC might not be operated immediately after the tsunami, hence promoting necessary arrangements. However, the report "the IC lineup was complete and the water injection restarted" at 18:21, along with other reports, have supposedly led them misbelieve that IC resumed its function.

(2) **Discussion on "Poor handling of alternative water injection at Unit 3"**

(a) Operations executed and ones written in the operational procedures: In the wake of the SBO (Station Black Out) by the tsunami attack, the operators on duty put the RCIC (Reactor Core Isolation Cooling) system in service at 16:03 on March 11 and provided the DC load separation from around the evening (Reference [49] p. 95, 96 ditto). The RCIC system was automatically tripped at 11:36 on March 12 and the HPCI (High Pressure Coolant Injection) system automatically started up at 12:35 due to low reactor water level. In order to avoid a rapid increase of the rector water level due to the large amount of water supplied by the HPCI system, they decreased and adjusted the flow from the HPCI system. They also suppressed consumption of DC power and secured time to restore the AC power (Reference [49] p.96, 170 ditto). Thus, the operators on duty struggled to follow the SBO procedures written in AOP (Abnormal Operating Procedures) while paying attention to suppress consumption of the DC power.

At 20:36 on March 12, the DC power of the reactor water level indicator was dried up, which resulted in unavailability of monitoring of the reactor water level. This satisfied the conditions to introduce "reactor water level unknown" in EOP (Emergency Operating Procedures). However, they had continued the injection by the HPCI system with a little increasing of the flow rate and they also tried to recover the reactor water level indicators (Reference [49] p. 170, 171 ditto). Subsequently, the discharge pressure of the HPCI pump became almost the same level as the reactor pressure, which had the operators unconvinced the injection to the reactor. In order to start the alternative injection through the FP system with the D/DFP, they put the D/DFP for spraying in the S/C (Suppression Chamber) into the reactor injection line, and removed the HPCI system from service at 2:42

before opening a SR (Safety Relief) valve in order not to break the HPCI system. Then they tried to open the SR valve (2:45, 2:55) but failed. As a result, the injection to the reactor finally stopped (Reference [49] p. 170, 171, 172, 173 ditto).

(b) Discussion on the operations after 20:36 on March 12: The operators on duty had continued the injection by the HPCI system even after 20:36 when monitoring of the reactor water level was no longer available. There are some descriptions in the Accident Analysis Report by TEPCO as the following (Reference [50] p.202 ditto).

- The HPCI system was more reliable than the D/DFP which was small in capacity for fire extinguishing.
- The situation did not allow them to shift to the injection by the D/DFP of Unit 3, in parallel with Unit 1.
- The reactor water level may rapidly decrease due to decompression boiling with accelerating risk of fuel exposure.

The following information was obtained from what was inquired to TEPCO, especially, on "The HPCI system was more reliable than the D/DFP which was small in capacity for fire extinguishing".

The FP system including the D/DFP, the outdoor piping system, etc. was designed as seismic design class C. As was concerned by the General Manager of the power station, and that really was the case (Reference [49] p. 122 ditto), rupture and water leakage were found at the outdoor piping system from the filtrated water tanks installed away on the hill (capacity: 8,000 t × 2 tanks) to the turbine building. The FP system was not in the condition enough to make it work as designed. Also, it was necessary to supply fuel to the kerosene tank of the D/DFP to keep its running. However, the fuel supply system was unavailable due to the loss of the power, and the kerosene tank had small capacity as well (the capacity is calculated to be equivalent to run the pump for about 20 h, from what written in the business diary [53] of Unit 3). They had to access down to the hard hit area by the tsunami and manually supply fuel in order to keep it running. From these circumstances, it was the common recognition among the ERC and the MCR to continue the injection by the HPCI system which is seismic design class S and had been stably running. Rather than using equipment for fire extinguish which might not fully work, they chose the option to use equipment then active and used for water injection. It seems to be sufficiently rational. From the viewpoint of driving source of the HPCI system and the D/DFP, it would be reasonable to run the HPCI system as long as possible and set the diesel fuel apart for the later use, which could result in prolonging the water injection hours. And at around 2:40 on March 13, as the injection by the HPCI was no longer functioned, they switched to the D/DFP for the alternative injection based upon the common recognition between the ERC and the MCR. Generally speaking, operational procedures are prepared under assumptions on operational situations, events, etc. Those must be complied with as much situations allow. However, this accident was far beyond the situation contemplated when the operation procedure had been prepared. Under this circumstance, those in the ERC and the MCR examined the feasibility of what written in the procedures rather than only following them and made their own decisions based upon their knowledge and experience in order to continue injecting water to the reactor as long as possible.

(c) Discussion on the action at around 2:40 on March 13

(i) "it must be confirmed if the alternative injection line was completed before the HPCI system was removed from service" It would be agreeable as suggested above in non-urgent situations. However, this time no communication means were available for the operators who went to the turbine building to change the lines. Also, the lines had to be urgently switched. The HPCI system was not automatically tripped, despite the reactor pressure was lower than one designed to trip the HPCI automatically. The risk of HPCI damage was getting larger. Also, it was unknown if the injection had been available because the discharge pressure of the HPCI pump was almost equal to the reactor pressure. It seems to be risky but inevitable to remove the HPCI from service before confirming the injection line.

(ii) "SR valve should be operated open before removing the HPCI system from service" It is also agreeable if there is no risk of the HPCI damage due to the reactor decompression. However, opening the SR valve (decompression operation) would further increase the rupture risk of the HPCI system and also lower the reactor water level through decompression boiling. On the other hand, the DC power which had supplied the electricity to the HPCI and the RCIC systems for many hours, run the oil pump (5,600 W) of the HPCI system and put the SR valve status indicator on immediately before stopping the HPCI system. Beyond any doubt, they might have thought "the SR valve was available for open". Their decision seems to be inevitable under the circumstances where they had to hurry to inject water to the reactor and avoid damages to the HPCI system.

(3) **Lesson learned though the analysis**

In "Misjudgment of operational situation of the IC at Unit 1", they lost the functions of the MCR control panels and SPDS, essential to monitor the plant status. In "Poor handling of alternative water injection at Unit 3", the operation procedures were no longer available to be applied to. Thus, they were forced to face the situation where they had to make their own decisions and act on their own, based upon their knowledge and experience. From this accident we got an insight that, although the assumptions and hypothesis against unexpected situations were insufficient, human can take flexible approaches. We should remember that "things do not go as expected", "things that could not happen do actually happen" and "only human can cope with unexpected situations" to help improve safety.

6.11.1.4 Education and Training

(1) **Purpose and method to examine**
In this section, the details and problems of the education and training which had been provided before the accident in Fukushima Daiichi Nuclear Power Station are discussed. Also a review will be made on what the BWR Operator Training Center Corporation (BTC), Nuclear Power Training Center Ltd. (NTC), and the electric power companies implemented and planned in the wake of the accident, together with a discussion on the direction of the education and training based on these studies and the results.

(2) **Requirement of regulations for simulator training before the accident**
Education and training using plant simulators was a must for nuclear power plants before the accident. Article 35 of the former Nuclear Reactor Regulation Act and Article 12 of the former Commercial Reactor Regulations provides provisions on "operators in nuclear reactor operation", "operating conditions", and "operation supervisor ". Also, Notification No. 200 of the Ministry of Economy, Trade and Industry stipulates knowledge and skills on nuclear reactors.

Also, JEAG4802-2002 summarized the education and training policy for nuclear power plant operators, including the item, "SBO by all AC power loss" which was occurred in the accident.

Appendix of JEAG4802-2002 suggests that, for the "education and training on the core damage accident and on the maintenance of the important safety functions", the curriculum should include accident cases in/out of the country, emergency core cooling function, important parameters for monitoring, recognition of core damage, hydrogen gas generation, and accident management.

(3) **Discussion on education and training before the accident**
Before the accident in Fukushima Daiichi Nuclear Power Station basic education and training were considered to be performed on the events occurred in the accident. However, they did not provide effective simulator training assuming long-term SBO (station blackout by all AC power loss) and nuclear reactor core meltdown, which is considered to be a major cause for delaying restoration from the accident.

The Government's Investigation Committee mentioned "they hardly seem to have acted based on that kind of knowledge," at the Accident in Fukushima Daiichi Nuclear Power Station (Government Investigation Committee on the Accidents at the Fukushima Nuclear Power Station [54] p. 402). Furthermore, "they are expertized in their own field. While, on the contrary, it is hard to say that they have enough knowledge on another field even if that is closely related to their specialty", suggests problem of the vertically segmented organization (Government Investigation Committee on the Accidents at the Fukushima Nuclear Power Station [54], p. 403). From this viewpoint, the chairperson who directs the entire commission commented that "(7) Understand that it is important to judge and act after seeing and thinking for oneself, and develop such ability" (Government Investigation Committee on the Accidents at the Fukushima Nuclear Power Station [54] p. 447).

On the other hand, they failed to collect enough knowledge from operators of the main control room, local support staff and Emergency Preparedness Headquarters members, which could be effective to avoid the accident. This fact reveals the poor communication between members and teams, even taking account of the limited information collection and communication means.

Also, some of the operators seemed to lose their sense of composure because of the frequent big aftershock quakes, no light condition in the main control room and the wretched working conditions wearing radiation protection suits, etc. Before the accident education and training assuming such a harsh scenario were not performed.

Emergency operating procedures were classified into 3 classes: event base, symptom base and severe accidents. Because the "severe accident" procedure assumes the presence of power supply, the procedure was not effective under the condition of long-term power loss in the Fukushima Daiichi accident.

(4) **Approach to improve education and training after the accident**

(a) Approaches by Operator Training Centers: BTC and NTC are engaged in improving education and training based upon the lessons learned from the accident in Fukushima Daiichi Nuclear Power Station. For example, BTC developed " review for the accident in Fukushima Daiichi Nuclear Power Station and performing countermeasures training" and started training from August, 2012 [55]. This course aims at experiencing the simulated accident in Fukushima Daiichi Nuclear Power Station events, and understanding the aims and effects of emergent safety measures. Also, preparing for an expansion to severe accident training programs, they added an severe accident simulator models to simulate damages of the core and the Primary Containment Vessel.

(b) Approaches by electric power companies: Electric power companies have promoted to improve education and training based upon the experiences learned from the Niigata Chuetsu Earthquake and other trouble events. In addition to enhance training simulator to be able to simulate long-term SBO training under situation of the accident such as Fukushima Daiichi Nuclear Power Station, they plan to improve the education and training program, develop education and education-support tools, and introduce analysis models for nuclear reactor core meltdown. Also, they adopt CRM (Crew Resource Management) training [52] to reinforce the team performance, and reorganize the organization of shift operators in order to respond to multiple unit accidents at the same time.

Also, the Abnormal Operating Procedures was reviewed to cope with a long-term SBO as well as other procedures on severe accidents.

(5) **Future direction of education and training**

(a) Future direction of education and training: In and outside the country, the lessons learned from the accident in Fukushima Daiichi Nuclear Power Station have been actively discussed, and some institutions propose ideas aiming at safety enhancement for a nuclear plant such as manual operation training in the field as well as simulation straining programs to enhance an ability to recognize the situation, communication and top management skills. It is specifically recommended to improve both hardware and software aspects, that are developing professionalism, courage, and sense of cooperation, and to promote an activity to cultivate a further safety cultures.

(b) Addressing 30 items recommended by the former Nuclear and Industrial Safety Agency: The former Nuclear and Industrial Safety Agency (NISA) concluded 30 technical findings as lessons learned from the accident [56] to be reflected in the regulations. In regards to education and training, it is required to prepare necessary information including operation procedures, design drawing and documentations, etc., staffing at emergency, construction of on-call system, training programs under high-radiation circumstances, night or bad weather, etc.

(c) Addressing 28 Recommendations to the International Atomic Energy Agency: Recommendations based upon the accident analysis and lessons learned from the accident in Fukushima Daiichi Nuclear Power Station are reported to the International Atomic Energy Agency (IAEA) [57]. As for education and training programs, they require enhancement of the training programs to respond to severe accidents.

(d) Recommendations and proposals by peer review of stress test in Europe: The stress tests and peer review on the accident in Fukushima Daiichi Nuclear Power Station conducted as requested by the European Council include summarized recommendation and proposals. In the training section, (i) facilities inspection and training programs, (ii) severe accident management (SAM) training, etc., are listed.

(e) Addressing to the Institute of Nuclear Power Operations report: The Institute of Nuclear Power Operations (INPO) studied and analyzed the Fukushima Daini Nuclear Power Plant, and provided the lessons learned based on the results [58]. Those concerning education and training include the following: as needed to prepare for more unexpected events and priorities in operational response, core cooling in the initial stage, development of clear strategies and communication at restoration activities, guidance to primary containment vessel venting, enhancement of nuclear safety culture.

6.11.1.5 Problems and Measures in Communication and Information Sharing

(1) **Objective and method of the study**

This section aims at analyzing problems and proposing measures to cope with communication and information sharing in the case of the accident at Fukushima Daiichi Nuclear Power Station with the following conditions:

(a) because the "sites" had a significant influence on the accident situation and the operations were restricted by time, the subjects of this study are set to the Main Control Room (MCR) and Emergency Preparedness Headquarters, (b) the main data used was the detailed descriptions included in the TEPCO's report [50, 59] and the Investigation Committee on the Accidents at the Fukushima Nuclear Power Station Report [49, 54], (c) if any measures were proposed in the reports, then the validity was evaluated in this study. If it is necessary, additional measures were proposed as the part of this study. The following paragraphs show the analysis results in three situations: information sharing between the MCR and the Emergency Preparedness Headquarters; within the MCR; and in the Emergency Preparedness Headquarters.

(2) **Information sharing between the Main Control Room and the Emergency Preparedness Headquarters (information sharing between two groups)**
Information on Unit 1 operation status and situations was not fully shared within the MCR operators. The communication from the MCR to the Emergency Preparedness Headquarters should follow the process as below: [MCR]—hot line (oral communication) → [Emergency Preparedness Headquarters]—oral communication → [Chief of Power Generation Team of the Headquarters]—oral communication → [entire Headquarters]. The information about detailed operations by operators, e.g., valve controls, and the sound they heard, e.g., generation of steam, which might prove IC was functioning, was communicated to the entire Emergency Preparedness Headquarters. However, the information that denies IC's functioning was not transmitted from the MCR to the entire Headquarters for some reason. As the result, there was a period of time that the operators in the MCR understood "IC's not functioning", while the members of the Headquarters recognized the situation as "IC's functioning" (Reference [50] p. 323, Appendix 8–10).

TEPCO proposed four measures in the report: (a) to understand the situation visually, communication form, e.g., simple diagram, should be used for communicating plant and system status, (b) the common template should be set in the Emergency Preparedness Headquarters and the MCR (e.g., dedicated sheets on white boards), (c) communications should be made whenever information is updated, (d) the use of these methods should be trained through disaster drills (Reference [50] p. 344–345, Appendix 16–3).

In situations where old information was recorded on the template at the Headquarters, it is difficult for the members to find out what the operators forgot to inform. To address this problem, they need to compare records on the both templates. A feasible measure is using a hardware that the Headquarters can visually confirm the template in the MCR with. A software measure is stationing of staffs in charge who perform a periodical report about the information in the MCR.

A strategy that demands to measure and communicate everything with many items might be not feasible at the time of emergency. These items must be selected based upon importance assessment. More flexible strategy may be appropriate, which ask to handle only essential information according to the situation.

(3) **Information sharing in the Main Control Room (information sharing in a working group)**
The problem was that information on the operation status immediately before SBO could not be shared by the operators in the MCR of Unit 1. One operator testified, "Valve 3A was closed before power source was lost. I told the information to another operators" (Reference [50] Appendix 8–10). But no similar testimony was obtained from other operators. When external memory such as a control panel is not available, information that should be stored in the memory rapidly increases. This easily causes a memory failure.
Although not written in the reports, the measures with the block diagram and template mentioned above can be also applicable to information sharing in a group. This may help address the problem.

(4) **Instructions and directions at the Emergency Preparedness Headquarters (information sharing between commanders and order takers)**
In the reference [54] (pp. 403–404), it was reported that an instruction by General Manager (preparation of water injection by fire engines) was not promptly accepted by the members of each function teams and groups at the Emergency Preparedness Headquarters. It also pointed out that because the roles of the teams and groups are fractionated, they lack a way of thinking which is recognizing the situation in a comprehensive manner, designing their roles, and providing necessary support service.
No measures were proposed in the reports. To address this problem, it may be a good idea to visualize the details and allocation of the tasks, and ongoing status on a white board. This allows the commanders and order takers to clearly share the information about the task. Furthermore, if the display of the MCR can be seen from the Headquarters, it may be effectively used to develop necessary support and advice to the MCR.

(5) **Notes for considering the measures**
For a practical use of the measures, each license holder should evaluate the effectiveness and feasibility at sites in details. One of the requirements should be satisfied is "do not interrupt the task on the site." The first priority should be assign to the control tasks in the MCR, where resources are limited. The information sharing task should be allotted to the Emergency Preparedness Headquarters to inhibit the interruption to the task process of the operators of the MCR.
Actually it is reported that workers in charge were allocated in the Fukushima and Tokai Daini Nuclear Power Plants. On the other hand, there is no such report about the case of the Fukushima Daiichi Nuclear Power Plant.

6.11.1.6 Analysis on the Emergency Response Capability by Organizations

(1) **Analysis method**
Based upon analysis methods used for various accident reports [60], and new methods advocated in recent years such as Resilience Engineering [61] and

High Reliability Organization [62], we extracted successful and failure cases in regards to how they responded to the Accident in Fukushima Daiichi Nuclear Power Station, from individual via. organizational, and to the external response levels. At the analysis, based upon the report from TEPCO, we discuss the timeline on water injection at Unit 1, especially on the judgment to continue seawater injection.

(2) **Methods used for the analysis**

(a) According to the definition of Resilience engineering (RE) [61], it is a strategy to control the state steady by adopting human situational awareness when the change of a system status is severe, in contrast with the concept to design a robust system against disturbance to avoid a conventional human error. Resilience (flexible and robust) refers to an capability to adjust the function, which an organization inherently has, in responding to the environment and disturbance before, amid and after it, which includes (i) studying ability, (ii) predicting ability, (iii) monitoring ability, and (iv) responding ability.

(b) High reliability organization (HRO) [62] studied organizational capability, and refers to "honesty" (report any small indication), "prudency" (to be very careful), and "sharpness" (sharp sense about operation), at ordinary time, and then "agility" (to fully respond to problem-solving) and "flexibility" (to entrust authority to the most suitable person), at the time of emergency. HRO is a concept to review a successful case from the standpoint of an organization, which has a common objective to alleviate accident trouble, in line with the present direction of RE.

(c) Risk literacy (RL) is an capability to examine the background of a risk, and to understand and deal with the influence of the risk. To ensure an effective risk management of an organization, it is important for the organization or risk manager to have a risk literacy [63]. This capability includes analysis capability (collection, understanding and predictive ability), communication capability (networking and communication ability), and practice capability (response and applied ability).

(3) **Analysis results of organization factors**

Water injection timeline of Unit 1 from the viewpoints of RE, HRO, and RL are analyzed. As an example, the analysis result is shown from the viewpoint of RL in Table 6.36. The horizontal axis shows suggested emergency response capability, and the vertical axis shows each individual, organizational (newly divided into the site and the managerial section), and external response level.

Also, successful cases are written in green Gothic style, while failure cases in red Italics.

(4) **Discussion on the accident response capability**

As shown in Table 6.36, a difference in accident response capability is found between individual & organizational levels, and national & industrial levels. Many successful examples of resilience were found on individual and organizational basis. The operators on the site seem to have a sense of duty, a

Table 6.36 Water injection analysis results of Unit 1 associated with Risk Literacy

Risk Literacy / Analysis level		Normal time				Emergency situation		
		Analysis capability		Communication capability			Practice capability	
		Collection capability	Understanding	Prediction	Networking (information sending)	Communication capability	Response capability (now available emergency response)	Applied capability (fundamental measures)
Individual		Tsunami damage case	Tsunami damage risk awareness	Loss of power risk awareness	-	-	Judgment on continuing sea water injection	Emergency training
Organization	Site	Collection of cases: Jogan tsunami	Consequence assessment on earthquake/ tsunami PSA implementation	Awareness on accident magnitude	Sharing of site information	-Command system (site) -Base-isolated building (unification) *-MCR-EHQ communication*	-Use base-isolated building as EHQ - Use of fire engines -Freshwater/sea water injection -Venting	-Command system -Tsunami operation -AM operation -Prevention of tsunami damage
	Management division	Accident case collection: Jogan tsunami, JNES tsunami PSA, La'bruie tsunami, Madras tsunami	*Tsunami damage risk awareness*	*Loss of power Source false recognition*	*Information sharing at Headquarters /site*	-TV conference system (2nd floor) *-Disturbance of headquarters/ site*		-Base-isolated building installation -Fire engine installation -Education /training system review
Liaison (PM residence, etc.)		-Collecting overseas terrorism cases: USA *-9.11 terrorism b.5.b Importance false recognition*	-Classification of external event importance -False recognition of earthquake and tsunami risk	-Importance of external event *-Infrastructure damage risk false recognition*		*-Media, municipality and overseas PR* *-Official residence/head quarters/site direction system*	*-Delay in initial response* *-Government's command system*	-Manufacturer, cooperative firm, external support -Radical measures/organizational revision(regulation/ power) *-Revision of insurance system*

Note: Successful cases: green Gothic, failure cases: red Italics

critical mind for usual work, and an experience of accident training programs, which seemed to have worked effectively at the situation of emergency. This is the significance of safety culture development. In this context, it is important to "establish study (feedback) system as an organization" on daily basis.

On the other hand, there are many flaws in crisis response of managerial and country levels. In the management division, trainings is dispensable that focus on emergency responsibility allotment, evaluation of severe situation assessment, and mode shift from normal time to emergency. Failure cases are concentrated on rare event recognition and challenges in organization culture, in the national level and industrial base. According to bounded rationality [64], they used the limited information to make a rational decision in the limited environment, which may have been a failure in the sight of God. It is suggested that it is important to destroy bounded rationality, or to "establish the system which prioritize judgment on the site (allows violation of order). The typical example was seen in the judgment to continue seawater injection despite the order from the official residence and the headquarters. A higher priority was placed on the conclusion on the site. Also, rebuilding of the safety concept integrating unexpected responses is designed in order to eliminate errors in risk recognition.

Analyzing documents including lessons learned from the Fukushima Daini Plant accident as shown in Reference [65], the causes of such difference were due to the severity in damage and the availability of power source. In the Fukushima Daini, the damage of the whole system was less than the Fukushima Daiichi, and the total power source was not lost. Considering the four capabilities of Resilience Engineering, the response was not greatly different between Fukushima Daiichi and Daini Plants.

TEPCO proposed, in the accident summary newly submitted [66], in addition to the hardware measures by the Investigation Committee on the Accidents at the Fukushima Nuclear Power Station of Tokyo Electric Power Company (Investigation Committee) [50], such means to avoid a negative chain of organization as "to improve safety awareness by the top management" and "to introduce incident-command system" for addressing the challenges of the organization suggested in this paragraph.

6.11.1.7 Factors That Inhibited Operation of the Reactors and Smooth Accident Response from the Aspect of Site Operation and the Improvement Plan

(1) **Purpose and method**

In this section hardware factors that may have inhibited the operators and workers from responding smoothly are analyzed and improvements of hardwares for them are proposed. Target hardwares are monitoring/control and recording system, site related operations system, Main Control Room and Important Anti-seismic Buildings related operations system. The method is by analyzing the related incidents after extracting the incidents from references [49, 67]. The areas of analysis are the post-SBO reactor cooling operation for Units 1 through 3 in the wake of tsunami, pressure containment venting operations, and alternative water injecting operations.

(2) **Analysis about monitoring/control system and record related matters**

SBO caused loss of lighting of the Main Control Room, function of monitoring/control system and SPDS functions of important anti-seismic building. Operators in the Control Room were not able to gain necessary information which caused trouble to operate the Nuclear Plant. The director and the supporters in the important anti-seismic building were not able to support the operators precisely due to loss of SPDS functions. Another problem was the poor instrumentation system when the severe accident relating to a nuclear reactor such as fuel exposure occurred. So power source and service systems that maintain power source for a long time even at the time of abnormality and natural disaster, the functions to identify the plant status at the time of severe situation and the computer support system for the cooling operation should be needed

(3) **Analysis about operations on the site**
Reactor operations are based upon monitoring and control system; then, operations including important IC, RCIC, HPCI, pressure containment venting are not supposed to have been readily available when the functions were lost. Manual operation on the site, not too dependent on remote control, should be included in the system design.

(4) **Operations in relation to the sites, including the Main Control Room and the important anti-seismic building**
It is required to eliminate the following factors that inhibited accident response from the operation aspect.

   (a) Main Control Room related matters: (a) insufficient power source, (b) weakness of communication system (telephones, etc.), (c) insufficient radiation protection performance, and (d) insufficient livability for operators.
   (b) Important anti-seismic building related matters: (a) weakness of communication system, (b) insufficient radiation protection performance, (c) insufficient radiation management facilities, and (d) insufficient livability for operators.
   (c) Work site related matters including reactor building (R/B), turbine building (T/B) and road: (a) Weakness of lighting outside the buildings, (b) weakness of the communication system, (c) lack of quake/tsunami resistance of work space in/outside the building, (d) insufficient radiation protection performance inside the buildings, (e) debris on the roads in/outside the buildings, (f) insufficient radiation protection performance of heavy machinery, (g) vulnerability of gate power source.

6.11.2 *Human Resources in Nuclear Field*

6.11.2.1 Object and Method of Investigation

This study considers the issues related to human resources for nuclear power and some countermeasures to the issues from viewpoints of the cause of the accident at Fukushima Daiichi Nuclear Power Station (hereinafter referred to as "Fukushima Daiichi"), the cause of inappropriate responses against the Fukushima Daiichi accident (hereinafter referred to as "the accident"), as well as the cause of failure of the regulatory system which did not work well in the accident. Since the scope of our own investigation to the accident was limited, this study extracts the issues with reference to disclosed reports concerning the accident and "Promotion of Measures to Secure and Develop Human Resources for Nuclear Energy (Statement)" issued by Japan Atomic Energy Commission.

In addition, skilled workers have been dispersed due to the sharp decrease in demand for maintenance and inspection jobs, because most nuclear power plants in

Japan have suspended operations due to the effect of the accident. Further, new graduates are also avoiding career choice in nuclear-related fields due to the accident and the current situation around nuclear power. Therefore, this report covers relevant countermeasures, since securing human resources and fostering their careers are also important.

6.11.2.2 The Issues Related to Human Resources for Nuclear Power in Case of the Accident

(1) **The issues regarding the cause of the accident**

(a) Lack of awareness of the potential risk, overconfidence and complacency on technologies and nuclear safety by nuclear plant operators: Nuclear plant operators publicly announced that nuclear power plant was safe for responding lawsuits [68], and they themselves believed that "severe accidents would never happen in Japan" [69], so such background can be said to have prevented the uptake of new knowledge.

(b) Insufficient commitment by top management to safety, and lack of a sense of safety as the top priority in Tokyo Electric Power Company (hereinafter referred to as "TEPCO"): In 2008, TEPCO had estimated the potential for a giant tsunami to strike at Fukushima Daiichi. But sufficient countermeasures for the potential disaster had not been taken by the time of the Great East Japan Earthquake [69]. Moreover, the low performance of nuclear power plants had continued for an extended period, due to the revelation of falsifying self-inspection records in 2002 and the occurrence of the Chuetsu-oki earthquake in 2007. As a result, senior management decided against investing in measures for severe accidents such as the tsunami, whereupon this lack of action has taken the company to a serious situation, which might lead to go out of business [69].

(c) Lack of questioning and learning attitude in TEPCO: In the United States, in response to the terrorist attacks September 11, 2001, NRC mandated nuclear plant operators to implement appropriate measures against potential attacks on nuclear facilities in 2002. There may have been some opportunities to recognize the need for these measures in Japan, but they were overlooked [70]. In addition, considerable available information which could predict the possibility of the accident at the Fukushima Daiichi has been obtained, so work was underway to study some measures against severe accidents in Japan. However, these issues were not able to wake the "awareness" in TEPCO, in terms of the need to respond urgently to take measures beyond safety regulation [69].

(d) Lack of imagination and expertise for designing plants by suppliers and for operating plants by TEPCO: Fukushima Daiichi Unit 1 was installed by GE Corporation with a full turnkey contract and was designed with placing the emergency diesel generators, emergency site-power supply systems and

DC power supply systems on the basement of the turbine building. The plant suppliers and TEPCO followed the same layout for Units 2 to 6, as well, hence these were not designed to avoid the risk of power loss due to internal flooding or tsunamis.

(2) **Issues regarding the cause of inappropriate severe accident response**

(a) The lack of recognition of the special properties of nuclear power: It is needless to say that the reactor must be monitored the condition, even after shutdown, and it is also essential to ensure to continuously remove decay heat. In Japan, as for the severe accident training on the external power loss, the accident scenario assumed that external power would be restored after a certain period of time. Therefore, more severe situations like the loss of all AC power at multiple units were not simulated [68–70]. Moreover, much more severe situations, such as a loss of all AC and DC power for an extended period, leading to a meltdown, were not assumed in simulator training for operators [68–70]. Consequently, it is considered that plant personnel were unable to enhance their knowledge and skills sufficiently to respond to severe accidents. In addition, conventional disaster prevention drills were impractical, so problems during practical evacuation events were not found and elucidated, which led to some confusion during the evacuation from the accident [68–70].

(b) Insufficient knowledge and skills of the individual facilities in TEPCO

(i) Isolation condenser of Unit 1 In regard to the isolation condenser (IC) installed early boiling water reactors (BWRs), IC was not included in the simulator training program in the training center for operators [68, 71]. In addition, information about the operating condition of IC was not shared between the main control room and the emergency response facility, which led to discrepancy of recognition [69, 72].

(ii) Emergency alternative water injection at Unit 3 The operators stopped the high-pressure coolant injection system (HPCI); anticipating to be able to depressurize the reactor by the safety-relief valves (SRVs) and to be able to inject water into the Unit 3 reactor by a diesel-driven fire pump (DDFP) which had been switched from HPCI. However, SRVs could not be opened due to loss of power, which prevented efforts to reduce the reactor pressure and inject water into the Unit 3 reactor [69, 71]. These incorrect mindsets about the priority of the emergency situation, and wrong decisions taken had exacerbated this problem.

(iii) Indicated reactor water level As for the fact that the indicated value of the reactor water level remained unchanged for an extended period, it suggested that the operators had not recognized that the reactor level indication was not correct if the reactor water level had been gone below the reactor inlet pipe [70].

(c) Insufficient communication within TEPCO: In Fukushima Daiichi, communication apparatus between the main control room and the emergency response facility was lost, with a few exceptions, which severely limited the communication. Accordingly, misunderstandings by either party were very likely.

Moreover, it had been pointed out that communication between TEPCO and the central and local governments was insufficient, and the essential information was not sufficiently transmitted. Although there might be some insufficiency of the emergency response telecommunication apparatus as well as some on-site confusion, TEPCO's training and confirmation of information capability was still considered insufficient [69].

(d) The roles of the chief engineer of reactor were unclear in severe accidents: The duties of the chief nuclear reactor engineer include directing safety issues concerning the reactor operation as specified in the Act on the Regulations of Nuclear Source Material, Nuclear Fuel Material and Reactors. Conversely, in the Act on Special Measures Concerning Nuclear Emergency Preparedness, the nuclear emergency response manager, vice managers and disaster prevention personnel are specified, but the chief reactor engineer is not. In the revelation of falsifying the seawater temperature data at the condenser inlet and outlet in nuclear power plants in 2006, TEPCO revised the role of the chief nuclear reactor engineer. In this revision, the independence of the chief reactor engineer against the plant manager was enhanced and he/she had a new role of reporting to the company president, which meant the chief nuclear reactor engineer was no longer in a position to lead plant safety management under the supervision of the plant manager [69]. However, there is no report about the verification result of the function of the new role. Moreover none of the company president or chief nuclear reactor engineers might have recognized the role.

(3) **Regulatory issues in Japan**

(a) Insufficient expertise of regulatory staff: In the United States, nuclear power plant personnel and NRC resident inspectors established a good relationship with mutual respect as professionals, which helped reduce trouble and ensure over 90 % capacity factor.

Conversely, in Japan, nuclear plant operators tended to adopt an attitude that regulation would be sufficient if they got permission. They urged to limit the scope of regulatory inspections in order to escape from the troublesome regulations as far as possible. Therefore, it can be said that deep expertise is no necessary requirement for the inspectors. Moreover, this fact was suitable to the personnel system of the government, because government officials regularly change their post in a few years.

(b) Insufficient advice from the national emergency organization: Only some experts on nuclear power were convened following the Fukushima Daiichi accident, and the regulatory agency did not dispatch any other expert than staff members to the site. Consequently, the helpful and comprehensive advice was not supplied from the national emergency organization, which also meant it could not support the site [71, 72].

6.11.2.3 Countermeasures to Issues Related to Human Resources for Nuclear Power in Case of the Fukushima Daiichi Accident

(a) Enhancing awareness of potential risks and concern about technologies and nuclear safety: As lessons learned from the Fukushima Daiichi accident, there is a need to eliminate complacency and overconfidence about nuclear safety and to share and ingrain the concept of nuclear safety-first, throughout the nuclear power plant operators.

(b) Increasing the commitments of the top management and enhancing the concept that the nuclear safety is top priority: The top management must recognize the specific risk of nuclear power and also fulfill his/her commitments to nuclear safety. Since the top management may not necessarily be an expert in nuclear power, it is important to ensure the opportunity to enhance his/her awareness about nuclear safety. For example, he/she should positively participate in the meetings of the World Association of Nuclear Operators (WANO), the top seminars of Japan Nuclear Safety Institute (JANSI), and the peer reviews of WANO and JANSI. In addition, the role of nuclear expert staff supporting the top management during decision-making process is important.

   Plant managers must supervise plant safety and be responsible to all efforts in the event of an emergency. Accordingly, they must have the professional knowledge of the facilities and have the proper abilities, including supervision, decision-making, problem-solving and leadership. To enforce these abilities, there is a need to ensure crisis management training and practical exercises simulating severe accidents at periodic intervals. While captains of ships and airplanes need official qualifications for responsibilities to human lives, no official certification is required for nuclear plant managers in Japan. Efforts should be made to clarify the eligibility requirements of plant managers, from the perspective of the importance of the emergency response of a nuclear power plant.

(c) Improving a questioning and learning attitude: When nuclear safety-first concept is shared in a nuclear power plant operating company, it will emerge in individuals as a questioning and learning attitude. The meaning of a learning attitude is to appropriately analyze and evaluate lessons learned from past events and operating experiences in nuclear power plants at home and abroad; based on these, the necessary improvements to facilities and the modifications of operating procedures will be made. Conversely, a questioning attitude means ensuring a constantly self-assessing attitude with regard to daily work which involves design, construction, operation and maintenance. Such attitude would be crucial in establishing the highest levels of safety and reliability of nuclear power plants.

To allow the top management to confirm such attitude to be established, it is effective for middle management to check the daily work performances of their staff members and to have a periodic review by external experts.

(d) Improving imaginable ability and expertise in nuclear plant design: It is important to improve professional knowledge by accumulating practical work experience regarding design, operation and maintenance, etc., but "awareness" would not occur solely by engaging in day-to-day operations. In Fukushima Daiichi, the emergency diesel generators, emergency power supply systems and DC power supply systems of all 6 power plants were installed in the basement of the turbine buildings, based on the original design concept without any change.

As mentioned above, by rooting "learning and questioning attitude", awareness occurs, and improvements to facilities and operations can be made possible.

(e) Improving recognition of the specific properties of nuclear power: Of course, the risk related to nuclear energy depends on the radioactive materials involved. In case of a severe accident, radioactive materials may have great impact on the residents around a site.

It is important to improve knowledge and skills by repeating practical emergency drills assuming a severe accident and also to improve the awareness of all those who may be involved in any aspects of nuclear power.

(f) Promoting visualization to improve knowledge and skills: The nuclear plant operators in Japan have their own education and training system. There is no common education and training guideline between the operators like in the U.S. Since the rule which determines the competence of staff members differs from each other, it is difficult to evaluate the personnel of each operator. Therefore, to improve the knowledge and skills of staff members based on the evaluation of human resources, using common evaluation rules, it is important that the nuclear power industry and academia collaborate to standardize the knowledge and skills required for each of the human resources involved in the design, construction, operation or regulation of nuclear power plants. In other words, the visualization of the required knowledge and skills is important.

Acknowledging the above, we can easily show Japanese abilities ensuring nuclear safety, regarding the design, construction, operation and regulation of nuclear power plants both domestically and overseas.

Incidentally, for the purpose of visualization, utilizing the official qualification system should also be considered. In Japan, specific professional qualifications in the nuclear power field include the chief engineer of reactor, the chief engineer of nuclear fuel, radiation protection supervisor, and the chief reactor operator. In addition, there is a certification system of maintenance skills. As for eligible qualifications for expertise, general judgment abilities, problem detection abilities and problem-solving abilities, there are professional engineer, PhD, Master of Business Administration (MBA)/Technology Management Master (MOT), etc. However, certification is not a goal, and there is a need to gain more experience as well as striving to improve expert knowledge and skills. There is a need to consider incentives and to impose a duty of continuous efforts after certification acquisition.

Under the present life-time employment system in electric power companies in Japan, it is possible to maintain a system of continuous evaluation for expertise knowledge, skills and abilities, including questioning and learning attitude, through in-house training or qualification systems, or through evaluation of day-to-day operations, without depending on the official qualification systems.

To encourage the acquisition of qualifications, it is effective to link to some incentives such as promotion, award, etc. However, it is more desirable that qualification acquisition would be based on recognition of the importance for self-development, because it will be linked to subsequent improvements.

Academia and nuclear power industry are expected to cooperate to study the effective use of qualification to ensure the competence of nuclear power personnel.

(g) Improving communication: To improve internal communication between the head office emergency response center and the nuclear power plant emergency response center as well as between the emergency response center and the main control room at the nuclear power plant site, there is a need to carry out practical "exercises" repeatedly, assuming a severe accident.

In addition, there is a need to allocate company communicators to improve external communication. Communicators are hoped to play an important role for restoring social confidence in nuclear power. There is a need to collaborate with various professionals, including from fields of sociology and psychology in order to train such communicators. Accordingly, through practical "exercises" in which communicators participate, it is expected that the effectiveness of external communication is verified, and it is also expected that the improvement of communication is achieved, too.

(h) Clarifying the role of chief nuclear reactor engineers in severe accidents: Chief nuclear reactor engineers have a national qualification and have the duty to oversee nuclear safety defined by the nuclear regulation law, so the nuclear plant operators must reconfirm these roles and improve their authority and responsibility.

The roles of licensed chief electric engineers and licensed chief boiler-turbine engineers in emergency situations should also be clarified and be improved, too.

(i) Development of regulatory personnel with enhanced professional expertise, judgment capability and global outlook: Regulatory authority, being entrusted by people, is supervising the safety of nuclear installations by examining, licensing and inspecting. Based on the experience of the accident, it should be necessary for regulatory authority personnel to improve expertise, problem detection ability, problem-solving ability and decision-making ability. In addition, communication skills on explaining to the society about the risks and safety of nuclear facilities are also important. The U.S. Nuclear Regulatory Commission sends staff members to universities to enhance their expertise and technical abilities and to acquire Ph.D. It should also be desirable to encourage regulatory staff members to obtain official certification, such as professional engineers, chief nuclear reactor engineers, Ph.D., etc., in Japan.

In addition, collecting a wide range of the latest global knowledge and information, to be reflected in regulations, is important as a part of the mission of regulatory authority staff members. From this perspective, cooperation and collaboration with international organizations and foreign regulatory bodies are important.

Incidentally, the Japan Nuclear Energy Safety Organization (JNES) published a report concerning the role of human resource development in JNES in June, 2013, which reported the subject about human resource development in JNES as follows: (1) clarifying the required expertise and ability; (2) enhancing JNES staff expertise and ability; (3) promoting continuing professional development effort. Embodiment of the report would subsequently be discussed.

(j) Enhancing advice from the national emergency technical advisory organization: In case of a severe accident, the emergency technical advisory organization must convene numerous nuclear experts so as to examine the accident and advise about the countermeasures from a comprehensive perspective. Therefore, even under circumstances where communication and transportation are stopped, it is necessary that a framework is being established in order to promptly summon such nuclear experts and to start their activities without awaiting an official convocation. Hence "to establish and exercise the framework" is important to achieve these purposes.

6.11.2.4 Measures for Securing and Fostering Human Resources for Nuclear Power

There are some issues in regard to securing and fostering human resources for nuclear power. The following countermeasures are required to deal with them:

(a) Securing younger generation interested in nuclear power: In the course of elementary school, junior high school and high school, pupils learn scientific literacy and study the basics of nuclear and radiation. Scientific literacy makes it possible to reasonably evaluate the benefit and risk of radiation or its use, and also to be justly afraid of the risk of radiation. As a result, the interest in science and technology may be increased and lead to choosing a nuclear-related profession.

The basics of nuclear and radiation are taught in the third year science class in junior high school, after Japanese curriculum guidelines have been revised. However, after the Fukushima Daiichi accident, information concerning the effects of radiation is being confused. Hence, the urgent needs to educate on radiation issues, regardless of junior high school. In particular, most of the elementary school and junior high school teachers have not learned about radiation. This is why a training course on nuclear and radiation issues for teachers is necessary in the first place. The government, academic societies, universities, research institutions, and industry must actively support these training courses for teachers, including providing teaching materials and dispatching lecturers, etc. Furthermore, academic societies should act as coordinators to implement these activities smoothly and effectively.

Education in universities is what precedes actual practices and is likely to offer an opportunity for students to learn about the basic knowledge and skills for expertise, and to learn the ability to detect and solve problems, the responsibility for ensuring safety and the ethics for the engineering.

Nuclear Engineering is an integrated system with deep expertise, experiences and skills based on various technological fields. This feature makes it possible for students of nuclear engineering to develop the ability to detect, solve and manage problems in other fields. Therefore, personnel, who have learned nuclear engineering, can work not only in the nuclear engineering field but also in many others. It might be possible to increase the best students choosing nuclear engineering courses if this feature is acknowledged. This might be considered the role of universities and the Atomic Energy Society of Japan.

Conversely, the nature as an integrated system of nuclear engineering provides opportunities; not only for students of nuclear engineering, but also students in various other courses, such as mechanical engineering, electrical engineering, chemical engineering, civil engineering, architectural engineering, etc. Therefore, related information to students not specializing in nuclear engineering is important to motivate superior students from other fields so as to choose nuclear industry as their profession. To do so, it is important to provide opportunities for them to get to know nuclear power. These might be mainly provided by industry.

To increase the degree to which the younger generation aspires to nuclear-related occupations, it is important to show that the nuclear-related professions deserve challenging and attractive, as well as to provide opportunities for such internship, in which they would experience real work. Industries are expected to get proactively involved in this area as well.

(b) Ensuring skilled workers: In Japan, a periodic inspection and maintenance outage of a nuclear power plant is performed at a certain interval, with more than 2,000 workers engaging. Most workers belong to subcontracting companies which have deep relationships with the area where the nuclear plant is located. However, in the event of a prolonged plant shutdown like at present, this drastically reduces the amount of work and forces some companies to withdraw from businesses related to nuclear plant periodic inspection and maintenance works. Amid aging of skilled workers, maintaining local businesses and skilled workers to ensure the safety of nuclear plants has become an issue.

Major plant suppliers have been striving to ensure technical capabilities through implementing seismic improvement works and safety margin improvement works to comply to the new regulatory guidelines. However, inspection and maintenance works involved in areas such as turbine generator systems, instrumentation systems, etc. have suspended. Moreover, some small- and medium-sized companies with special technologies and skills are leaving from the nuclear-related business. As a result, there has been losing some skilled workers. This is a serious problem from assuring human resources perspective.

Education and training by utilizing the maintenance training center of each nuclear plant operator could be considered, as a temporary measure to securing the required level of skills. Governmental large-scale R&D projects are also expected for human resource development.

Nuclear industries have to do their best for resuming nuclear power plants operation, as well as clarifying the future.

(c) Ensuring education and research personnel: Education and research personnel have duties to participate in councils of Japan's nuclear regulation authority as experts, and provide professional advice. They have a fundamental role to raise younger people who have to play a leading role in future. Hence, it is most important to ensure education and research personnel covering all fields.

In Japan, it is getting difficult to ensure education and research personnel of all nuclear related fields in a single university. Therefore, certain strategies are needed to prevent human resource shortfalls; not only in terms of the age structure but also in specialized fields. Accordingly, collaboration among universities is important. In Europe, the European Nuclear Education Network (ENEN) was established in 2003 and focuses on Master-level education and training in the nuclear sciences. In Japan, collaboration of nuclear education has been carried on in some universities and is expected to be enhanced in future.

6.11.2.5 Conclusions

The Fukushima Daiichi accident revealed various issues related to human resources for nuclear power.

Since humans are the most important resource in Japan, human resource development is a key issue that should foresee about 10–20 years and should be carried out intentionally and steadily. With the Fukushima Daiichi accident in mind, we can say that humans indirectly caused the accident, but humans fought against and stabilized the accident. Even if there is no scenario, humans can think and determine how to settle the accident and can cope with the accident, so securing and fostering the right human resources is essential.

Human resource development for nuclear power should be performed based on visualizing the required competence in order not only to restore confidence in nuclear power but also to comply with exports of nuclear power plants to emerging countries.

6.11.3 Responsibility and Duty of the Chief Reactor Engineer

In the Fukushima Daiichi Nuclear Power Station, a chief reactor engineer with a national qualification has been assigned. How the chief engineer played his/her role in the accident have been studied, with details described on the desirable roles.

6.11.3.1 Legal Definition of the Chief Reactor Engineer

For all reactors operated in Japan, operators must select a chief reactor engineer and assign him/her to an appropriate position in compliance with the law. The chief reactor engineer is selected from among those qualified by a national examination, and has the role of supervising safety reactor operation. The national examination is intended to confirm advanced expertise and practical experience, regarding collective aspects of the reactor, including theory, design, operational control, fuel and materials, radiation protection and related laws and regulations, and those with such qualifications are considered senior experts on reactor operation. According to the laws and regulations, those who operate the reactor must comply with the instructions of the chief reactor engineers regarding safety.

Conversely, no obligation is defined for the work of chief reactor engineer equivalent to the chief engineer of other systems (licensed electricians, chief gas engineers, etc.). In general, the assigning of a chief engineer is a requirement to evaluate the technical competence of the operating organization.

However, pursuant to the laws and regulations mentioned above, the nature of activities required for chief reactor engineers had not been clarified, during normal times and accident times. In general, a corporate officer not in the chain of command for normal operation of the power plant, would be assigned as the chief reactor engineer, and is usually considered to have the role of routinely monitoring compliance with safety regulations.

Incidentally, in the nuclear disaster prevention system based on the Special Measures Concerning Nuclear Emergency Preparedness Act, no specific role was assigned to the chief reactor engineer licensed under another law (the Nuclear

Reactor Regulation Law), and he/she was not assigned as a member of the disaster prevention staff, for the disaster prevention organization of the operator at a time of emergency.

6.11.3.2 Accident Response of the Chief Reactor Engineer at the Fukushima Daiichi Nuclear Power Station

A chief reactor engineer is basically assigned at each reactor/unit, but there is also scope for one chief reactor engineer to be assigned to multiple reactors/units and in the case of the Fukushima Daiichi Nuclear Power Station, one person was concurrently assigned to Units 1–4. In response to the accident, the chief reactor engineer in charge of Units 1–4 had been waiting in the emergency room in the Main Anti-Earthquake Building, together with the chief reactor engineer in charge of Units 5–6, even though no specific accident response actions had been identified.

6.11.3.3 The Role of the Chief Reactor Engineer for Serious Accidents

The nature of responsibilities which should be assigned to the chief reactor engineer in the event of serious accidents have been studied.

In the United States, a "Shift Technical Advisor" system was introduced after the Three Mile Island nuclear power plant accident. This is a system which allows advice to be passed on, following the complication of an accident, based on a deep knowledge of the reactor, by assigning individual experts to each of the reactor operation teams. Unlike operators who target familiarity with operations for various assumed situations, it has scope to assign experts capable of making case-by-case decisions based on professional and technical expertise in response to circumstances not assumed in training.

Considering such cases, issues regarding the chief reactor engineer would be clarified.

(1) **Is it necessary to station an engineer with expert knowledge of nuclear reactors in power plants in the event of a serious accident?**
Staff of the operator of the nuclear power plant must have undergone repeated training, and have operational proficiency, including the procedure of starting and stopping the reactor, and methods of coping with assumed serious situations, etc. Conversely, to cope with circumstances which were not assumed beforehand, as in this accident, the ability to understand the underlying principle and its significance is required, as well as the ability to organize appropriate countermeasures by themselves. Engineers with the relevant expertise are therefore considered necessary to handle serious accidents. Regarding decisions as to whether such engineers should be assigned to each of the operation teams, to each reactor, or to each power plant; these decisions should depend on

timing and how promptly they could respond, in the event of an accident. However, at least one person must be assigned to each reactor, to cope with simultaneous accidents occurring in multiple reactors.

(2) **Has the chief reactor engineer acquired the required expertise and knowledge to a sufficient extent**?
The chief reactor engineer is a highly sophisticated professional with a national qualification, and is considered a qualified expert in the event of a serious accident. However, knowledge concerning how to cope with serious accidents has not been explicitly confirmed in qualification examinations to date. Confirmation of the capacity to manage serious accidents would also subsequently be necessary. Incidentally, to fulfill his/her responsibilities as director of safety, it is important to have experience, not only in individual areas of expertise but also experience in practical exercises. Accordingly, 3 years of experience were established as the required qualification for the chief reactor engineer, in the revised law for furnaces in July, 2013, and revised commercial reactor rules.

(3) **Should the chief reactor engineer be the person responsible for supervision or a mentor in the event of a serious accident**?
Examples of occupations requiring national qualifications include medical doctors and airplane pilots. These occupations are characterized by the high expertise and skills consistently required to participate in normal operations. Meanwhile, reactor operation normally goes without a problem, whereby the specific steps are processed smoothly in compliance with manuals, hence the need to foster operators with a proficiency course. In the event of any abnormality, the role of the chief reactor engineer is considered important as the key person to analyze the situation calmly from the perspective of not being chased for an immediate response, and making a decision from an engineering and expertise perspective. Therefore, unlike medical doctors and airplane pilots, his/her stance might be that of a mentor, and the operator must follow that technical decision.

It is inappropriate for a chief reactor engineer to adopt a role as disaster prevention administrator, since in the event of a serious accident, the on-site disaster prevention administrator must command various tasks overall, including directly dealing with the accident, off-site measures, and making contact with relevant organizations. Accordingly, the advice of the chief reactor engineer becomes even more significant, and there is a need for the disaster prevention administrator to follow that engineering decision. Consequently, it would be appropriate for the chief reactor engineer to adopt a role of substantial responsibility and command the necessary instructions, etc., regarding with measures to ensure the safety of the nuclear reactor, in the capacity of nationally qualified senior expert. For this purpose, some staff members supporting the chief reactor engineer are also necessary.

(4) **What is the role of the chief reactor engineer during normal periods**?
To date, the role of the chief reactor engineer has been assumed under normal circumstances, and is not necessarily always clear. Accordingly, assigning to multiple reactors was not against the law or regulation. From now on

continuous efforts to improve safety by the utility would subsequently be important. In this sense, the utility should make the best use of the chief reactor engineer, as the person in charge of advancing voluntary safety measures at the job site of operation, including coaching the operator and making improvements based on accident information.

6.12 Relationship with International Society

(1) **Background**
Though Japan was investigating international trends and involved in developing safety standards of the International Atomic Energy Agency (IAEA),[20] both licensees and regulators were reluctant to introduce international standards. Namely, though Japan felt the need to do so in a long term, handling of short-term issues kept it low-priority. To prevent any recurrence, there is a need to investigate what countries and relevant international organizations think of the lessons from the Fukushima Daiichi Nuclear Power Station Accident, and, based on the results, arrange and analyze measures for international standards. We should also refer to international standards for new reactors (Generation III + and Generation IV), which had been considered before the accident. Therefore, this chapter describes post-accident international trends and the introduction of international standards for existing reactors based on reports of institutions issued by 2012.

(2) **International trends**

(a) Concepts of safety for new reactors before the Fukushima Daiichi Nuclear Power Station accident: New reactors consist of Generation III + and IV reactors. The former, Generation III+, is an extended light water reactor, while the latter is an innovative reactor using no light water as coolant. Currently, as well as Western countries, Russia and developing countries are also designing new reactors, which target improved safety and economics compared to existing reactors. This section reports safety concept in Western countries, which most countries still focus on.

In Europe, the Western European Nuclear Regulators Association (WENRA) was established in 1999 to establish common nuclear safety and regulations. International standards for the safety objectives of new reactors proposed (i) a reactor core meltdown accident should be practically eliminated as it will lead to early or large radioactive release and (ii) the design should be provided that damage from a reactor core meltdown accident, which cannot be practically eliminated, can be limited to the minimum off-site response (emergency evacuation of citizens near the nuclear reactor facilities) [73].

[20] Government Accident Investigation Committee Report Chapter V.

Currently, next-generation light water reactors (Generator III + reactors) EPR under construction in France, Finland, and China have emphasized reducing the probability of severe accidents with safety enhancement effort. The design of reactor containments includes countermeasures against aircraft crash and reactor core meltdown. The EPR design guide [74] prepared by German and French experts said that defence in depth should be reinforced, the design should avoid the need for any evacuation in the event of a reactor core meltdown accident whenever possible, and "practically eliminate" events leading to unacceptable consequences.

In the U.S., the NRC issued a policy statement on new reactors (Generation III + reactors) [75] showing a policy to protect the environment and public health, safety, common protection, and security at least equivalent to the current-generation light water reactors. In consideration of the September 11, 2001 terrorism attacks, to exercise safety and security functions, a large commercial aircraft crash should be considered. Further requirements were safety margin extension, simplified, passive, and other revolutionary measures provision.

Conversely, in comparison with conventional reactors, the Generation IV International Forum (GIF), the international project led by the U.S., has promoted the development of Generation IV reactors which excel in economics, safety, sustainability, and nuclear nonproliferation in comparison to conventional reactors, with construction targeted for around 2030 [76]. The scope of Generation IV reactors includes sodium-cooled fast reactors (SFR), gas-cooled fast reactors (GFR), lead-cooled fast reactors (LFR), very high temperature reactors (VHTR), molten salt reactors (MSR), and supercritical water-cooled reactors (SCWR). SFR is the most feasible and promising of all. GIF established technology goals of safety and reliability in 2002: SR-1 "to excel in safety and reliability in operation," SR-2 "to have a very low likelihood and degree of reactor core damage," and SR-3 "to eliminate the need for offsite emergency response." To achieve these goals, basic safety principles common to six type reactors were shown in 2008, and safety design criteria for SFR were established in 2013. In conclusion, they have been making efforts to improve the safety of new reactors in comparison with t existing reactors. Namely, they tried to prevent any reactor core meltdown accident to eliminate the need for public evacuation in areas near the nuclear reactor site as well as taking measures to mitigate any potential accident consequences. For new reactors, measures against reactor core meltdown accidents are studied in advance, and after the Fukushima Daiichi Nuclear Power Station Accident, the importance of reactor core meltdown accident countermeasures and accident management are reaffirmed.

(b) International assessment of the Fukushima Daiichi Nuclear Power Station accident: The International Atomic Energy Agency (IAEA) [77] report shows 15 conclusions and 16 lessons. One of these conclusions describes that the on-site responses to the accident were not faulted in consideration of severe situation for personnel during this accident. Lessons to improve nuclear safety are derived.

The WENRA [78] report says the initial design did not provide for proper anti-tsunami measures, but proper improvement was not conducted during the periodic safety review. It also says safety culture and organizational factors, including decision-making capabilities, did not help establish any proper protection and hindered accident management. It also indicates the importance of properly implementing defence in depth to ensure safety. In particular, it cites the need to provide adequate protection against external hazards and confirm the same with regulators during periodic safety reviews, as well as the need for comprehensive safety analysis using adequate deterministic and probabilistic methods and to consider multi-unit sites and long-term countermeasures. Moreover, it also deems it necessary to adequately protect the Emergency Response Center from external hazards.

According to a report of the Institute of Nuclear Power Operations (INPO) [79], nuclear power generation licensees worldwide focused on continuous improvement as an effort to learn lessons from reactor core meltdown accidents at the TMI-Unit 2 and the Chernobyl Power Station. The report pointed out that the Fukushima Daiichi Nuclear Power Station, however, proved the need to prepare for unexpected events (beyond design basis). In addition, it says TEPCO and the wider nuclear power industry were not prepared to maintain important safety functions or implement effective emergency response procedures and the accident management plan under the extreme conditions faced in the accident.

The Electric Power Research Institute (EPRI) [80] report analyzes the root causes and concludes that geographic properties were not considered when determining the design basis tsunami height and that the protection against tsunamis beyond the design basis and mitigative capacity were insufficient for the actual event.

The American Nuclear Society (ANS) [81] report organized the event progression, health physics, cleanup, and urgent issues. It should be noted that unlike other institutes, the event progression, pollution, and other aspects were also analyzed. The accident analysis points out that the DC power failure prevented the steam relief safety valve from being operated, while unclear feed-water actuation due to shortages in the instrumentation system showed the importance of depressurization, water supply and containment vessel venting, etc., and cites the need to review accident management. In addition, the ANS thinks human errors and regulatory oversights exacerbated the accident and points out that these problems should be tackled before modification of the facilities.

As mentioned above, while the response to the accident under extremely difficult conditions due to the extreme external hazards should be highly evaluated, it is commonly pointed out that safety assurance efforts were insufficient, despite opportunities for improvement.

(c) Lessons learned from the Fukushima Daiichi Nuclear Power Station accident: The IAEA Mission Report [77] shows 16 lessons. Lesson 1 deals with external hazards, Lessons 2 to 9 severe accidents, Lessons 10–13 the off-site emergency response, Lessons 14–15 the on-site emergency response and Lesson 16 the independence of regulation and transparency of roles. In addition, after Director General Amano made a statement in June 2011 that they should strengthen the IAEA safety standards, the IAEA studied gaps between the lessons from the Fukushima Daiichi Nuclear Power Station Accident and the IAEA safety requirement documents from fall 2011 to spring 2012. Consequently, the current IAEA safety requirement documents essentially lack no overarching requirement but points of enhancement for the safety standard were checked and revision was progressed.

The WENRA report, March 21, 2012 [82] highlights the following issues based on aspects of organization, system, culture, and technology. As for the organization and system aspects, the report shows the independence of the regulatory authority, clarifies the roles and responsibility of governments, regulators, and licensees, and reveals the need for a periodic safety review and timely implementation of reasonably executable improvement, and mutual assistance for the accident response between regulators. As for the cultural aspect, the report concludes a high safety standard and continuous improvement should be enhanced, and efforts to explore improvement should persist. As for the technical aspect, WENRA should prepare a guideline to identify and assess natural disasters (cliff-edge effects), review the safety reference level in light of measures against pressurizing containment vessels and a safety reference level relevant to the accident management.

The U.S. INPO [79] reports 26 lessons from the perspective of preparing for unexpected events, reactor core cooling, containment vessel vents, accident response, personnel arrangement, personnel limits, preparation for emergencies, roles and responsibilities, communication, radiation protection, off-site support, design and facilities, procedures, knowledge and technology, operating experience, and nuclear safety culture according to its own review.

The American Nuclear Society (ANS) [81] report suggests the regulation to use risk information, extreme natural disaster hazards, consideration of multi-unit sites, hardware modification, severe accident management guidelines, accident diagnostic methods, the chain of command in the event of accidents, emergency plans, health physics, social risk comparison, ANS risk communication and crisis communication. These were specified in the regulatory measures proposed by the short-term task force of the U.S. Nuclear Regulatory Commission (NRC), The American Society of Mechanical Engineers (ASME) [83] reports the root causes of the accident as the tsunami, flood, and defective design basis for accident management and that the severe accidents significantly impacted on the politics, society, and economy. The report summarizes lessons learned including enhancement of the design basis, the defence in depth system using risk information, human performance management, accident management, emergency response preparation management, communication and public trust. In addition, a new atomic safety system is proposed, including planning and coordination, a system to be implemented to ensure the design, construction, operation, and management of nuclear power stations to prevent wide-ranging social confusion based on radiation releases.

The Massachusetts Institute of Technology (MIT) [84] report discusses technical issues listed as lessons from the accident and makes proposals for existing and future reactors. The issues are organized in six sections: "Emergency Power following Beyond-Design-Basis External Events", "Emergency Response to Beyond-Design-Basis External Events", "Hydrogen Management", "Containment", "Spent Fuel Pools", "Plant Siting and Site Layout".

From the above, the institutions present valuable lessons extracted from their own various perspectives. Major lessons include external hazards, severe accident management, design (consideration of multi-unit sites, etc.), facilities (power supply and containment vessels, etc.), emergency response, improvement of organization and system, and others.

(3) **Analysis of safety improvement measures**

The European Council decided to conduct comprehensive and transparent risk and security evaluation results and so-called "stress tests" for 143 nuclear power plants in a total of 17 European countries on March 25, 2011. Based on the stress test specifications proposed by WENRA, three phases of nuclear operator assessments, national regulator reviews and multilateral peer reviews were conducted from June 2011 to April 2012. The peer review pointed out the following items for improvement in Europe [85].

- Development of guidance on natural hazards and margin assessments. In addition, the development of guidance on margins beyond the design basis and cliff-edge effects.
- Implementation of reviews of natural disasters and related countermeasures for power stations (periodic safety review). Natural hazards must be re-evaluated at least once a decade.
- Urgent implementation of measures required to ensure the integrity of containment vessels. These measures include equipment, procedures, and accident management guidance to depressurize the primary circuit, prevent hydrogen explosions, and prevent high-pressure reactor core meltdown.
- National regulatory bodies examine the implementation of measures to allow accident prevention and minimize their impacts in case of extreme natural hazards. Typical measures include fixed equipment like instrumentation and communication means, mobile equipment protected against extreme natural hazards, emergency response centers protected against

extreme natural hazards and contamination and rapid-response rescue teams and equipment to support local operators during long-term events.

The final stress test report in October 2012 [86] recommends the following further improvements:

- The earthquake risk calculation is insufficient in 54 of 145 reactors, while the flood risk calculation in 62 reactors does not meet the standard. The report recommends that the risk calculation should be based on a 10,000-year time frame.
- An instrumentation system to measure and alarm earthquakes should be installed in each station. These must be installed or improved in 122 reactors.
- To maintain integrity of the containment vessel in the event of a severe accident, filtered vent equipment should be installed, which has not yet been done for 32 reactors.
- Equipment to manage severe accidents should be stored in protected locations, which are shielded even in the event of general devastation in an accident, and rapidly retrievable. The report points out this has not been done for 81 reactors in the EU.
- A backup emergency control room should be available in case the main control room becomes uninhabitable due to an accident. This had not yet been done for 24 reactors.

National action plans were prepared by the end of 2012 and underwent peer reviews at the beginning of 2013. A stress test recommendation will be reported in June 2014, which emphasizes security is within the scope.

The Autorité Administrative Indépendante (ASN) of France implemented stress tests. Its June 2012 report says the safety of nuclear facilities is fully ensured, no facilities should be immediately shut down, and further safety improvement measures are required [87]. Concurrently, ASN proposed the concept of a "hardened safety core" to ensure the robustness of the facilities and organization in response to unexpected events [88]. A "hardened safety core" helps prevent severe accidents or stops events progressing, and limits massive radioactive releases as crisis management responsibilities of operators for the exceptional scale of natural disasters beyond design basis or periodic safety reviews and their combination, and the loss of heat sink influencing all on-site facilities and causing long-term power loss.

Electricity of France (EDF) [89] will take flood-control measures like water-tight doors as well as improving the robustness of facilities and equipment against earthquakes and floods. In addition, to reinforce the water and power supply, they consider the need to reinforce the steam generator, primary system, and equipment to supply water to a fuel pool as well as equipment with an emergency diesel generator, emergency pump, and instrumentation. They consider improvements to the filter efficiency of vents with containment vessel filters introduced to limit the massive release of radioactive materials. In addition, they will develop a Nuclear Rapid-Response Force (FARN) by the end of 2014 as a special unit to take action within 24 h in the event of a severe accident and cool and restore nuclear reactors.

Conversely, in the United States, following the September 11 attacks, NRC issued a temporary safeguard inspection order EA-02-026 in 2002 and requested the development of influence mitigation measures and response procedure in the event of an aircraft impacting on the nuclear power station in B5b [90]. Moreover, the regulatory standard 10CFR50.54(hh)(2) requires guidance and an anti-fire/explosion plan should be prepared and implemented to maintain or restore the cooling of the reactor core, containment vessel, and a pool of spent fuel, even if most of the plant is lost due to explosion or a fire. The short-term task force report summarizing the lessons learned from the accident at the Fukushima Daiichi Nuclear Power Station recommends that the equipment installed according to 10CFR50.54(hh)(2) should be properly protected so that it is available after an external event beyond the design basis occurs.

The Nuclear Energy Institute (NEI) proposed safety measures entitled "Diverse & Flexible Industry Strategies" (FLEX) in December 2011 [91]. FLEX is defined as "variable and flexible influence of mitigation ability to provide backup to normal service equipment, which may be unavailable due to severe or extreme natural phenomena or malicious acts." The basic actions involve providing the electricity and water required for the important safety functions of the reactor by preparing multiple sets of portable equipment and protecting the same from natural phenomena.

In summary, Europe conducts stress tests to reflect on lessons learned from the accident, recommends improvement to plants, and tries to improve comprehensive safety. In particular, France introduced a new concept of the "hardened safety core" to strengthen the resilience of facilities and organizations, while the U.S. is introducing flexible strategies with portable equipment.

(4) **Proposed Future Direction**

While Western countries adopted safety improvement measures for severe accidents according to the defence in depth protection measures as stipulated in the international safety standard before the accident at the Fukushima Daiichi Nuclear Power Station, the introduction of these standards into Japan was delayed. In consideration of the above-mentioned international status, Japan should immediately introduce the international standard. Licensees promptly implement countermeasures, and regulatory authorities should introduce such international standards into Japan.

Japan participated in the revision of the IAEA safety standard and actively strove to reflect on lessons from the accident at the Fukushima Daiichi Nuclear Power Station in the IAEA international safety standards. The IAEA international safety standards are important for efforts to improve nuclear safety and are used by Japan for this purpose. It is essential to systematically introduce IAEA safety standards to Japanese safety standards.

In addition, there is also an urgent need to enhance security measures including B5b terrorist countermeasures adopted in the U.S.

Japan should consider establishing an organization like the Nuclear Rapid-Response Force to be established in France.

The periodic safety review is a globally established approach, which Japan also implements, but it failed to prevent the accident. We should review its effectiveness and consider improvements to create a more effective system.

In consideration of the fact that an approach to proper improvement of safety measures has been globally established based on comprehensive risk analysis using a probabilistic method, Japan should positively utilize the probabilistic risk assessment method.

In future, there is a need to close examination on individual security for various reactor types and new reactors, referring to countermeasures after the accident adopted abroad.

(5) **Summary**

Trends in international studies on the Fukushima Daiichi Nuclear Power Station Accident have been investigated, reviewed efforts to improve the safety, and analyzed lessons from the accident and safety improvement measures. Based on the results, discussions and proposals have been made on the future direction of progress in Japan. In the future, lessons from the accident should be reflected and the concept of ensuring safety against new reactors should be considered.

6.13 Information Dissemination

The responses of the Government and TEPCO following the occurrence of the Fukushima Daiichi Nuclear Power Station accident showed various problems. Examining the aspect of communication only, various problems in terms of information collection and decision-making by the Governments and other bodies, information transmission, and communications between the bodies concerned. There are not only fundamental deficiency and a lack of crisis awareness by the parties concerned but also many errors in basic understanding and responses to the crisis in the background. This section deals with the communication of the body or between bodies, including the Government and TEPCO after the Fukushima Daiichi accident. The first section shows where a deficiency was in the crisis management system and the second section shows what to do with the crisis management framework based on the result of the first item.

(1) **Discovered deficiency in the crisis management framework**

Immediately after the accident in the Fukushima Daiichi Nuclear Power Station, the parties responsible for crisis management in the Government centered upon the Prime Minister's official residence and TEPCO received a huge volume of emergent and important tasks and their responses were very confused. Failures in their responses from March 11 to 15, 2011 can be organized as follows from the communication perspective.

(a) Failures in information collection, distribution, and integration: The video conference system which links the head office and the site of TEPCO and its information had not been extended to or shared by the Prime Minister's official residence and the Nuclear and Industrial Safety Agency (NISA). During the time, the information obtained by the Prime Minister's official residence and NISA was fragmented and delayed. Therefore, the Prime Minister's official residence, etc. often added some information collection routes on an ad hoc basis, but the information needs could not be fully satisfied. As a result, the Prime Minister's official residence intervened in TEPCO without fully getting information on venting, seawater injection, and withdrawal problems, which TEPCO was to implement.

    Moreover, various failures occurred in the information analysis, evaluation, and integration in emergencies. For example, knowledge of experts, including the Nuclear Safety Commission members, special committee members, and JNES members was not fully used for decision-making and as the basis for decisions at the Prime Minister's official residence, The biggest reason for preventing was given the lack of full information to experts, they could not sufficiently analyze it.

(b) Failure in information dissemination: The Government's explanation to people was improper in terms of both the amount and quality of information. For example, it was not easy to understand the explanation of the radiation effects, with many examples which those were not information along to the needs of receivers. In addition, immediately after the accident, the Government kept avoiding the use of "reactor core meltdown" contrary to the words of experts. Though NISA determined it was very likely that the reactor core had melted down as of March 18, the Government kept avoiding the expression, and on April 10, changed the expression to "fuel pellet meltdown". It was May 16, 2 months after the accident, that the Government formally accepted the reactor core meltdown. This is an intentional restriction of information. Further, in the background, the Government might have wanted to arbitrarily underreport the plant status, rather than report it as-is.

    The elite panic means the elites themselves went into a panic, afraid the people would also panic during the disaster. Caron Chess [92] said it was elites rather than ordinary people who were put into a panic in emergencies. Elite panic is unique because it is caused by elites' imagination that general people would go into panic and that elites may intentionally restrict public relations. Elite means a high-ranking group in the social system as well as one with strong social influence. The parties that responded to the accident were the Government including the Prime Minister's official residence and NISA, the top management of the electric power company, and some academic experts. It was the Prime Minister's official residence that was in the highest rank when responding to the Fukushima Daiichi accident. It is likely that the elite panic resulted in the Prime Minister's official residence avoiding the use of the term reactor core meltdown as shown above [93].

(2) **Crisis management and crisis communication**

The crisis communication is communication among parties, people, and administrative organs in the event of a crisis (severe accident), and the crisis is the occurrence of an unprecedented and unexpected dangerous state [94]. Communication here is interactive distribution of information; involving both its transmission and reception. In addition, crisis management is a series of actions by the parties concerned to minimize damage in the event of an emergency. Risk management is analysis and assessment and responses to prevent the occurrence of a crisis itself. Moreover, the crisis management includes both crisis and risk management.

Immediately after the Fukushima Daiichi accident, great confusion arose in crisis management by the Prime Minister's official residence, NISA, and TEPCO, for the following reasons:

(i) Safety measures for the use of nuclear power in Japan, focusing on preliminary measures to prevent any accident, and measures to minimize accidental damage on the presupposition that an accident can occur were not prepared. Similarly, resilience [95] measures were not prepared to recover from the accident.

(ii) As the bureaucrats in charge are reshuffled every few years, the crisis-management knowledge inherited and passed on was insufficient.

(iii) TEPCO, the main party responsible for crisis management at the Fukushima Daiichi accident, knew the site better than the Prime Minister's official residence and NISA, and had the technology, knowledge, and resources for the crisis management. Still the Prime Minister's official residence exerted overly strong leadership for crisis management. TEPCO was subject to excessive interference, which might have influenced the awareness of the parties concerned in the crisis management of TEPCO.

(iv) Risk manager did not correctly understand panic. Panic does not grip the general people when they are informed of an imminent serious crisis. Because of the psychological mechanism of normalcy bias, they are convinced that "they will not fall into a dangerous situation" and that "someone will properly help them" to deal with the crisis. In fact, empirical research in sociology and social psychology revealed few cases where panic spread in a disaster. However, in the Fukushima Daiichi accident, the crisis managers fell into an elite panic and failed to engage in sufficient crisis communication.

Based on these, some specific items to improve crisis communication are suggested.

In the Fukushima Daiichi accident, the emergency advisory organization for crisis managers rarely functioned. It is necessary to build a system to reflect experts' analysis and related advice in policy initiatives and decisions as well as the mechanism of providing information to experts in emergencies.

In addition, in Japan, mechanism and training for information transmission in emergencies, including a nuclear power station accident, were not fully realized. It is necessary to build a mechanism to transmit unified supervisory information by a professional spokesperson as well as transmit information respectively and provide non-stop training by introducing training sessions equal to those of Western countries (for example, unannounced drills, training for full-time spokespersons, securing and training emergency call center staff, emergency response on the web site, etc.) and response capability training by those involved in disasters in the army, police, fire stations, hospitals, and municipalities based on the lessons learned from the Fukushima Daiichi accident.

Moreover, crisis communication with people through news media failed to distribute appropriate scientific information in some respects, including some harmful rumors. To prevent this, a process is required to enable professional engineers and groups of researchers like the AESJ and the Institution of Professional Engineers, Japan to ensure a "neutral stance", collect information from the Government, electric power companies and local governments, etc., and further clarify the progress of the accident and others from a technical perspective.

Conversely, non-experts/non-parties concerned on the street expressed their wish online to relieve unnecessary concerns from those around them and ensure correct information was transmitted. Today, community-based information is filled with new media including social media. The offering of the basis for our decision to these new media is required to prevent biased contents and directions of discussions and not to invite useless concerns including harmful rumors. Approaches are required to help non-experts/non-parties concerned on the street explain the situation to their surroundings in social media and daily conversation as well as enable experts and parties concerned to explain directly by themselves in order to realize the two step flow of communication. For example, the Government and AESJ can provide information to which the general public can easily refer and simply explain to those around them and the outline of explanation.

Appendix: Items Related to Accident Progression That Require Further Investigation and Consideration

Events progressed concerning several rector cores and spent fuel pools in the accident at the Fukushima Daiichi Nuclear Power Station, while various accident responses were conducted. Consequently, the development of the accident was very complicated.

In addition, due to high level of radiation particularly in the reactor building, detailed investigations including human access remain limited in some areas. Details of the status of the inside of the containment vessels in Units 1, 2, and 3 are unknown. From the above, this section summarizes topics concerning the accident progress which particularly need to be investigated and studied in detail. Such summary seems to be important during further investigation in future.

Facts related to the accident were checked focusing on 14 reports [96–109] released as reference materials. Perspectives which are important from an R&D perspective are summarized as remarks. Topics to be studied include (1) those that have not been fully investigated yet, those with room for further examination, or those for which reasonable explanations are currently difficult, and (2) those that seem reasonable but evidence is lacking so far.

Summarized topics are listed in Table 6.A.

Table 6.A Items requiring more detailed investigation in future in the events of the Fukushima Daiichi nuclear power station accident

Number	Objective (Unit)	Date and time	Category	Unresolved issues	Remarks	Source
1	1, 2, 3, 4, 5	In the event of a tsunami	Tsunami	Tsunami route, inundation start time of rooms (sections) in the reactor building, the control building, and the turbine building, the maximum inundation depth, and the time variations of inundation depth	For example, it is reported that as for the Unit 2 RCIC room, the reactor building started to be inundated by around 2:00 A.M. on March 12. Conversely, the Government Accident Investigation Committee Final Report says it is difficult to reach a deduction on the inundation of the Unit 2 reactor building in consideration of the inundation status of the Unit 4 reactor building. Moreover, the National Diet of Japan Fukushima Nuclear Accident Independent Investigation Commission report points out the emergency DG might stop before the attack of the tsunami. Information on the inundation route and time variations are required to check details in chronological order of various equipment and power sources installed in the lower positions of the reactor building	Technical Knowledge, p. 58 Government Accident Investigation Committee Final Report, Attachment p. 133 The National Diet of Japan Fukushima Nuclear Accident Independent Investigation Commission, p. 31
2	1, 2, 3, 4	–	Tsunami	Validation of tsunami damage conditions of the building and equipment	It is important to see whether the damage of the building and major facilities can be reproduced by numerical simulation to verify the tsunami simulation to determine the tsunami wave power. In addition, it is important to confirm the effects of floating debris	

(continued)

Table 6.A (continued)

Number	Objective (Unit)	Date and time	Category	Unresolved issues	Remarks	Source
3	1, 2, 4	In the event of a tsunami attack	Power source	Time gap between AC and DC power losses	The point in time when DC and AC power were lost is unclear. Difference may arise in the subsequent plant behavior when DC power is lost before AC power and when DC power is lost after AC power, which is important in designing safety	The Interim Report of the Gov't Accident Investigation Committee, p. 92
4	1	At around 1:48 on March 12	Cooling	Reason why a diesel-driven fire pump stopped working	The reason is said to be a malfunction, but was not obvious. As the diesel-driven fire pump is regarded as a piece of equipment with AM countermeasures.	The Interim Report of the Gov't Accident Investigation Committee, p. 129
5	1	In the event of tsunami	Cooling	Opening of isolation valves in the containment vessel of IC piping	As the isolation valve in the containment vessel is driven by AC power, the timing of the loss of DC power controlling the control signal and AC power for driving the valve determines the final opening. This information is necessary to check the IC cooling effects in detail while analyzing the reproduction of the accident progress	Technical Knowledge, p. 13
6	2	Tsunami attack—at around 13:25 on March 14	Cooling	RCIC drive mechanism in the status of DC power failure	It is true that the DC power failure led to the RCIC going out of control. Under the circumstances, it is pointed out that the feed water system was balanced by the level to which the water rose in the reactor, the flow from the main steam pipe, which turned into a dual-phase flow, into the RCIC turbine, and the	Technical Knowledge, p. 69

					worse drive efficiency of the turbine. It may be possible to perform verification tests using RCIC turbines at experiment facilities overseas where steams from thermal power stations can be used. It is important from the perspective of analysis to reproduce the accident progress	
7	2	At around 13:25 on March 14	Cooling	Reason for the RCIC Shutdown	The RCIC shutdown can be assumed by the tendency toward a lowered water level, but the mechanism leading to the shutdown remains unclear. However, according to the information provided by TEPCO, the water level indicator is shown at around the upper limit in the measurement range. If the actual water level exceeds the upper limit, the downward trend of the water level may not be measurable until the level reaches the measurement range even if the water level starts to be lowered. It is important from the perspective of analysis to reproduce the accident progress	Technical Knowledge, p. 69
8	1, 2, 3, 4	In the event of a tsunami attack	Cooling	Status of function loss of injection pumps (HPCI, CS, CCS, MUWC, CRD, SLC, RHRS) due to water damage and water injection	It is necessary to classify function losses into those due to the loss of power or the support system and those due to water damage and submergence. The status is summarized in Table IV-2-1 Technical Knowledge. According to the Interim Report of the Gov't Accident Investigation Committee, the first basement was submerged. It cannot be confirmed whether the equipment installed on the basement could have been used if the power source had not been lost. It is necessary to classify the equipment into two to further study	Technical Knowledge, Table IV-2-1, The Interim Report of the Gov't Accident Investigation Committee Reference II-12, p. 24

(continued)

Table 6.A (continued)

Number	Objective (Unit)	Date and time	Category	Unresolved issues	Remarks	Source
					accident responses: those that could be used if AC power supply was not lost and others that could not be used if the AC power supply had not been lost.	
9	1	Around March 11 night	Cooling	Unit 1 had no record of SRV opening operation, and the reactor pressure was lowered. This reactor pressure reduction mechanism and time variations, and time variation of the containment vessel pressure at the same time period	Molten fuel debris might damage the dry tube in the phase before PRV damage and lead to depressurization. In addition, the exposed fuel might cause the gas (steam) in the pressure vessel to become extremely hot, damaging the flange gasket in the main steam piping flange, which led to the gas leak. This depressurization process is important to precisely reproduce the accident progress	Technical Knowledge, Table IV-2-2 The Interim Report of the Gov't Accident Investigation Committee, p. 143 the National Diet of Japan Fukushima Nuclear Accident Independent Investigation Commission report, p. 31
10	3	11:36:00 March 12, 2013	Cooling	Reason for RCIC automatic shutdown	No reason for the shutdown of RCICs in both Units 2 and 3 is revealed, but this is important information to correctly understand the accident progress. Further, it is desirable to study it when considering application to other plants. In Unit 3, some drops of oil fell from the ceiling to the latch of RCIC and there is a description that the latch was detached	The Interim Report of the Gov't Accident Investigation Committee, p. 164
11	1, 2, 3	When alternate water injection started	Cooling	Time variations of water injection immediately after the start of alternate	As the water injection quantity was determined by a combination of discharge pressure, core pressure and leak quantity of alternate water injection line, the quantity injected to the core	IAEA Report, pp. IV-48, 60, 73

				water injection by fire engines	at the start of alternate water injection was unclear. The possibility the water injection was diverted to some by-pass line. The alternate water injection quantity was important to accurately analyze the core damage progress	
12	1, 2, 3	–	Cooling	Amount of salt accumulated by sea-water injection in the pressure vessel and the containment vessel	The amount is necessary to assume the influence of cooling effect of alternate water injection, cooling inhibitor effect of salt, and corrosion over the system. However, the leakage in liquid phase from the pressure vessel might have occurred before the precipitation of salt became problematic	
13	4	–	Cooling	Time variations of water flow from the reactor well into the fuel pool	The starting time and the time variations of water flow from the reactor well to the fuel pool are important information to accurately reproduce the change in water level. Structurally, if the water level of the reactor well exceeds that of the fuel pool, the water levels of the reactor well and fuel pool change nearly almost as the same	Technical Knowledge, p. 67
14	1, 2, 3	–	Core damage	The damage state of fuel, melted and dropped fuel debris in the pressure vessel, and the distribution of the containment vessel	Results differ depending on analytical codes and institutions, anyway, the damaged fuel might drop on the bottom of the pressure vessel, and further, part of fuel might fall into the pedestal of the containment vessel, Under the pressure vessel, lie the control rod drive mechanism and the large-sized equipment, to which the molten fuel might adhere. At present, there is considerable uncertainty of analysis, and detailed observation is required in future when removing fuel debris to enhance	JANTI report p. 2–72, Attachment 7-1

(continued)

Table 6.A (continued)

Number	Objective (Unit)	Date and time	Category	Unresolved issues	Remarks	Source
					knowledge. This information is important for studying the detailed process of taking out of fuel debris	
15	1, 2, 3	–	Core damage	Detailed Progress of the Accident	The reproduction analysis for the following by the simulation is implemented: changes in the reactor water level, the starting time of fuel damage, fuel damage and movement process, the pressure in the pressure vessel and the temperature change, the temperature change of the containment vessel, the damage process of the containment vessel, etc. The uncertainty of analytical result was considerable. In the Government Accident Investigation Committee Final Report, the analytical results vary widely and some events could not be fully reproduced. Through future observation of inside containment vessel, removing fuel debris, etc., it is important to obtain information to reduce such uncertainty and improve the accuracy of producing severe accidents	IAEA Report, p. IV-39 Government Accident Investigation Committee Final Report, Attachment p. 23
16	1	15:36:00 March 12, 2013	In the event of hydrogen explosion	A relationship with the PCV vent at around 14:00 on the same day	It may be possible to directly assume a relationship with vent by quantitatively evaluating hydrogen leakage from the containment vessel. In addition, it is related to the effect of flow control damper by separating SGTS from the vent line in the power source loss. This is important information to understand the accident progress process and AM measures	Technical Knowledge, pp. 32–33

17	3	11:01:00 March 14, 2013	In the event of hydrogen explosion	A relationship of PCV vent conducted several times on March 13	It may be possible to directly assume a relationship with PCV vent by quantitatively evaluating the hydrogen leakage from the containment vessel. In addition, it is related to the effect of gravity damper by separating SGTS from the vent line in the power source loss. Government Accident Investigation Committee Final Report assumed the vent operation at Unit 3 did not succeed except the first and second operations	Technical Knowledge, pp. 32–33 Government Accident Investigation Committee Final Report, Attachment p. 179
18	1, 2, 3	–	In the event of hydrogen explosion	Time variations of hydrogen concentration in the containment vessel and details of the building and the progress of the explosion	Reproduction analysis by the simulation is implemented. In future, it will be important to improve the accuracy of reproduction analysis while using the knowledge obtained by the progress of decommissioning measures to understand the more detailed effects of hydrogen explosion	Technical Knowledge, p. 36
19	4	9:38:00 March 15, 2013	Fire	Cause of fire on 4 F of the reactor building	It seems there was a relationship with the hydrogen explosion occurrence at about 6:00 in the morning of the same date, but the cause was not revealed. This was information to determine whether the fire was associated with the hydrogen explosion	IAEA Report, p. IV-77
20	1, 2, 3	–	Containment	Damage state of structures in the reactor core and the pressure boundary including the pressure vessel and the control rod drive mechanism	The fuel melted in the high temperature might damage the dry tube in the reactor or the fuel debris dropped on the bottom of the pressure vessel might damage the penetrating portion of the control rod drive mechanism, etc. Details must be observed when the fuel debris will be removed. In addition, the core pressure when the pressure vessel is damaged influences the	Government Accident Investigation Committee Final Report, Attachment p. 166

(continued)

Table 6.A (continued)

Number	Objective (Unit)	Date and time	Category	Unresolved issues	Remarks	Source
					distribution of the molten fuel debris. According to the Government Accident Investigation Committee Final Report, it is assumed that the depressurization was triggered by molten fuel dropping into the water remaining at the bottom of the pressure vessel and leading to pressure spikes. This is important information for the analysis to reproduce the accident and removal of fuel debris	
21	1, 2, 3	–	Containment	Damage status of deviation of D/W, pedestal, and S/C	If the fuel drops on the pedestal of the containment vessel, the molten core concrete interaction might damage the floor area of the pedestal. This is important information to confirm the confinement of the containment vessel, and it is necessary to observe the details when fuel debris is removed in future	JANTI Report, Attachment 7-1
22	1, 2, 3	–	Containment	The possibility of DCH. shell attack, water vapor explosion, etc.	It seems that DCH shell attack, water vapor explosion, etc., which were important containment vessel damage modes, did not occur. It is important to study why these events were prevented to consider severe accident measures in future	
23	1, 2, 3	–	Containment	The leak mechanism and route of gas (including hydrogen and steam). Time variation of leakage	The dose distribution in the building shows efforts were made to identify the leak route and the flange and penetration seal might be deteriorated. The time variations of leakage are related to the timing of the release of radioactive materials outside the station and the	Technical Knowledge, p. 27

					released amount. It is necessary to investigate the containment vessel flange and the penetration area in detail. This information is important from the perspective of estimating the accuracy improvement of the release of radioactive materials and AM measurement planning	
24	1, 2, 3	–	Containment	The leak mechanism and route of liquid from the containment vessel. Time variations of leakage	The water level at Unit 2 was only around 60 cm from the pedestal and it was assumed that there is a leak at the bottom of the containment vessel, including the connection pipe to the S/C. It is important to check whether the bellows piping connects the drywell and the S/C. There are certain to be some leaking portions in containment vessels Units 1 and 3, but the locations have not been identified. It is necessary to identify leaking portions by remote detection technology and others. In addition, such investigation may also identify the leakage mechanism. The information is important from the perspective of improving the estimate accuracy of the amount of radioactive materials released and planning AM measures	
25	1, 2, 3	–	Containment	The quantity of radioactive materials, time variations, and chemical forms of radioactive materials released as gas phase from the	The information is important to evaluate the diffusion of radioactive materials into the air. The general values were evaluated in past analyses, but it is important to strive to improve the accuracy of analysis to reproduce the accident to reduce the uncertainty whenever possible. The information is important from the	TEPCO Released Amount Measurement

(continued)

Table 6.A (continued)

Number	Objective (Unit)	Date and time	Category	Unresolved issues	Remarks	Source
				containment vessel to the reactor building, and further, into the environment	perspective of improving the estimate accuracy of the amount of radioactive materials released and planning AM measures	
26	1, 2, 3	–	Containment	The quantity of radioactive materials, time variations, and chemical forms of radioactive materials released as liquid phase from the containment vessel to the reactor building, and further, into the environment	The information is important to evaluate the diffusion of radioactive materials into the sea. The general values were evaluated in past analyses, but it is important to strive to improve the accuracy of analysis to reproduce the accident to reduce uncertainty whenever possible. It is directly connected to improving the accuracy of the release of radioactive materials	
27	1	At around 17:50 on March 11	Containment	Reason why a higher radiation level than usual was detected at the double door of the reactor building	If IC did not operate after all the power sources were lost, it was assumed that the reactor water level reached TAF at around 18:00 on March 11. At the time, different results were obtained from data analysts about whether the fuel damage was reached. In addition, it is necessary to confirm whether the leak that raises the radiation level outside the reactor building or the increase in direct radiation is possible. The Government Accident Investigation	The Interim Report of the Gov't Accident Investigation Committee, p. 103

					Committee report says no other abnormalities were found, including increases in the radiation level within and around the building	
28	1	Early morning of March 12	Containment	Plant parameter changes including PRV, the internal pressure of the containment vessel, temperature, etc., and changes of monitor post readings	To improve effectiveness of severe accident measure, it is important to check over temperature of the containment vessel, the progress of excessive pressure damage, and time variations in the amount of radioactive material leakage from the containment vessel	IAEA Report, p. IV-44 Estimated TEPCO Release
29	1,2,3	–	Containment	The emission rate of rare gas released from the containment vessel, the diffusion direction, and the exposure dose of rare gas	The current emergency preparedness guidelines evaluate most effective doses are caused by rare gas. In the accident, it seemed that the almost the entire quantity of rare gas were continuously released from three reactors. This is likely to impact significantly on doses in surrounding areas depending on the diffusion directions, so it is important from the perspective of exposure dose evaluation at an early stage to evaluate the dose in the site surroundings by overlapping then detailed climate conditions as well as revealing the rare gas emission rate. In addition, the Nuclear Safety Commission, NISA and TEPCO have different opinions in estimating the emission source. Further study is required to fill the gaps in future	Estimated TEPCO release

(continued)

Table 6.A (continued)

Number	Objective (Unit)	Date and time	Category	Unresolved issues	Remarks	Source
30	1, 2, 3	–	Containment	Radioactive material removal factor in the S/C in the wet ventilation	To accurately evaluate radioactive materials released by vent, it is necessary to accurately evaluate the DF value by the S/C. The information will help ensure more effective assessment of AM measures	IAEA Report, p. IV-44
31	1, 2, 3	–	Containment	Cause of the peak of high radiation level shown in the monitoring result, particularly causes of high peaks at 10:00 on March 15, at around 23:00 on March 15, and at around 11:00 on March 16	The following estimations can be made with a certain rationality, but it is difficult to confirm as of now At around 10:00 on March 15: It seems there is some relation with the depressurization of the containment vessel in Unit 2 At around 23:00 on March 15: It is estimated as the release from Unit 2 At around 11:00 on March 16: It seems there is some relation with the massive release of vapor from Unit 3 The information is required to segregate the changes in wind direction and emission rate. The information is important from the perspective of deepening understanding of the accident progress and the estimate accuracy of the amount of radioactive materials released	Technical Knowledge Fig.V-1-1 Estimated TEPCO release, p. 9
32	2	–	Containment	Status of rupture disk	If the rupture disk did not operate, it is necessary to find the reason. It may be because the containment vessel pressure led to over temperature damage, not reaching the rupture disk	Technical Knowledge, p. 34

					working pressure, or because the vent valve might have not opened when the containment vessel pressure was high. In the measurement data, when the vent valve was open, the pressure in the suppression chamber (S/C) did not reach the working pressure of the rupture disk, The measurement data at around 22:10 on the 14th was lower than the drywell pressure, and after that, the gap got bigger. The data reliability is a problem. The information will help ensure more effective assessment of AM measures	
33	2	At around 6:00 on March 15	Containment	Reason why the pressure reading of the S/C suddenly dropped	At the same time, the D/W pressure lacked a large decrease. The Government Accident Investigation Committee report says the trouble of the electric system downscaled the S/C pressure, which might be erroneously recognized as 0 MPa (abs). The relationship with hydrogen explosion at the Unit 4 reactor building, which almost simultaneously occurred, was limited	TEPCO report, p. 79 Government Accident Investigation Committee, p. 65
34	2	0:00–6:00 on March 15	Containment	Reason for deviation of D/W and S/C pressure	The drywell and suppression chamber are not isolated structures. The pressure should have changed similarly The Government Accident Investigation Committee Final Report assumes that it was an error signal of the pressure measurement created by the S/C pressure measuring device for AM, but that the detailed reasons were unknown	IAEA Report, Fig. IV-5-5 Government Accident Investigation Committee Final Report, Attachment p. 111

(continued)

Table 6.A (continued)

Number	Objective (Unit)	Date and time	Category	Unresolved issues	Remarks	Source
35	2	7:20 to 11:25 on March 15	Containment	Causes of D/W pressure reduction	It seemed that the leakage from the flange or penetration might reduce the D/W pressure. At present, neither leakage points nor leakage mechanism have been identified. The information is useful in studying more detailed mechanism of the containment vessel damage	IAEA Report, Fig. IV-5-5
36	2	8:25:00 March 15, 2013	Containment	Cause of white smoke, which looked like steam, from the 5 F of the reactor building (blowout panel opening)	It is estimated there is some relation with the D/W depressurization, which seemed to occur at the same time, but the cause is unclear. The information is useful in studying the amount of radioactive materials released and the damage mechanism of the containment vessel	IAEA Report, Table IV-5-2
37	3	At around 8:30 on March 16	Containment	Source of a large amount of water vapor released from the reactor building	The fuel pool and inside the containment vessel had large amounts of water and a heat source that could produce water vapor, but the source has not yet been identified. In considering the lack of leakage found in the spent fuel pool and the relation between the decay heat of the spent fuel and holding water quantity, the spent fuel pool at Unit 3 was unlikely to be the source. The D/W pressure tended to drop from March 15–17, but it is not determined whether it has any correlation with the occurrence of water vapor. The information is useful in studying the amount of radioactive materials released and the damage mechanism of the containment vessel	IAEA Report, Table IV-5-3

38	4	At around 10:30 on March 14	Containment	Reason why the dose rose in the reactor building	The inside of the building was shielded from the outside, and there was no reason why the air dose rate rose. The possible reasons for high dose in the Unit 4 reactor building include: (1) the low water level of the Unit 4 spent fuel pool was and the increase in direct radiation, (2) the shutdown of the air-conditioner made it possible for the outdoor air to intrude into the reactor building, and (3) the reverse flow from the Unit 3 vent line released gas containing radioactive materials (and hydrogen) into the Unit 4 building. However, the actual reason was not found. At 4:08 on March 14, the temperature of the Unit 4 pool was measured, and the dose in the Unit 4 building might have increased from 4:00 to 10:30 on the 14th. Unit 3 building exploded at 11:01 on the 14th. This is important supporting evidence for the Unit 4 hydrogen explosion	TEPCO Final Report, Attachment 2, p. 111 The Interim Report of the Gov't Accident Investigation Committee, p. 217
39	3	–	Containment	Reason why the damage mechanism of equipment hatch and the biological shield were moved	At Unit 3, the equipment hatch was investigated and the leak of water was confirmed. What was a cause of this leakage and why the shield plug at the hatch entry part was moved? It is necessary to consider these questions	
40	1, 2, 3	–	Spent fuel pool	State of the inside of Units 1 and 2 fuel pools	The insides of Units 1 and 2 have not been observed. It is necessary to do so to remove the fuel	Technical Knowledge, Attached Reference 2
41	1, 3, 4	In the event of hydrogen explosion	Spent fuel pool	Water level of the fuel pool after the hydrogen explosion	Especially in Units 1 and 3, it is likely that debris of reactor building fell and the water level was lowered due to the hydrogen explosion. In addition, the accurate water level of	Technical Knowledge, Attached Reference 2

(continued)

Table 6.A (continued)

Number	Objective (Unit)	Date and time	Category	Unresolved issues	Remarks	Source
					Unit 3 was unknown when the helicopter on March 17 and then, water cannon trucks started to inject water into the fuel pool on March 19. Originally, the decay heat of the spent fuel in the fuel pool at Unit 3 was around a quarter of that at Unit 4, and they had time before the water injection if the water did not drop in the early stage. This is useful information for the analysis to reproduce the accident progress	
42	1, 3, 4	–	Spent fuel pool	The changing water level since the start of alternate water injection (including the water discharge from the ground)	As for Units 3 and 4, the accurate water level was not measured for 1 month until mid-April. As for Unit 1, the water was not measured until the end of May. This is useful information for analysis to reproduce the accident progress	Technical Knowledge, Attachment Reference 2
43	4	Before the explosion of building	Spent fuel pool	Reason why the preparation for injection into the pool is delayed	There is a description that the work was given up due to a high radiation level in the building. This is useful information to study the effectiveness of AM measures	TEPCO Final Report, Attachment 2, p. 111 The Interim Report of the Gov't Accident Investigation Committee, p. 216, p. 217

44	1	Early morning of March 13	Spent fuel pool	The cause of white smoke from the reactor building observed early morning of March 13	According to a preliminary calculation, as of March 13, the water temperature of the spent fuel pool was 40 °C or less, it was unknown that the white smoke could be observed. This can be the information useful to more accurately estimate the amount of radioactive materials released	The Interim Report of the Gov't Accident Investigation Committee, p. 215 Technical Knowledge, Attachment Reference Fig. ii-1
45	1	–	Instrumentation	Cause of gap between A-system and B-system reactor pressure readings	In future, it can be the reference information in developing high reliability pressure gauge	IAEA Report, Fig. IV-5-1
46	1, 2, 3	–	Instrumentation	Reliability of measurement value of PRV, drywell (D/W), suppression chamber (S/C) pressure and temperature	The measuring instrument was exposed to the severe condition after the accident, and it was difficult to assess how much measurement error was due to this influence. What the qualitatively reasonable trend is shown and what is not are mixed together. For example, there was a gap between D/W and S/C pressures at around 0:00 on the 15th, which is physically implausible. The analysis to reproduce the accident progress was performed based on limited measured values. It is important to assess the uncertainty of measurement values. In future, this may be reference information when developing high reliability measuring instruments	The Interim Report of the Gov't Accident Investigation Committee, p. 231

(continued)

Table 6.A (continued)

Number	Objective (Unit)	Date and time	Category	Unresolved issues	Remarks	Source
47	1, 2, 3, 4	–	Earthquake ground motion	The influence of the earthquake ground motion on equipment	As for important safety equipment, the plant data until the station blackout seemed to show the influence was not serious enough to interfere with the safety function, but for the present, small leakage has not been fully accounted for. The status of equipment less important for safety has also not been directly checked yet. To enhance knowledge in future, the check should be promoted	Technical Knowledge, p. 50 The National Diet of Japan Fukushima Nuclear Accident Independent Investigation Commission report, p. 30 The Final Report by the National Diet of Japan Fukushima Nuclear Accident Independent Investigation Commission, p. 27, etc.
48	1, 2, 3, 4	–	Others	The relation between the subsurface water quantity in inflow path into the reactor and the turbine building and the groundwater level	The inflow of groundwater seems to increase to about 400 tons of contaminated water per day. It is important to assess the relation between the inflow path and the groundwater level and the correct value of the increase to consider countermeasures for the contaminated water	

References

1. IAEA, *Basic Safety Principles for Nuclear Power Plants 75-INSAG-3 Rev.1, INSAG-12* (IAEA, 1999)
2. IAEA, *IAEA Safety Standards, Fundamental Safety Principles No. SF-1* (IAEA, 2006)
3. AESJ-SC-TR005, *Fundamental Concept on Nuclear Safety – Nuclear Safety Objectives and Fundamental Safety Principles, Technical Report by Standards Committee, Atomic Energy Society of Japan* (AESJ-SC-TR005, 2012)
4. Special Committee, Nuclear Safety Commission, *Interim Report on the Investigation and Review on Safety Goals* (Special Committee, Nuclear Safety Commission, 2003)
5. USNRC, *Recommendation for Enhancing Reactor Safety in the 21st Century, The Near Term Task Force* (USNRC, 2011)
6. IAEA, *Safety Reports Series No. 46, Assessment of Defence in depth for Nuclear Power Plants* (IAEA 2005)
7. IAEA, *Defence in depth in Nuclear Safety, INSAG-10* (IAEA, 1996)
8. IAEA, *Basic Safety Principles for Nuclear Power Plants 75-INSAG-3 Rev. 1, INSAG-12* (IAEA, 1999)
9. Interim and Final Reports of the Japanese Government Investigation Committee on the Accident at the Fukushima Nuclear Power Stations, Investigation Committee on the Accident at the Fukushima Nuclear Power Stations of Tokyo Electric Power Company, Medialand (2012)
10. IAEA Mission Report, *The Great East Japan Earthquake Expert Mission, IAEA International Fact Finding Expert Mission of the Fukushima Dai-ichi NPP Accident Following the Great East Japan Earthquake and Tsunami* (2011)
11. The Diet Accident Investigative Committee Report, *TEPCO Fukushima Nuclear Power Station Accident Investigation Committee report* (Tokuma Shoten Publishing, 2012)
12. H. Miyano, N. Sekimura, K. Demachi, S. Arai, M. Matsumoto, *10th Academic Lecture Summaries*, Japan Society of Maintenance (2013), pp. 329–335
13. Nuclear Regulation Authority Project Report, *Nuclear Regulation Authority, FY 2004 Deterioration Technology Evaluation Enhancement Project* (Nuclear Regulation Authority, 2004)
14. Japanese Standard Association, *Method of Determining the Repassivation Potential for Crevice Corrosion of Stainless Steels, JIS G 0592* (Japanese Standard Association, 2002)
15. M. Akashi, G. Nakayama, T. Fukuda, Corrosion **98**, 158 (1998)
16. M. Sakiya, R. Matsuhashi, T. Mastuhashi, A. Takahashi, Mater. Environ. **58**, 378 (2009)
17. M. Kobayashi, T. Narabayashi, Maintenance Technology. Jpn. Soc. Mainten. J. **12**(2), 84–93; T. Narabayashi, Japan Society of Maintenology publication "Maintenology" (2013)
18. UNSCEAR, *Sources and Effects of Ionizing Radiation* (UNSCEAR, New York, 2010)
19. ICRP, *The Recommendations of the International Commission on Radiological Protection, ICRP Publication 103* (ICRP, 2007)
20. R.R.K. Nair et al.. Background Radiation and Cancer Induce in Kerala, India-Karunagappally Cohort Study. Health Phys., **96**(1), 55–66 (2009)
21. Act on Special Measures concerning the Handling of Environment Pollution by Radioactive Materials Discharged by the NPS Accident Associated with the Tohoku District – Off the Pacific Ocean Earthquake That Occurred on March 11 (Act No. 110 of August 30, 2011), posted on the website of the Ministry of the Environment together with relevant government and ministerial decrees http://env.go.jp/jishin/rmp/.html
22. The Ministry of the Environment, Decontamination Policy in the Decontamination Special Areas (the Decontamination Roadmap) (2012), http://www.env.go.jp/jishin/rmp/attach/josen-area-roadmap.pdf
23. ICRP, *A Application of the Commission's Recommendations to the Protection of People Living in Long-term Contaminated Areas after a Nuclear Accident or a Radiation Emergency, ICRP Publication 111* (ICRP, 2008)

24. The MOE, Guideline relevant to decontamination, 2nd edn. (MOE, 2013) http://www.josen.env.go.jp/material/index.html
25. The MOE, Basic policy on interim storage facilities required for response to environmental contamination caused by radioactive materials associating to accident of the Tokyo Electric Power Co., Inc., Fukushima Daiichi Nuclear Power Station (MOE, 2011) http://www.env.go.jp/jishin/rmp/attach/roadmap111029_a-().pdf
26. Nuclear Suffers Life Support Team, Cabinet Office, Japan Atomic Energy Agency(JAEA), Reference, Meeting reporting on results of decontamination model demonstration projects, (MOE 2012) http://www.jaea.go.jp/fukushima/kankyoamzen/d-model_report/app_2.pdf
27. The MOE, Policy on transportation to interim storage facilities, Deliberation committee on safety countermeasure of interim storage facilities, second meeting, reference 9 (2013) http://www.josen.env.go.jp/area/processing/pdf/safety_measure_02.pdf
28. The MOE, Outline of interim storage facilities, Deliberation committee on safety countermeasure of interim storage facilities, first meeting, reference 4 (2013) http://www.env.go.jp/area/processing/pdf/environmental_protection_01.pdf
29. The AESJ Website, Decontamination catalogue version 1.0 (relevant to the Clean-Up Subcommittee), (2011) http://www.aesj.or.jp/information/fnpp201103/chosasenmoniinkai.html
30. Nuclear Safety Commission, Environmental Radiation Monitoring Guideline (Nuclear Safety Commission, 2010), p. 51
31. M. Kayano, At. Energy Soc. Jpn. J. **55**, 220–224 (2013)
32. Nuclear Regulatory Commission, *Nuclear Disaster Countermeasure Guideline* (Nuclear Regulatory Commission, 2012)
33. Committee on investigation and verification on TEPCO Fukushima Daiichi Nuclear Power Station accident, "Final Report", July 23, 2012 http://www.cas.go.jp/seisaku/icanps/post-3.html
34. I. Kataoka, J. Nucl. Sci. Technol. **50**(1), 1–14 (2013)
35. Atomic Energy Society of Japan, Atomic Energy Society of Japan standard, *Implementation Standard on Probabilistic Safety Assessment Targeting Output Operation of Nuclear Power Plant* (Level 2, PSA): 2008, AESJ-SC-P009: 2008 (Atomic Energy Society of Japan, 2009)
36. G.E. Wilson, B.E. Bouyak, Nucl. Eng. Des., **186**, 23–37 (1998)
37. D. Magallon, A. Malliat et al., Nucl. Eng. Des., **235**, 209–346 (2005)
38. S. Kinase, M. Kimura, S. Takahara, T. Homma, Evaluation of averted doses to infants by tap water restrictions after the Fukushima Daiichi nuclear power plant accident. Trans. At. Energy Soc. Jpn. **10**(3), 149–151 (2011)
39. Ministry of Health, Labor and Welfare, *Measures for Radioactive Material in the Tap Water, Interim Report. Study Meeting of Measures for Radioactive Material in the Tap Wate*r (Ministry of Health, Labor and Welfare, 2011)
40. Discover Twentyone, Independent Investigation Committee on the Fukushima Daiichi Nuclear Accident, *Report of Investigation and Verificatio*n (Discover Twentyone, 2012), p. 342
41. Independent Investigation Commission on the Fukushima Daiichi, *Nuclear Accident Report of investigation and verification.* (Discover Twentyone, 2012), p. 343
42. Nuclear and Industrial Safety Agency, Japan Nuclear Energy Safety Organization, Guideline for the Regulatory Authority to Evaluate Licensees' Efforts to Prevent Degradation of Safety Culture and Organizational Climate deterioration (Nuclear and Industrial Safety Agency, Japan Nuclear Energy Safety Organization, 2007)
43. IAEA, *Safety Culture* (1991)
44. IAEA, *Nuclear Security Culture* (2008)
45. Advisory Committee for Natural Resources and Energy Nuclear and Industrial Safety Subcommittee Nuclear Emergency Preparedness Subcommittee, *Concerning response to internal threat(insider) in the nuclear facilities* (Advisory Committee for Natural Resources

and Energy Nuclear and Industrial Safety Subcommittee Nuclear Emergency Preparedness Subcommittee, 2005)
46. Atomic Energy Commission Nuclear Energy Advisory Committee on Nuclear Security, Strengthening of Japan's Nuclear Security Measures (Atomic Energy Commission Nuclear Energy Advisory Committee on Nuclear Security, 2012)
47. IAEA, INSAG-24 Interface between the safety and the security of the nuclear power plant (2011)
48. Example of Regulation document about the measurement control: NRC:PART 74, DOE-STD-1194-2011
49. Interim report by the Government Investigation Committee on the Accidents at the Fukushima Nuclear Power Station of Tokyo Electric Power, Study Committee (Main ver.) (2011)
50. TEPCO, *Investigation Committee on the Accidents at the Fukushima Nuclear Power Station* (Main ver.), (Attached documents) (2012)
51. Information from the Nuclear and Industrial Safety Agency, former organizations; TEPCO Fukushima Dai-ichi NPS, plant related parameters, (http://www.nsr.go.jp/archive/nisa/earthquake/plant/2/plant-2-2303.html)
52. A. Ishibashi, D. Karikawa, M. Takahashi, T. Wakabayashi, M. Kitamura, J. At. Energy Soc. Jpn., **9**(4), 384–395 (2010)
53. TEPCO, *Work operation transfer diary for Unit 3* (http://www.tepco.co.jp/nu/fukushima-np/plant-data/f1_4_Nisshi3_4.pdf) (2011)
54. Government Investigation Committee on the Accidents at the Fukushima Nuclear Power Station of Tokyo Electric Power Company report, verification committee and final report (main-text ver.) (2012)
55. BTC Training Newsletter, No.89 (April 2012), No.90 (2012)
56. Technical findings of Nuclear and Industrial Safety Agency, Fukushima Dai-ichi Nuclear Power Plant (2012)
57. Emergency Preparedness Headquarters, *Additional Report on International Atomic Energy Agency by Japanese Government- Fukushima Nuclear Plant Accident-*(2nd Report) (2012)
58. Japan Nuclear Technology Institute, Lessons Learned from Nuclear Accident at the Fukushima Daiichi Nuclear Power Station; Institute of Nuclear Power Operations, Fukushima Dai-ichi NPS, INPO 11–005 Addition (2012) (translated by Japan Nuclear Technology Institute)
59. TEPCO, *Accident Analysis Report by the Investigation Committee on the Accidents at the Fukushima Nuclear Power Station* "Appendix 2: Response Status to the Fukushima Dai-ichi and Dai-ni Nuclear Power Stations" (TEPCO, 2012)
60. J. Reason, *Managing the Risks of Organizational Accidents* (Ashgate, 1997)
61. E. Hollnagel, *Safety culture, safety management, and resilience engineering, ATEC, International Aviation* Safety Forum (2009)
62. A. Nakanishi, *Requirements for High Reliability Organization* (Seisansei Publishing Co., 2007)
63. Y. Hayashi, *Case studies: Introduction to Risk Literacy* (Nikkei BP Publishing Co., 2005)
64. K. Kikuzawa, *Unreasonable Organization* (Diamond Publishing Co., 2000)
65. TEPCO, Response to the Pacific coast of Tohoku Earthquake and Tsunami, Survey and Lessons Learned from Extracted; Japan Nuclear Safety Institute (Proposal) (2012)
66. TEPCO, *Summary of the Fukushima Nuclear Accident and Nuclear Safety Reform Plan* (2013)
67. Investigation Committee on the Accidents at the Fukushima Nuclear Power Station of Tokyo Electric Power Company, Report by the national Diet of Japan Fukushima Nuclear Accident Independent Investigation Commission (2012)
68. The National Diet of Japan Fukushima Nuclear Accident Independent Investigation Commission, Report, July 5, 2012

69. Tokyo Electric Power Company, *Fukushima Nuclear Accident Summary and Nuclear Safety Reform Plan* (Tokyo Electric Power Company, 2013)
70. Government Investigation Committee on the Accident at the Fukushima Nuclear Power Stations of Tokyo Electric Power Company, "Final Report," July 23, 2012
71. Government Institute of Nuclear Power Operations (INPO), Lessons Learned from the Nuclear Accident at the Fukushima Daiichi Nuclear Power Station, Special Report INPO 11–005 Addendum, August 2012
72. Investigation Committee on the Accident at the Fukushima Nuclear Power Stations of Tokyo Electric Power Company, "Interim Report," December 26, 2011
73. WENRA RHWG, Safety Objectives for New Power Reactors (2009)
74. GPR/German experts Technical Guidelines, Oct. 19 and 26, 2000
75. U.S. NRC, NRC-2008-0237, *Federal Register*, vol 73 (2008), p. 199
76. U.S. DOE NERAC&GIF, GIF-019-00 (2002)
77. IAEA Mission Report, 24 May – 2 June 2011
78. WENRA RHWG, Safety of New NPP designs (2012)
79. INPO, INPO 11–005 (2012)
80. EPRI, 2012 Technical Report (2012)
81. ANS, Fukushima Daiichi: ANS Committee Report (2012)
82. WENRA report, March 21, 2012
83. ASME Presidential Task Force report (2012)
84. J. Buongiorno, R. Ballinger et al., MIT-NSP-TR-025 (2011)
85. ENSRG, Peer review report, Apr. 25, 2012
86. EC, Communication on Stress Tests, Oct. 4, 2012
87. ASN, Complementary Safety Assessments of the French Nuclear Installations (2011)
88. IRSN, Translation of IRSN report No. 708 (2012)
89. A. Keramsi, N. Camarcat, ATOMEXPO 2012 Int. Forum, Moscow, Russia, June 4–6, 2012
90. NEI, NEI 06–12 (2006)
91. US NRC, JLD-ISG-2012-01
92. L. Clark, C. Chess, Soc. Forces **87**(2), 993–1014 (2008)
93. T. Sata, J. At. Energy Soc. Jpn. **54**(7), 434–440 (2012)
94. S. Tsuchida, J. At. Energy Soc. Jpn. **54**(3), 181–183 (2012)
95. M. Kitamura, J. At. Energy Soc. Jpn. **54**(7), 721–726 (2012)
96. The Atomic Energy Society of Japan (AESJ), the Research Committee of "Nuclear Safety," the Technology Analysis Subcommittee, Lessons learned from the accident at the Fukushima Daiichi Nuclear Power Plant (2011)
97. The Government Nuclear Emergency Response Headquarters, Report of Japanese Government to the IAEA Ministerial Conference on Nuclear Safety. The Accident at TEPCO's Fukushima Nuclear Power Stations, June 2011 [IAEA Report]
98. The Government Nuclear Emergency Response Headquarters, IAEA Additional Report, Additional Report of Japanese Government to IAEA – Accident at TEPCO's Fukushima Nuclear Power Stations- (Second Report), September 2011 [IAEA Additional Report]
99. The JANTI Fukushima Daiichi Nuclear Power Station Accident Committee, The JANTI Report, Examination of Accident at Tokyo Electric Power Co., Inc.'s Fukushima Daiichi Nuclear Power Station and Proposal of Countermeasures (2011)
100. TEPCO Fukushima Nuclear Power Station Accident Investigation Committee report, *The Interim Report of the Gov't Accident Investigation Committee, TEPCO Investigative Committee on the Accident at the Fukushima Nuclear Power Stations of Tokyo Electric Power Company*, Interim Report (2011)
101. TEPCO, *Fukushima Nuclear Accident Analysis Report* (Interim Report) (2011)
102. The Nuclear and Industrial Safety Agency (NISA), *Technical Knowledge of the Accident at Fukushima Daiichi Nuclear Power Station of Tokyo Electric Power Co., Ind.* (Interim Summary) (2012)

103. Team H2O Project, *What should we learn from the severe accidents at the Fukushima Daiichi Nuclear Power Plant?* Final Report (2011)
104. Independent Investigation Commission on the Fukushima Daiichi Nuclear Accident, Private Accident Investigation Committee, *Investigation and Verification Report, Independent Investigation Commission on the Fukushima Daiichi Nuclear Accident*, Discover 21 (2011)
105. TEPCO, *TEPCO Determined Released Amount Fukushima Daiichi Genshiryoku Hatsudensho Jiko ni okeru Hoshasei Busshitsu no Taiki chu heno Hoshutsuryo no Suitei ni tsuite (Estimated Radioactive Material Release into the Air in the Accident at the Fukushima Daiichi Nuclear Power Station)* (TEPCO, 2012)
106. TEPCO, *Fukushima Nuclear Accident Analysis Report (Final Report)* [TEPCO Accident Report] (2012)
107. The National Diet of Japan Fukushima Nuclear Accident Independent Investigation Commission, *TEPCO Fukushima Nuclear Power Station Accident Investigation Committee, TEPCO Fukushima Nuclear Power Station Accident Investigation Committee report* (the National Diet of Japan Fukushima Nuclear Accident Independent Investigation Commission, 2012)
108. TEPCO Fukushima Nuclear Power Station Accident Investigation Committee, *TEPCO Fukushima Nuclear Power Station Accident Investigation Committee* (Final Report), July 2012 [Government Accident Investigation Committee]
109. Nuclear Safety Subcommittee of the Atomic Energy Society of Japan (AESJ), *Seminar on the Fukushima Daiichi Nuclear Power Station*, Report "What was Wrong and What to Do in Future" March 2013

Chapter 7
Analysis and Issues on Nuclear Safety System

Abstract In this chapter, the Government, industry, and the academia, particularly the AESJ, are respectively analyzed to show where the problem lay in the system and the nature of the problem to maintain and improve safety in Japan before the accident and how to improve the same problem in future based on analysis and evaluation results of the accident described in Chap. 6.

Keywords Global nuclear safety system • Industrial community • Nuclear safety system • Role of AESJ • Safety regulatory system • Safety research system

In this chapter, the Government, industry, and the academia, particularly the AESJ, are respectively analyzed to show where the problem lay in the system and the nature of the problem to maintain and improve safety in Japan before the accident and how to improve the same problem in future based on analysis and evaluation results of the accident described in Chap. 6.

Nuclear technology is basically handled differently from general industrial technology in terms of the regulation system, since the industry handling nuclear source material and fuel material needs to get a license from the Government before starting a business. If an accident should occur, it may severely impact on the socioeconomic activities of general citizens and environment. The Government only allows those who are recognized to have the ability and economic resources to keep nuclear safety. Naturally, the licensee has a primary responsibility to ensure safety while the Government plays a key role in recognizing the ability of the licensee to safely promote its business and issues a business license accordingly. The academia and research institutions also have important roles in promoting R&D of safety technology, which is the benchmark for the judgment of the Government.

This chapter analyzes whether issues of countermeasures against tsunamis, the cause of the accident, and severe accidents were recognized as problems in each system, what prevented the issue from being solved, and what each system should do in future to take advantage of this lesson and play a role in ensuring safety at nuclear power stations. Figure 7.1 "Role Sharing to Ensure Safety among the Government, Industry, and Academic World" shows the sharing of safety roles among the Government, industries, and academia.

The Fukushima Daiichi Nuclear Accident: Final Report of the AESJ Investigation Committee, DOI 10.1007/978-4-431-55160-7_7

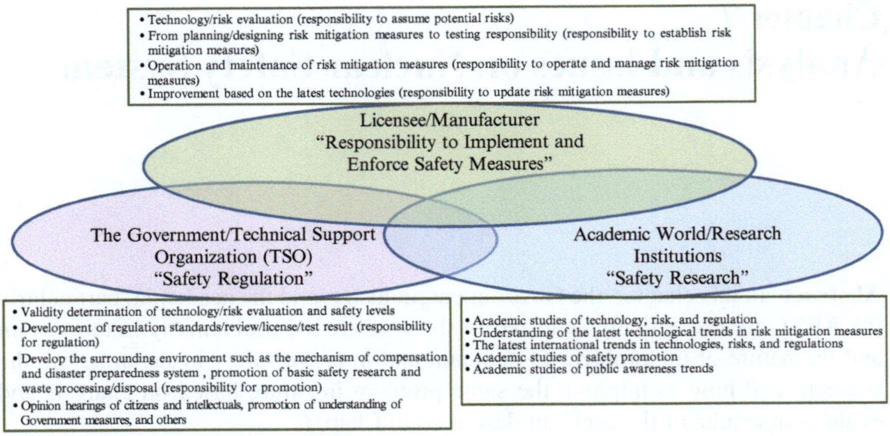

Fig. 7.1 Role sharing to ensure safety among the Government, industry, and academic world

The biggest lesson learned of the Chernobyl accident in 1986 was that if an accident should occur in a nuclear power station, as well as the country itself, its neighbor countries would also be affected. Accordingly, safety standards common to countries using nuclear technology were developed and complied with, and an international safety standard for the International Atomic Energy Agency (IAEA) was developed. The analysis in Chap. 6, however, reveals that Japan did not always fully comply with the IAEA safety standard. Where the problem lay in the international system and how it should be improved will be analyzed.

This chapter particularly emphasizes our own role, which involved carrying out a questionnaire survey of all AESJ members as experienced directors/auditors. The survey committee asked all members for their opinions on the analytical results prepared by the committee and then, based on many opinions and views, the committee analyzed past problems and future improvement measures.

7.1 Safety Regulatory System

The Japanese safety regulatory system has been greatly improved by drawing on from the lessons learned on the accident. This section analyzes the inherent problems in the previous safety regulation system, which contributed to the accident.

The safety regulation issues revealed in the accident can be classified into the following three categories: The first is the problem of specialization—namely the inability to provide proper governance in emergencies, while the second involves the accident response and insufficient preparation for all aspects of the legislative system, hardware, and management for such severe accidents. The third is that of organizational frameworks whereby while the safety regulatory administration was splintered into many administrative organs, the regulatory administration was not independent of the nuclear promotion administration.

(1) **Problem of competence**

The key lesson learned in the safety regulation was the problem of competence. As well as residents who were forced to evacuate, many citizens who were also afraid of their own well-being and other people worldwide closely focused on the daily information issued by the Government of Japan, which was unfortunately unable to meet such expectations.

The primary cause was the fact that the specific Japanese job rotation system was applied to safety regulatory organs and the regulators lacked sufficient expert knowledge. The Act for Establishment of the Nuclear Regulation Authority, which was enacted in June 2012 (hereinafter called the "AENRA"), specifies that the no-return rule should be applied to senior-level regulators of the Nuclear Regular Authority (NRA), who should be excluded from job rotation. In addition, it stipulates as present measures, the Japan Nuclear Energy Safe Organization (hereinafter the "JNES") should be integrated and measures considered to utilize highly specialized human resources. However, this is just a temporary strategy. In response to major issues occurring in the future, medium- and long-term drastic measures should be taken to enhance the competence.

(2) **Problem of unexpected accident response**

The immediate cause of the accident was the tsunami, which exceeded all expectations, but as the analysis in Chap. 6 shows in detail, measures taken to guard against the risk of such accidents exacerbated the impact of the accident, given their insufficiency in all aspects of the legislative system, facility design, and operation management. After the accident, the enactment of the AENRA improved the legislative system. The Act on the Regulation of Nuclear Source Material, Nuclear Fuel Material and Reactors (hereinafter called the "Reactor Regulation Act") was revised, and unexpected accident measures were legally defined as "serious accident measures" and make into regulatory requirements. In addition, based on the lessons learned from the accident, national guidelines and plans for emergency readiness were drastically revised.

The issues to be discussed in future in the national guidelines and plans include improvement and reinforcement of severe accident management (hereinafter "SAM") and clarification of shared roles between the Government and licensees, which were insufficient at the time of the accident and caused confusion. A major example was the authority over the decision to inject sea water or to vent the gas from the pressure vessel. In addition, the traffic line of evacuees and that of materials and equipment procured by the licensee for emergency measures went in opposite directions and neither past disaster prevention plans nor nuclear disaster countermeasure drills took this into consideration. In future, guidelines and plans for not only "disaster readiness" but also "SAM support" should be urgently developed.

As the Reactor Regulation Act incorporates severe accident measures, these will be reinforced by licensee, including filtered vent, and the Government will check them in the NRA safety regulation process.

The key severe accident measures are, as mentioned in Sect. 6.3, SAM for situations beyond the design assumption, since merely improving the legal system and hardware countermeasures are insufficient as accident countermeasures. This is because nobody clearly assumes what is required for situations beyond the design assumption. When accidents occur, management ability is required to promptly understand the situation, select the prepared hardware properly and strive to mitigate the consequences of the accident. To achieve this, as well as desk plans, repeated maneuvers and drills to improve skills are required.

(3) **Problem of organizational frameworks**

The most important lesson learned for organizational frameworks is the weakness in the independence of the regulation. The National Diet of Japan Fukushima Nuclear Accident Independent Investigation Commission pointed out "regulatory capture", which is likely to have been a factor behind the first and second problems described above. The AENRA, which aimed to improve this point, reinforced independence and consolidated regulatory authority. In other words, the Nuclear Regulation Authority, which was established in the Cabinet Office as a council, and the Nuclear and Industrial Safety Agency (NISA), which was established as an administrative organization in METI, would be integrated into the NRA as the Article 3 Committee with high independency in the Ministry of the Environment. All "the Three Ss" (Safety, Safeguard, and Security) regulatory functions, which were formerly distributed among MEXT, METI, and others, would be consolidated into the NRA.

Now, ten recommendations received from the IAEA in the "Integrated Regulatory Review Service (IRRS) to Japan" in 2007 are listed. The legal revision in 2012 satisfied R1 and R9. By assigning authority over personnel management and budget, an environment to improve the issue in R3 was created. Other items also seem close be being solved but it is desirable to continuously follow these recommendations and receive another Integrated Regulatory Review Service (IRRS) from the IAEA.

IRRS 10 Recommendations in 2007

R1. Recommendation: The role of NISA as the regulatory body and that of NSC, especially in producing safety guides, should be clarified.

R2. Recommendation: NISA should enhance its training requirements and programmes to ensure that all aspects of inspection requirements, such as attributes of quality management systems, and knowledge and awareness of licensees' operational requirements and practices are adequately included.

(continued)

(continued)

R3. Recommendation: NISA should produce a workforce plan that clearly identifies its minimum staffing needs to discharge the functions and tasks required to secure effective nuclear safety regulation in Japan against the elements of its 5-year strategic plan. Future staff number/budget requests would then be based on these minimum resource needs plus any supplement required for additional work/tasks. (The workforce of the regulatory system JNES/NISA and NSC should be ensured considering respective functions –mandates, completeness, fairness, neutrality, etc.—for this issue.)

R4. Recommendation: NISA should define its expectations more clearly concerning reporting of minor inspection findings and events, to screen them for early identification before they become problematic.

R5. Recommendation: NISA should leverage inspection and enforcement to ensure licensees have efficient processes for learning lessons from other domestic facilities and foreign facilities.

R6. Recommendation: NISA should continue to review and revise its regulatory requirements to provide assurance that licensees' operational safety programmes are comprehensive and address all safety-related elements in operation, including human and organizational factors.

R7. Recommendation: NISA should ensure that its inspectors have the authority to perform inspections at the site at any time and on an ongoing basis. This would ensure that inspectors have unfettered access to the site to interview people, and request the review of documents at any time rather than just prescribed inspection times as by law. This applies to both the construction and the operational inspection programmes.

R8. Recommendation: NISA should clarify the basis for authority to shut down a nuclear power plant in instances of poor performance, in addition to the existing clear law for shutting down due to hardware type problems.

R9. Recommendations: As the regulatory body in Japan, NISA should take major responsibility for developing and endorsing safety regulations and guides.

R10. Recommendation: NISA should continue to develop its comprehensive Quality Management System (QMS); focusing on practical implementation rather than the philosophical and conceptual rationale. As a first step the QMS should take account of the five-year strategic plan when formulating the Divisional Annual Plans.

7.1.1 *Analysis on Safety Regulations*

It was revealed that Japanese regulations had many problems while reflecting on lessons learned from the accident in the Fukushima Daiichi Nuclear Power Station. A new system of Japanese nuclear regulation started by establishing the Nuclear Regulation Authority (hereinafter "NRA") on September 19, 2012. It is expected that the NRA will improve regulations based on reflection on the lessons learned from accident. This section describes the expectations of improved regulations as well as analysis of what was inappropriate or insufficient. As the first, the problem of the insufficiency in "continuous improvement" about regulation before the accident is handled in Sect. 7.1.1.1. In particular, insufficient consideration of external events such as natural phenomena and the insufficiency in establishing regulations for unexpected accidents (severe accidents) are handled in Sects. 7.1.1.2 and 7.1.1.3.

In consideration of what future regulatory improvement should be, first of all, "it should be emphasized that the primary concepts to ensure safety, which have been evaluated highly, are important after all". The importance of "defence in depth" is re-recognized globally and already covered by this report. Here, particularly important ones among other principles, "scientific and rational regulation" problems and "Industry/academia/government cooperation and regulatory independence" are covered in Sects. 7.1.1.4 and 7.1.1.5.

7.1.1.1 Continuous Regulatory Improvement is Important

The Fukushima Daiichi Disaster was triggered by the tsunami following the earthquakes. The licensees have somehow been striving for "continuous improvement", reflecting new knowledge on tsunamis. Namely, they re-evaluated tsunamis based on new knowledge and proposed means of evaluation; gradually raised the design basis of the tsunami height, and considered various tsunami-resistant designs accordingly. As "primary responsibility belongs to the reactor licensee", such activities are matters of course and reviewing the results shows that the licensee's responses were not necessarily sufficient. As a result of these improvements, NPSs other than the Fukushima Daiichi managed to prevent the occurrence of severe accidents (SA).

For regulatory bodies, continuous improvement is crucial and as a rule, regulation must be enacted based on the latest knowledge. The framework of a periodic safety review (PSR) was developed. When new knowledge was acquired, particularly when the standard and guideline were revised, existing reactors also had to respond to the revision.

The guidelines themselves related to tsunamis, but were very simply briefed at the end of the earthquake-proof guidelines, simply referring to the Japan Electric Association JEAC4601 Technical code for seismic design of nuclear power plants and to the latest evaluation method by the Japan Society of Civil Engineers (JSCE) in JEAG4601. Consequently, the PSR did not reach the safety level required.

As for SA countermeasures, when the accident management (AM) was developed for all light water reactors in Japan as part of independent safety measures for the licensee in the 1990s, the common perception of most regulatory parties was that the next regulatory requirement would be SA countermeasures, but this was not practically implemented.

"Reflection of operating experience" was insufficient. The importance of reflecting operating experience; not only at Japanese facilities but also elsewhere, in regulations had long been recognized. Japanese regulators were aware of the inundation of the Madras reactor in India in the event of the Indian Ocean Tsunami, but just thought "This was an accident in India and has nothing to do with Japan". The above-mentioned September 11 attacks were also not reflected in Japanese regulations. Japanese regulators should have learned from these examples more humbly.

In future, improvement based on the lessons learned from the Fukushima Daiichi accident is crucial, which will require understanding "what actually happened during the accident", identifying what was wrong, and striving to reform according to understanding and identification. NRA has already reviewed the design standards for earthquakes, tsunamis, and SA and will back-fit existing reactors in accordance with this new safety standard in future.

Still, the Fukushima Daiichi accident does not justify blindly adopting such improvement measures. As mentioned in Sect. 7.1.1.4, reasonable improvement is required with the effectiveness of risk mitigation taken into consideration.

Basically, such problems should comply with the internationally common concepts. IAEA is drafting a back-fit standard (DS414) as follows:

> Measures for new-generation NPSs in severe accidents are now included in the plant design. However, it may not be realistic to apply all design requirements in this safety requirement document to existing NPSs in operation and those under construction. Moreover, it may not be feasible to revise the designs approved by the regulatory authority. As for such design safety analysis, it is expected to determine whether the safety operation of the station should be further improved as part of periodic safety reviews for the station, for example, by comparing it with the current standard and feasible measures to enhance safety.

7.1.1.2 Issue of Design Standards for External Events Including Natural Phenomena

Looking at the results, the Fukushima Daiichi accident shows the design basis is lower than the tsunami, which is a natural phenomenon. The current design and regulation did not always fully take specific measures for external events like natural phenomena other than earthquake ground motions, including the problem of the design basis hazard (DBH) setting, into consideration.

At the beginning of the "Regulatory Guide for Reviewing the Safety Design of Light Water Nuclear Power Reactor Facilities" (hereinafter the "Safety Design Guideline"), response requirements for various events are listed.

> In "Guideline 2. Design Consideration of Natural Phenomena", there is a need, in consideration of the importance of safety functions, for structures, systems, and components (SSC) with safety functions "to be designed to fully withstand such seismic design force as seem fit", "the safety of nuclear reactor facilities should be designed to be unhindered by predicted natural phenomena other than earthquakes", and SSC with crucial safety functions "should be designed in consideration of conditions considered to be the severest of predicted natural phenomena or cases where natural forces combine with the accident load". Accordingly, the interpretation is as follows: "Conditions considered the severest of predicted natural phenomena" are of equivalent or greater severity to past records of the target natural phenomena, with their reliability in mind and are considered statistically valid.

As a result of this "explanation", it seems that the design basis hazard was determined based solely on short-term records when the natural phenomena were rarely recorded in a fully reliable way previously and its strategic validity could not be confirmed.

As for tsunamis, licensees obtained the predicted tsunami heights using a method of "tsunami evaluation technology in NPSs" by the Japan Society of Civil Engineers (JSCE) developed in February 2002 and voluntarily reinforced the anti-tsunami design. The JSCE method is based on historical tsunamis from 1611 to 1978 (recorded tsunami height: previous tsunami). This is the method along with the above-mentioned "explanation". The tsunami height calculated in this method should be equal to the maximum height of historical tsunamis from 1611 to 1978. The result obtained shows we should assume the maximum height of tsunamis over the past four centuries. Though some tolerance was included, tsunamis occurring once a millennium (10^{-3}/year) or so were predicted and no countermeasures were taken for those beyond such level. Under these circumstances, there was little hope of meeting the following performance goals in Japan:

- Core damage frequency (CDF): 10^{-4}/year
- Containment failure frequency (CFF): 10^{-5}/year

It was also quite natural for the JSCE to develop an evaluation formula based on the historical tsunami, although the significance from a nuclear security perspective seems not to have been discussed. This shows a problematic lack of communication, but the nuclear community, which is responsible for controlling risks, should highlight this perception gap under normal circumstances. The community should feel deep remorse for being unable to do this while at the same time, certain measures should be taken to prevent any recurrence of such event.

The NRA is currently developing a more explicit standard of individual external events, based on which it is necessary to (i) "establish the design basis hazard" and (ii) "require a proper protection design for the design basis hazard (for example, tide embankment for tsunamis". When assuming the design basis hazard, there is a need to remember records of historical earthquake and tsunami have only been kept for a certain period. For the period without any record, the occurrence frequency should be evaluated based on experts' study and the corresponding hazard should be used as the design basis.

However, if the design basis hazard is reviewed in this way, the potential for beyond-design-basis hazards (DBH) remains. In fact, before the March 11 earthquake and tsunami, the related parties had no common understanding that an earthquake with such a big magnitude occurred off the northeast coast of Japan and that the tsunami associated with several seismic sources is exacerbated due to overlapping. (iii) It is necessary to "consider a beyond-design-basis event may happen and prepare for it".

Moreover, in the Regulatory Guide for Design, there is a problem of protection for the design basis accident, the third layer of the defence in depth. When assuming the design basis accident, the Regulatory Guide for Design affirmed "it is not necessary to consider ... long-time power loss", but in the Fukushima Daiichi station, a station blackout (SBO) occurred for an exceptionally extended period. Such improper description of the guideline was due to complete dependency on the data of past experiences.

Incidentally, as aforementioned, the IAEA Integrated Regulatory Review Service to Japan in 2007 recommended that "the Nuclear and Industrial Safety Agency" (hereinafter the "NISA") should, assuming responsibility for the regulatory body, prepare the safety review and assessment by themselves". Reviewing the Fukushima Daiichi accident, this recommendation is crucial.

In Japan, basic designs were reviewed according to the NSC safety review and assessment, and the NISA rarely prepared review standards by themselves with a few exceptions including the "Aircraft Fall Evaluation Standard". Therefore, the mechanism to constantly review the guideline system was not necessarily sufficient.

The safety review and assessment standard should be continually reviewed, and an organization responsible for this task is required. The NRA has already established an organization responsible for unitarily reviewing the standard, which is expected to constantly gage an overview of the reference system and continually strive to improve it.

7.1.1.3 Regulated Requirements for Severe Accident Measures

As mentioned above, it took time before regulations required severe accident measures. Japan developed comprehensive AM measures for all domestic light water reactors in the 1990s, which was a prompt response to global trends at the time. The regulation related parties commonly thought that in the next step, AM measures would constitute the regulatory requirement. However, it was not discussed whether the AM would really function effectively. It is essential to establish measures for severe accident regulatory requirements as well as confirm their effectiveness.

Incidentally, many "unpredicted events" occurred during the Fukushima Daiichi accident. As "predictable events should be properly predicted" and firm measures prepared, there is naturally a need to "change the basis if the predicted design basis is insufficient" and assume "what would happen beyond the design basis". However, unpredicted problems occurred: hydrogen explosions occurred in the reactor building, the hydrogen produced in the Unit 3 reactor flowed into the Unit 4 reactor building and triggered an explosion there. There is a need to "consider flexible measures, assuming unpredictable events may occur".

The U.S. has developed measures against unpredicted terrorism events since the September 11 attacks. Such trends were not reflected in Japanese regulations. Considering what can happen in terrorism and other external events, Japanese licensees should consider flexible response measures, including off-site support and Japan requires regulatory authorities capable of responding to the measures.

7.1.1.4 Scientific and Reasonable Regulations is Required as Expected

After the Fukushima Daiichi Disaster, the sole aim pursued was to enhance regulations. Originally, "the regulation had to be scientific and rational". This remains true, but to achieve this, a probabilistic risk assessment (PRA) is key factor. The so-called "graded approach", whereby resources are allocated according to the risk level, should be naturally applied. In addition, regulatory "requirements" are determined by the regulatory authorities, but the "standard performance requirements" whereby detailed provisions to achieve the same were used following a proper review of academic society's standard, remain unchanged in principle.

However, before the Fukushima Daiichi Disaster, the use of risk information was limited to the "Revision of Seismic Design Guidelines". This revision simply "reinforced the seismic resistance requirements because the results of the earthquake risk were large", and the risk information was sufficiently reflected.

In the Fukushima Daiichi Disaster, there were major differences between the impact of the earthquake and that of the tsunami. The earthquake was generally within the design basis range, although with some exceptions. Moreover, despite the fact actual safety-critical SSC at Fukushima Daiichi Units 1–4 could not be confirmed, such SSC at the Fukushima Daiichi Units 5 and 6, Fukushima Daini, Onagawa, and Tokai stations, were neither damaged nor failed. As explained in the above-mentioned Guideline 2 of the Regulatory Guide for Design "in response to the target natural phenomena and with the reliability of past records in mind, at least equivalent reliability". It was indeed problematic to have earthquake beyond the design basis, but as a result of the response on each occasion following full examination.

Meanwhile, the tsunami impact was enormous, because there was insufficient consideration of tsunamis.

Previously, PRA targets were merely so-called internal events (to be specific, random phenomena) and earthquakes. As for aircraft falls, only the hazard evaluation standard was developed to calculate the collision probability and determine whether or not protection was required. No other probabilistic evaluations were developed specifically as guidelines and/or standards. The standards for tsunamis, fire, and terrorism, etc. were missing or insufficient and no individual plant examination for external events (IPEEE) was conducted for each external event to identify weak points.

During both the regulation and safety research, related parties tend to focus on known aspects or peculiarities. To improve safety, however, it is important to find and tackle insensible problems. For example, many people considered "security" problems important but did not tackle them sufficiently.

In a regulation utilizing the risk information, there was a delay in the "application to the safety goal regulation". The safety goal essentially involves producing the "logical regulations". It can be said, however, that almost nothing was done to spearhead the PRA results and the safety goal to improve regulations, except to reinforce the above-mentioned earthquake-proof guidelines. What was actually done was a proposal to "compare the safety goal with PRA absolute values and mark symbols of circle (yes) and cross (no)", which is far from the PRA concept and the actual condition. This is a typical example of delay in introducing scientific and logical technology to the regulation.

Reconsideration of the safety goal index may also be required. The current safety goal index was established solely with the human health influence in mind. However, environmental pollution also exerted a significant influence. The "prevention of environmental pollution" should be added to the safety goal. Previously, the Nuclear Safety Commission reported this as follows:

> The event had various influences such as the direct personal and post-incident mortality risks due to the accident, the number of direct and aftereffect group fatalities, and economic damage caused by the accident. The safety goal only deals with "personal mortality risk" among these risks. Therefore, the safety goal is not set at all risks but for the most important and quantificational risk. In this sense, the safety goal on this occasion was the "first step of the beginning".

The safety goal remained as the first step of the beginning. In future, the NRA is expected to resume discussion of the safety goal and promote regulation using the risk information.

7.1.1.5 Industry/Academia/Government Cooperation and Regulatory Independence

The regulatory system was not fully independent before the accident, but improved with the establishment of the NRA. However, after the Fukushima Daiichi accident, acute awareness of independence drastically limited the scope for dialog between the NRA and industry. However, cooperation among industry, academia, and Government and regulatory independence must all be retained. The regulation must not be isolated and licensees should be responsible for ensuring safety. The regulation is responsible for monitoring the same, and licensees for its execution. The regulation must not inform safety without knowledge possessed by the site licensees. Therefore, the regulation must ensure transparency but the regulatory authority and industry must exchange opinions and information quite closely. NRA is expected to improve these issues properly in future.

The following shows what the NISA targeted to establish as a regulatory organ, quoting the first report summarized by the Nuclear and Industrial Safety Subcommittee, formerly the supreme advisory committee, in 2001, the year of its establishment:

(1) Trust of Citizens
(2) Scientific/logical regulations, effective/efficient regulations
(3) Crisis management skills
(4) Knowledge base
(5) Foundation of human resources

Actually, not all of these were achieved as expected. In particular, "trust from citizens, the first target listed by the NISA" was lost in the Fukushima Daiichi accident.

The regulatory targets themselves remain unchanged. In future, NRA and the regulation authority are expected to target the NISA goals, particularly to win the trust of citizens, and fully achieve them in future.

7.1.2 Conditions and Future Approach on the Regulatory System

This section discusses what the Japanese safety regulations should be based on the lessons learned from the accident. Originally, in Japan, very strict regulations were imposed on the safety regulation of nuclear power technology from the initial development stage under principles of independence, democracy, and openness to the public as stipulated in the Atomic Energy Basic Act. A representative example is the prohibition of private free use, whereby the Government reviews the technical ability and economic power, etc. of the individual and the licensee applying for use. Only when the Government judges there is no hindrance to the atomic use of the applicant is the use permitted. Moreover, compliance with actual standards, etc. as stipulated by the Government is regularly checked during actual use.

So, who is liable for ensuring safety? The first of ten safety principles provided by the IAEA is that "the prime responsibility for safety must rest with the licensee". However, as aforementioned, the Government is responsible for permitting the licensee to run the business and check its safety activities. If there is any flaw in the safety activities of the licensee, the Government is undeniably responsible for pointing it out in advance and requesting the licensee to correct it. Still, it would be too hasty to assume the Government should check every move of the licensee. The ideal nature of national regulations varies depending on countries: ranging from establishing detailed rules and checking every move of the licensee to focusing on risks that threaten nuclear safety and monitoring activities of the licensees from a risk perspective.

When discussing Japanese safety regulations, one important constraint to recognize in full is the limited nature of the resources to be input (funding and human resources, etc.). The roles of Government should be limited, and there must be consideration of an optimal mechanism to perform the mission as well as cost-effectiveness.

When considering what regulations should be, the following three points are particularly important:

(1) Analyzing the importance of roles the Government should fulfill, and allocating resources to important activities.
(2) Promoting voluntary continuous improvement activities among licensees.
(3) Presenting scientific reasons when conducting instructions. The Government is accountable for judgments to stakeholders.

7.1.2.1 Roles to be Fulfilled by Government

The roles to be fulfilled by Government can be effectively considered using the level of risk of impairment to nuclear safety as a yardstick. It is effective to consider the risk of the Government check mechanism. The potential for an accident to occur can be reduced if the input resource, which was input into the activity involving less significant risk, is allocated to that with greater risk. Vice versa, the risk will be higher and severe accident measures were the most important lesson learned from the accident. When the discussion finally started 20 years after foreign countries commenced such dialog to make severe accident measures a mandatory regulatory requirement, the accident occurred. Consequently, the twenty-year postponement was one of the causes. The report on severe accident measures issued in 1992 [1] was not discussed for 20 years. When considering safety measures, it is also essential to evaluate individual event risks to consider which roles the Government should take as well as examining safety measures to take. As essential indicators to evaluate safety, technology using the risk levels must also be urgently legislated.

Specifications of regulatory standard performance by the Government are also important. There is a problem in the specifications standard. However, the activities of licensees were formalized and the Government requires licensees for compliance, which hinders efforts to improve safety in the mid- to long term. Originally, it was important to continuously improve the manual so that activities could focus on important risks. With this in mind, the Japanese regulatory standard performance should be specified and a detailed manual should be improved at the discretion of the licensee. Continuous improvements should also be made to allocate resources, which were previously set for activities involving less significant risks, to those with greater risks.

(1) **Issues on regulations before the accident**
The regulation using risks as a yardstick is called Risk Informed Regulation (RIR); research into which has been progressing over the years. RIR has been studied for diverse aspects of design, construction, operation, and maintenance, and such research has been implemented to improve regulations using risks as a yardstick. The maintenance area is where RIR can be used effectively. Nuclear power stations are complicated entities, comprising hundreds of thousands of devices and systems, but only a small portion are important from a safety management perspective. Even if the resources are unchanged, safety can be considerably enhanced by noting the maintenance frequencies of parts at high risk, switching parts with low risk for subsequent maintenance, and other improvements.
In 2009, the maintenance program was introduced to the regulation based on the maintenance database. However, unfortunately, because RIR has not been introduced, the maintenance program failed to vary the pace of maintenance according to risk, and the method was applied to maintain parts at both high and low risk in the same way. As mentioned above, allocating resources to parts with low risk similarly is dangerous in that it hinders resource allocation to parts with high risk. It is safe to say that introducing the maintenance program without using RIR conversely increases the risk of facilities. In fact, regulation without using RIR cannot prevent accidents because it involves allocating resources uniformly without varying the pace of maintenance, regardless of risk levels, and a lower priority was given to intensive resource allocation to parts at high risk.
(2) **Improvements and regulatory problems after the accident**
Unfortunately, post-accident regulations still do not use RIR and require uniform compliance with regulatory standards regardless of the risk level. Moreover, it seems the emphasis on documents is stronger than before the accident, which prevents both regulators and licensees from focusing resource allocation on places which are most at risk. It is difficult to say that limited resources are effectively allocated to improve safety. The RIR should be introduced to immediately improve the situation.

7.1.2.2 Promotion of Continuous Improvement by Licensees

One important lesson learned from the accident is our deep awareness that there is no safety myth. As mentioned in Chap. 6, the accident risk is not zero, and we must prepare for a possible accident. It is insufficient to reflect this lesson by enhancing the accident response and although the accident risk is undeniable, the relevant parties must strive to continuously alleviate the risk. It is no exaggeration to say that the use of nuclear power is permitted only by continuing efforts to minimize risk. It is insufficient to merely comply with regulatory standards stipulated by the Government. The science and technology advances step by step. If the licensee just complies with the national regulatory standard and abandons efforts to improve, it can be said that safety will deteriorate on a relative basis. We should remember that the minimum requirement for licensees is to reduce accident risk on a continual basis through voluntary improvement.

In the U.S., ROP[1] was introduced as a mechanism urging licensees to implement voluntary improvements, followed by the introduction of PI/SDP[2] 30 years ago as an improvement indicator. This is a mechanism to evaluate licensees' activities from seven perspectives and objectively check them to see whether improvement has progressed. In this PI/SDP, the risk of the aforementioned accident is used as a yardstick. Since introducing this mechanism, in the U.S., licensees have activated voluntary improvement, safety has improved, and the operation rate has soared. Attention should be paid to the mechanism of activities of the local inspectors in the U.S. Here, local inspectors inspect licensees' activities from dual perspectives of risk and improvement. A database called CAP[3] is used in the inspection, in which licensees register daily activity issues and promote improvements based on the register. The local inspectors evaluate the extent to which risk has decreased by improvement activities of licensees, which is another yardstick. In the U.S., the regulations urged licensees to engage in such voluntary improvement activities and Japan should learn lessons from this.

(1) **Regulation before the accident**

The Japanese regulatory body was planning to study the aforementioned U.S. ROP mechanism, combine it with RIR, and regulate activities from a bird's-eye perspective [2]. This promotes improvement efforts by licensees and targets efforts to enhance nuclear safety. Japan planned to experimentally introduce PI/SDP, following the example of the U.S., and create a mechanism to encourage improvement. Japan, however, introduced a method quite different from that of the U.S., which merely urges licensees to strictly observe manuals, regardless of the risk level.

There was a movement to use PI/SDP as a risk-based indicator, but Japanese PI/SDP is ultimately unrelated to any risk and Japanese PI/SDP has not been implemented after the accident. It is strongly desirable to improve it into risk-based indicators and implement it.

In addition, the operational safety program audits whether activities are performed according to the operational safety program on a quarterly basis. If this audit is conducted on a risk basis, it will be effective, but actually, only documents are checked to check that the inspection is implemented in accordance with the provisions of the operational safety program. The hurdles are considerable to improve the operational safety program. In addition, regardless of the risk levels involved with the program, items linked with both high and low risks are handled in the same way when performing an audit, which means

[1] Reactor Oversight Program; Where regulatory technology reflects the quality of operation performance in terms of operational frequency, etc.

[2] PI: performance indicator; SDP: Significance Determination Process;

[3] Corrective Action Process

priority allocation to assign resources to items of higher risk is not available. In other words, the regulation has actually been adopted only to hinder voluntary improvement.

(2) **Regulation after the accident**

Utilizing the lesson learned, namely the importance of licensees promoting improvement activities, and some licensees promoting voluntary safety improvement activities before the Government determines new regulatory requirements. Conversely, some licensees still target efforts to meet the national regulatory standard and do not strive for further improvement. It can be said that the aforementioned trend before the accident has had a lasting effect in some way.

It is important that all licensees share an understanding that voluntary improvement by licensees is key to safety. To achieve this, improvements such as risk-based revisions of the operational safety program and reducing the frequency of lower-risk auditing items are effective.

7.1.2.3 To Present Scientific Reasons When Conducting Instructions

NPPs are very complex systems and improvements based on partial effects alone often fail to trigger system-wide improvements. Therefore, when improving the safety, we must always confirm that the system-wide risk has been lowered. This is done using the PRA method, and it is important to a means of risk-evaluation method including its operation.

Risk determination must always be based on scientific grounds and the circumstances and grounds for such determination should be explained with transparency to stakeholders. Moreover, explanations should also be given to the whole nation including local governments, licensees, and manufacturers.

In the U.S., the site inspectors are given free access rights and all on-site activities are evaluated through the CAP database, using risk as a yardstick. Moreover, risk mitigation measures, if regarded as necessary, will be taken after consultation between licensees and the parties concerned. Site inspectors have skills to use simple risk-evaluation tools and can engage in a risk-based discussion.

(1) **Regulation before the accident**

Japanese regulators, especially on-site inspectors, did not always have a professional capacity. In comparison with U.S. inspectors, who always form judgments based on risk evaluation, there is a strong tendency for Japanese inspectors not to emphasize risk evaluation but to check whether the activities comply with the articles of the operational safety program.

To make scientific judgments and conduct a comprehensive risk evaluation, a broad range of knowledge and experience is required. As the related education was not strongly stressed by the Japanese regulatory body, Japan was forced to rely on external experienced resources. It can be said that given the lack of risk tools and insufficient knowledge on risk, regulators' professional capacity was insufficient.

In addition, in Japan, some facilities did not fully develop the CAP database, while there was also a lack of tools to determine the status of facilities. Communication between licensees and regulators was also substandard, and regulators were unable to get sufficient information to evaluate risk, which was another factor behind the lack of professional capacity.

(2) **Regulation after the accident**

The professional capacity of site inspectors remains unchanged from before the accident, but a positive development is the fact that regulatory bodies have started focusing on education to strengthen capacity. In future, regulatory bodies should ensure penetration of the risk-based concept, correctly understand the importance of the operational safety program, and ensure trained inspectors who can provide proper directions as early as possible.

To enhance safety, the key is to know the site effectively. To do so, it is crucial for the licensees and regulator to engage in close face-to-face communication and dialog to achieve this goal.

There were judgments of regulators where the scientific basis was occasionally not clarified, and other examples where accountability as the basis for judgment was not completely fulfilled. Considering the fact that the key role of the regulatory body is to earn citizens' trust, this situation is cause for concern and seems attributable to the fact that the regulatory bodies remained in their infancy. It is important to eliminate judgments based on unclear scientific information and ensure accountability to citizens in future.

7.1.2.4 Summary

As mentioned above, Japanese regulations were one of the causes of the accident. Every time the licensees caused some trouble or acted wrongly, Japanese citizens sought regulatory enhancements, and regulators repeatedly strengthened examination tests for many low-risk items in response, which meant the regulators were unable to form a priority allocation to assign regulated resources to improve high-risk safety. Consequently, severe accident measures and other responses could not be taken in time.

As regulated resources were uniformly wasted on everything, regardless of the risk level, to prepare documents as alibis, regulations which should have promoted voluntary improvement among licensees instead hindered such efforts and elicited adverse results.

Of course, this did not justify licensees' failure to make voluntary improvements. They should not have delayed high-risk safety improvement measures such as anti-tsunami measures and should have worked on the same diligently.

Based on the aforementioned lessons, it will be essential in future to achieve risk-evaluation regulation based on scientific grounds, which does not hinder continuous improvement and which ensures accountability. Naturally, licensees must comply with Japanese regulatory standards. Without resting on their laurels, they must voluntarily promote continuous improvement and prioritize efforts to resolve issues with higher risk as well as fulfill accountability obligations.

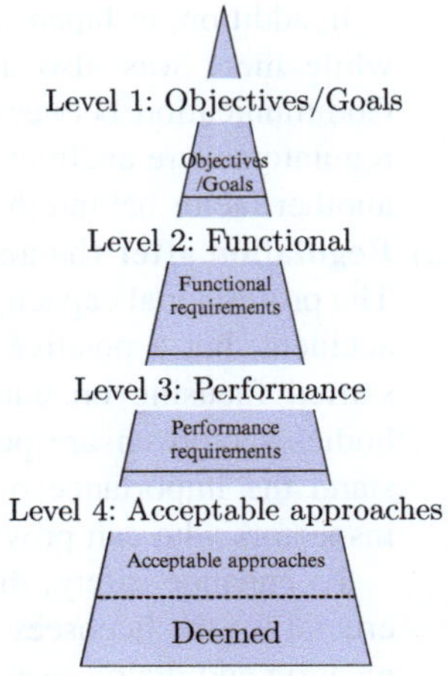

Fig. 7.2 Concept on safety regulatory framework

7.1.3 Regulatory Framework for Ensuring Nuclear Safety

7.1.3.1 Regulatory Framework on Nuclear Power Generation

Safety regulations on nuclear power generation, or rules to ensure safety prioritize the rules and guides defined by the IAEA. Conversely, the international standard primarily for manufacturing is the global ISO (the International Organization for Standardization) standards, which prevail over the American Society of Mechanical Engineers (ASME) and European Norm (EN) standards, as well as various Japanese subordinate standards. Recently, a consistent and systematized framework on these standards have been promoted, the relationship on which is shown in Fig. 7.2. Namely, key "objectives/goals" of various rules provided by international organizations and countries are set at the top of the hierarchy (Level 1), followed by a number of "functional requirements" set forth in meeting the objectives/goals (Level 2), and the quantitative criteria on the functional requirements (Level 3). Various measures to achieve this required performance level are determined in the form of detailed provisions (Level 4), under which various industrial standards are determined and implemented by commercial enterprises. The SSCs of nuclear facilities are manufactured and nuclear power stations are constructed based on these rules.

7.1.3.2 Life Cycle and Safety Regulations on Nuclear Power Stations

The life cycle of nuclear power stations comprises the siting, design, construction, operation, and decommissioning phase. How is safety regulation applied to this life cycle? The following shows how nuclear power plant safety was ensured under the Japanese safety regulations (regulation as applied at the time of the accident).

- During the siting phase, land for site construction is selected and environmental inspections is conducted in accordance with siting evaluation guideline and the Regulatory Guide for Reviewing the Safety Design of Light Water Nuclear Power Reactor Facilities.
- During the design phase, assessments are conducted for authorization such as safety assessment, nuclear facility installation assessment, construction plan, and fuel assembly design according to safety assessment review guide and Ordinance Nos. 62, 187, and 123, etc.
- During the construction phase, fuel assembly inspection, safety management inspection on welding, pre-use tests, and authorization of operational safety program according to the Regulatory Guide for Reviewing the Seismic Design of Nuclear Power Reactor Facilities, Ordinance No. 62, and the operational safety program

A nuclear power station starts operation after undergoing the above safety regulatory procedures.

- During the operation phase, regular inspections, periodic safety reviews, regular safety management inspections, regular licensee's inspections, and operational safety inspections are conducted. Recently, aging technical evaluation has also been included. The standards related to these reviews and inspections are established based on Rules for the Installation, Operation, etc., of Commercial Power Reactors (commercial reactor rules), Ordinance No. 62, safety assessment examination guide, Ordinance for the Enforcement of the Electricity Business Act, Fire Service Act, and the Act on Special Measures Concerning Nuclear Emergency Preparedness, etc.
- During the decommissioning phase, authorization of decommissioning plan, regular facility inspection, and confirmation of the end state of decommissioning are conducted according to Ordinance No. 77 and Act on the Regulation of Nuclear Source Material, Nuclear Fuel Material and Reactors, etc.

Specific and detailed specification standards need to be formulated in accordance with the regulatory standards shown above for actual work execution. Internal documents and technical documents by regulatory support organizations and affiliated organizations have been utilized as supporting information on the detailed specifications, as well as standards by professional societies such as the AESJ Standards Committee (SC), the Japan Society of Mechanical Engineers (JSME) Committee on Power Generation Facility Codes, and the Nuclear Standards Committee of the Japan Electric Association (JEA), etc., have been instrumental as the bases to these specifications.

Subsequently, specification requirements should be formulated utilizing standards developed by professional societies, and continuously updated with latest insights and knowledge. Although a policy on reinforcing functional performance requirements had been issued, related regulatory codes had not yet been organized. The regulatory safety framework shown in Fig. 7.2 is still in the development process, not fully effective with some parts not entirely defined.

7.1.3.3 Issues in the Japanese Safety Regulatory Framework

Following are the issues on the Fukushima Daiichi accident in light of the above circumstances.

The regulatory body and licensees had so far applied regulatory standards by making adjustments and ensuring coherency, which underline the key issues on Japan's regulatory safety framework – its complexity and the numerous ambiguities in the regulatory standards.

(1) In principle, the framework on laws and ministerial ordinances provides the basis to establishing regulatory standards.
Level 2, or the regulatory safety framework designates the minimum required functions, while Level 3 determines the quantitative performance requirements, based on which Level 4 should deal with the acceptable method of using standards established by professional organizations as specification standards (functional enhancement). However, in reality, quantitative performance requirements were not shown clearly and left to individual discretion. In some cases, relevant functional requirements determined based on performance goals have been formulated by commercial codes and standards.
(2) Criteria on safety requirements for ensuring "nuclear safety" is supposed to be shown clearly in the guidelines established by the Nuclear Safety Commission. However, it is assumed that because the safety requirements were not linked to laws, ordinances and public notices related to "nuclear safety", the judgmental basis on ensuring "nuclear safety" remained ambiguous.
Development of a safety framework founded on IAEA's global standards was not making progress. Regulatory framework on "nuclear safety" based on guidelines by the Nuclear Safety Commission was also not organized and needed to be addressed.
Under these circumstances, the Fukushima Daiichi accident occurred.

7.1.3.4 Effective Nuclear Safety Regulatory Framework

The nuclear safety regulatory framework based on guidelines established by the Nuclear Safety Commission has been outlined so far, which has remained static to date.

However, it is now commonly recognized in Japan that safety starts by presenting the aim and the approach on ensuring "nuclear safety". The Japanese version of the fundamental concept on the nuclear safety in reference to the IAEA's fundamental safety concept has been developed and established by the Standards Committee, Atomic Energy Society of Japan (AESJ) in the Technical Report, "Fundamental Concept on Nuclear Safety—Nuclear Safety Objectives and Fundamental Safety Principles" (AESJ-SC-TR005). On the basis of this fundamental safety principles, a legal framework on nuclear safety should be developed. The former "Electricity Business Act" comprised of "rules on ensuring power (electricity) supply", "rules on general safety" and "rules on safety other than nuclear safety". Under the newly established legal framework, "nuclear safety" is presented clearly in "The Act on the Regulation of Nuclear Source Material, Nuclear Fuel Material and Reactors" and "rules on ensuring nuclear safety" and "rules on the handling of nuclear material" have been separated from the "Electricity Business Act". A legal framework founded on functional and performance requirements formulated as government ordinances on the basis of national "Guidelines" should be established for the achievement of "nuclear safety". Under this framework, functional and performance requirements, as well as standards by professional societies may be formulated on the basis of specific safety "Guidelines". All regulatory standards and standards developed by professional societies may be integrated into a framework for ensuring "nuclear safety" in such a manner.

Figure 7.2 explains the co-relationship between each level defined in reference to safety regulatory framework. Levels 1 and 2 define the overall scope of safety, whereas Level 3 defines the minimum level of safety warranted with considerations given to uncertainties. The minimum level is determined on the basis of consensus reached on, following considerable discussions and verifications. The verification includes for example, demonstration and qualification tests which enable confirmation on performance reliability of the total system, and do not require visual verification of individual components. Standards by professional societies are intended to ensure safety based on facility maintenance criteria with considerable margin on safety limit. Consequently, in addition to the required safety margin, products are designed and manufactured with a further margin over variances that may occur. Hence, from safety limit perspectives, significant margin is provided in the criteria on the design and manufacturing of products.

As described, an adequate and clearly defined framework on ensuring "nuclear safety" must be established.

7.1.3.5 Role of the Regulatory Body

Significance of nuclear safety relates to society. Nuclear safety is defined under the societal context because there is no end point in ensuring safety. The level of safety will depend on the consensus established in society on what level of safety is acceptable. The meaning and significance of nuclear safety objectives and fundamental safety principles is closely tied to the values and consensus on safety established in society. To date, severe accident management had been left to the voluntary discretion of the operators, and defence in depth-based approach in severe accident management and emergency preparedness and response measures had not been established in nuclear plants in Japan. The Fukushima Daiichi Plant accident highlighted the issues that should have been addressed, which has been pointed out strongly by various organizations.

What were the factors that induced the severe consequences? First, it was the insufficient understanding on the system, organization, framework, and the interactions between these elements necessary for fulfilling responsibility for safety;

secondly, lack of understanding and resolve in utilizing nuclear power and the risks involved; and thirdly, lack of focus in the application of the defence in depth concept, by all individuals and organizations involved in nuclear power generation. On the basis of the lessons learned, the organizations in public and private sectors in Japan must jointly strive to achieve the highest standards of safety in nuclear power generation. To this end, nuclear safety codes and standards for the enhancement of nuclear power generating facilities should be established.

Was there a shared perception of "responsibility"? Responsibility, including obligations associated with assigned roles of all parties involved should be given consideration. Nuclear power generation has been promoted as part of the state policy in securing energy sources—ensuring nuclear safety is not only the responsibility of the operators. The local site is at the forefront of ensuring safety, and has the prime responsibility for ensuring safety. The operators must not only adhere to the rules, but are expected to make the best achievable efforts in ensuring safety. Whilst in technical aspects, the operators have developed design and operating procedures on nuclear facilities, have conducted safety reviews, and carried out construction and operation of nuclear facilities in compliance with the regulatory rules. The regulatory body is responsible in this respect, which should also be taken into account.

Responsibility for the accident does not rest solely on the operators. It is essential for all stakeholders in the nuclear power generation community, including central and local governments, academia, utilities, manufacturers, etc., to recognize responsibilities for ensuring safety in the event of emergencies commensurate with the assigned roles. In addition, the involvement of the mass media and the public in the process perhaps should be given some thought.

On the basis of the lessons learned, dialogue should be established between the regulatory body, operators, supporting organizations (manufacturers, etc.), the academia, and the general public. Subsequently, various regulatory standards and rules that determine activities involving nuclear power generation should be reviewed for re-establishing an adequate operational framework. For this purpose, interactions and role sharing between all related parties is essential for ensuring a solid framework on nuclear safety.

Finally, as shown by the report, "Nuclear Safety Objectives and Fundamental Safety Principles" established by the Atomic Energy Society of Japan (AESJ), it is essential to utilize the breadth of accumulated expertise of professional societies.

7.2 Nuclear Safety in the Industrial Community

7.2.1 The Role of Licensees

Principle 1: Responsibility for safety pursuant to the 10 principles shown in the IAEA Basic Safety Principles (SF-1) says "The prime responsibility for safety must rest with the person or organization responsible for facilities and activities that give rise to radiation risks". Needless to say, licensees play the key role in nuclear safety.

7.2.2 *Licensees' Response to Nuclear Accidents*

7.2.2.1 Overseas Accidents

In December 1979, after the TMI accident, U.S. power companies targeted the "highest level of safety and reliability" lest a similar accident should recur and established the Institute of Nuclear Power Operations (INPO). Subsequently, U.S. nuclear power generation significantly improved both the quality and operational level in the 1990s and 2000s, which heralded the bright prospects for what was a previously stagnant industry. Public support for the nuclear power generation also soared remarkably. A good example of significant growth is the voluntary safety improvement activities in the industry.

Following the Chernobyl accident, the need to share safety-related knowledge; not only among electric power companies in a single country but also among those worldwide was strongly recognized. In May 1989, the World Association of Nuclear Operators (WANO) was established and around 130 or more electric power companies in 35 countries joined.

In response to such overseas trend, Japanese industry acknowledged the importance of voluntarily striving to improve safety. The Japan Nuclear Technology Institute (JANTI) was established; recruiting 110 companies in the nuclear generation industry as members by March 15, 2005, and strove to improve safety by targeting the following five goals: (1) collection, systematization, and effective use of technical information, (2) traction/business management functions, (3) development of human resources and organizational culture, (4) support upon request from members, and (5) cooperation with relevant organizations. Major JANTI activities included peer reviews of members and sharing of technical information.

7.2.2.2 Establishment of the Institute of Nuclear Power Operations (INPO) in the U.S.

On March 28, 1979, a severe accident occurred at the Three-Mile-Island nuclear power plant (TMI). In the general report by the President's Commission that investigated the accident (the Kemeny Commission) said "we found problems with the people who operate the plant, with the management that runs the key organization, and with the agency that is charged with assuring the safety of nuclear power plants. In the testimony we received, one word recurred over and over again. That word was "mindset". It also pointed out "The industry should establish a program that specifies appropriate safety standards, including those for management, quality assurance, and operating procedures and practices, and that conducts independent evaluations". In response, the U.S. industry established the Institute of Nuclear Power Operations (hereinafter the "INPO") as a voluntary regulatory organ in 1979.

INPO's vision is that "setting the global standard in nuclear safety, pursuing excellence of ourselves and expecting it of others" and started activities with the mission "to promote unrivaled safety and reliability—and promote excellence—while operating commercial nuclear power stations". The initial issue was to "win trust, gain competent and capable staff members, take correct actions, and establish an excellent culture". Initially, it was difficult to achieve the expected purpose. To date, however, the INPO has achieved good results and remarkably improved the operation rate of nuclear power stations, although it took more than a decade for the INPO to achieve strong leadership. The key to success involves building and maintaining a mechanism to lead electric power companies, fulfill their accountability obligations, and ensure independency. In addition, the top management of electric power companies spearheaded such efforts.

INPO now has significant influence over licensees and organizations relevant to nuclear power; not only in the U.S. but also worldwide. More than 20 countries participated in the INPO international programs and the INPO has influence over 75 % or more of nuclear power stations worldwide. The nuclear industry has a peculiar industrial structure, whereby even a single accident at a nuclear power station influences nuclear power generation. Under these circumstances, the INPO exercises strong leadership and spearheads activities to improve nuclear safety.

7.2.3 *Lessons Learned from the Fukushima Daiichi Accident*

First, organizational factors are extracted from the factors which triggered the severe accident on the Fukushima Daiichi NPS based on reports published by TEPCO. Next, based on reference materials, organizational factors in the industry are extracted, how the nuclear power industry addresses these issues is analyzed, whereupon subsequent issues are compiled.

7.2.3.1 Extraction of Organizational Factors of TEPCO in Triggering the Fukushima Daiichi Accident

TEPCO established the "Nuclear Reform Special Task Force" in September 2012. The task force analyzed the technical causes, organizational factors and background to the accident, and prepared a "Report on the findings of the Nuclear Reform Monitoring Committee to the Board of TEPCO on the Reassessment of the Fukushima Nuclear Accident and Nuclear Safety Reform Plan"; supervised by the Nuclear Reform Monitoring Committee (Released on March 29, 2013).

The following summarizes reflections on inadequate preparations for a severe accident, inadequate tsunami countermeasures, lack of preparation for an accident response, and publicity coping in the event of accident, etc. from the perspective of safety awareness, engineering capabilities, and communication ability.

(1) **Inadequacies in severe accident measures**

From 1994 to 2002, TEPCO developed measures involving the containment vessel vent system, an interchange of emergency diesel generator among units, etc. (hereinafter "AM (accident management) measures"). However, deeming existing measures to be sufficient to fully secure safety and evaluate core damage risks during periodic safety reviews (PSR), TEPCO also believed their measures compared favorably with those overseas and focused on accumulating daily activities rather than new AM measures. There was a fact that while some experts expected the impact of external events to exceed that of internal events, specific measures were not taken and this point was regretted by TEPCO.

Conversely, elsewhere, AM measures were steadily improved and reinforced based on external events, i.e. flood like the flooding of Blayais NPS in France in 1999 and terrorism like the September 11 attacks in 2001. Operational experience in NPS shows that if AM measures had been taken with long-term SBO and the loss of ultimate heat sink, the accident might have been immediately and properly mitigated, regardless of the initiating events. We should deeply reflect on why we failed to learn from successful cases overseas.

Analyzing the root causes why our investigation of countermeasures against the severe accident was slower than other countries, we classified the issues into three categories: (a) safety awareness, (b) engineering capabilities, and (c) communication ability.

(a) Safety awareness: TEPCO lacked a common understanding that continuous safety improvement was an important business challenge and was overconfident that current safety measures were sufficient.
(b) Engineering capabilities: TEPCO failed to share understanding that the severe accident risk caused by external events could not be ignored and lacked sufficient engineering capabilities to determine useful measures by harnessing overseas information, etc.
(c) Communication ability: TEPCO lacked communication ability, believing that if it acknowledged the severe accident itself, it would be difficult to explain the current safety.

(2) **Factor of inadequate tsunami countermeasures**

TEPCO lacked sufficient humility to carefully respond to the risk of natural disasters, thought it would be adequate if the legal requirements, standards, and criteria were satisfied, and lacked the ability to carefully study the tsunami risk by themselves. Moreover, in general, to secure safety according to the principle of prevention, TEPCO was reluctant to acquire new knowledge and prepare a conservative design. Analyzing the root causes why tsunami countermeasures were lacking based on these facts, we could classify issues into three: (a) safety awareness, (b) engineering capabilities, and (c) communication ability.

(a) Safety awareness: TEPCO did not make sufficient efforts to take countermeasures after recognizing numerous uncertainties and failed to fully see the need to take measures against the third and fourth layers of defence in depth; regardless of the possibility of occurrence.

(b) Engineering capabilities: TEPCO lacked sufficient efforts to conduct additional investigations, judge by itself and leverage flexible engineering capabilities for feasible and cost-effective measures in a short time.

(c) Communication ability: TEPCO lacked sufficient communication ability to communicate with the regulatory authority and the local society.

(3) **Lack of preparation for accident response**

Analyzing the root cause why TEPCO failed to fully prepare for the tsunami countermeasures based on these facts, we could classify the issues into three categories: (a) safety awareness, (b) engineering capabilities, and (c) communication ability.

(a) Safety awareness: Confident that no severe accident would occur, TEPCO had insufficient and formal training plans and failed to fully prepare the necessary materials and equipment.

(b) Engineering capabilities: the necessary works in emergencies were determined, but since the implementation guidelines were not developed, TEPCO could not respond promptly. As TEPCO did not prepare responses when the SBO prevented staff members from obtaining information on the state of the plant, they could not estimate it. Given the shortcomings of the mechanisms and training in information sharing during emergencies, the parties concerned could not smoothly share information among themselves, or arrange information for off-site inquiries and instructions, and the chain of command on site was also unclear.

(c) Communication ability: TEPCO lacked sufficient communication ability to report on the progress status of the accident promptly and accurately to the relevant organs and local governments.

(4) **Organizational issues**

TEPCO analyzed the primary causes of the accident, identified problems from perspectives of insufficient "safety awareness," "engineering capabilities," and "communication skill," and concluded that the structural problem, which promoted these issues, is the "negative chain," which took route in the nuclear power section. In addition, it was not only the negative chain of problems in the nuclear power section that triggered the accident. TEPCO revealed the inability on the part of their senior management to manage risk properly.

7.2.3.2 Responses of the Utilities

(1) **Experiences of NPS incidents**

Since Unit 1 at the Tsuruga Nuclear Power Station, which started commissioning in 1970, Japanese nuclear power generation has accumulated over 40 years of operating experience. Previous examples are listed in descending order of the International Nuclear Event Scale (INES) introduced in 1992, the JCO

Criticality Accident (Level 4, 1999), the fire explosion in the former PNC asphalt solidification facilities (Level 3, 1997), and SGTR accident with the Kansai Electric Power Company, the Kansai Electric Power Company Mihama Unit 2 (Level 2, 1991).

Unlike major accidents in other countries, such as the Chernobyl NPS accident (Level 7, 1986), the Windscale NPS accident (Level 5, 1957), and the Three-Mile-Island NPS accident (Level 5, 1979), the impact of Japanese accidents has been quite limited. The only major NPS accident was the Mihama Unit 2 SGTR accident. Levels 1 and 0 of the NPS accidents or those not covered by INES were regarded as quality problems affecting high reliability and the accidents were not perceived as opportunities to review the definition of design-basis accidents (DBA). Nuclear power generation licensees devoted themselves to strictly observe detailed routines.

Following the Chernobyl accident, research into severe accidents and the probability safety evaluation was accelerated, based on which the above-mentioned WANO was established to improve performance by collectively conducting peer reviews and benchmarks, trying to improve performance through mutual support and information exchange, and learning the best practice for optimal operational safety and reliability. Japanese nuclear power generation licensees joined WANO and participated in the peer review, operating experience, technical assistance, technical exchange programs and activities.

Analyzing the relationship with the Government, we note that even nation-wide, advance preparation was insufficient.

In 1992, the then MITI determined AM measures were not regulatory but voluntary measures of nuclear power generation licensees and requested that licensees develop AM measures. They were determined as voluntary measures because (a) Strict safety regulations ensure the safety of Japanese NPSs and severe accidents are very unlikely to occur for engineering reasons, (b) the measures to date sufficiently reduced the likelihood of accidents, and AM measures further reduced the risk, (c) AM measures are "knowledge-based", depending on technical knowledge of nuclear power generation licensees. It is desirable for nuclear power generation licensees to use such knowledge flexibly according to circumstances [3], This means we used to ignore overseas accidents as unrelated to us and as proof that the nuclear power parties concerned had the "mindset" that no severe accidents would occur in Japan.

This raises the potential that the fire at the Kashiwazaki Kariwa NPS caused by the Niigata Prefecture Chuetsu-Oki earthquake shows the complexity of the natural disaster and nuclear accident and the potential for a severe accident exceeding the design standard. Once it happens, a natural disaster may trigger a bigger accident, but if safety has been prioritized, preparation for natural disasters should also have been enhanced. The lesson learned from the Niigata Prefecture Chuetsu-Oki earthquake was only reflected in the reinforcement of seismic design, and no tsunami countermeasures were reinforced.

(2) **Reflection and response of the entire licensees**
Power companies established the Fukushima Support Headquarter of the Federation of Electric Power Companies of Japan (FEPC) and developed a system to efficiently support TEPCO. All companies dispatched support staff members to Fukushima prefecture for environment monitoring, decontamination guidance, and provided radiation measuring instruments and other materials and equipment. The number of staff members dispatched from other electric power companies in the 10 months since the accident reached about 60,000.

The direct causes of the Fukushima Daiichi accident included the loss of all power, the loss of the ultimate heat sink, and the inundation of important equipment. Therefore, after the Fukushima Daiichi accident, NPSs all over Japan reinforced "redundancy" and "diversity" as measures to eliminate these direct causes. Emergency safety measures included the deployment of mobile power supply vehicles as backup measures for emergency power supplies, deployment of conveyable pumps and hoses in preparation for circumstances where seawater pumps were inundated and reinforcement of building inundation protection measures to prevent water from coming above tide embankments and damaging facilities. In addition, as diversification measures for emergency power generating devices, external power supply measures, measures for electric facilities within the power plant, and measures for cooling and water injection equipment, including air-cooled emergency power generating devices, which do not need cooling water, were deployed, spare parts of seawater pumps motor were deployed so that the motor could be immediately restored even if damaged by water, raise the tide embankment, and establishment of key seismic buildings.

In addition, the nuclear power industry reflected on the fact that the voluntary safety improvement activities implemented by JANTI before the accident were insufficient. The earthquake and big tsunami are natural phenomena, which can have a significant impact, even if the probability is minute. For such phenomena, insufficient efforts were made from the perspective of responding to the situation beyond assumption and the need to investigate, consider, and then reflect on overseas safety improvement activities. Moreover, while the past stable operation results and experience of misconduct ensured compliance with the rules, TEPCO should have striven to pursue activities to enhance safety without settling for the current situation. It was pointed out that JANTI, which was established to support safety improvement activities of electric power companies, was not fully utilized and issues concerning the mechanism for organization utilization were identified. With these reflections, the nuclear power industry positively dissolved JANTI in November 2012 and newly established the Japan Nuclear Safety Institute (JANSI) using the U.S. INPO as a model for the organization design.

In addition, if an accident should occur, the "Nuclear Emergency Support Organization" will be established within 2015 to make diversified and advanced disaster responses available, and before that, the Nuclear Emergency Support Center, whose execution entity is the Japan Atomic Power Company was established in January 2013. Following a request for mobilization from the damaged licensee, it conducted reconnaissance of the site subject to a high dose, measured the air dose rate, removed rubble, etc., and transported materials and equipment, including robots, to minimize the radiation exposure to licensees and in the surrounding area as a means of supporting the activities intended to cope with emergencies.

7.2.4 Future Issues of the Nuclear Power Industry

7.2.4.1 Future Issues

The circumstances and analysis revealed the following issues which the nuclear power industry, including electric power companies, must face:

(1) From the accident, the lesson was learned: once the accident occurs, nuclear safety problems not only impact on the station but also worldwide. This should be accepted as an industry-wide problem for the nuclear field as well as one of TEPCO.
(2) The nuclear power industry must recognize lessons learned from the accident and face up to the organizational issues extracted from the perspective of safety awareness, engineering capabilities, and communication ability as common issues facing the nuclear power industry.
(3) A warning should be issued in response to the mindset: "the accident will not occur because the measures against recurrence of past accidents are conducted". The licensee should always explore any accident factor and try to penetrate the safety culture, which means retaining a policy to continuously enhance safety, from senior management to the very bottom of the organization.
(4) In the peaceful use of nuclear energy, once a severe accident occurs, which is rare, it has an immense socioeconomic impact. With this in mind, the entire nuclear power industry should continuously strive to enhance safety. The transient effort is not desirable.

7.2.4.2 Addressing the Issue: Establishment of JANSI and the Role of the Nuclear Power Industry

Based on reflection of TEPCO in response to the Fukushima Daiichi accident, the Japanese nuclear power industry established JANSI to reinforce voluntary safety improvement activities more strongly; adamant that no severe accident should recur.

Securing the transverse network, the Japanese nuclear power industry must cooperate closely with relevant bodies in other countries. This enables the nuclear safety to be steadily secured by (1) unitarily collecting information from overseas and safety improvement measures, which used to be individually collected, (2) collecting the latest knowledge domestically and overseas, (3) making prompt proposals and recommendations to electric power companies and the industry, and (4) supporting safety improvement activities, focusing on severe accident countermeasures. JANSI is responsible for this realization and release efforts, and asked society to evaluate the results.

Presidents of electric power companies reflected on the failure of JANTI to fulfill its goals and expressed strong commitment. They announced they would treat the evaluations, proposals, and recommendations from JANSI seriously and ensure efforts to enhance safety with strong resolution and determination and clarified their position to actively enhance safety improvement. This should be neither superficial nor transient.

The nuclear power industry must ensure technical independence, which is not influenced by nuclear power generation licensees, utilize commitments of senior management in power companies, clearly demonstrate the execution of nuclear safety measures for citizens and commit to their realization.

7.3 R&D and Safety Research System

It is pointed out that the causes of the severe accident at the Fukushima Daiichi Nuclear Power Station were insufficient tsunami and severe-accident countermeasures. While tsunamis and severe accidents have been studied in Japan, here, we consider whether or not the research results have been reflected in measures, and further, the causes of the problems, and how to overcome them.

(1) **Whether research results are reflected in measures**
Tsunami research developed and dealt with the Jogan earthquake and the tsunami earthquake along the ocean trench offshore Fukushima as the wave sources [4]. In particular, though few documents remained on the Jogan earthquake, the tsunami deposit investigation started to reveal the tsunami-inundated area. In addition, as for the tsunami wave height evaluation on the coast, a technique to calculate the tsunami propagation from the tsunami wave source by solving a two-dimensional shallow water equation and thus determining the tsunami wave height was almost established [5, 6]. TEPCO calculated the tsunami propagation with the wave source of the Jogan earthquake and the earthquake along the ocean trench offshore Fukushima, and obtained the calculation results, which are equivalent to the tsunami wave height on March 11, 2011, in advance. However, TEPCO did not voluntarily take tsunami countermeasures based on these results. Moreover, neither regulatory authorities nor researchers aggressively persuaded TEPCO to do so.

Japan has a long history of severe accident research. The former Japan Atomic Energy Research Institute (JAERI), the former Nuclear Power Engineering Corporation (NUPEC), and universities, etc. have actively researched the area, example results of which are summarized in [7]. This report is highly regarded for almost covering the entire scope of current issues of thermal hydraulics of reactors with severe accident countermeasures and the passive functions after listing the U.S.WASH-1400 (so-called Rasmussen Reports), TMI-2 accident, and the Chernobyl accident in the former Soviet Union in "1. Introduction" in the document [7], describing how severe accident countermeasures have actively progressed worldwide, the Nuclear Safety Commission and the MITI take the lead and request that electrical power suppliers develop the accident management in Japan.

The progress of severe accidents at the Fukushima Daiichi plant can be explained based on past researches after the accident. In other words, water could not be supplied due to the power source loss, cooling water in the pressure vessel gradually declined, whereupon the core was exposed and melted. Subsequently, molten materials fell to the bottom of the pressure vessel. Once the bottom was damaged, the molten materials reached the floor of the containment vessel, causing molten core-concrete reaction. Conversely, no large-scale steam explosion or direct heating of the containment vessel, which could have triggered a significant release in the early stage, occurred. Past researches did not deal with large-scale hydrogen explosions in the reactor buildings due to the leakage from the containment vessels.

As total DC and AC power loss due to the tsunami was not expected, measures prepared beforehand had little effect. TEPCO had to depend on on-site contingency actions. For example, removing vehicle batteries and connecting them for the opening operation of the safety relief valves (SRV), alternative water injection by fire engines to the inlet nozzle of the fire protection system, using seawater as a water source, etc. Further, while advanced reactors equipped with water injection and heat removal systems without power were developed and constructed worldwide, there was no movement to construct such reactors in Japan; despite the fact that the current status is summarized in the document [7].

The probabilistic risk assessment (PRA) is useful for considering severe accident measures. As for PRA, while methods for internal events were developed and standardized, the Standard [8] for the earthquake as one of the external events was published, but the Standard [9] related to the tsunami was only issued after the Great East Japan Earthquake, and the development of methods for other external events were delayed. However, in Japan, it was recognized that the risks of core damage in external events exceeded those in internal events [4]. Though the U.S. adopted severe accident measures based on the PRA involving external events, Japan did not implement an accident management strategy for external events [10].

(2) **Problems and issues**

Although knowledge of the Jogan earthquake and the earthquake along the ocean trench offshore Fukushima was gained, specific measures were not taken, for the following reasons; (i) an earthquake bringing a high tsunami to the Fukushima Daiichi Nuclear Power Station was being investigated, (ii) the licensee did not take measures at its disposal based on the latest research results, and (iii) the regulatory authorities did not request the licensee to take measures based on the latest research results.

Opinions were divided among researchers on the earthquake along the ocean trench offshore Fukushima, as of March 11, 2011. Some researchers thought such earthquake would not occur as there was no previous record, while others thought it might occur because of the mechanism of tsunami earthquakes. Conversely, the research results of tsunami deposits in the Sendai plain seemed to reveal almost all details of the Jogan earthquake. Though opinions were divided among researchers and nobody knew who was correct, the latest research results should have been reflected in tsunami countermeasures from the safety viewpoint for nuclear power stations. Learning a general lesson from the accident, the licensees and regulatory authorities should have reflected incomplete research results in measures from the safety viewpoint for nuclear power stations.

Before March 11, 2011, a trial analysis of the probabilistic risk assessment (PRA) of tsunami was conducted; the results of which said a tsunami exceeding the expected tsunami wave height would cause an extremely serious situation. In addition, these results could be easily anticipated by PRA experts. Therefore, considering the PRA as well as the tsunami research, tsunami countermeasures should have been conducted, even though the academic society had differing perspectives on tsunami research knowledge.

The primary responsibility for safety measures should go to licensees, followed by the supervisory regulatory authorities. However, researchers involved in nuclear safety would have recognized the need for tsunami countermeasures if they had known both the latest knowledge of tsunami and risk assessment results. It was the harmful effect of the specialization and subdivision of research that prevented it. In addition, the top managers of licensees and research organizations, who are in a position to organize separated specialists, should have recognized the need for tsunami countermeasures. We believe the evil of specialization and subdivision can be overcome by active exchanges among specialists in different fields and the attitude of top managers to consider nuclear safety comprehensively.

With regard to the severe accident, there was a delay, not in research but in reflecting research results in actual severe accident countermeasures. The first reason for this was an inexplicable pride, which meant even researchers believed no severe accident would occur in Japan. The Japanese could not accept in reality that a severe accident like the Three Mile Island Nuclear Power accident in the U.S. and the Chernobyl nuclear power plant accident in the former Soviet Union might occur in Japan. For example, although advanced reactors with static safety systems that could be operated without any power supply were developed in China and Korea, as well as Europe and the U.S., Japan did not go out of its way to adopt the same.

The second reason was the harmful effect of the vertical sectionalism of research organizations. The former Japan Atomic Energy Research Institute

(JAERI) (the current Japan Atomic Energy Agency) was the main body for the severe accident research. The competent authority was the Ministry of Education, Culture, Sports, and Science and Technology while the competent authority for commercial reactors was the Ministry of Economy, Trade and Industry. The system made it difficult to use the research results of the former JAERI to improve the safety of commercial reactors. There are some examples, including a reactivity accident, where research results were utilized, but generally, there was no direct close relationship between safety research and commercial nuclear reactors, which is a consequence of vertical sectionalism of organizations.

The third reason is the weak mechanism whereby the results of safety research are adopted for commercial nuclear reactors in the safety regulation framework. The Nuclear Safety Commission and the Nuclear and Industrial Safety Agency supervise safety regulations and new measures could not be adopted for commercial nuclear reactors without any approval from these regulatory authorities. It was difficult for the licensees to voluntarily adopt the latest research results into safety measures. Basically, the licensees thought if the safety regulations were satisfied, safety would be secured. Voluntary efforts to improve safety were insufficient.

Water injection from fire engines and seawater injection were implemented as on-site responses in the severe accident at Fukushima Daiichi, but were not in manuals of severe accident countermeasures and no training was given. These responses have been imagined in advance as one of severe accident countermeasures for on-site personnel; not those conceived after the accident occurred. However, such imagined measures were not documented nor studied during actual training. Many issues arose. For example, when conducting water injection by the fire engine, it took time to find the inlet nozzle, the fuel for the fire engines ran out along the way, which meant supplying water was suspended, and the number of fire engines was insufficient from the start. In addition, for the sea water injection, two fire engines had to be connected to supply seawater from the ocean to the inlet nozzle because one was not able to pump water sufficiently. This could have been recognized in advance if specific plans had been made and actual training had been given along with the plan.

A mechanism was required to directly connect the latest safety research results with safety improvements in commercial nuclear reactors. Aware that safety is the top priority, we must create a mechanism for licensees to voluntarily improve safety. Preparation for technical standards in academic societies, exchange of accident information by licensees themselves and peer reviews, etc. were specific examples. It is true that severe accident countermeasures will be reinforced by imposing a regulatory requirement for severe accident countermeasures by regulatory authorities, but there is also a need to develop a mechanism whereby licensees always voluntarily adopt the latest research results and strive to improve the safety.

An important underlying cause of the severe accident at Fukushima Daiichi was presumption that no severe accidents would occur in Japan due to explanatory ability to the local society, lawsuit measures, and consistency in safety regulation. Conversely, researchers have studied severe accidents based on the presupposition that a severe accident might occur. It is important to establish a mechanism of continuous improvement of comprehensive risk assessments of various external events and reinforcement of severe accident countermeasures. Moreover, to ensure this can function in actual society, scientific and reasonable concepts of safety should be prioritized and not changed due to political or social reasons. Nuclear safety researchers are greatly responsible and must recognize this.

7.4 International System

Nuclear power was used in an international context from the very beginning, whereupon international activities involving treaties and others activated various forms of multinational/bilateral cooperation, including joint research project and information exchanges. Japan actively participated in this international framework and was treated as an experienced country with high technology and skill in nuclear power. This section considers why this international framework did not function to prevent this accident and why the accident occurred in Japan.

(1) **International nuclear safety framework**

Initially, nuclear power technology was adopted for military use in the form of the nuclear bomb. However, President Eisenhower's speech at the United Nations in 1953 enabled this huge energy brought by atomic reactions to be used for peaceful purposes. At the same time, to prevent the proliferation of nuclear weapons, the International Atomic Energy Agency (IAEA) was established with the Nuclear Nonproliferation Treaty (NPT) and Safeguards (surveillance) as pillars. As an institution to promote the peaceful use of nuclear energy as well as nuclear nonproliferation, the IAEA supported member countries in their efforts to harness radioactivity and energy. While the use of the nuclear technology was mainly focused on nuclear power generation more than ever, an international cooperative framework of nuclear safety was established, including the Convention on Nuclear Safety, the Convention on the Early Notification of Nuclear Accidents, and the Convention on Assistance in the Case of a Nuclear Accident or Radiological Emergency through the Three Mile Island Nuclear Power accident and the Chernobyl nuclear power plant accident.

In particular, important international activities for nuclear safety include the development of nuclear safety standards and peer review services within the IAEA. The purpose of the nuclear safety standards is to develop systematically and comprehensively the standards required to secure safety. In particular, international knowledge, including defence in depth and other basic concepts, is collected to continue to develop and revise standards. Meanwhile, peer review is an activity whereby review teams, centering on experts from various countries, visit countries to be reviewed, evaluate safety measures in specific fields based on international safety standards and specialists' knowledge etc. and find issues to be improved, etc. The IAEA also develops standards for nuclear security and enhanced review activities and tries to persuade Western countries and others to use it.

In recent years, countries, especially those planning to newly start nuclear power generation, have adopted the IAEA safety standards. In addition, until recently, Europe, the U.S., and other nuclear power advanced countries have tended to prioritize their own safety standards and made light of the IAEA. But now, all of them changed their policies and tried to adjust their domestic/regional standards to the IAEA safety standards. In particular, European countries developed and revised the EU Directive simultaneously with IAEA safety standards in a manner that both standards are effectively correlated with each other at almost the same time.

The safety research is essential to establish firm basis of nuclear safety. As countries can bring their knowledge and use the research facilities together to reduce the expense necessary for research, they actively cooperate in safety research activities under the framework of international cooperation. OECD Nuclear Energy Agency (OECD/NEA) works as the core of such international cooperation. Since its foundation, it has established the framework of an "international joint research project" and has created a wide range of joint research projects, in which Japan has also actively participated and worked as project lead hosting some of them.

The top regulators of countries with similar experience and scale of nuclear power capacity periodically convene together and exchange opinions on mutual concerns. Japan joins the International Nuclear Regulators Association (INRA), alongside the U.S., France and some other countries. Moreover, national regulatory authorities who are currently or will be tasked with the review of new reactor power plant designs have established the Multinational Design Evaluation Program (MDEP) to study common issues on the safety review and assessment of advanced new reactors.

As for the radiation safety field, the United Nations Scientific Committee on the Effects of Atomic Radiation (UNSCEAR) scientifically evaluates new knowledge of radiation effects, the results of which are then used as the basis for study at the International Commission on Radiological Protection (ICRP). The ICRP is a voluntary organization comprising experts participating as individuals, which has a significant international influence due to its highly professional expertise and past achievements. It is no exaggeration to say that the ICRP virtually sets out the basis of international standards for radiation protection. ICRP recommendations are discussed by related international organizations, including the IAEA, and then reflected in a jointly established safety standards for radiation protection, whereupon they become formal international standards.

(2) **Initiatives toward severe reactor accidents and nuclear disasters**

The IAEA established "Basic Safety Principles for Nuclear Power Plants", which prevails over all other safety standards, in 2006 (signed by eight international organizations, including the NEA, the World Health Organization (WHO), the International Labor Organization (ILO), and the Food and Agriculture Organization of the United Nations (FAO). It includes important perspectives reflecting experiences of the Chernobyl accident. The safety goal is to "protect humans and the environment", reflecting the problematic awareness of protecting against damage caused by radiation contamination. The ALARA principle of radiation protection is applied to ensure facility safety during the life cycle of nuclear facilities, and basic concepts related to retrofitting, which have long been discussed, which will lead to apply more advanced safety standards if a new reactor can achieve higher safety than an existing reactor at the same cost.

Based on this idea, Europe considers higher safety standards for new reactors focusing on measures against severe accidents, which is an attempt to elaborate the concept of defence in depth and effectively apply measures including the upgrade of the plant design to ensure the confinement function of radioactive materials in the event of a severe accident under the concept of the "design extension condition" and virtually eliminate the release of radioactive materials to achieve higher safety for new reactors. It was established as a safety requirement "Safety of Nuclear Power Plants: Design (SSR-2/1)" at the IAEA general Conference in September 2011. The requirement includes stipulation to show a guarantee that the cliff edge effect will be prevented by the probabilistic method.

Globally important consideration was given to radiation protection. The ICRP published new recommendations in 2007 and the IAEA, FAO, ILO, NEA, Pan American Health Organization (PAHO), WHO and other international organizations jointly established the requirements as a publication of "Radiation Protection and Safety of Radiation Sources: International Basic Safety Standards (Revised BBS, GSR Part 3 (Interim)" in 2001, which takes account of findings of ICRP new recommendations. The important point of newly adopted requirements is a means of responding to unexpected situations, including accidents. Classifying actual exposure statuses into three different exposure situations, namely planned, emergency, and existing, relevant standards were developed from the perspectives of justification, optimization of protection, and dose limits.

As for the design requirement for severe accident measures, and the revised BBS, which can be applied in the event of a nuclear disaster, were internationally approved after the occurrence of the Fukushima Daiichi NPS accident, however, lessons learned from the accident were not included. Both were approved as a result of a long-term review and have already been adopted by some European countries in national safety regulations.

(3) **Response to the evolving global standard**

Japan has the third largest nuclear power generation installed capacity in the world. As a country providing epidemiological survey information in Hiroshima and Nagasaki, which is the basis of the radiation protection standard, Japan has been globally contributing to both the safety of nuclear facilities and radiation protection. However, what about involvement in global discussion on severe accident measures directly related to the accident?

First of all, Japan was obsessed with the former notion that the IAEA standard was for countries whose domestic standards had not been fully developed or those whose resources to develop standards were limited, while other countries changed their attitude toward the IAEA safety standard system. For example, Western countries strove to actively incorporate the IAEA system into domestic regulatory standards. Consequently, Japan generally refrained from actively trying to make proposals to lead preparation for international standards or adopting the IAEA system into domestic standards, excluding exceptional efforts like a study of aging in NEA. Japan normally focused on negative checks to avoid any disadvantage for Japan. For example, "Basic Safety Principles for Nuclear Power Plants", should prevail over all safety standards, and Japan should have examined the desirable nature for Japanese safety regulations and standards based on these principles. Looking at the safety goal of protecting "human and the environment", there is the possibility of awareness of the problem and need to prevent damage by radiation contamination to be reflected in Japanese regulations. Still, as the principles are not specific regulatory details in character, the action on this matter was late. The Nuclear Safety Commission showed a policy to adopt them, but it was just before the accident.

As mentioned before, in a case of a new reactor which can achieve greater safety than existing reactor at the same cost, more advanced safety standards should be applied. Based on this idea, standards with higher safety requirements for new reactors were studied and established as SSR-2/1. However, the Japanese regulation eliminated the scope to change regulatory scheme between new and existing reactors due to the principle of uniform regulation. It was difficult to even join the global discussions on this issue. During this period, the concept of defence in depth was developed and severe accident measures were continuously discussed in the international nuclear society.

Concerning the revision of radiation protection standards, Japan actively discussed with the ICRP but our response was slow. The concepts of the emergency and existing situation of radiation exposure did not fully spread; even among specialists. Accordingly, the latest international knowledge was not always properly used for evacuation and return home as guidelines.

(4) **International review**

IAEA review service is important international scheme along with IAEA safety standards. However, Japan did not actively accept the IAEA review services except the operational safety review for electrical power suppliers (the IAEA Operational Safety Review Team (OSART)). Moreover, few experts were dispatched to review teams for other countries.

In this accident many issues of regulatory bodies were revealed. IAEA has a review service specially designed for regulatory body, Integrated Regulatory Review Service (IRRS). While major nuclear power advanced countries accepted IRRS, the Nuclear and Industrial Safety Agency (NISA) requested the IAEA and accepted the IRRS review team in 2007. Some items pointed out in the report of review results overlaps with the lessons learned of the accident. First, the report made a suggestion that the independence of NISA from the Agency for Natural Resources and Energy (ANRE) should be demonstrated more clearly in future. The human resource problem, including an appropriate personnel plan, was also raised in both recommendations and suggestions. As for severe accident measures, suggestions stated that NISA, which started to study severe accident management measures, should continue such efforts. Considering the circumstances that these review results were not used effectively to prevent the accident, it seems that systematic approach is necessary, such as a third party organization which has an authority to improve the organization of regulatory bodies.

The more comprehensive and obligatory international review is carried out based on the Convention on Nuclear Safety. The status of handling safety measures of signatory countries is reviewed every three years. Before the accident, a review is conducted in September 2010. At the time, the Government of Japan reported the above-mentioned responses to the IRRS review results and other issues. Although such review meetings are not open to the public, many countries publish their reports, which include the status of response to items highlighted in the previous meeting. Japan expressed the mid- and long-term efforts, including the enhanced application of risk-informed regulations. These safety measures function effectively in the sense of encouraging efforts to enhance safety measures through international discussion. Conversely, it is difficult to handle problems related to administrative systems which vary from one country to another, like reviews of regulatory organs. Therefore, in the IRRS, the description of independence of the organization seems to be stated as a suggestion.

The Japanese licensees accepted the IAEA's OSART for operational management and also actively promoted mutual review activities by the World Association of Nuclear Operators (WANO) as a major member of this international nongovernment organization. Consequently, given the fact that this review did not lead to the prevention of the accident, we think the system must be improved.

(5) **Summary**

Various discussions, which might have prevented the accident, had been held globally involving Japan. As shown in the responses to severe accident measures and risk-informed regulations, a movement toward reviewing the system gradually started, but the accident occurred before the reform had been completed. The international framework is effective in encouraging safety improvement efforts, but the strong will of each country and individual approaches to using the framework effectively are essential for achieving the goal.

For Japan, contribution to global society entails a great burden compared to Western countries due to its linguistic and geographical constraints. To promote the use of nuclear power in future, we should face up to such burden, actively participate in international activities, and create a system to reflect the discussions there in Japan effectively. In future, it seems likely that more countries will start using nuclear power. Japan should actively leverage its experience including nuclear disasters, to help them create systems to secure nuclear safety. From these perspectives, we must develop human resources who play a role in leading international discussions.

If Japanese plant manufacturers aim to develop global business, the industry must take part in works to develop a global nuclear framework.

7.5 The Role of Atomic Energy Society of Japan

The Atomic Energy Society of Japan is the only organization in Japan that aims to contribute towards progress in the development of atomic energy by seeking academic and technological advances pertaining to the peaceful use of atomic energy.

To achieve this goal, the AESJ established an action guideline and Code of Ethics, as well as implementing various activities alongside the same. For example, the action guidelines stipulate an "active involvement in policy recommendations on nuclear power technology" and "support for activities to maintain and improve the safety and reliability of nuclear facilities". In the Code of Ethics, the AESJ asks members for "efforts to secure safety, learn safety knowledge and information, prioritize commandment of efficiency, economy, efforts to improve safety, the requirement for caution, commandment for overconfidence in the maturity of technology, commandment of each member's sense of security, acquisition of new knowledge, learning from experience and technology inheritance, and disaster prevention among member organizations".

However, these action guidelines and Code of Ethics did not fully penetrate among Society members and may lose substance. During the operation, the AESJ may not have made sufficient efforts to penetrate the above-mentioned details provided in the action guideline and the Code of Ethics among Society members.

Therefore, AESJ Investigation Committee on the Accident at the Fukushima Daiichi Nuclear Power Station of Tokyo Electric Power Company heard opinions on the Fukushima Daiichi accident from society members and determined the issues of the AESJ operation, and the role to be played for the AESJ. The following shows the results:

(1) **Analysis of the roles played by the AESJ**

First, AESJ Investigation Committee performed a questionnaire for past and current Society members and those who experienced directors, etc. This questionnaire was performed to study what the AESJ could do to prevent accidents or alleviate their influence or cover what should have been done. The survey respondents were current or past members who were experienced directors or subcommittee chiefs, and those in other positions whom the AESJ could contact via e-mail or post, comprising 289 individuals in total. The AESJ e-mailed or posted questionnaires and responses were received from 102, representing a response rate of 35. 3 %. The questionnaire requires a narrative form, the essence of which was analyzed and summarized.

(a) Why we could not prevent the accident

- We did not strive to "learn from others and the past".
- The reduction of safety research may reduce the SA response activity.
- Defence in depth was not sufficiently understood and implemented.
- The collaboration and cooperation in the AESJ were insufficient
- Lack of knowledge in terms of a bird's-eye view and control of the whole.

* Example response

"Our study and reflection with the systematization of ethics concerning overall nuclear safety were insufficient".

(b) What caused the problem and what was our problem?

- We were conceited and overconfident in our technology.
- Our environment prevented free and frank exchanges of opinion.
- The members of the AESJ should have widely shared awareness of the responsibilities which it should take.

* Example response

"We had the wrong mindset that our power stations were safe, or at least not subject to urgent risks".

"We had no time/space to frankly exchange opinions on safety".

"We should build a space where members with rich imagination can discuss, share and evaluate risks of which they are afraid. We should establish a system to express our concerns to society".

"We should have a space where individuals can focus on research fields other than their own and comprehensively understand issues".

(c) What to do in future

- Controlling the accident and implement measures to investigate the causes, and reflect them in as many lessons as possible,
- Strive for reconstruction in Fukushima, and
- Investigate and implement a strategy to improve the AESJ in future.

Example response

"We should examine the fundamental question of what nuclear safety should be in greater depth".

"We should establish a vision of AESJ's responsibility for society and how to realize it".

"We should consider a code of conduct, share a vision of AESJ's duty and responsibility and what to target and develop a code of conduct".

"As a specialist group, we should develop an approach, system, and skills to properly advise administrations, licensees, and other parties concerned as real experts with specialized skills that enable us to understand the entire nuclear system".

"We aim to form a strong influential organization capable of attracting abundant human resources with views, desire, and leadership from various societies, and create a space that integrates the roles shared by various societies; centering on the sound promotion of the use of nuclear power".

(2) **Roles to be fulfilled by the AESJ**
AESJ Investigation Committee compiled a draft "Strategy to Improve What the AESJ Should Be in Future", which was released on the AESJ website in July 2013, and suggestions invited from Society members, whereupon eight members advanced their opinions. AESJ Investigation Committee studied these opinions and compiled them as follows:

(a) Re-acknowledge duty to be exercised by the AESJ : With the following basic recognition, the AESJ shows its duty in its articles of incorporation, action guideline, and Code of Ethics, etc. Society members should reaffirm the duty to be exercised as a member of a specialist group of scientists and engineers engaged in nuclear power technology.

Nuclear power technology is the edifice of knowledge painstakingly established to emphasize rationality and verification and a valuable asset to be shared by all mankind as well as for the benefit of society. Therefore, AESJ activities can gain social recognition only subject to the trust and mandate of society. The AESJ, which is engaged in intellectual activity, assumes the grave responsibility to fulfill a mandate from the society as experts as well independence from certain authorities and the interests of organizations with academic freedom; enjoying the right to seek out the truth based on its own professional judgment. Particularly at present when scientific activities and their results exert a wide and profound influence on mankind, society requires scientists to constantly make ethical judgments and actions. Moreover, there are social requirements for roles to be assumed by science in the process of forming policies and public opinions.

The Fukushima Daiichi Nuclear Power Station accident forced the AESJ to review their past actions by reflecting on whether they had really fulfilled the trust and mandate from the society and showed issues to be resolved by the AESJ; targeting the reconstruction of the disaster region and revitalization of Japan. Accordingly, the AESJ revised articles of incorporation at its general assembly in June 2013 and widely shared the sense of responsibility as the Society members.

(b) Free discussion in the AESJ: The AESJ will take the suggestion seriously that is the AESJ "had little atmosphere of free discussion" previously. The Society members reconfirm the following basic duties and develop an atmosphere that enables the free and frank exchange of opinions.

(i) The AESJ members are responsible for maintaining expert knowledge and technical quality which they themselves have created, and further, contributing to human health and welfare, social safety and public peace, and sustainability of the global environment using our own expert knowledge, technology, and experience.

(ii) The AESJ members shall make judgment and take actions honestly and sincerely, try to maintain and improve expert knowledge, abilities, and skills, and strive their utmost to scientifically show accurate and legitimate knowledge.

(iii) AESJ members shall be convinced that scientific autonomy can be founded on the trust and mandate from the society, understand the relations between science/technology and the society/natural environment from a wide perspective, and take actions properly.

(iv) AESJ members shall be responsible for meeting expectations of society for discovering the truth and achieving various issues. When developing the research environment and using research funds, the AESJ shall remain aware of the existence of wide social expectations.

(v) The AESJ members shall try to elicit the meaning and role of our research, actively explain it, evaluate the influence of this research on human beings, the society, and environment, and changes that can be induced by such research, and neutrally and objectively release the results as well as establishing a constructive dialogue with society.

(vi) The AESJ members shall recognize the potential for their research results to be used for wrong purposes, and when implementing the same, select a proper channel and ways accepted by society.

(c) Reinforcement of safety research: Obviously, it is true that nuclear safety research has been declining for years in Japan, which has seen the number of researchers and engineers engaged in safety plummet. We can also cite this as one of the causes of the accident.

In future, to sustain the use of nuclear power, we must rebuild the safety research system by trying to develop and implement the concept of safety culture and revising the mechanism to continuously implement safety improvement research, which will be the basis for restoring public confidence in nuclear energy.

The AESJ should adopt a leading role in developing a road map and continuous revisions, etc.

(d) Reinforcement of interdisciplinary efforts: The AESJ shall establish a space for bird's eye discussions and cooperation with other academic societies of nuclear safety and similar and play a leading role.

Nuclear power is a comprehensive form of science technology encompassing various special fields. To secure nuclear safety, a comprehensive perspective is essential to avoid gaps in boundaries between these special fields. To date, the AESJ has been trying to reinforce functions. In future, it will cross over fields, including other academic societies, and make and reinforce continuous comprehensive efforts; the results of which shall be released as suggestions from the AESJ.

(e) Contribution to continuous improvement of safety regulations: The deviation of the Japanese safety regulation mechanism from the international standard is the key matter on which we should reflect. There was significant progress with the establishment of the Act to Establish the

Nuclear Regulation Authority (AENRA) in June 2012, the establishment of new regulations in July 2013, and like the regulations, continuous improvement is required for safety regulations. In response, the AESJ reinforces research and standard development activities, which give evidence of the regulatory system, and properly releases the results to society. Research should be conducted to cover not only technical but also social aspects of safety regulation. Emergency plans, including the disaster prevention plan, and risk research, which show how society should face up to an influential risk with low frequency, are example important issues of social aspect research.

References

1. The Nuclear Safety Commission, *Research Report on Accident Management Against Severe Accidents. Measures Against Reactor Containment.* Round-table conference of common issues, Special Committee on Reactor Safety Standard (The Nuclear Safety Commission, 1992), Japan
2. Japan Nuclear Energy Safety Organization, Anzen Jisseki Shihyo (PI), Anzen Juyodo Hyoka (SDP) Shuho Seibi Donyu Kentokai (Chukan Hokoku no Gaiyo (Committee on Development and Introduction of Safety Performance Indicator (PI) and Safety Significance Determination Process (SDP) Technologies (Overview of Interim Report)) (2008). http://www.meti.go.jp/committee/materials/downloadfiles/g80408a09j.pdf
3. Rebuild Japan Initiative Foundation (RJIF), *The Independent Investigation Commission on the Fukushima Nuclear Accident* (RJIF, 2012), Japan
4. Government Investigative Committee on the Accident at the Fukushima Nuclear Power Stations of Tokyo Electric Power Company, *Interim Report* (2011)
5. The Tsunami Evaluation Subcommittee of the Japan Society of Civil Engineers (JSCE), *Tsunami Assessment Method for Nuclear Power Plants in Japan* (The Tsunami Evaluation Subcommittee of the Japan Society of Civil Engineers (JSCE), 2002), Japan
6. The Japan Electric Association (JEA), *Genshiryoku Hatsudensho Taishin Sekkei Gijutsu Shishin (Guideline on Seismic Design Technology for Nuclear Power Station).* JEAG4601-2008 (The Japan Electric Association (JEA), 2008), Japan
7. The Special Expert Committee on Survey and Research of Thermal-Hydraulics in Containment Vessel of the Atomic Energy Society of Japan (AESJ), *Sibia Akusidento to Shingata Keisuiro no Netsu Ryudo Kyodo Kenkyu no Genjo (Current Status of the Research on Severe Accidents and Advanced Light Water Reactors)* (1994)
8. The Atomic Energy Society of Japan (AESJ), *Genshiryoku Hatsudensho no Jishin wo Kiin toshita Kakuritsuronteki Anzen Hyoka Jisshi Kijun (Standard for the Probabilistic Safety Assessment on Earthquakes for Nuclear Power Stations)*, AESJ-SC-P006:2007 (The Atomic Energy Society of Japan (AESJ), 2007)
9. The Atomic Energy Society of Japan (AESJ), *Genshiryoku Hatsudensho nitaisuru Tsunami wo Kiin toshita Kakuritsuronteki Risuku Hyoka nikansuru Jisshi Kijun (Standard for the Probabilistic Risk Assessment on Tsunami for Nuclear Power Stations),* AESJ-SC-P006:2011 (2011)
10. Government Investigative Committee on the Accident at the Fukushima Nuclear Power Stations of Tokyo Electric Power Company, *Final Report* (2012)

Chapter 8
Root Causes of the Accident and Recommendations

Abstract In Chaps. 2–5, the Accident Investigation Committee grasped the facts of the accident process. Based on this, in Chap. 6, it analyzed and assessed where problems lay in the accident at the Fukushima Nuclear Power Stations, followed in Chap. 7 by analysis of the nuclear safety organizations.

This analysis and assessment aims to determine the root causes of this accident and boost efforts to prevent any repeat nuclear disasters by drawing lessons and providing recommendations to nuclear power stakeholders.

In this Chapter, the committee presents its view on the root causes and provides recommendations based on them. The main focus of the root cause analysis in 8.1 is on organizational issues. Recommendations are made in Sect. 8.2, in addition to those based on the results of root cause analysis and those from other perspectives, drawn from the analysis and assessment of Chaps. 6 and 7.

Keywords Direct cause • Recommendation • Root cause analysis • Underlying cause

8.1 Root Cause Analysis

8.1.1 Direct Causes

The cause-effect relationship based on the facts in the severe accident at the Fukushima Daiichi Nuclear Power Station and the consequent damage on residents is that much of the equipment rendered unusable due to the ground motion and tsunami generated by The Great East Japan Earthquake, which damaged the reactors and released a large amount of the radioactive materials to the environment. In particular, the loss of almost all power sources due to the tsunami flooding exacerbated the condition. There were three direct causes of this accident and the damage caused to residents as follows:

- Inadequate countermeasures for the tsunami
- Inadequate severe accident measures
- Inadequate emergency measures, accident response measures and various mitigation and recovery measures

The Fukushima Daiichi Nuclear Accident: Final Report of the AESJ Investigation Committee, DOI 10.1007/978-4-431-55160-7_8

With regard to the tsunami, despite fresh knowledge accumulated of two findings before The Great East Japan Earthquake, no action was taken against them. The first finding was the Jogan earthquake, which was recorded in old literature. Tsunami deposits corresponding to this earthquake were discovered; mainly in Miyagi prefecture and details of the tsunami wave source reproducing the earthquake was published in a scientific paper. The second finding was the tsunami earthquake along the Fukushima prefecture offshore trench. The Headquarters for Earthquake Research Promotion of the Ministry of Education, Culture, Sports, Science and Technology highlighted the possibility of occurrence. In 2008, based on simulation related to each of these tsunamis, TEPCO obtained calculation results of 9.2 and 15.7 m, respectively, for the maximum wave heights at the Fukushima Daiichi Nuclear Power Station. These results far exceeded the wave height of 5.7 m assumed by TEPCO in its tsunami countermeasures. However, TEPCO postponed countermeasures for these tsunamis for the following judgements; (1) there was no consensus on these tsunami wave sources in academic society and (2) the occurrence probability was not high enough to require countermeasures. Instead, if considering countermeasures using probability, TEPCO should have prioritized the so-called cliff edge in which the core damage probability sharply rises for tsunamis exceeding the postulated wave height.

Severe accidents are accidents beyond the design basis. As design basis accidents are based on certain assumptions, probabilities of severe accidents are not zero. Responding to this in advance, measures to prevent damage to the core and primary containment vessel (PCV) in severe accidents were prepared as accident management (AM) measures. Meanwhile, voluntary efforts by utilities included implementing AM measures in all nuclear power stations by 2002, although there was little follow-up review of AM measures since. In particular, no countermeasures were implemented for severe accidents related to natural disasters such as earthquakes and tsunami. In the United States, based on the lessons of the September 11 attack in 2001, counterterrorism measures at nuclear power stations were developed. Though these were part of the countermeasures for severe accidents, no equivalent counterterrorism measures were taken in Japan. We believe that counterterrorism measures would also be effective against natural disasters.

When a large amount of radioactive materials is released into the environment, urgent evacuation of residents is necessary. Such emergency response plans were prepared in advance by municipalities in which nuclear power stations were located and neighborhood municipalities. The plans assumed the evacuation of residents living within 10 km of the nuclear power stations. However, since the actual situation was more serious, an instruction was issued requiring residents living within 20 km of the plant to evacuate. The off-site center was located approximately 5 km away from the Fukushima Daiichi Nuclear Power Station. With virtually all communication methods of the center disabled by the earthquake, it was no longer able to function as the Local Nuclear Emergency Response Headquarters. Regarding the administration of iodine tablets, related instructions were not fully spread. These failures during the resident evacuation were attributable to the inadequate implementation of emergency plans in advance.

Delayed decontamination was also caused by the failure to assume such circumstances in advance.

With regard to the onsite response by TEPCO personnel, several examples of inadequate handling were recognized such as delayed observation of the isolation condenser (IC) in Unit 1 being out of function, and the extended delay of low-pressure water injection implemented after manual termination of high pressure water injection in Unit 3. However, these failures were not the misoperations for those required in accordance with the operation manual but the operations which are concluded as inadequate in the later validation even though the situation was extremely severe and far beyond the experience. As for the Unit 1 IC being inoperative, it emerged that the inner isolation valves were open in System B. If the staff member had entered the Reactor Building at an early stage and manually opened the external isolation valves, the reactor could have been cooled by System B. However, this was identified in retrospective investigation and analysis and we cannot argue that such a response should have been made at the time of the accident. As for the manual termination of high pressure coolant injection system in Unit 3, the operation continued under the pressure out of the operation range and it was likely to stop due to failure. The success in the PCV vent at Units 1 and 3 was achieved by bringing portable generators and compressors on line, though AM measures did not assume any loss of power sources. Water injection using fire engines could not prevent core damage, but established the decay heat removal at a later stage. The extended delay of the PCV vent and water injection by fire engines was attributable not to an inappropriate onsite response but inappropriate preparation for such accident in advance. Accordingly, inappropriate onsite response was not deemed to be the direct cause of the accident.

8.1.2 Underlying Causes

In this section, we analyze the underlying causes that led direct causes to the accident. In particular with reference to the organizational underlying causes, we focus on academic specialists, utilities and regulatory authorities. Given the present document as a report of the accident investigation by the Atomic Energy Society of Japan (AESJ), a group of academic specialists, we emphasize the underlying cause related to academic specialists in the initial part of this section in particular.

(1) **Lack of awareness of their own roles at academic specialist**
With academic specialists having confined themselves to their own specialist areas, oversights occurred in safety. Although tsunamis had been recently discussed; mainly by specialists in the field, efforts to determine the risks tsunami would impose on nuclear power stations were insufficient. Many nuclear safety specialists, possessing deep insight into the plants, were less aware of the risks of natural disasters.

Regarding tsunami countermeasures and severe accidents, there had been many opportunities for academic specialists to provide their views and one of presentation materials at an AESJ conference highlighted the high risks of natural disasters. However, it was not taken into account when planning the actual safety measures for nuclear power stations.

The framework to utilize the research and alerts by academic specialists in the communities was not enough, and generally speaking, few alerts by individuals would be employed in the communities. At the same time, as there are various opinions in communities and accepting these different opinions is healthy for both natural science and communities, it is difficult to implement the entire range of individual opinions in the safety measures. With this in mind, a framework to utilize the opinions agreed in the specialists group in the safety measures as technical standards made by the specialists themselves. Experts must do the efforts of continuous revisions for the latest knowledge to be reflected in the technical standards. Moreover, safety measures by utilities and safety regulations by regulatory authorities must also be implemented based on this latest knowledge. In the AESJ, the standard committee is responsible for creating and revising technical standards. Regarding the countermeasures for natural disasters and severe accidents, although the AESJ standard committee had been developing the technical standards, it was unable to prevent the accident occurring at the Fukushima Daiichi Nuclear Power Station.

Within academic societies and associations participated by specialists, various academic activities are conducted, including not only creating technical standards but also organizing academic seminars and investigation committees. Communities expect these academic activities to be neutral, which rules out any activities conducted in the interests of particular organizations. These academic societies and associations must be managed not to be suspicious from the communities. Insufficient efforts by the academic societies and associations undermined the reliability to the academic specialists.

(2) **Lack of the sense of safety and efforts for safety at utilities**

As a utility, TEPCO cannot complain about criticism that it failed to face the risks identified by new knowledge about tsunamis and severe accidents and postponed required safety measures. TEPCO overlooked the risks of the accident in the managerial judgment.

The utilities lacked the attitude to drive voluntary safety measures which would be even more stringent that regulatory requirements. They never voluntarily improved tsunami countermeasures and severe accidents where not required to do so by regulatory authorities. Traditionally, since nuclear safety regulations have been stricter than those for other industries, it has spawned the concept that safety could be maintained simply through compliance with the regulatory requirements.

Severe accidents at nuclear power stations not only seriously damage communities but also significantly burden the business management of the electricity utility itself. As risk control is part of the business management, we would have to say that the utilities lacked comprehensive management ability to prioritize safety.

(3) **Lack of awareness on safety of the regulatory authorities**
As a regulatory authority, the Nuclear and Industrial Safety Agency (NISA) acquired information about postulated tsunamis from TEPCO. However, NISA never instructed TEPCO to take safety measures. We would have to say that NISA lacked the sense of safety as a regulatory authority responsible for safety.
Safety regulations concerning countermeasures for severe accidents and nuclear emergency also fell far behind the international standards. However, the regulatory authorities never improved such safety regulations promptly.
Management in the event of emergency actions had not been established, which explains the numerous inadequate responses to the accident by the Nuclear Emergency Response Headquarters and the Local Nuclear Emergency Response Headquarters.

(4) **Lack of efforts to directly learn from international activities and joint works**
Few efforts were made to directly learn about countermeasures for severe accidents and natural disasters based on overseas experiences and international activities such as those by the IAEA. In the earthquake in the Indian Ocean off Sumatra with a magnitude 9.1 in 2004, for example, a destructive tsunami occurred, which caused flooding in a nuclear power station located on the opposite side of the Indian Ocean. However, no one assumed an earthquake and tsunami on such scale would occur in the sea around Japan, nor was there any forecast that nuclear power stations would be flooded, and no action was taken.

(5) **As a huge complex system deeply related to communities and the economy, there was a lack of human resources with a comprehensive perspective to ensure the safety of nuclear power plants and foundation for organization management**
A nuclear power station is a huge complex system, which is not just a huge and complex engineering system but also an entity deeply related to the communities and the economy. Safety measures, for example, do not work just by installing safety equipment, but are critically linked to human management such as maintenance and emergency operations. As a common reason of the underlying causes discussed above, we would like to point out the failure to develop human resources with a comprehensive perspective to ensure the safety of nuclear power plants as huge complex systems and establish a foundation for organization management.

8.2 Recommendations

Based on the analysis in Sect. 8.1, it was identified that this accident was triggered by a natural phenomenon of unexpected tsunami generated by the earthquake and expanded to a nuclear disaster due to the direct causes. In the background, there were various complex issues mainly related to organizational aspects.

About how we should address these complex issues, the Investigation Committee provides respective recommendations corresponding to the direct causes and underlying causes related to organization as well. These recommendations include those based on the results of the root cause analysis, as well as those from other perspectives drawn from the analysis evaluation in Chaps. 6 and 7. As shown below, firstly we provide recommendations concerning basic items for nuclear safety that had wide-ranging impact on the direct and underlying causes (Sect. 8.2.1). Secondly, we provide recommendations according to the direct causes (Sect. 8.2.2), those according to the organizational causes for each of the three sectors (academia, industries and government) among the underlying causes (Sect. 8.2.3) and those related to common items (Sect. 8.2.4). In addition, as recommendations for restoration, we discuss recommendations for environmental restoration in Sect. 8.2.5 and provide the conclusion in Sect. 8.2.6.
(Items of the suggestions)

8.2.1 Recommendation I (Basic Items of Nuclear Safety)
(1) Initiatives for setting goals for nuclear safety and forming a framework
(2) Deepening understanding of defence in depth and encouraging its application

8.2.2 Recommendation II (Items Related to Direct Causes)
(1) Strengthening the measures against external events
(2) Strengthening the measures against severe accidents
(3) Preparation for emergency and strengthening the response framework
(4) Enhancement of the nuclear safety assessment technology

8.2.3 Recommendation III (Items Related to Organizational Causes Among Underlying Causes)
(1) Initiatives of AESJ and academia as expert groups
(2) Initiatives of the industries
(3) Initiatives of nuclear regulatory authorities

8.2.4 Recommendation IV (Common Items)
(1) Strengthening the nuclear safety research foundation
(2) Strengthening the international cooperation framework
(3) Developing human resources related to nuclear power

8.2.5 Recommendation V (Items Related to Restoration)
(1) Initiatives toward future environmental restoration

8.2.6 Conclusion

We hope these suggestions will lead to concrete activities in future for the various parties concerned such as the government, including regulatory authorities, the industries, and academic and research organizations. These recommendations include items which the AESJ itself should address. Not only sincerely responding to these items, the AESJ will continue to encourage related organizations aiming to make the recommendations effective.

As we are in a position to prioritize the transparency of nuclear-related information above all, we are confident that these recommendations should be broadly shared with all people concerned in nuclear power generation. All of the organizations and experts involved in nuclear activities should regard these recommendations as the requests to them and tackle with the recommendations seriously. Any of the organizations or experts should be aware that they would lose the qualification to work in the nuclear field if they could not tackle with these recommendations.

8.2.1 Recommendation I (Basic Items of Nuclear Safety)

(1) Initiatives for setting goals for nuclear safety and forming a framework

(a) The reason why the accident at the Fukushima Daiichi Nuclear Power Station was expanded to a "disaster" is that the core melt could not be prevented, following which adequate countermeasures to block the release of a large amount of radioactive materials were not taken, resulting in the long-term evacuation of residents and contamination of the surrounding environment. In addition to inadequate accident management, emergency responses to be taken under such circumstances did not work appropriately to minimize the damage. We are certain that the reason is omission to consider an event whereby the worst case scenario of release of radioactive materials into the environment might actually occur. Consequently, no management measures or practical disaster prevention plan were prepared corresponding to severe accidents, which exposed many problems.

We think this was primarily due to inadequate understanding of the safety goals to be achieved and their importance which should be positioned as the foundation of nuclear power safety. In Europe which experienced radioactivity pollution caused by the Chernobyl accident, protecting the environment as well as human beings has been considered an essential goal. The regulatory authorities set common safety goals to enhance the measures to prevent environmental pollution. In the Safety Fundamentals (SF-1) formulated by the IAEA on top of safety standards, the objectives of fundamental principles for nuclear safety include protecting human beings and the environment from hazardous radiation generated from nuclear-related facilities and their activities, and the perspective of protecting the environment is clearly stipulated. Based on these fundamental principles, specific safety requirements as lower-level regulations are currently being reorganized.

The safety goals to be achieved could be set as the risk probability. In many countries, quantitative risk levels are set, while in Japan, although related reviews have been conducted, they have not been included in regulations. In principle, risk analysis and assessment are effective methods of discovering vulnerabilities of plants and improving them. However in Japan, these methods have not appropriately been used in analysis related to the impact and progress of external events. Although these methods have not yet been completely established, the event progress scenario like the accident at the Fukushima Daiichi Nuclear Power

Station could have been extracted if comprehensive risk assessments had been conducted.

Based on the above insight, we provide the following recommendations:

- Quantitative safety targets show the acceptable level of risks for the communities. To share this with the communities, continuous efforts for dialog should be made. Along with these safety targets, risk information should be proactively utilized. Regulatory authorities should improve the transparency, predictability, rationality and consistency. Operators should voluntarily and continuously strive to reduce the risks related to activities using nuclear power.

(b) Regarding the design and operation of nuclear power facilities, as safety measures, assuming the accidents in advance, preventive measures are taken. However, in this accident, failure to prepare for cases of events beyond the design basis based on that assumption resulted in the disaster. To date, from the perspective of preventing similar accidents based on related experience, nuclear safety has been improved by drawing lessons that have led to effective measures.

The task force established within the US Nuclear Regulatory Commission (NRC) to examine the lessons of the accident at Fukushima Daiichi Nuclear Power Station reviewed the status of safety regulations in the United States, and concluded that while effective measures have been taken as a result of constant improvement efforts, a more balanced application of the Commission's defence in depth philosophy using risk insights would provide an enhanced regulatory framework that is logical, systematic, coherent, and better understood. Such a framework would support appropriate requirements for increased capability to address events of low likelihood and high consequence, thus significantly enhancing safety.

We are certain that one of the essential approaches to be able to respond to unpredictable events is to establish a basic concept on safety systematically, which would increase the possibility of acquiring comprehensive safety policies covering all loopholes which would be able to respond to various events. In particular, for measures to which fixed procedures cannot be applied like accident management, the importance of higher safety philosophy will increase. Examination on SF-1 and the new defence in depth by the IAEA is part of such systematic initiatives. Regarding individual facilities and systems, to optimize not only individual parts but also system-wide combinations in various events, we must consider a comprehensive safety system, covering all aspects from design to management. In addition, to promote the peaceful use of nuclear power, it is essential to ensure nuclear security as well as nuclear safety. However, despite common elements in both areas, they have been addressed independently. Therefore, these two areas must be consistently implemented in future.

Japan had worked to advance individual technology that realizes nuclear safety, which has been highly recognized internationally. However, it had put little emphasis on activities to consider safety systematically and conducting in-depth studies on safety concept. Accordingly, after the IAEA had developed SF-1, Japan had not yet positioned a higher level of safety concept corresponding to the fundamental principles for nuclear safety in its regulatory system. On the contrary, to find inconsistencies and omissions in safety requirements related to each facility, it is vital to create a framework working as a compass to examine them comprehensively. Based on this idea, the AESJ formulated the fundamental principles for nuclear safety based on SF-1 in November 2012. We must continue such efforts.

Based on the above insight, we will make the following recommendations:

- We must strive to develop and deepen higher-level thought concerning safety such as the fundamental safety principles in cooperation with the international communities. When doing so, knowledge in fields other than nuclear power should be proactively introduced. The regulatory authorities should clearly position the higher-level thought for safety such as the fundamental principles for nuclear safety in the regulations. Based on this, the regulatory authorities should organize systems for regulatory standards, etc.
- To ensure nuclear safety measures and nuclear security measures are consistently implemented and generate a synergistic effect, organizations which supervise each of these measures should enhance their information sharing and opinion exchange, while paying attention to the treatment of their classified information.

(2) **Deepening the understanding of defence in depth and strengthening the application**

In the accident at the Fukushima Daiichi Nuclear Power Station, the event actually occurring exceeded the design basis scale. Due to this serious damage, the expansion of the accident could not be prevented. Though measures assuming severe accidents such as main steam line breaks were included in the design before the accident, a situation where all power and coolant was lost simultaneously was not assumed. The facilities for venting as countermeasures for severe accidents could not work appropriately. The reasons for this include the fact that remote control for the vent valve of the primary containment vessel (PCV) did not work during the loss of all power. Another reason is that because the vent valve was located very close to the PCV, due to the high level of radiation, sufficient operation time could not to be taken.

The reasons why preparation for the event exceeding the design basis are as follows; As is commonly known, reactor design technology in Japan was introduced from the United States, and the safety designs were also based on design concepts in the United States. In addition to inherent safety, quality control and safety systems for postulated accidents, safety measures have been strengthened through thorough implementation of recurrence prevention measures against accidents and troubles experienced previously. Such approach had been regarded as adequate measures.

The above concept was globally widespread until the 1980s. However, the Chernobyl accident in 1986 changed the concept of safety design dramatically, since people became aware of the importance of countermeasures for events beyond design basis. Amid growing mutual concerns to clearly stipulate a safety concept and share it worldwide, the IAEA formulated "Defence In Depth" as the INSAG report (INSAG-10) in 1996. Until this report was published the concept of defence in depth was differently by each county, but a common definition worldwide was given defence by the report. However, considering that "an accident like that in Chernobyl would never occur with light water reactors adopted in this country", Japan had never incorporated the IAEA's defence in depth concept, which included countermeasures for events beyond the design basis assumptions, into its safety regulations. Therefore, the implementation of these countermeasures relied on voluntary efforts by the operators. As Japan had participated in the formulation of INSAG-10, it should have reflected these standards in its safety regulations at the time of its publication.

The Nuclear Regulation Authority (NRA) already stated its decision to adopt the IAEA's defence in depth in safety regulations in Japan and formulated various countermeasures for events beyond design basis. We hope the issue mentioned above will be improved toward resolution. Considering the importance of this issue, we make the following recommendations:

- The "Fundamental Safety Principles" proposed by the AESJ based on SF-1 should be used to formulate regulatory documents that stipulate the basic safety design concept.
- The concept of IAEA's defence in depth and relevant guidelines for implementing the defence in depth concept should be formulated as regulatory documents.

8.2.2 Recommendation II (Items Related to Direct Causes)

(1) **Strengthening measures against external events**

One of the direct causes of the accident at the Fukushima Daiichi Nuclear Power Station was inadequate design and preparation for tsunami, a natural phenomenon that hits the power station from outside. Due to the tsunami, a lot of equipment malfunctioned. In particular, the electric facilities were damaged by flooding and almost all power sources were lost, which further exacerbated the situation.

Though countermeasures for tsunami were steadily prepared, the area was hit by a tsunami on a scale exceeding what was supposed. In addition, in this accident the cliff edge, in which the impact sharply rises when exceeding postulated conditions, clearly appeared. In other words, when a tsunami exceeding a certain height flooded the facilities, many of the safety facilities lost their function, which compounded the situation. Based on this accident, under new regulatory standards, a tsunami exceeding the maximum level of the largest recorded flood in the past was included as a "reference tsunami", for which measures such as tsunami protection facilities e.g. tide embankments would be mandated.

It is important to strengthen countermeasures for tsunamis. At the same time, based on the experience that by focusing on countermeasures for earthquakes only, adequate countermeasures for tsunami may be overlooked, we must prepare for other events, in addition to countermeasures for earthquakes and tsunamis, which may cause malfunctions of all safety facilities simultaneously due to common causes just like in this accident.

To implement this thought, regarding various external events such as future earthquakes and tsunamis, the effective method involves quantitatively evaluating risks using a probabilistic risk assessment (PRA) and checking the resistance against huge natural disasters. It may be difficult to measure the risks of external events comprehensively and quantitatively. However, using the PRA impact assessment, the vulnerability of the plant can be identified, and by continuously improving such vulnerability based on results, the safety of the plant as a whole can be enhanced. At the same time, as evaluation of external events includes major uncertainty, it is important to use the PRA and act based on the concept of defence in depth. Preparation for artificial incidents such as terrorism would be also vital.

Based on the above insight, we make the following recommendations:

- External events to be assumed include earthquakes, tsunami, fires (forest fires, etc.), strong winds (typhoons, tornados), floods, avalanches, volcanoes, freezing, high temperatures, low temperatures, transport/factory accidents, airplane crashes, etc. An envelope assessment of these external events, identification of vulnerabilities in each plant, and definition of the responses of each plant based on these results must be mandated. When doing so, to prepare for uncertainty, measures should be taken considering the defence in depth as well as the PRA evaluation.
- For external events, the vulnerabilities should be identified through the detection of the cliff edge, identification of plant behavior and possible responses in case of the loss of safety function. Then appropriate actions for the vulnerabilities should be conducted.
- For artificial causes such as terrorism, to proactively utilize overseas knowledge, we should join international discussions, develop human resources and enhance the preparedness.

(2) **Strengthening measures against severe accidents**

Regarding the installation of nuclear power stations, to prepare for potential abnormalities and accidents, a safety assessment is conducted considering the representative events that cause major impacts called "design basis events", based on which safety facilities are designed. At the Fukushima Daiichi Nuclear Power Station, as events far exceeding these design basis events occurred, severe damage to the core (severe accident: SA) occurred, which caused the loss of functions for confinement, cooling of PCVs and probing the reactor condition.

Measures to prevent the progress toward SA and mitigate the impact at the time of SA are called accident management (AM). At the Fukushima Daiichi Nuclear Power Station, AM measures had been formulated by May 2002 as voluntary efforts of the electricity utility. However, little review had been conducted of the AM measures since then. At that time, there were repeated discussions and continuous improvement efforts globally with regard to the AM measures. Given these circumstances, at the time of the accident at the Fukushima Daiichi Nuclear Power Station, we have to say Japan was far behind the global standards. These inadequate countermeasures for severe accidents are one of the direct causes and reasons why the accident triggered a disaster.

SA is an accident where an event occurs on a scale exceeding the design basis. The important issue when planning preparation for such events is to consider comprehensive risks for the plant, including operation and maintenance. However, we must also take into account that adding measures (in particular for facilities) does not always reduce risks. At the same time, given that the nature of such events cannot be identified, without depending on the particular accident scenario, we must be ready for whatever events may occur. Therefore, it is important for plant operators to know the plant well, and be able to manage it using all resources through exercises. Moreover, equipping the necessary hardware and software for this management is also important without just relying on hardware. In addition, the management must consider the measures for each, not only to Level 3 of the defence in depth within design basis events but also to Level 4 beyond the design basis events and Level 5 for disaster prevention measures. Assuming the case of a complex disaster in which an accident at the nuclear power station and natural disasters proceed concurrently, social infrastructure has been destroyed, and multiple plants are subject to a severe accident situation concurrently, it is also important to enhance facilities, materials and machinery, procedures, education training, human resources for response, and organizations.

Based on the insight above, we make the following recommendations.

- In SA the event may not proceed as assumed in the scenario, hence a flexible response ability is required to address the event as the management. To foster this ability, continuous improvement activities should be conducted through exercises/drills.

(3) **Strengthening emergency preparedness and the response framework**

The legal basis of the nuclear emergency response system in Japan is based on the "Disaster Countermeasures Basic Act" and the "Act on Special Measures Concerning Nuclear Emergency Preparedness". The Basic Disaster

Management Plan is a planning basis document that describes the roles and responsibilities of the relevant organizations. The "Regulatory Guide on Emergency Preparedness for Nuclear Installations" issued by the former Nuclear Safety Commission (NSC) should be considered an important technical document by national and local governments and utilities for use in establishing an emergency plan and for implementing protective actions. The framework and related documents seemed to be well prepared. However, the basic concept and clear operational procedures for measures to protect the residents in emergencies were not specified. In emergency response drills, a decision-making scheme to determine urgent protective actions were established over-dependent on calculation and forecast systems—ERSS (Emergency Response Support System-source term predictions for accident progress and released amount etc.) and SPEEDI (System for Prediction of Environmental Emergency Dose Information)

In the emergency response to the accident at the Fukushima Daiichi Nuclear Power Station, various problems occurred just like the case of the JCO accident such as confusion over the initial response, lack of collaboration between responsible authorities, and unclear decision-making schemes. However, the concerns and discussions simply focused on the utilization of such tools and announcement of the results gained by tools. From the perspective that the Level 5, emergency preparedness and response, is the last resort for the IAEA's five-level defence in depth, and to achieve the emergency response target of how we could protect residents against the radiation impact, the Accident Investigation Committee analyzed the issues in implementing urgent protective actions, and those related to emergency management and operation including identification of the responsibilities and roles of operators, municipalities and the national government to draw the lessons and recommendations.

When preparing for emergent situations and developing response measures, the worst situations must also be taken into consideration. To ensure the radiation risks are reduced against rationally predictable events, emergent situations must be examined including the worst scenario that operators can anticipate based on the assessment of possible events for facilities, and complex disasters, such as a combination with general emergency situations such as earthquakes. During the crisis management of the response phase, the response should be undertaken in accordance with predetermined criteria, then subsequently responding more flexibly when deviation from the pre-established arrangements is required. There is also a need to develop such capability during the ordinary time.

To achieve this, we make the following recommendations. It is necessary to review the responsibility and role of the organizations concerned at each level - operator, regional, national, and international—based on the recommendations. Moreover, it is necessary to make an agreement between organizations, including arrangements to coordinate a unified response, and re-examination following training to ensure that it functions effectively.

- At the initial crisis management stage when little information can be obtained and there are many uncertain elements, a scheme must be established based on an assessment performed on the facilities beforehand, so that urgent protective actions within a predetermined zone can be implemented promptly before a release of radioactive material to the environment in collaboration between operators and municipalities.
- Responsible parties such as the national government, municipalities and operators should consult, decide on and document the roles and responsibilities for on-site and/or off-site emergency response. In principle, operators should be responsible for the response to on-site matters and municipalities for off-site matters, and the national government should support them.
- Regarding the crisis management, detailed response policies such as various procedures and urgent protective actions should be examined and clarified in advance through exercises/drills.
- Regarding the diffusion analysis information of radioactive materials by SPEEDI or any other similar system, relevant organizations should understand the limit of such systems which cannot be used for decision-making in evacuation at the early stage of the accident. The handling method for such system should be clarified.
- The activities of each relevant organizations in nuclear emergency—municipalities implementing disaster management measures, police standing at the forefront of the resident protection, firefighters and Self-Defense Forces, and the national government—are basically the same as those in other general disasters. Therefore, activities in nuclear emergency and those in other general disaster must be integrated into a common platform with reference to the cases of other countries.
- With regard to measures for radioactivity, one of the unique characteristics of nuclear emergency management, all staff members working for the accident response should have sufficient knowledge of the principles of radiological protection and impact of radiation exposure, and increase the ability to respond to nuclear emergency.

(4) **Enhancement of the nuclear safety assessment technology**

Due to the significance of its potential risks and the uncertainty related to events exceeding the assumption, the nuclear power field requires technology to analyze and assess the progress of an accident and uncertainty related to the same progress and its impact, efforts to appropriately reflect new knowledge and data in the safety assessment, to avoid any omission or gap, and to further ensure quality in all areas. As risks of rare event are dominant cases, in addition to combining the existing data and design methods related to the postulated events, efforts should also be made to estimate the progress and magnitude of a broad range of scenarios while constantly increasing the quality of estimate.

Regarding natural phenomena such as earthquakes and tsunamis, it is important to establish methodologies to appropriately assume the scale and combination of natural phenomena through for constantly collecting and analyzing the latest insights. When considering natural phenomena, it is essential to check for any accompanying phenomena or secondary events, and to take into account the frequency of occurrence, magnitude and margin for responding time.

Regarding the seismic design and analysis on tsunami wave propagation and flooding (upstream behavior), based on recent and near-future development of computers, progress is expected in analytical methods for multi-dimensional finite element method, time history response analysis and numeric calculation methods using the latest computing performance such as large scale calculation technology. By using such simulation technology to calculate the safety factor and safety margin, to analyze uncertainty in data and scenarios and to make quantitative assessment of risks, the precision and quality of future safety assessment technology are expected to greatly increase.

Though the reconstruction analysis of the accident at the Fukushima Daiichi Nuclear Power Station has been conducted in several organizations, transient details of the plants had not yet been fully reconstructed due to the reliability of measured values at plants, lack of information concerning operation and equipment handling while responding to the accident, and the limits of analysis code models. At the same time, by correctly understanding these technical issues/limits, essential and effective information can be obtained to analyze the accident progress and future responses. For the phenomenon model and source terms, the Phenomena Identification Ranking Table (PIRT) showing the severity from the perspective of these accident impacts has been performed using all knowledge and up-to-date data. However, due to the lack of validated models and data, no precise analysis codes have yet been developed. Given that progress of simulation technology is vital to deepen understanding of the progress process of the accident, validation of analysis and accumulation of data by experiments must be promoted to make the analysis code more complete.

Based on the above insight, we provide the following recommendations:

- To increase the quality of estimate on natural phenomena, development and application of probabilistic risk assessment which considers the uncertainty of natural phenomena and plant system durability should be prioritized.
- Regarding the seismic design and analysis on tsunami wave propagation and flooding, numeric calculation methods always using the latest computing performance should be applied. At the same time, aware of the complexity in natural phenomena and the limits of our knowledge, validation of simulation technology and proper application should be maintained.
- By correctly remaining aware of the issues and limits when applying, the simulation and risk assessment can be utilized effectively in the safety assessment. In addition to proactively using them, the government, industry and academia should collaborate to make the technology more complete, collect new knowledge, and assure quality.

- International cooperation in nuclear safety assessment technology which creates benefits to both parties, should be promoted proactively and continuously.

8.2.3 Recommendation III (Items Related to Organizational Causes Among Underlying Causes)

(1) **Initiatives of AESJ and academia as expert groups**

Despite the potential risk was suggested by experts from the academic perspective before the accident, such knowledge was not used to prevent the accident. The academic research on nuclear science/engineering should have originally aimed not only to contribute to pure science but also improve safety for real facilities and safety regulations. What kind of research results would be utilized in design and regulations in what way should be determined by each organization of the government, industry and academia. However, the Atomic Energy Society of Japan (AESJ), which has been closely involved in nuclear safety research, must play a major role in this field such as issuing alerts not to disregard important point. To fulfill this role, it is essential for the AESJ to continuously revitalize its research activities concerning safety improvement, and strive to foster a safety culture to ensure research achievements are reflected in the real design and safety regulations, collaborating with government and industry.

Reconsidering above lessons learnt, extensive discussions have also been conducted in academia. In particular, being aware of its responsibilities and recognizing its obligation to review its activities as an expert group, the Investigation Committee provides the AESJ with the following recommendations. We hope these recommendations will be examined at the AESJ, reflected in the activities of the AESJ and its members, and spawn further discussions in the academia. The AESJ should also make the efforts to use the outcomes actively.

(a) Reminding responsibilities the AESJ should assume: The AESJ has to fulfill mandate from the communities and regain public confidence. In particular, being aware that the nuclear power technology has serious impacts on human beings on particular case, there is a need for the AESJ to always make decisions and act in an ethical manner. In addition, at the AESJ General Meeting in June 2013, the activities toward restoration of the affected regions and assistance in decommissioning work at Fukushima Daiichi Nuclear Power Station were stipulated in its articles of association. Now we, AESJ members, must recognize that these activities are our responsibilities.

(b) Unrestricted discussions at the AESJ: Being aware of the importance of independent activities from the objective and fair perspective, the AESJ must strive to foster an atmosphere in AESJ that allows exchanges of free and honest views.

(c) Enhancing the safety-related research: The framework to continuously conduct research to improve safety must be restored and the system for nuclear safety research must be reconstructed. In such research activities, the AESJ must assume the leading role by formulating a roadmap and its continuous amendment.

(d) Enhancing the interdisciplinary initiatives: In addition to setting up a "place" for comprehensive discussions and collaboration with other fields of academia related to nuclear safety, the AESJ must play a leading role there.

(e) Contribution to continuous improvement of safety regulations: The AESJ must enhance research that supports the regulatory system and the activities for developing standards, as well as appropriately disseminating achievements to the communities, including the research results on social dimension, as necessary.

(2) **Initiatives of the industries**

Major underlying causes of the accident at the Fukushima Daiichi Nuclear Power Station include issues of the operator, such as lack of awareness and recognition of safety, inadequate technological skills, and lack of communications with regulatory authorities and the communities. Some of representative cases of the direct causes related to industries are as follows: the decision was delayed and failed to implement the measures before the accident though increasing the height of tide embankments was discussed as the countermeasures for tsunamis, the risk of potential flooding was pointed out within the company but the emergency power generators had never been moved from the underground floor of the turbine building and the same concerns were raised for the direct current power supply and power panels. The quality management system needs to be appropriately operated through both top-down and bottom-up approaches.

In the IAEA's "Fundamental Safety Principles (SF-1), ten principles of safety are specified. Principle 1 for the "Responsibilities for safety" defines that "The prime responsibility for safety must rest with the person or organization responsible for facilities and activities that give rise to radiation risks". Needless to say, this means that the operators (licensee) have the most important roles in nuclear safety. In the same IAEA's safety principles, Principle 5 for the "Optimization of protection" defines that "Protection must be optimized to provide the highest level of safety that can reasonably be achieved". This means that the operators should not be satisfied with just "meeting the regulatory requirements" but are required to continue efforts for improving safety within the range that can reasonably achieved. We hope for the operators to declare that "We have not only complied with the national requirement but also continue to increase the safety beyond the required level within the range that we can do".

The industries established the Japan Nuclear Technology Institute (JANTI) as an organization promoting peer reviews and sharing of technical information, to improve the safety of nuclear facilities. Based on the regret that it was not workable to prevent the accident at the Fukushima Daiichi Nuclear Power Station, the JANTI was abolished and a new organization named the Japan Nuclear Safety Institute (JANSI) was established as a driver for strengthening the safety measures to renew the framework related to the safety. We hope that this will mark the first step for reform. However, based on the reflection of the underlying cause or organizational cause of this accident pointed out by the Investigation Committee that "The overall operators lacked awareness of safety and recognition on safety and failed to utilize the JANTI which was responsible for the advice on correction", we make the following effective reforms:

- At the time of an accident, the safety issues of nuclear power stations are beyond the boundaries of the relevant power station and have impacts on the entire communities and the world. This lesson is the issue not only of TEPCO, which is the relevant party, but also of all operators. The industries should be reminded of this lesson and face the organizational issues extracted from the perspective of safety awareness, technology and communication ability as common issues of themselves to work toward resolution.
- The entire nuclear industries should remain aware of the unique risks of using nuclear power and strive to improve the safety continuously, not temporarily.
- It is essential for the senior management to be committed to prioritizing nuclear safety. The senior management should eliminate overconfidence in safety, proactively participate in opportunities to raise awareness of nuclear safety, and infiltrate the safety culture that promotes rigorous commitment to continuously enhancing the safety, into the organization.

(3) **Initiatives of nuclear regulatory authorities**

As an underlying cause, we pointed out in Sect. 8.1 the lack of awareness of safety by regulatory authorities of the national government as represented by the fact despite new knowledge related to the tsunami assumption acquired in advance, they did not order actions. In addition, regarding the safety regulations concerning countermeasures for severe accidents, Japan had been significantly delayed. Though examinations were underway, no prompt actions had been taken. Regarding emergency preparedness, no effective measures had been implemented, nor had the workable management system been established to unify related organizations in emergencies and carry out appropriate measures.

These issues were attributed to the organization. First, as the nuclear regulatory organization had been incorporated into the framework of HR rotation, a unique system in the Japanese bureaucratic organization, regulatory officials had less expertise. As a result when implementing the quality assurance system, there were several cases of superficial inspection. In addition, the Nuclear and Industrial Safety Agency (NISA), the regulatory administrative organization of nuclear power stations which belonged to the Ministry of Economy, Trade and

Industry, was less independent with limited rights over personnel matters, budget and license permission.[1] In addition, they should have made more efforts to improve regulatory systems to boost safety.

As we have mentioned above, they failed to introduce the internationally standardized defence in depth such as countermeasures for severe accidents. As the regulatory authority had built a regulation system on the assumption that "no accident would occur", the introduction of regulatory methods using risk information which had been introduced overseas was delayed.

After the accident, the first thing the government addressed was the reform of regulatory systems mainly for organizational aspect. By the Law for Establishment of the Nuclear Regulation Authority enacted in June 2012, the national regulatory organization was drastically restructured. The safety regulation system of the government, which had been divided into small segments, was integrated into the Nuclear Regulation Authority (NRA) which was granted the status of "Article 3 committee" without any interference from politics and other government organizations as well as stakeholders. By this reform, in a way the controlling organization of nuclear safety which had not been clearly defined before then was made clear and a system equivalent to that of western developed countries was established. As a measure to improve professional ability of regulatory officers, the system included a rule to exclude senior managers from the Government's general personnel rotation. In new regulatory system the concept of IAEA's defence in depth was introduced. The use of probabilistic safety analysis was expanded in fields other than seismic design, and the assessments of external events such as tsunamis were required in the scope of safety design. Thus the regulatory issues we mentioned above had been improved. To establish firmly these initiatives over the mid- and long-term periods and continuously improve the practice of the regulations, we provide the following recommendations:

- The priority should be placed on recovering confidence in safety regulations lost by the accident at the Fukushima Daiichi Nuclear Power Station. To build confidence, regulatory practices based on scientific and rational judgment should be accumulated as track records. When doing so, efforts to fulfill accountability about such judgment process and results are needed. Dialogs with licensees, local residents near the nuclear power facilities, general public, academia, and international communities should be proactively promoted.
- The regulatory authorities must also continuously improve their own organization and system as it requests such efforts to licensees. They must strive to closely communicate with the licensees to access the latest on-site information, and by eliminating self-complacency figure out challenging issues in the regulatory system and its operational practice. In addition, they

[1] The authority of government licenses before the accident was not the Director General of the Nuclear and Industrial Safety Agency but the Minister of Economy, Trade and Industry.

should use international peer review services and consider the implementation of audit system.

- From the perspective of prioritized allocation of regulatory resources to facilities and management activities with high-risk of accidents, the introduction of risk-informed regulation effectively contributes to safety improvement under limited regulatory resources, which should be addressed proactively. We are confident that these efforts will help regulatory officers develop the ability to assess the risks, which leads to practical safety improvements.
- The regulatory authorities are requested to shift conventional regulations which had major concerns on the mechanical performance of hardware to a regulatory system focused on software, which means a regulation emphasizing basic concept of nuclear safety, the performance and functions of whole system, and management of operation. Also it is requested for them to make efforts to foster human resources for implementing above mentioned shift of the regulation.
- To continuously maintain and improve nuclear safety, it is important to encourage voluntary safety improvement efforts by operators. To do so, measures to prevent the operators from falling into the trap of thinking that "all we must do is to follow the regulations". From this perspective, the risk-informed regulation is an important method to encourage these efforts by operators. Japan should also proactively use standards of the private sector like practice of regulatory systems in western countries. These measures will enhance technologies for safety standards in the private sector, expand the base of engineers for standards, and consequently contribute to long-term safety upgrading.
- Nuclear power technology is broad and complex and requires maximum use of relevant expert knowledge in a balanced manner when regulating it. Accordingly, in operating the Examination Committees, appropriate consideration should be made in the structure of the Committees to prevent imbalance of experts, using academic organizations such as the AESJ.

8.2.4 Recommendation IV (Common Items)

(1) **Strengthening the nuclear safety research foundation**

To ensure nuclear safety, it is important to clarify the basic concept of nuclear safety, set safety goal, effectively contribute to safety improvement based on the probabilistic risk analysis, and continuously pursue appropriate application of defence in depth concept to plant design, accident management, and disaster prevention measures. Research on nuclear safety is a foundation of safety measures and underpins these continuous efforts for safety.

In general, the achievements of research not only foster the development of scientific technology, but also increase technological options in society. The safety research is expected to increase the flexibility for upgrading the safety level, point out potential issues and sound alerts based on discussions regarding new scientific technological knowledge.

To ensure safety by the concept of defence in depth, cause events of all possible accidents that may occur must be considered. In addition, by identifying the whole picture based on the risks, the requirements for equipment design must be shown, the facilities must be maintained, and appropriate management at the time of accidents is required. We are certain that in the safety research, by applying the probabilistic risk assessment method to identify the whole picture, the approach toward cause events would be enhanced. We are also confident that applying probabilistic risk assessment to research on security would be effective.

Researchers tend to deepen their own specialized fields. At the same time, safety may fail through the gap in many technological fields and areas. Keeping a comprehensive perspective, research plan need to be formed, and achievements must be utilized effectively.

Based on the above insight, we provide the following recommendations:

- Safety research should become a driver for deepening the understanding to overview the approach to safety and continuous advance of various software and hardware to improve safety.
- Safety research is also important to maintain and develop advanced human resources for nuclear field, which should be conducted seriously while promoting international cooperation.
- The government, industry and academia should recognize that they are responsible for pursuing safety research through information exchange and discussion among various levels in society.
- Regarding the probabilistic risk assessment to identify the whole picture, the application range should be expanded to the research on safety for accidents caused by external events such as tsunamis and fires. From this perspective, deeper and wider research related to security should be introduced in addition to the safety research.
- By discussing an ideal status to achieve the goal of nuclear safety and facing the current technology, a map of comprehensive technological issues to be addressed should be prepared, and to resolve these issues, mid- and long-term roadmaps as well as short-term ones should be developed. These roadmaps should be widely published together with the evaluation perspectives and continuously revised through communications with society.

(2) **Strengthening the international cooperation framework**

The peaceful use of nuclear technology commenced within the international framework from the beginning, where various international collaboration such as obligatory activities specified in the treaties, joint research project and information exchange programme had been vigorously conducted multilaterally and bilaterally. Japan, which has also proactively participated in this international framework, is regarded as a country with high-level technology and a wealth of experience related to nuclear science/engineering. Internationally, cases of damage to nuclear power stations flooded by huge tsunamis and submerged plants due to floods were reported, which could be used as lessons. In addition, measures for facilities and management responding to severe accidents have been introduced mainly in European countries, and discussions to position these measures as regulatory requirements have been made internationally. As shown above, international talks which may have prevented this accident had been made in various forms, in which Japan had also participated. The United States implemented measures to enhance power supply based on B.5.b as counterterrorism measures and informed such measures to Japan. However, the information provided through international cooperation was not effectively used to prevent this accident.

The reason for the failure is considered to be the lack of strong commitment to introduce these measures into Japan and the effective system to use that framework. We provide the following recommendations based on the assumption that Japan will use nuclear power stations as electricity source facilities:

- Japan should proactively participate in international activities and establish an effective system to reflect the discussions there in its national measures.
- As an increasing number of countries are launching nuclear power, Japan should proactively provide its experience including that of nuclear power accidents to contribute to building systems toward ensuring nuclear safety. From this perspective, developing human resources who lead international meetings is required.
- If Japanese nuclear reactor vendors target international business expansion in future, the industries should proactively participate in international frameworks.

(3) **Developing human resources related to nuclear power**

In the accident at the Fukushima Daiichi Nuclear Power Station, "humans" committed the underlying causes, "humans" responded to the accident and "humans" settled the accident. Based on this lesson, we are certain that even in the emergency situations of an unpredicted accident, the priority should be given to developing "humans" who are able to think, judge and respond. At the same time, to complete the decommissioning work over the next 40 years to the final stage, it would be vital to develop experts over many years. Based on the above insight, we provide the following recommendations:

- Regarding development of the nuclear power human resources, the sense of value to put top priority to the nuclear safety, "safety first", should be continuously enhanced. It is necessary to constantly eliminate overconfidence or arrogance, settle the "attitude to learn" and the "attitude to ask", and regularly review the level of such settlement. In particular, it is vital for the top management in the organizations to demonstrate a strong commitment to nuclear safety. They must provide instruction by themselves to increase awareness of nuclear safety on every occasion. In addition, by clearly institutionalizing that safety-related knowledge and experience inherent in nuclear technology such as radiation protection is essential for works in this field, the necessary education and training should be rigorously provided to the workers.

- Knowledge and skills necessary for human resources in the nuclear power field should be more clearly set by establishing qualification systems. Specific issues are as follows: qualification requirements for directors and operation managers of nuclear power stations should be clearly defined considering the emergency responses; roles of the chief reactor engineer, which requires a national license, should clearly specified to ensure that they are able to take responsible actions both during ordinal time and at the time of accident; and expertise, global view and judgment ability of regulatory personnel should be improved. In addition, it is also important for relevant organizations to give incentives to a staff by establishing human resource management system to place a high value on the staff acquired above mentioned abilities and careers.
- To continuously secure human resources in the nuclear power field for which a high level of knowledge, skills and management ability are required, it is important to enhance education on nuclear engineering in universities. At the same time, focus should be also placed on human resource development for research. To introduce the latest research achievements and maintain the nuclear safety at the highest global standard, it is essential to keep the research in the most advanced level. Accordingly each of the national government, regulatory authority and the industries are required to be actively involved in the research for safety.
- From the perspective of continuous development of human resources, there is a need to increase the interest of the younger generations in nuclear field. To achieve this, education on radiation should be urgently enhanced. Relevant organizations must cooperate in study and training about nuclear power and radiation for teachers in elementary/junior-high/high schools, while providing information to increase interest in nuclear science/engineering.

8.2.5 Recommendation V (Items Related to Restoration)

(1) **Initiatives toward future environmental restoration**

Based on the response by the national government following the accident, we provide the following recommendations for radiation monitoring, legislation and guidelines, designation of areas for decontamination, decontamination and its technology, and deposit and storage of decontamination waste:

- **Radiation monitoring**: Regarding the monitoring at the time of future emergencies, a system for centralized collection and storage of data from the initial stages must be established so that it would be workable in emergencies.

In addition, there is also a need to conduct long-term assessments of radioactive dose for residents including infants. New methods for individual radiation monitoring should be developed to establish a system for managing continuous assessments.

- **Legislation, regulations and guidelines**: As installation of facilities such as temporary storage facilities of decontamination waste is delayed and there are cases where the decontamination is not notably effective, guidelines for decontamination methods should be enhanced by including the latest knowledge and ensuring a flexible and practical implementation to actual decontamination work. The generation of contaminated soil, debris and grasses/trees are the same both in and outside the nuclear power station site. Therefore, the relation between the Act on Special Measures concerning Handling of Radioactive Pollution and the existing Law for the Regulations of Nuclear Source Material, Nuclear Fuel Material and Reactors should be reviewed to make them more effective. For this purpose the upper level concept of these laws should be shown.
- **Designation of decontamination areas**: The national government uniformly designated areas where the additional radiation dose is 1 mSv/year or more as decontamination areas. One mSv/year should be the long-term target. More practical decontamination targets and decontamination areas should be set, considering the cost, time and effect related to decontamination as well as annual individual effective residual dose based on the optimization principles of ICPR. Regarding decontamination, not using the "average individual" in dose control but it should be reviewed based on the individual dose measurement results.
- **Decontamination and technology**: Regarding the decontamination implemented by municipalities, to ensure flexible decontamination according to the circumstances, prompt decision-making near the site should be made possible. When implementing decontamination, the responsible parties should make their utmost efforts to ensure cooperation and participation by the local residents. Decontamination methods should be selected case by case basis according to the characteristics of their locations and subjects. Early establishment of a one-stop service should be recommended where the framework to systematically organize and organically links the achievements of efforts by each responsible organizations as well as effectively reflecting the achievements in the guidelines and manuals for decontamination should be built by the national government and municipalities in an integrated manner,
- **Deposit and Storage of decontamination waste**: As installation of the temporary storage facilities has a direct impact on the progress of decontamination, the responsible parties must actively engage in dialog with local residents, and proactively encourage participation by the local residents in site selecting process. Contaminated waste is to be stored initially at temporary storage facilities, then at Intermediate Storage Facilities, and in the end transferred to final disposal facilities. In this flow, minimization of transported materials volume greatly boosts smooth transportation.

Therefore, the volume reduction and reuse of contaminated waste are essential. Responsible parties should implement necessary measures to ensure that these measures are taken promptly.

As shown above, we have made some suggestions aiming at prompt progress of environmental recovery. To further advance the environmental recovery, the understanding, cooperation and participation of residents in the vicinity is essential. The AESJ must continuously hold and co-organize forums, host dialog meetings with local communities, and continue to be proactively involved with the existing Decontamination Information Plaza. As a neutral academic organization, the AESJ assumes the role of digesting the thought of the residents, serves as a contact with the administration, and requests the necessary measures to national government and responsible authorities as needed. For this reason, the AESJ would like to proceed to talk with local residents through assisting activities such as dispatching the latest information in plain words, and proactively offer opinions to relevant organizations.

8.2.6 Conclusion

In this Chapter, we have made broad recommendations based on the results of the root cause analysis. To ensure nuclear safety, there are a wide range of issues such as identifying the basic concept of nuclear safety, using probabilistic risk assessment (PRA), and setting safety goal, as well as correctively understanding the concept of defence in depth, and applying it to plant design, accident management, and disaster prevention. Here, we would like to re-emphasize the importance of continuous development of research on nuclear safety as the foundation for these initiatives. Research activities expand the area of human knowledge, deepen our understanding of intrinsic nature of the issues, and lead us to the optimal resolution measures. The AESJ is committed to sincerely pursuing research, developing human resources and contributing toward resolution of challenging issues related to the use of nuclear technology.

Chapter 9
Post-accident Management in Progress

Abstract Post-accident measures that are ongoing can be classified into three tasks, namely on-site decommissioning work, off-site environmental remediation, and the healthcare of residents and employees. Off-site environmental remediation has been already discussed in Sect. 6.7 in detail. The future tasks for on-site decommissioning work and the healthcare of residents and employees are described in this chapter.

Keywords Contaminated water • Decommissioning • Handling of damaged fuel • Healthcare • Radioactive waste • Stable storage of components

Post-accident measures that are ongoing can be classified into three tasks, namely on-site decommissioning work, off-site environmental remediation, and the healthcare of residents and employees. Off-site environmental remediation has been already discussed in Sect. 6.7 in detail. The future tasks for on-site decommissioning work and the healthcare of residents and employees are described in this chapter.

In decommissioning work, the following operations will be performed for an extended period: spent fuel removal, in-plant investigation, removal of rubble, work to stop leaks, and fuel debris sampling and removal. During this period, one of the key tasks is to maintain the current cold shutdown state. Conversely, it is important to note that the declaration of a nuclear emergency as issued on March 11, 2011 for the Fukushima Daiichi Nuclear Power Station (hereinafter referred to as Fukushima Daiichi NPS) remains in place. Related parties must note that the ongoing decommissioning work is still being carried under such emergency and pay careful attention to the work. The Fukushima Daiichi NPS was designated as special nuclear facilities by the Nuclear Regulation Authority (NRA) on November 7, 2012. Related parties must be adequately aware that the NRA has instructed them to perform decommissioning under special safety management unlike that required for normal operation. The designation of the Fukushima Daiichi NPS as special nuclear facilities will be lifted after such designation is no longer necessary, whereupon the NPS will enter into the next decommissioning stage.

The Fukushima Daiichi Nuclear Accident: Final Report of the AESJ Investigation Committee, DOI 10.1007/978-4-431-55160-7_9

9.1 Treatment and Cleanup of Contaminated Water

(1) **The current status of the contaminated water cleanup and treatment system**

In Units 1–3 of the Fukushima Daiichi NPS, cooling water is being continuously injected to cool the nuclear fuel inside the reactors (it is estimated that most of the fuel have been damaged and completely misshapen). The cooling water injected into each reactor has leaked into the reactor pressure vessel (RPV), the primary containment vessel (PCV) and further into the reactor building (R/B). Finally, the injected cooling water is accumulating in a large area and reaching the turbine building (T/B) basement. The accumulated water (hereinafter referred to as contaminated water) is contaminated by substances such as radioactive fission products (FPs). Therefore it is anticipated that if the contaminated water were left unattended, it could overflow from the turbine building basement, leak outside the T/B, and be partly released into the sea. Subsequently, the contaminated water is being blocked inside the T/B to prevent its overflow. As shown in Fig. 9.1, the wide area ranging from the RPV to the T/B basement is now regarded as a cooling water storage container. To prevent the total cooling water volume inside the entire cooling water storage container from increasing, the contaminated water is being pumped out continuously or as required. After impurities such as FPs are removed from the contaminated water, the purified water is re-injected into the reactor. Thus each fuel assembly is being cooled in such a large-scale cooling water circulation system. The leakage locations from each PCV must be identified; the leakage sealed, and perform cooling inside the small-scale circulation system ranging from the RPV to the PCV as early as possible.

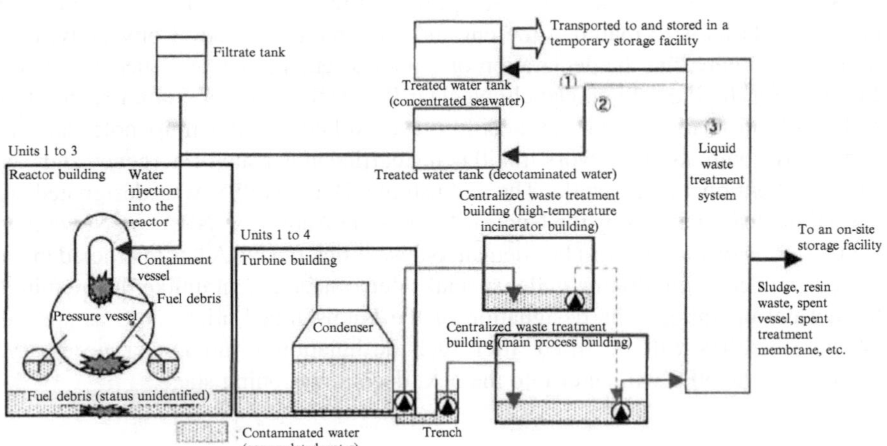

Fig. 9.1 Treatment of contaminated cooling water (pooled water) circulating in the Fukushima Daiichi NPS

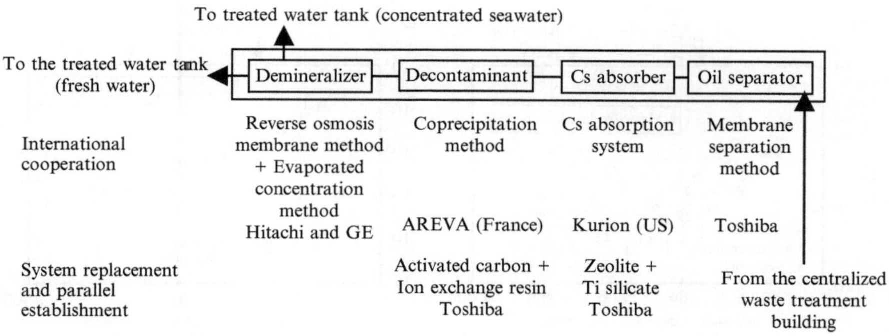

Fig. 9.2 Contaminated water treatment system

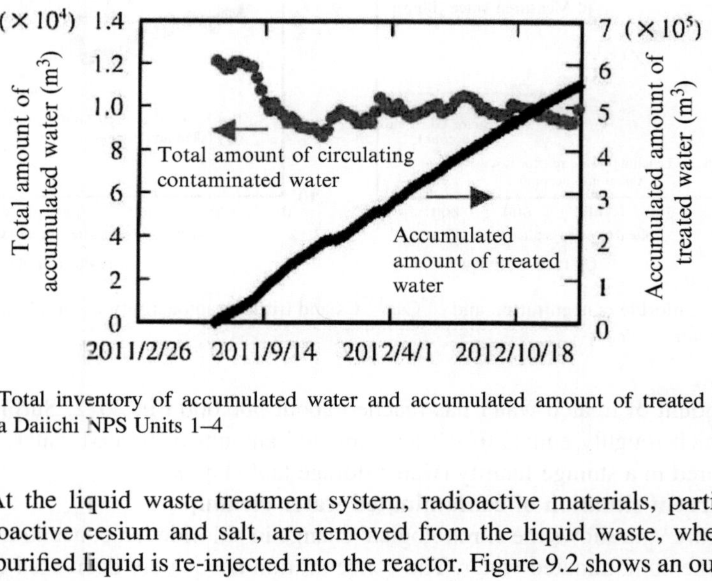

Fig. 9.3 Total inventory of accumulated water and accumulated amount of treated water at Fukushima Daiichi NPS Units 1–4

At the liquid waste treatment system, radioactive materials, particularly radioactive cesium and salt, are removed from the liquid waste, whereupon the purified liquid is re-injected into the reactor. Figure 9.2 shows an outline of the system.

Initially, two domestic manufacturers provided oil separators and demineralizers, while as part of international cooperation, Kurion (US) and AREVA (France) provided Cs absorbers and coprecipitation-type decontaminants respectively, since which time domestic devices have gradually been introduced instead of those manufactured overseas.

Water is being continuously injected into the reactors of Units 1–3 at a rate of about 20 t/h to cool the fuel. Groundwater is flowing into the turbine building and the accumulated water is increasing by the amount of groundwater inflow (almost equal to the amount of water injected into the reactors). Figure 9.3 shows the total amount of contaminated water circulating in Units 1–4 and the accumulated amount of treated water. The total amount of circulating contaminated water has reached 100,000 t, while the accumulated

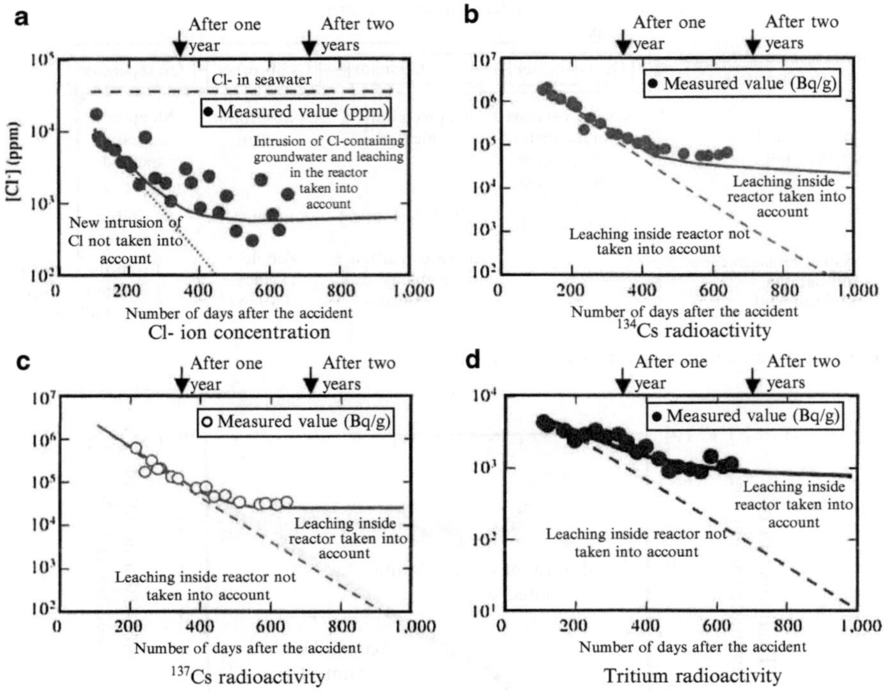

Fig. 9.4 Chloride concentration and ^{134}Cs, ^{137}Cs and tritium radioactivity in circulating contaminated water

amount of treated water has reached about 500,000 t in 2012. Surplus water, which roughly equals to the accumulated amount of treated water, has been stored in a storage facility (water storage tanks) on site.

(2) **Current situation of contaminated water cleanup**

Figure 9.4a shows the chronological change in chloride ion concentration (Cl^-) in the contaminated water. The concentration adequately decreased compared with Cl^- (3.5 %) in injected seawater and has been now almost leveled off (stopped decreasing). Because the salt concentration in groundwater mixed with the contaminated water is high and due to leaching inside the reactors, efforts to reduce Cl^- in the polluted water are being hampered, despite the adequate salt removal by demineralizers using a reverse osmosis membrane.

Figure 9.4b shows the chronological change in cesium-134 (^{134}Cs) radioactivity in the contaminated water. As in the case of Cl^-, ^{134}Cs concentration initially decreased due to Cs removal by the Cs removal tower of the liquid waste treatment system. Eighteen months after the accident, however, the concentration has almost stabilized (stopped decreasing). The main source of the ^{134}Cs is its leaching inside the reactors or containment vessels and the ^{134}Cs seems still leaching. Conversely, Fig. 9.4c shows a chronological change in ^{137}Cs radioactivity.

Said chronological change is almost the same as in the case of Cl- and ^{134}Cs.

Tritium, which is a hydrogen isotope, cannot be removed by the liquid waste treatment system currently used. However, the tritium concentration decreased, albeit moderately, due to the diluting effect of groundwater mixed into the contaminated water shown in Fig. 9.4d. The tritium concentration was also almost leveled off (i.e. stopped decreasing) about eighteen months after the accident. As in the cases of chloride ions and Cs radioactivity, the tritium concentration seems to have become constant (i.e. stopped decreasing) due to the balance between in-reactor leaching and its dilution by groundwater.

As aforementioned, leaching inside the reactors is ongoing and is basically hampering efforts to reduce the radioactivity concentration in contaminated water. Even improving the performance of cleanup and treatment devices will not reduce the radioactivity concentration in the contaminated water provided the current recirculation-type cleanup continues. However, to adopt a once-through-type treatment to enable release from the system, it is essential to improve the performance of cleanup and treatment devices.

(3) **Multi-nuclei removal system**

In early 2013, ALPS, a multi-nuclei removal system capable of removing radionuclides, including cesium shown in Fig. 9.5, started commissioning. The ALPS is expected to effectively remove radionuclides from the on-site excess contaminated water [1].

The multi-nuclei removal system ALPS does not replace the cyclic-type contaminated water treatment system currently used but is used as a once through-type treatment system. It is important to optimally exploit its decontamination performance. Its decontamination factor is 6×10^6 (outlet concentration: $1/(6 \times 10^6)$), which means the treatment device's inlet radioactive concentration can be reduced to below the specified concentration limit. To effectively utilize the multi-nuclei removal system ALPS, accelerating the water sealing of containment vessels and downsizing the cooling water circulation system are essential. The leaching of radionuclides inside the reactors is inevitable, but blocking the increase in contaminated water accumulating on site, purifying it as far as possible and restricting its leakage into the environment are all essential.

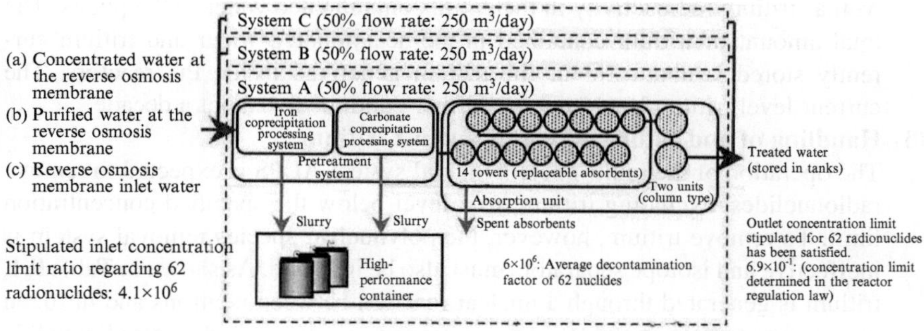

Fig. 9.5 Multi-nuclei species removal system ALPS

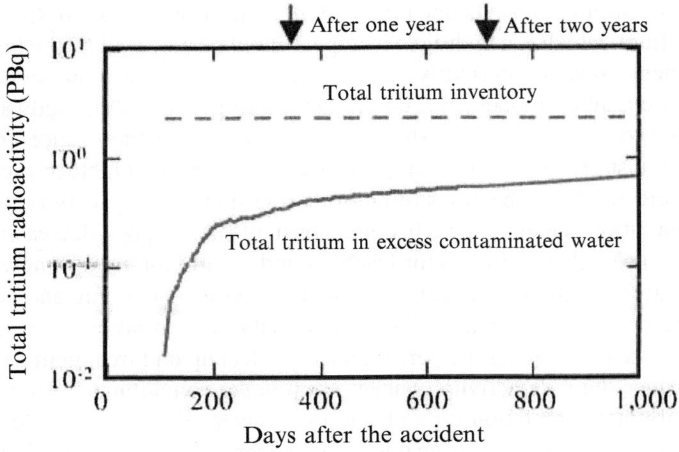

Fig. 9.6 Total tritium radioactivity in excess contaminated water

Table 9.1 Characteristics of tritium

Half-life	12.3 years
Biological half-life	12 days Small accumulation in ecosystem
Representative formation reaction	^{16}N (n, ^{3}t)^{12}C
Amount generated on earth	1 EBq/year (10^{18} Bq/year)
Natural background	0.01 Bq/g

(4) **Sources of tritium and the current situation of radioactivity in the contaminated water**

In the case of BWRs, tritium is generated through ternary fission among other uranium fissions (ternary fission: three fission products are generated). The probability of ternary fission is 0.9 to 1.2×10^{-4} per nuclear fission and can be determined based on the total number of uranium nuclear fissions calculated from in-reactor fuel burnup [2]. Figure 9.6 shows the tritium inventory (total generated amount) calculated from average fuel burnup at Reactor Units 1–3 as well as tritium radioactivity in the entire contaminated water in the plants. The total amount of tritium contained in the accumulated water and tritium currently stored is about 1/3 of the tritium generated inside the reactors. The current-level tritium leaching may further continue for about a decade.

(5) **Handling of and future measures against tritium**

The operation of the multi-nuclei removal system ALPS is expected to remove radionuclides, excluding tritium, to a level below the specified concentration limit. To remove tritium, however, the polynuclear species removal system is ineffective and isotope separators must also be installed. As shown in Table 9.1, tritium is generated through a nuclear reaction between neutrons and nitrogen or oxygen, or between protons and nitrogen or oxygen and is a radionuclide naturally present on Earth. Its background level in the environment peaks at 0.01 Bq/g and with a half-life of 12.3 years. Tritium is a hydrogen isotope and easily discharged from the body through the metabolism. Its biological half-life is also short, i.e. 12 days.

(6) **Cleanup and treatment of tritiated water**

Since tritium is a hydrogen isotope, it cannot be removed via coprecipitation or ion exchange but can be removed using an isotope separation method. After the removal of other radionuclides through the multi-nuclei removal system ALPS is complete, the ultimate problem with contaminated water is tritium, which cannot be easily removed by methods other than isotope separation.

To process tritiated water in bulk and at a low concentration, several methods can be used, examples of which actually exist. All the processes center on isotope separation before dilution and release of the tritiated water. Details are not provided here, but the isotope exchange methods shown in Table 9.2 are applicable. In any of these methods, after the removal of other nuclides, chemical species and salts, processing using tritium-containing pure water is necessary. Each process requires only minimal energy consumption except for electrolysis. However, the transport of bulk water and heating requires considerable energy.

Application examples similar to the bulk processing of relatively low-concentration tritium include water distillation using the ordered packing (stacked packing) method (by Sulzer) and a process centering on gas-phase chemical exchange at the Darlington Tritium Removal Facility (Canada). At the International Thermonuclear Experimental Reactor (ITER), a combination of a solution-layer chemical exchange and an electrolytic method is adopted to process 3.7×10^{11} Bq/kg of tritium water at a rate of 20 kg/h. These examples have problems inherent to the isotope separation process: Although these examples make it possible to reduce the tritium concentration in treated water, their decontamination efficiency is low and even in an attempt to reduce the concentration to 1/10, the need for multi-stage processing in the cascade arises, which requires significant processing. Moreover, reducing the tritium concentration requires equipment and energy consumption on an unrealistic scale.

Either tritium water treatment process mentioned above targets tritium concentration several digits higher compared with tritium water expected to be generated at the Fukushima Daiichi NPS (1 to 5×10^6 Bq/kg), making it difficult to apply them to the Fukushima Daiichi NPS' system. A processing capacity of up to 1 million tons within a few years requires them to be upscaled by 100 times or more.

(7) **Tasks and proposals to clean up and treat contaminated water**

To restrict the generation of contaminated water, identifying the locations of containment vessel leakage, accelerating their sealing, and shifting from the current large-scale circulation system, including turbine buildings, to a smaller-scale circulation system limited to containment vessels is necessary. Leakage sealing is essential to start removing the fuel debris and also important to

Table 9.2 Representative tritium water treatment processes

Method	Principle	Advantage	Disadvantage	Contaminated water treatment (special note)	Others
Water distillation	Utilization of isotopic difference in vaporization temperature (light water H_2O boiling point: low)	Effective impurity resistance Tritium resistant Simple operation High safety Bulk treatment possible	Separable with small vapor pressure differential/Small coefficient/Large tower height necessary	Low efficiency/Large-scale system	Basic experiment for fuel reprocessing system (Nagoya University)/Past achievement of providing CANDU reactors (Sulzer Corporation: about 30 units)
Water electrolysis	Utilization of isotopic difference in electrolytic potential (light water H_2O electrolytic potential is low. H_2O is collected at low electrical potential.)	High isotope separation factor at a single stage	Water electrolysis energy consumption is relatively large, making it unsuitable for bulk processing	Effect of impurities such as metal ions	Past achievement of providing Fugen degraded heavy-water refinery system
Gas-phase chemical exchange—cryogenic hydrogen distillation	Utilization of differences in the steam/hydrogen isotope exchange chemical equilibrium (catalyst required). (^3T tends to migrate to the liquid side.)		Utilization of cryogenic hydrogen distillation to collect ^3T. Cryogenic technique and refrigerator are required	Maximum use of hydrogen. Catalyst poisoning under the presence of impurities	Past achievements of removing ^3T from heavy water (France and Canada)

Water-hydrogen sulfide dual temperature exchange process (GS method) + water distillation	Utilization of equilibrium-constant temperature change (utilized for early-stage condensation and coupled with the latter-stage water distillation method)	Catalyst not required. Small energy consumption. Many past achievements at heavy-water production plants	Handling of H_2S Material corrosion and environmental pollution caused by H_2S	Handling of bulk H_2S	Past achievements of producing heavy water (U.S. and Canada)
Ammonia-hydrogen method + water distillation	Exploiting the tendency of heavy hydrogen isotope tending to migrate to NH_3. (utilized for early-stage condensation and coupled with latter-stage water distillation method)	Many past achievements at heavy-water production plants	Catalysts need to be added into NH_3 manufacturing system. NH_3 decomposition device is necessary, which requires high equipment and operation cost	Handling of bulk NH_3	Past achievements of producing heavy water (India and Argentine)
Water-hydrogen exchange + water electrolysis	Exploiting the tendency of the heavy hydrogen isotope to migrate to water (coupled with water electrolysis)	High separation performance	A large amount of hydrophobic catalyst required. Unsuitable for bulk processing. Large water-electrolysis energy consumption	Catalyst poisoning under the presence of impurities. Electrolytic-cell impurity effect	Applicable to ITER water treatment. Past achievements of providing Fugen degraded heavy-water refinery system

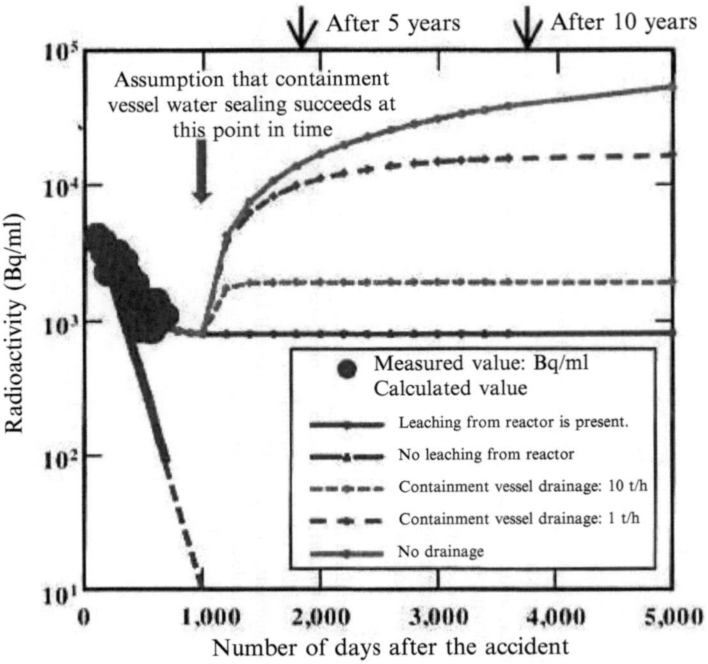

Fig. 9.7 Estimated tritium radioactivity in circulating contaminated water (tritium radioactivity after containment vessel water sealing)

minimize the amount of contaminated water. However, shifting to such smaller-scale circulation systems may increase the tritium concentration in cooling water. Figure 9.7 shows the anticipated tritium concentration values [2]. If said smaller circulation system is realized with an unchanged tritium release rate from the fuel debris, the tritium concentration in the circulation system may increase, reaching a concentration about 50 times the current value. However, it has been reported that the tritium concentration in the heavy-water system of the advanced thermal reactor (ATR) Fugen and Canadian deuterium uranium reactors (CANDU) is in the order of 100 MBq/g. A tritium atmosphere in the order of 10,000 Bq/g is not anticipated to cause significant radiation exposure during debris removal work [3]. Judgment as to whether the tritium concentration should be lowered by installing an isotope separation device to remove tritium, or whether circulating water should be partially removed to reduce tritium concentration, or fuel debris should be removed immediately to reduce tritium concentration caused by such contamination source seems to depend on the progress of future work.

It is anticipated that the operation of the multi-nuclei removal system ALPS will remove radioactivity below the specified concentration limit for radionuclides excluding tritium. To remove tritium, however, the aforementioned isotope exchange method must be adopted, but is difficult to adopt from an

Table 9.3 Tritium treatment—Contingency plans

Procedures	Outline	Subjects	Reliability for application	Risks for environment
1. Stored at the site	Storage without release	Possible contamination of underground water	High	High
2. Tritium removal or concentration	Isotopic exchanger	Engineering difficulty [reasonable DF: <10]	Low	Low
2.1 Release of diluted water	Accompanied by dilution process			
2.2 storage of highly ^3H enriched water	Decrease of total amount of tritium (^3H) in the release water		Low	High
3. Dilute and release	Remove other nuclei	Agreement with local community	High	Low
3.1 release into the ocean	^3H is diluted for release in the ocean			
3.2 Release into the air	Tritiated water is evaporated to release ^3H in the air (same as TMI)	Neighboring contamination due to rain	High	Middle

Note: Reactor regulation (Operational safety provision) for Fukushima Daiichi NPS prior to the: concentration; <20 Bq/g

engineering perspective. Table 9.3 summarizes measures deemed feasible for tritium in contaminated water under current circumstances.

The most effective method adopted at reprocessing plants to date to process bulk high-concentration tritium has been dilution and release into the sea.

La Hague Reprocessing Plant in France has the past achievement of releasing 14 PBq/year, which would enable the contaminated water at the Fukushima Daiichi NPS to be discharged within a month or so.

At the Fukushima Daiichi NPS, it is technically possible to release tritium using a condenser cooling pump (2.8×10^5 t/h) and reducing its concentration immediately before release to below its environmental concentration limit.

Conversely, regulations on total emissions (total amount control) of radionuclides are in place in the form of a local agreement. Regulations on total emissions for the Fukushima Daiichi NPS during its operation cite a figure of 22 TBq/year, which may restrict emissions more strictly than the concentration limit. Because such value is unsuitable for the current circumstances of the facilities with accident-damaged reactors, it must be modified to a more reasonable value. Meanwhile, emissions with a low concentration at the release point and almost equivalent to the natural background level are possible.

However, such emission level is easily detectable in the environment using the latest measurement technology. Therefore in terms of social receptivity and harmful rumors, it is essential to provide adequate explanations in advance to the effect that "emissions into the sea at a low level, almost the same as the environmental background level, do not result in biological concentration and a rapid dilution effect is expected".

During evaporated release performed at the TMI NPS (U.S.), contaminated water was evaporated by saturated water vapor pressure. Emissions of radiation lower than the natural background level such as from radon are possible, but even such radiation can be detected in the surrounding environment when it is deposited e.g. via rainwater. This may arouse harmful rumors concerning farm products, even when environmental standards are complied with. Unlike the TMI NPS located along a river and equipped with cooling tower-type condensers, the Fukushima Daiichi NPS requires the installation of additional evaporators.

In conclusion, the tritium-contaminated water accumulating in the Fukushima Daiichi NPS should be diluted, after other radionuclides are removed by the multi-nuclei removal system ALPS, until its concentration approximates its natural background level and then released into the sea. This is a method to reduce the risks of unanticipated radiation exposure and environmental contamination due to, for example, accidental contaminated water leakage due to on-site storage and is better than adopting a cleanup and treatment system for isotope separation or similar.

(8) **Summary**

This section has described contaminated water generation, past processing achievements and problems and countermeasures; particularly related to tritium. Storing all the radioactive waste water on site might be one solution. However, considering the long-term integrity of storage tanks and the potential for groundwater contamination due to leakage, the choice of diluting the contaminated water utilizing the sea under appropriate control and surveillance seems the most realistic solution.

9.2 Handling of Damaged Fuel

In Units 1–3 of the Fukushima Daiichi NPS, most of the in-core fuel is considered to have been meted and relocated down to the reactor pressure vessel or the containment vessel. At Unit 4, no fuel was loaded in the reactor vessel, and all of the fuel was in the spent fuel pool. As a result of inspecting the unused fuel in the pool, the integrity of almost all of the fuel in the pool is considered to be maintained. With regard to Units 1–3, however, the integrity of fuel in each spent fuel pool has not yet been confirmed. Because the fuel in each spent fuel pool and molten fuel (fuel debris) in each reactor containment vessel of the accident plants are the largest source of radiation, they must be removed as soon as possible and then properly managed. The current situation and future plans to cope with the fuel in the spent fuel pools and fuel debris inside the reactor containment vessels are shown in *The Mid-and-Long Term Roadmap for Decommissioning the Tokyo*

Electric Power Co.'s Fukushima Daiichi Nuclear Power Station, Units 1–4 (hereinafter referred to as Roadmap) presented at the 5th meeting for promoting the decommissioning of Units 1–4 of the Fukushima Daiichi NPS, Tokyo Electric Power Co. (TEPCO) held on June 27, 2013.

This section discusses implementation plans and research & development (R&D) plans for unloading fuel assemblies from the spent fuel pools and their storage as well as fuel debris removal and storage, and summarizes related issues and proposals, especially from technical and academic perspectives.

9.2.1 *Unloading of Fuel Assemblies from Spent Fuel Pools and Their Storage*

9.2.1.1 Implementation Plan

According to the Roadmap, the fuel stored in the spent fuel pools of Units 1–4 will be transferred to an on-site common (shared) pool and stably stored. To secure an area for storing fuel in the common pool, fuel retaining integrity and stored since before the accident will be transferred to dry cask temporary storage facility to be newly built. To transfer all the spent fuel assemblies of Units 1–4 into the common pool, the extension of the storage facility and secure procurement of dry casks are necessary.

Rubble on the operating floor of Unit 4 reactor building was removed in December 2012. Currently (as of June 2013), the installation of equipment for removing fuel in the spent fuel pool is ongoing. Removing the fuel from the pool will start in November 2013 and be completed around at the end of 2014.

The damage and contamination situations of reactor buildings differ by Units, therefore, an appropriate plan for removing the fuel from each spent fuel pool of Units 1–3 will be selected for each Unit, considering the quake-resistance of the reactor building, the progress in decontamination, and the availability of equipment for removing the fuel in the pool. The spent fuel removal is scheduled to start in FY 2017 for Unit 1, FY 2017 at the earliest or FY 2024 for Unit 2, and FY 2015 for Unit 3, depending on the plan to be selected.

9.2.1.2 Research and Development Plan

A R&D plan shown in Supplement of the Roadmap describes the following R&D plan to establish technologies for long-term storage at the common pool and dry storage of the spent fuel and to investigate technical issues in reprocessing the spent fuel.

(a) Assessment of the long-term integrity of fuel assemblies removed from the spent fuel pools (FY 2011–2017)

Evaluation of the integrity of the spent fuel before and after its transfer to the common pool by non-destructive inspections and strength tests, establishment of methods to assess long-term integrity of the spent fuel by corrosion and strength tests, and assessment of fission products leachability from damaged fuel by leaching tests with irradiated fuel pellets, and basic tests related to long-term integrity of the spent fuel.

(b) Study to treatment of damaged fuel removed from the spent fuel pools (FY 2013–2017)

Domestic and overseas case studies related to the damaged fuel, a study of the effects of damaged fuel on chemical processing, a study concerning damaged fuel handling, and development of a classification index for the damaged fuel.

9.2.1.3 Proposals

In the R&D plan, considering the importance of the long-term integrity assessment of the spent fuel assemblies after their transfer to the common pool, the effects of the seawater injection and the rubble that fell on the spent fuel assemblies will be investigated. With regard to damaged fuel, its long-term integrity during storage including fission product leachability will be appropriately taken into consideration in the plan.

It should be noted that fission product leachability have been studied, e.g., as shown in Katayama et al. [4]. The R&D plan for the leachability should be improved by referring to such existing data.

While study on the chemical processing of the damaged fuel is also planed, its implementation at this stage is too early in terms of the prioritization of research resources, although such study is important as a future task. Before completion of the decommissioning, technical development for secure storage and monitoring of the intact and damaged spent fuel assemblies should be prioritized.

9.2.2 Removal and Storage of Fuel Debris

9.2.2.1 Implementation Plan

Because the damage and contamination situations of reactor buildings differ by Units, an appropriate plan for removing the fuel debris from each reactor pressure vessel or containment vessel of Units 1–3 will be selected for each Unit, considering the quake resistance of the reactor building, the progress in decontamination, and the availability of equipment for fuel handling. The target period for removing fuel debris depends on the plan selected for each unit. Fuel debris removal is scheduled in FY 2020–2022 for Unit 1, FY 2020–2024 for Unit 2, and FY 2021–2023 for Unit 3, respectively.

The locations and properties of fuel debris and the damaged locations of the reactor containment vessels and pressure vessels remain unclear. However, as in the case of TMI-2, removing the debris in a water-flooded condition is regarded as the most secure method in the plan in terms of minimizing radiation exposure for workers. For this purpose, the following are planned: radiation dose reduction in the reactor buildings, investigation and repair for water-flooding the containment vessels, investigation of the inside of the containment vessels and the pressure vessels, developing technologies to remove, pack, transfer and store fuel debris, integrity assessment of containment vessel and pressure vessel, fuel debris criticality control, determining the damaged core situation by improving accident progression analysis, assessment of fuel debris properties, and preparation for fuel debris treatment and storage, and nuclear material accounting of the fuel debris. Technologies for water-flooding the containment vessels affected by the severe accident are challenging and need multi-stage development, so alternative methods for removing fuel debris without flooding the containment vessels are also being studied.

9.2.2.2 R&D Plan

A long-term is envisaged before starting removing fuel debris, depending on the situation of each unit and progress in on-site work. However, R&D required to remove fuel debris is scheduled as a project common to all units. Because there are many technical challenges before removing the fuel debris, the R&D plan may be modified significantly in future. In the R&D plan, therefore, schedule control and step-by-step approach will be taken, considering the future site condition, R&D results, and safety requirements. The outline of the general R&D project is as follows.

(a) The development of technologies for remote decontamination in the reactor buildings (FY 2011–2014)

Investigation of contamination situation, the study and manufacturing of remote control devices, the development of decontamination techniques and concept, decontamination tests for simulated contamination, the verification of decontamination techniques combined with remote control devices, and the remote verification of the validity of radiation shield installation.

(b) The development of a comprehensive dose reduction plan (FY 2012–2013)

The identification of work areas, such as locations to be investigated inside the containment vessels, explosion-damaged floors and common access routes (e.g. staircases); the assessment of the situation of the work area; the development of a dose reduction plan combining decontamination, shielding and flushing.

(c) The development of investigations and repair (for stopping water) technologies for water-flooding the reactor containment vessels (FY 2011–2017)

The study of leakage-detection techniques, fabrication of leakage-detection devices and their on-site verification, the study of repair techniques (for stopping water) and fabrication of repairing devices, the assessment of applicability to actual equipment, and the study of alternative techniques without water-flooding.

(d) The development of technologies to investigate the inside of the containment vessels (FY 2011–2016)

The assessment of the situation inside the containment vessels, the development of investigation techniques and devices for removing fuel debris, the compilation of existing technologies based on the understanding of the situation inside the containment vessels, and the study of methods for preventing radioactive materials from scattering out of the containment vessels.

(e) The development of technologies to investigate the inside of the pressure vessels (FY 2013–2019)

The determination of an investigation plan, the development of technologies to gain access to locations to be investigated, and the development of fuel debris sampling technologies.

(f) The development of technologies for removing fuel debris and reactor core internals (FY 2014–2020)

The cataloging and compilation of existing relevant technologies, studies regarding techniques for removing fuel debris, the development of devices for removing fuel debris and their applicability assessment, and verification of techniques by mock-up tests using full-scale test equipment.

(g) The development of technologies for canning, transferring and storing fuel debris (FY 2013–2019)

Survey of experiences of damaged fuel transport and storage, the study of fuel debris storage system, the development of the methodology to qualify fuel debris canister in terms of criticality control, radiation shielding, heat removal, sealing, structure, etc., the study of canning methods, canister fabrication, and the technology development for transfer and store fuel debris canister.

(h) The development of pressure vessel and containment vessel integrity assessment technologies (FY 2011–2016)

To assess the integrity of the pressure vessels and containment vessels about which corrosion is concerned because of their exposure to heated seawater, the following will be performed: structural material corrosion tests, corrosion control verification tests, the assessment of structure life and life extension, corrosion control system development and actual-equipment applicability assessment.

With regard to the integrity of the pedestals, deterioration of which is concerned about because of post-accident high temperature, the following will be performed: reinforced concrete degradation tests, and assessment of the effect of high-temperature fuel debris attack.

(i) The development of fuel debris criticality control technologies (FY 2012–2019)

To maintain subcriticality even after the change in fuel debris configuration and water volume during fuel debris removal work, the following will be performed: the study of criticality scenarios, the study of measures for mitigating radiation exposure in case of criticality, the technology development of criticality control for liquid waste treatment equipment and cooling equipment, the development of recriticality detection technologies, the development of criticality prevention technologies such as neutron absorber development, and the fundamental study of criticality control technology such as criticality calculation uncertainty assessment and analysis code development.

(j) The determining of the damaged core situation by improving accident analysis technologies (FY 2011–2020)

Because directly observing the damaged reactor cores is difficult, efforts to estimate and determine the actual situation will be continued using information obtained from decommissioning work, analysis of plant behavior under the accident, improved severe accident analysis codes and simulating test results.

(k) The characterization of simulated fuel debris, the characterization of actual fuel debris, and the development of fuel debris treatment technology (FY 2011–2020)

To prepare fuel debris removing devices and canisters based on fuel debris properties, simulated fuel debris will be fabricated and characterized. Actual fuel debris to be taken from a damaged core sample will also be characterized.

To have an insight into the long-term storage, treatment and disposal of the removed fuel debris, scenarios for coping with fuel debris (storage, treatment and disposal) and the feasibility of existing technologies for coping with fuel debris will be studied.

(l) The establishment of measures for material accountancy of fuel debris (FY 2011–2020)

Survey of experiences of nuclear material accounting in the case of the TMI-2 and Chernobyl accidents, the study of the current technology for nuclear material accounting, estimation of nuclear material distribution based on the results of the aforementioned e, j and k, and the establishment of measures for material accountancy of fuel debris.

9.2.2.3 Proposals

Since high radiation dose in the reactor buildings makes access into them difficult, several years will be required before sampling the fuel debris. The current implementation and R&D plans seem to be appropriate, where analysis and simulating tests will be implemented to obtain information on fuel debris distribution and properties until fuel debris sampling becomes possible. However, the following points need to be considered in future.

Water-flooding the pressure vessels and containment vessels is planned to remove fuel debris. The fission product leachability from fuel debris needs to be evaluated for the contamination assessment of fuel debris removal equipment and for the design of radiation shields and fuel debris canisters. The mechanical and chemical characteristics of fuel debris are basic information utilized not only for fuel debris removal but also for criticality control, nuclear material accounting, and storage of fuel debris. Therefore, comprehensive survey on relevant research performed domestically and overseas in the past is recommended to promote the R&D plan efficiently. When referring to such previous studies, various possibilities need to be considered based on the characteristics of the Fukushima accident. For example, the results of TMI Unit 2 (PWR) accident investigations show that fuel debris mainly comprises a ceramic phase, formed by the melting of zirconium oxide (generated by zircaloy fuel cladding oxidization due to water vapor) and the melting of uranium dioxide fuel, and a metallic phase consisting mainly of core internals and zircaloy; that the ratio of the ceramic and metallic phases differs depending on the in-reactor location; and that the size of ceramic phase crystals differs depending on the cooling rate. Because in the case of BWR the ratio of zircaloy in the reactor core exceeds that of PWR, the fuel debris characteristics, such as oxidization condition of in the ceramic phase and zirconium concentration in the metallic phase, may differ from those for TMI Unit 2. In the Fukushima accident, there is a possibility that part of the molten fuel has penetrated the pressure vessel and reacted with the pedestal concrete. In this case, the dissolution of concrete components in the molten fuel (fuel debris) results in a significant change in characteristics and properties of the molten fuel. It should be noted that the oxidization state in the molten fuel (the concentration of zirconium or uranium metals) affects the reaction between the molten fuel and concrete. To bring the accident under control, bulk seawater was injected into the reactors. Seawater contains various elements. Under circumstances where these elements contacted damaged fuel, various chemical reactions may have taken place. Table 9.4 summarizes the effects of the Fukushima accident features on fuel debris characteristics and properties [5].

The chemical form (composition and phase structure) of fuel debris can be estimated to a certain degree through thermodynamic equilibrium calculation assuming various conditions. Appropriate simulations can also clarify the relationship between fuel debris formation conditions and chemical form. Data and analytical results regarding fuel debris chemical form accumulated in this manner can provide important clues for clarifying the event sequence in the Fukushima Daiichi NPS accident from the results of chemical analyses of actual fuel debris samples. It should be noted that clarifying event sequence in the Fukushima Daiichi NPS accident will greatly contribute to understanding severe accident phenomena and provide a foundation for reasonable safety measures for light water reactors worldwide. Thus, as in the case of post-TMI-2 accident investigations, analyses of actual fuel debris samples should be performed in depth in cooperation with domestic and overseas organizations.

When developing technologies for canning, transferring and storing fuel debris, for which a research plan is being developed, it is important to design fuel debris canisters with consideration of conditions for removing fuel debris (e.g. additives in cooling water and methods of removing fuel debris) and fuel debris characteristics. In designing the canisters, considerations are also needed for measures against hydrogen generation due to water radiolysis, the control of

Table 9.4 Effects of the features of the Fukushima Daiichi NPS accident on fuel debris characteristics and properties

		TMI-2 (PWR)	Fukushima Daiichi NPS (BWR)	Anticipated effects on molten fuel characteristics and properties
Fuel and structural materials	Fuel structure	Spacer grid	Channel box	High zircaloy ratio in the core will lead to high Zr concentration in molten fuel
	Control rod	AG-In-CD/SS cladding	B_4C/SS cladding	Eutectic reaction between boron and iron may affect the behavior of noble metal fission products. The effect of CO_2 and H_2 gas generation due to reaction between B_4C and steam remains unclear
	Fuel	UO_2	Gd is contained. MOX is partially used	MOX and Gd may not significantly affect fuel debris properties
	Burnup	Three months after commissioning	Fresh fuel—high burnup	Due to a considerable amount of fission products in molten fuel, influence of water radiolysis needs to be considered
Core internals	Reactor vessel substructure	Core support plate, in-core monitor guide tube, etc.	Core support plate, control rod guide tube and drive shaft, etc.	Due to a larger amount of steel in the reactor vessel substructure, the Fe concentration in molten fuel at lower head is high. Significant fractions of noble metal fission products may have been involved in molten fuel
Event progress	Duration of melting	1–2 h	A few hours	The molten fraction of the core may be high. Volatile fission products may have been released significantly. Fuel debris may have been partially densified (consolidated). Molten fuel has partially fallen through the pressure vessel bottom and reacted with the concrete
	Pressure	>50 atm	Atmospheric pressure and higher pressure	Pressure may not significantly affect the metallurgical reaction
	Seawater injection	None	After meltdown	The behavior of seawater elements is unknown but may have caused fission product leachability and structural material corrosion. The effect on molten fuel properties is unknown

(continued)

Table 9.4 (continued)

	TMI-2 (PWR)	Fukushima Daiichi NPS (BWR)	Anticipated effects on molten fuel characteristics and properties
Time required to remove fuel debris after the accident	It took a decade to completely remove fuel debris	A decade is assumed to start removing fuel debris	If the cooling period is prolonged, the fuel debris properties may change. Effect on fission product leachability may arise

structural material corrosion due to seawater elements, safety margin to criticality, and radiation shielding. When developing technologies for canning, transferring and storing fuel debris, the design of canisters used to remove and store fuel debris at TMI-2 can be referred to. However, the aforementioned technologies must be carefully developed while taking into account the characteristics specific to the Fukushima Daiichi NPS such as high burnup, the possibility of seawater elements being mixed, and the possibility of molten core concrete interaction (MCCI).

9.2.3 *Fuel Inventory and the Likelihood of Recriticality*

9.2.3.1 Introduction

After the Fukushima Daiichi NPS accident there are several issues that drew attention, in terms of comparison with the scale of Chernobyl Nuclear Plant accident, i.e., "how much the radioactivity does exist inside the reactors?" "what rate of the radioactivity was released from the site?" and "will the damaged (molten) fuel reach the criticality again?"

In this paragraph, results of a study regarding the calculation of the inventory (amount of radionuclides) and recriticality problem are summarized and issues requiring future study are discussed.

9.2.3.2 Inventory Calculation

Radionuclides released from facilities and their radioactivity are called "source term" and it is important factor to evaluate the environmental effects of the severe accident. The source term is determined based on accident analyses by taking into account the conditions during the release of the radioactive materials. The basis for this evaluation is "core inventory" obtained by the burnup calculation. In normal fuel designs and core analyses, major actinides and fission products (FP) which have the long half-lives and the large neutron reaction cross sections affecting the neutron multiplication factor are the objective isotopes for the burnup calculation.

Conversely, in the post-accident inventory calculation, it is important to obtain the amount of the isotopes required for the dose assessment because of their short half-lives, which are not important for assessing the neutron multiplication factor. Therefore, a simple one point burnup calculation code is used rather than a detailed burnup calculation taking account of the change of the neutron spectrum during the burnup, which are necessary for the core analysis. In the inventory calculation performed by the Japan Atomic Energy Agency immediately after the accident to obtain the basic data for the accident assessment, ORIGEN2.2UPJ (http://www.oecd-nea.org/tools/abstract/detail/nea-1642) was used (ORIGEN2.2UPJ: the combined system of ORIGEN2.2 which is the latest version of the one point burnup calculation code (ORIGEN2) [6] most widely used in Japan, and the cross section data library developed in Japan). ORIGEN2 is characterized by the simple input. It is capable of calculating solely by inputting the amount and the composition of uranium and the operation history of the reactors. It has also an adequate function for the fast inventory calculation.

As a problem in post-accident inventory calculations, the difficulty in obtaining detailed data of the original fuel composition and the reactor operation history was pointed out and the effect of such data on the calculation results has been studied. With regard to the ORIGEN2, the effective cross section data, which changes depending on the burnup, is limited to built-in 20 or less reactions. Regarding this point, Nishihara et al. [7] calculated the amount of FP by taking into account the burnup dependency of its amount, using an integrated burnup calculation code system SWAT [8], which is the combination of ORIGEN2 and the neutronics calculation code SRAC [9]. Consequently, it was confirmed that the difference between the amount of FP calculated by the ORIGEN2 and by the SWAT is adequately small, given the purpose of the inventory calculation.

Moreover, Okumura et al. [10] performed an in-reactor three dimensional inventory calculation using a modular-type calculation code system MOSRA, which is an improved version of SRAC. Consequently, they pointed out that, with regard to the radionuclides generated through the neutron absorption reactions after generation of their parent nuclides by the fission reactions, the difference tends to be considerable between a simple analysis using the average burnup and a detailed analysis taking into account the power history of the fuels and the change of neutron spectrum. Okumura et al. performed this analysis assuming a generally presumed fuel loading pattern but it is an important suggestion to be considered in a detailed analysis.

Although the currently used inventory calculation based on ORIGEN2 are adequately precise, these indicate that it is important to ensure access to a set of required data to evaluate up-to-date inventory when an accident occurs, or to evaluate the inventory periodically as software-related future measures against the severe accident.

9.2.3.3 Recriticality

Recriticality means that the reactor reverts to the state of the criticality for some reasons after the nuclear reactor stops, i.e. stopping nuclear chain reaction and achieving subcriticality. During the work to bring the Fukushima Daiichi NPS accident under control, freshwater required for cooling the cores ran out. As an alternative, seawater injection was proposed, but the likelihood of recriticality due to the seawater was pointed out which resulted in confusion as to its appropriateness. Moreover, although measures to prevent recriticality such as boric acid injection had already been taken, there had been a concern that the recriticality may have already taken place or could possibly occur inducing radioactive material release. Thus, the discussions on recriticality had existed from immediately after the accident. It is deemed necessary to study "whether the recriticality could actually occur," "its possibility and what kind of condition is considered if the recriticality could occur," "the effects of the recriticality," and "how much the effects on the accident evolution is expected and how much the radioactive material release is expected."

9.2.3.4 Study on Recriticality Possibility

The possibility of the recriticality during each stage of the severe accident is as follows:

(1) **Reactor core**

(a) Process of core meltdown: During meltdown, water as the moderator basically does not present. Criticality does not occur for the low-enriched uranium system.

(b) Flooding process: During the loss of coolant accident, the control rods melt and fall down. If reflooding starts when the fuel is not melted and standing independently, criticality might occur. Moreover, the molten core may penetrate the pressure vessel and fall into the containment vessel. When cooling water is injected to cool the molten core, fragmented molten fuel could reach the criticality. The possibility of these scenarios was studied assuming certain conditions but it concluded that such possibility is significantly low [11].

(c) Cooling process: The size of the fuel debris generated due to the TMI-2 accident ranged from a large mass to granules. The fuel debris shape also varied; ranging from porous to crust. When molten fuel exists at the bottom as a large mass and there is no water intrusion, criticality is unlikely. Conversely, when molten fuel in a fine powder (or grain) state is distributed in cooling water or when water intrudes into large molten fuel masses, recriticality might occur. For example, unburned fuel with 4 wt% uranium enrichment may reach criticality when its degree of uranium concentration increases, namely about 0.4 g¯U/cm^3 (400 g¯U/L) or higher. In reality, it is hardly anticipated that uranium powder will uniformly disperse in water with such high concentration. However, recriticality risk arises when small fuel particles migrate via cooling water circulation and concentrate in a certain location such as pipe, hence caution is required. In addition, as shown by post-TMI-2 accident investigation, criticality might occur if the neutronics condition become adequate, when molten fuel turns into large masses of pumice form with porous at the bottom of the reactor pressure vessel and adequate water intrudes into the masses, or when molten fuel turns into relatively large particles and water permeates among the fuel particles.

(2) **Spent fuel pool (SFP)**
The fuel in the SFP could have been mechanically damaged due to the rubble generated after the hydrogen explosion but the integrity of most of the fuel is estimated to be maintained. If cooling shortage significantly damages the fuel and it accumulates on the pool bottom, the senario is almost equivalent to the case of the core meltdown, but the criticality risk may exceed than the case of the core because of the lack of the control rods in the SFP. Conversely, if the fuel storage rack is deformed with fuel integrity maintained, the possibility reaching the criticality is high because there are no control rods. Notably, it has been pointed out that when spent fuel subcriticality is maintained by the space between fuel assemblies rather than neutron absorption property of the storage rack, if the water level decreases due to boiling of cooling water, the neutron multiplication factor may increase [12].

9.2.3.5 Recriticality Effects

In general, the maximum scale of the supercritical event (critical accident event) depends on the degree of the excess reactivity and the system volume. In low-enrichment uranium system such as light water reactors, the criticality is unlikely to take place provided the system is small scale, and the degree of excess reactivity remains modest. Therefore, even if an accident occurs within such system, its scale will not escalate significantly. However, depending on the neutron slowing down condition (moderation condition), positive reactivity may be added due to the increase in the system temperature. In such cases, the criticality will continue until the bulk water evaporates and the system size diminishes, resulting in a large-scale accident. To date in the Fukushima Daiichi NPS, the recriticality has not been found during the cooling operation. If the fuel arrangement and shapes are maintained in future decommissioning operations, criticality is unlikely.

As a means of detecting the recriticality, a containment vessel gas management system has been installed, which monitors short-half-life xenon in the containment vessels and is capable of detecting spontaneous fission level concentration. In addition, boric acid water injection to stop the criticality is available at any time. However, given that cooling water is circulating inside and outside the reactors, maintaining subcriticality continuously by controlling and managing the boric acid concentration requires significant efforts.

9.2.3.6 Conclusion

Inventory calculation is possible using the aforementioned simple one point burnup calculation code. In inventory calculation after the accident, conservative estimation is required first. Finally, however, precise numerical values are required as far as possible. As the emergency countermeasures, it will be required at each plant to organize and manage data such as the basic core parameters, the operation history, the number of fuel assemblies in the spent fuel pool and their irradiation histories in order to ensure the data can be used at any time. Because the calculation speed of ORIGEN2 used for the inventory calculation is high enough, a new system could be established to perform the inventory calculatin at all reactors in Japan every day as the part of the system of nuclear emergency.

Moreover, the possibility of recriticality during the process of settling the Fukushima Daiichi NPS accident was discussed. It's possibility cannot be completely ruled out, hence preparation for prevention of the recriticality such as boric acid water injection and assuming the worst-case recriticality (with early detection and subcriticality measures taken) are necessary for promoting future decommissioning work and in terms of disaster readiness.

9.3 Decommissioning and Treatment and Disposal of Radioactive Waste

9.3.1 Introduction

Following the Fukushima Daiichi NPS (Units 1–4) accident, waste, cut down trees and rubble contaminated by radionuclides as well as secondary waste due to processing were generated. Moreover, various types of waste will be further generated in future decommissioning and related preparatory work. Basically, this waste will be processed and disposed of according to its characteristics and radionuclide concentration. Reducing the risk, compacting and stabilizing such waste according to the latest site circumstances are necessary. As in the case of back-end measures, optimizing each stage of decommissioning and radioactive waste processing & disposal does not necessarily result in the overall process optimization. It is essential for stakeholders to share how the whole decommissioning project should be managed, how the end state should be envisaged and flexibly handle technical issues and social receptivity.

While focusing on decommissioning considerations, this section summarizes the differences between normal and accident plants and proposals regarding related issues. With regard to the Fukushima Daiichi NPS, the Nuclear Emergency Response Headquarters has prepared *The Mid-and-Long Term Roadmap towards Decommissioning the Tokyo Electric Power Co.'s Fukushima Daiichi Nuclear Power Station, Units 1–4*, which details a 40-year plan comprising three periods and the necessary research and developments. The Atomic Energy Society of Japan established a special expert committee on radioactive waste processing and disposal at the Fukushima Daiichi NPS and issued a report *Concept for Identifying R&D Issues and Their Solution* [13]. This section avoids discussing contents which overlap with reports and describes the investigative committee's proposals made based on the current situation.

9.3.2 Decommissioning

After the fuel debris removal, decommissioning is scheduled for around 2022. Comparison between the Fukushima Daiichi NPS and normal nuclear power plants and considerations for decommissioning are provided here.

9.3.2.1 Decontamination of the Buildings

Unlike the decommissioning of normal power station, the buildings in the Fukushima Daiichi NPS need to be decontaminated below a certain radiation level to reduce radiation dose rate before removing the plant equipment and systems. Important points are (1) secondary waste generation, (2) selection of decontamination methods and (3) disposal classification. The disposal classification requires the following:

(a) On-site waste will be plotted in terms of α concentration and β and γ concentration relationship. Disposal types are classified into trench disposal, disposal with artificial barrier (concrete pit disposal), intermediate depth disposal and geological disposal (below-clearance also taken into account), all of which are reassessed via performance assessment.
(b) Subsequently, processing forms for stable storage will be examined.

9.3.2.2 Contamination Types

Contamination affecting in-building decontamination methods is classified into the following four types in terms of contamination source and objects: (1) surface contamination by volatile radionuclides, (2) penetration contamination by volatile radionuclides, (3) surface contamination by contaminated water (accumulated water and circulating water), and (4) penetration contamination by contaminated water (accumulated water and circulating water).

During work to bring the Fukushima Daiichi NPS accident under control, in-building decontamination to reduce the radiation dose rate will be the main target for the time being. To optimize radiation protection, there is a need to ensure the worker exposure dose caused by decontamination work does not exceed the specified level.

9.3.2.3 Decontamination Methods (Applicable Techniques)

When selecting decontamination methods, it is necessary to determine the above-mentioned contamination situation and level and select a method specific to each radiation exposure source so that decontamination can effectively reduce radiation exposure. Regarding secondary waste generated after decontamination work, less burdensome processing and storage methods should be selected. When it is difficult to decontaminate a little space or narrow area, its decontamination is not necessarily required and methods such as its dismantling, removal and shielding can also reduce radiation exposure. The environmental conditions of the accident reactor site are estimated to be severe, which means complicated decontamination methods using chemical agents or similar should be avoided as far as possible. Simple decontamination methods unlikely to cause re-floating (re-suspension) of volatile radionuclides should be selected. Specific building decontamination examples are as follows:

Water cleaning, high-pressure water jet cleaning, shot blast, dry ice blast, shaving, and gel decontamination (strippable coating)

It should be noted that using water for e.g. high-pressure water jet cleaning may spread contamination.

Selection of methods suitable for each type of contamination, a combination of multiple methods, integrated water treatment and the sequence of works must also be taken into consideration. An optimal decontamination plan should also be developed, to balance the exposure dose due to decontamination work and later-work exposure dose. In developing such plan, reflecting past examples and practices is important. Secondary waste generated due to decontamination work must be compacted as far as possible, separated according to contamination source (into liquid and gaseous waste), then stored on-site with information attached such as the name of the facility where the waste was generated, contamination status and contamination level that can be useful for later processing.

9.3.2.4 Considerations for Decommissioning (Differences Between Normal and Accident Plants)

Decommissioning of the Fukushima Daiichi NPS will be performed after fuel debris retrieval and is scheduled for around the middle of the 3rd period in the Mid- and Long-Term Roadmap. Multiple scenarios can be envisaged regarding how the contaminated facilities should be decommissioned, how the waste can be stabilized and how the end state should be. Selecting a scenario requires not only scientific and technological perspectives but also social perspective. Proposing decommissioning and waste disposal scenarios based on facts is important.

Table 9.5 shows differences in terms of decommissioning between normal and accident plants such as the Fukushima Daiichi NPS, where an accident involving core damage took place.

Table 9.5 Differences in terms of decommissioning between normal and accident plants

	Decommissioning of normal plant	Decommissioning of accident plant
Fuel	Fuel can be removed in the same way as during operation and transferred for processing	Collected in the form of fuel debris and stored for the time being (handling method must be considered.)
Facilities	Buildings can be utilized as shields	Buildings and facilities have been damaged
Determining situation	The contamination situation can be investigated in advance and a decommissioning plan can be developed	Determining the contamination situation in advance is difficult. The situation will be checked according to event progress
Environment	No environmental contamination	Soil, plants and beach sand, etc. have been contaminated
Radionuclides	Major radionuclide is Co-60 at structural materials around the reactor	Volatile radionuclides (Cs-134/137 and Sr-90) may be present in the air in addition to the structural materials in the left column. Heavy metals, fission product (FP) nuclides and fuel constituent nuclides may be present in the contaminated water
Penetrating Contamination	Almost no penetrating contamination into structures is found	Penetrating contamination into damaged facilities and basements must be considered
Amount of materials	Amount of materials regarded as radioactive waste is 10,000–20,000 t per unit	Amount of materials regarded as radioactive waste is estimated to be dozen tons to several million tons per unit
Disposal system	Disposal system is in place according to the current laws	The development of disposal system is required
Contaminated water	Existing facilities can process contaminated water	A large amount of contaminated water containing FP nuclides and salt is present

The following differences also exist:

(a) There is a possibility of deposition or incorporation (mixing in) of salt, boron, oil, or organic matters. Its chemical effect on processing or disposal must be taken into consideration.
(b) Structures to be dismantled have been damaged. In addition, the dismantling work environment is severe, meaning remote dismantling techniques are necessary.
(c) It is also important to reduce the amount of radioactive waste by limiting the sections to be dismantled and devising dismantling techniques.

Requirements for decommissioning are: (1) waste processing and disposal facility, (2) financial resource, (3) organization(s) in charge and (4) technologies. Particularly in the case of the Fukushima Daiichi NPS, the applicability of technologies for restricting waste generation should be considered.

Decommissioning the facilities requires interim storage facilities that can accommodate more than the amount of generated waste. In the case of Fukushima Daiichi NPS, the amount of waste is enormous. Therefore it is essential to optimize the handling of the waste without transferring it unnecessarily frequently. And when developing a decommissioning plan, outlook for entire waste processing and disposal activities must be established and specific scenarios must be selected for such activities. The following are essential points for decommissioning:

(a) The basic information required to develop a decommissioning plan is the current situation of the facilities and contamination distribution. Neither the current situation nor contamination distribution at the reactors and their vicinities is clear at the Fukushima Daiichi NPS. Thus, they should be qualitatively estimated based on certain assumptions and the decommissioning plan should be reviewed as necessary as the dismantling progresses.
(b) With regard to radioactive material contamination, the interrelationship between contamination sources and contaminated items will be determined and the types of contaminating nuclides and contaminated items and the presence/absence of contamination penetration must be checked. It is also important to check for the presence/absence of hazardous substances other than radioactive materials.
(c) As for dismantling techniques, it is desirable to apply a mechanical cutting technique to those contaminated, particularly by volatile nuclides, while the use of remote devices on an as-required basis should be considered and appropriate techniques suitable for the properties of those dismantled should be selected.
(d) No system for processing and disposing of the demolition waste of the accident facilities is currently in place. Therefore they must be roughly separated with the places from which they were generated and their dose rates taken into account.
(e) Decontamination to reduce workers' radiation exposure must be performed while considering its cost-benefit performance. Conversely, decontamination to reduce the amount of radioactive waste should be adequately reviewed at its planning stage as to whether it can be performed reasonably.

9.3.2.5 The Development of Laws Related to Accident Reactor Decommissioning

With regard to a legal system related to the decommissioning of accident reactors, *the Nuclear Reactor Regulation Law* was revised on June 27, 2012. Under this revision, the Fukushima Daiichi NPS was designated as special reactor facilities and tasked with the obligation to develop an implementation plan to perform measures stipulated by the Nuclear Regulation Authority (NRA). After the designation, the general stipulations, including decommissioning regulation, of the Nuclear Reactor Regulation Law are exempted and measures according to the implementation plan will be performed intensively. Specific contents of these measures will be decided based on the NRA judgment. Risks accompanying the decommissioning stage of the accident facilities will be reduced step-by-step.

Hereafter, rules to ensure the security of the Fukushima Daiichi NPS will be implemented. The contents of the rules are not so different from those for other general nuclear facilities. The rules need to be reviewed appropriately with decommissioning progress taken into consideration and in terms of how the rules should be to promote the stabilization and decommissioning of the Fukushima Daiichi NPS swiftly and safely.

9.3.3 Processing and Disposal of Radioactive Waste

Radioactive waste generated after the Fukushima Daiichi NPS accident includes not only the buildings but also cut down trees, rubble, and secondary waste generated from contaminated water treatment. The properties of the radioactive waste remain unknown. The long-term storage of this waste should be also considered. Its characteristics important for decommissioning, the processing and disposal of the waste are summarized, the importance of analyzing radioactive waste properties is discussed and proposals for the long-term storage of the radioactive waste are provided as follows.

9.3.3.1 Considerations of Waste Characteristics, Processing and Disposal

It should be noted that cut down trees and rubble may contain volatile nuclides such as iodine and cesium as well as fuel-derived nuclides such as TRU and fission products (FP). Salt and organic matters should be also noted.

(a) Cut down trees and others: Analyses and quantification of radionuclides deposited on driftwood are necessary for future waste package processing and disposal. Processing and disposal facilities must be constructed that can address bulk cut down trees, contaminated soil and the effects of organic matters, effects such as promotion of radionuclide migration due to complex formation (complexation), microorganism activation and gas generation, while developing techniques for assessing the safety of such facilities is important.

(b) Rubble: Rubble is broadly divided into concrete pieces and metals. With regard to rubble that will be stored in future, harmful materials must be removed and deposited radionuclides must be analyzed and quantified urgently, for which a system must be in place. It should be noted that, unlike normal decommissioning, the amount of highly contaminated rubble tends to be large.

(c) Secondary waste generated from contaminated water treatment: This will be processed according to the concentration of radionuclides contained. Therefore, before packaging, the radionuclides need to be analyzed and quantified.

Analysis of secondary waste already generated and stored is difficult. However, it is proposed that future contaminated water treatment facilities should enable the sampling of a representative portion for analysis the removed nuclides. In so doing, sampling representativeness must be ensured. Particularly in zeolite analyses, nuclides that migrate into the air during dissolution should be noted.

Before fabricating waste packages, the salt concentration and amount of boric acid in secondary waste must be analyzed. Before carrying out waste packaging, analysis of secondary waste generation chronology and quantification of hazardous substances other than radionuclides are also proposed.

Issues common to the waste in the Fukushima Daiichi NPS include: selection of waste packages while taking into account salt and other impurities such as oil, ferrocyanide, organic matter, and boron; barrier composition; and nuclide migration behavior and characteristics during disposal. Assessment of these with the whole system taken into account is proposed. Based on the conventional disposal forms, new forms also need to be studied.

9.3.3.2 Importance of the Property Analysis of Radioactive Waste and Proposals Regarding its Long-Term Storage

With regard to radioactive waste, as well as nuclides (inventory) contained in them and its concentrations, the disposal system performance must also be evaluated while considering its chemical properties therefore, before processing and disposal, analyzing and identifying the concentrations of radionuclides in the waste are important. Particularly in the case of Fukushima Daiichi NPS, unlike inventory assessments of normal nuclear power plant dismantling, the damage to equipment and components in systems due to hydrogen explosions and contamination (with fuel elements, fission products and corrosion products mixed up) via cooling water exposed to fuel debris need to be considered. Conventional scaling factor methods and average concentration methods cannot be applied directly to the power-plant waste in the Fukushima Daiichi NPS. Such waste must be handled as having properties similar to reprocessed waste. The radioactivity concentrations of the waste and nuclide compositions may vary, hence the urgent need to verify whether post-analysis statistical assessments are possible important. According to the Mid- and Long-Term Roadmap, the waste properties will be determined and the material amount assessed by around 2014; the applicability of the existing disposal concept to the waste will be checked by around 2017; and the safety outlook regarding waste processing and disposal will be confirmed by around 2021.

Based on these current conditions, the following are proposed:

(a) Currently, waste that can be analyzed is limited. To fulfill the plan, broad-range waste must be analyzed with priority taken into account. Moreover, to determine the applicability of statistical techniques and theoretical calculations to each type of waste, sample analyses should be immediately performed.

At the same time, to improve the quality of the analytical data, analysis standardization such as establishing an academic society standard is essential.

(b) An analytical facility should be established as soon as possible and is scheduled for the site vicinity. It is necessary to determine whether the currently planned number of samples (50 samples/year by the end of 2016 and 200 samples/year by the end of 2020) is adequate, and as required, the facility should be reinforced.

(c) Generated radioactive waste requires various types of long-term storage for 20–25 years. Issues common to the long-term storage of the diverse waste include inventory assessment and separation (based on radioactive concentration and recyclability) and measures against gas generation (due to radioactive decomposition of water, metal corrosion, organic matter decomposition, etc.). Currently, rubble that may result in the contamination expanding to the surrounding area is temporarily stored in containers. However, the selection of container functions and materials adaptable to future long-term storage is an important task. In particular, waste containing salt and boron must be stored in containers with corrosion resistance validated by corrosion assessment (if possible, such containers should also be reusable for transport).

9.3.4 Summary

Decommissioning and radioactive waste processing and disposal at the Fukushima Daiichi NPS are long term and wide-ranging tasks and should be performed while keeping in mind social receptivity, optimizing the entire process through appropriate management and flexibility according to the situation. Essential consideration is as follows.

(1) Various options not only in terms of technical perspective but also in terms of social perspective are possible for decommissioning, radioactive waste processing and disposal. With regard to how the end state of the decommissioning should be, study is needed with international expert opinions taken into consideration, and relevant information should be shared among stakeholders.

(2) Reviewing the regulatory system to swiftly and safely promote stability and decommissioning of accident reactors in terms of risk reduction is also important.

(3) In radioactive waste disposal, the assessment of disposal system barrier performance is necessary with radio-toxicity, chemical form, and the physical and chemical properties of solidification taken into account. As well as conventional disposal forms, new disposal forms should also be considered.

(4) In radioactive waste processing and disposal, analysis of the inventory, concentration and chemical properties is important. To accelerate decommissioning, such systems must be further strengthened.

(5) With regard to the long-term storage of radioactive waste on site, the maintenance, management and shipment of containers must also be considered.

9.4 Long-Term Stable Storage of Major Systems and Components

In work processes for decommissioning, the processing, disposal and storage of contaminated water and secondary waste as well as the removal and long-term storage of fuel assemblies and fuel debris are important. In this section, post-accident measures and issues and proposals related to the maintenance and management of major systems and equipment necessary to perform these work processes are provided, particularly in terms of the integrity of materials consisting these systems and equipment.

The structures of the accident reactors were significantly damaged by hydrogen explosions, so maintaining their structural integrity against seismic ground motions that may occur during long-term decommissioning is important. Appropriate aseismic performance evaluation in combination with the integrity evaluation of structural materials exposed to the severe environment due to the accident must be performed continuously.

9.4.1 Analyses and Countermeasures

9.4.1.1 Current Situation of Each System and Component

As described in Sect. 6.4.4, the reactors and surrounding facilities that underwent the accident have been exposed to environments not anticipated before. It should be noted that constituent materials have/had been exposed to seawater, fission product contact and irradiation. Under such environment, corrosion is likely to progress. Concrete that comprises the reactor buildings and trenches is under similar conditions. So are the facilities for various radioactive material removal facilities and contaminated water tanks installed after the accident. To decommission the disabled plants, appropriate maintenance and management of the reactors, buildings, various devices, equipment and pipes are necessary. The fundamental principle of preventing the leakage of radioactive substances requires adequate preventive measures against degradation. Progress of corrosion (uniform corrosion, crevice corrosion, and pitting corrosion) is presumed in various locations under the irradiation environment with seawater and/or fission products. The corrosion phenomena under the environment were yet to be dealt with, and the progress of corrosion is difficult to be estimated. However, the corrosion problem is crucial when discussing post-accident seismic safety. As described in Sect. 6.4.4, organizations such as Japan Atomic Energy Agency are proposing or implementing related research and development.

In the following paragraphs, the deterioration of equipment installed for post-accident processing in particular and measures against such deterioration are discussed. And in Sect. 9.4.2, research results concerning post-accident maintenance and management of the reactor pressure vessels (RPV) and the primary containment vessels (PCV).

9.4.1.2 Issues Related to Water Management Facilities

In the pre-accident Fukushima Daiichi NPS, wells called sub-drains were installed around the reactor buildings to pump up about 850 m^3/day of water, thereby restricting groundwater intrusion into the reactor building bottoms and preventing building buoyancy. However, the sub-drain function was lost by the post-quake tsunami, and now about 400 m^3/day of groundwater flows into the buildings. The influx routes are estimated to be the outer wall penetrations of the buildings. Conversely, as detailed in Sect. 9.1, cesium and salt are being removed from contaminated water containing radioactive material and removed from the buildings, whereupon the purified water is recirculated back to the reactors to cool them. This water contains the aforementioned groundwater. The same amount of water as that of the groundwater that has flowed into the buildings is collected to keep the in-building water level under the groundwater level. This is intended to reduce the risk of leakage of the in-building water which contains in-reactor radioactive substances. The collected water is stored in mid-to-low level contaminated water tanks and transferred to the Multi-nuclide Removal Equipment (ALPS), where the radioactive substances are removed by chemical reactions (coprecipitation) and adsorbent. TEPCO states the following three points as activity principles in the mid-and-long term roadmap: (1) removing the contaminants, (2) keeping the water away from the contaminants, and (3) no contaminated water shall be leaked. Based on the roadmap, activities such as decontamination of the water in buildings and trenches, construction of impermeable walls, and enhancement of contaminated water tanks and their monitoring have been ongoing.

Despite efforts by TEPCO and the government some troubles have occurred at the Fukushima Daiichi NPS to date. Data on these troubles have been reported through the TEPCO website each time and summarized in the database NUCIA [14]. Many of these troubles were caused by human and system errors but some were caused by material deterioration. Major troubles included (a) water leakage from salt-removal and reverse osmosis membrane (RO)-type concentrated water tanks (August 2013), (b) water leakage from the middle-to-low level contaminated water tanks (August 2013) and (c) a small amount of water leakage from the polynuclear species removal system (June 2013). Among these, the cause of (a) is under investigation (as of September 2013). Corrosion induced by the relatively highly concentrated saltwater environment in the RO concentrated water tanks is anticipated. Regarding the cause of (b), the possibility of a defective tank bottom structure or defective installation of the tanks has been pointed out. As for (c), local corroded pore formation due to pitting corrosion or crevice corrosion along the tank weld lines have been found, which suggests that under an environment where fission products are mixed and salt concentration is high, special attention is required for the vicinity of the weld lines. Another trouble was (d) contaminated water leakage from the basement storage tanks (April 2013), which is estimated to be attributable to defective impermeable-liner welding.

9.4.1.3 Need for Proactive Actions

It is significantly difficult to foresee the aforementioned troubles under the circumstances where various day-to-day decontamination work, pre-decommissioning work and others are required. Even though, since the social influence of leakage of radioactive substances is significant, reliable measures on "no contaminated water shall be leaked" is required as declared by TEPCO in its mid-and-long term roadmap.

As stated above, the main causes of contaminated water leakage are pore formation due to pipe and tank corrosion, or shoddy workmanship. Those types of leakage may occur in various locations in future.

We need to review that it was/is difficult to foresee troubles caused by materials and their degradations or not. As an example, corrosion problems in the seawater environment at the salt removal system and the multi-nuclide removal equipment are discussed here.

In the Fukushima Daiichi accident, seawater was intentionally injected. On the other hand, at the Chubu Electric Power Co.'s Hamaoka NPS Unit 5 after an earthquake, an event in which seawater flowed into the reactor occurred due to seawater flowed into the reactor due to piping damage which occurred during the reactor shutdown operation. As described in Sect. 6.4.4 in this report, this event did not result in a serious situation. In the Unit 5, salt removal work was performed relatively smoothly after the seawater influx event. During the work, leakage occurred due to the corrosion of the welded sections of the condensate recovery pump's recirculation piping (carbon steel). This corrosion mechanism is estimated as follows: Water containing seawater permeated and formed a corrosive environment in the piping, which resulted in general corrosion. Moreover, the continuous operation of the condensate recovery pump continuously distributed the seawater elements and dissolved oxygen and the corrosion progressed further. In general, the corrosion rate of a welded section exceeds that of the base material. Therefore the corrosion of the welded sections progressed selectively and resulted in leakage 3 or 4 months after the start of the salt removal work. By the time this event was found, the salt removal system and the multi-nuclide removal equipment in the Fukushima Daiichi NPS had been online or commissioning. The same events occurred in the systems after almost the same operation period. Other than this event, the chemical plant underwent the same event many times [15].

The TEPCO and Government's inadequate feedback activities regarding these findings cannot be denied. A framework should be in place to collect various knowledge and findings, including those of non-nuclear fields, nationwide so that measures can be taken before troubles surface. It is necessary for TEPCO to utilize such findings and knowledge by compiling them into a database, to foresee corrosion progress by exploiting non-destructive inspections, to improve techniques for detecting leakage from radioactive material tanks and piping in terms of hardware and software, and firmly promote activities for preventing contaminated water leakage as shown in (3), namely "no contaminated water shall be leaked" as stated in the mid-and-long term roadmap.

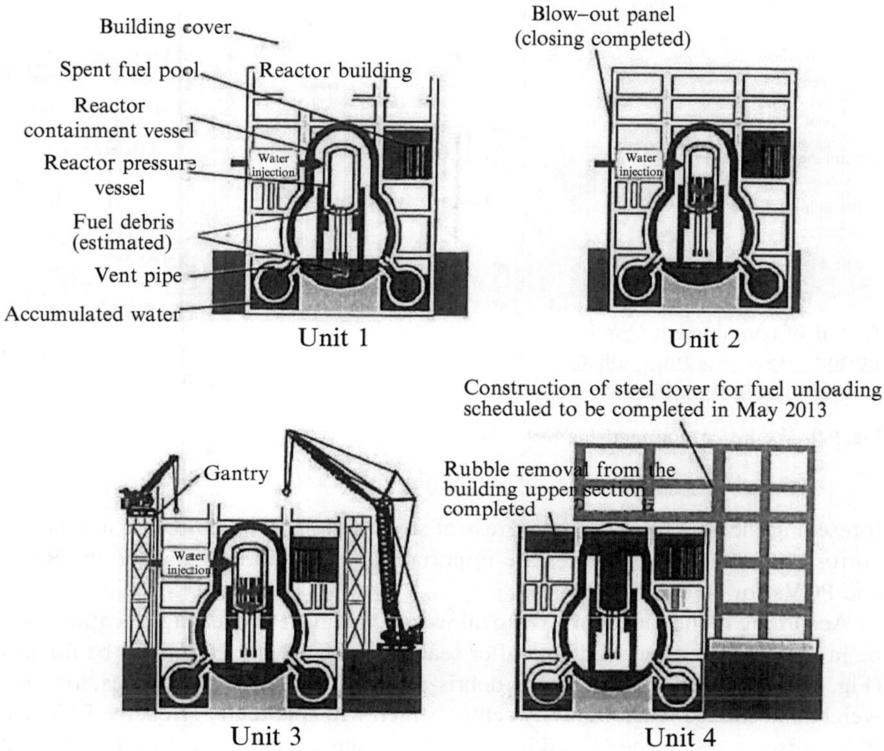

Fig. 9.8 Situation of each unit (June 2013)

9.4.2 Reactor Pressure Vessel and Primary Containment Vessel

Inside the Fukushima Daiichi NPS Units 1–3, fuel melted after the accident and reactor core components such as cladding tube are estimated to have dropped from the reactor pressure vessel (RPV) lower head onto the RPV pedestal concrete inside the primary containment vessel (PCV) and solidified (Fig. 9.8). When decommissioning of these reactors, preventing the RPVs and PCVs from fresh damage during the extended period until fuel debris removal from the RPVs and PCVs is completed and maintaining and storing the fuel debris stably are important. However, as described in Sect. 6.4.4, the RPV and PCV steel materials (RPV: low-alloy steel, and PCV: carbon steel) in Units 1–3, where seawater was injected, should be anticipated to be exposed to the water environment containing seawater elements and radiation for an extended period also in future. Given the potential for the RPVs, PCVs and their support structures to be damaged, there is concern that the materials of these vessels may be corroded and their structural strength and seismic performance may decrease. Therefore,

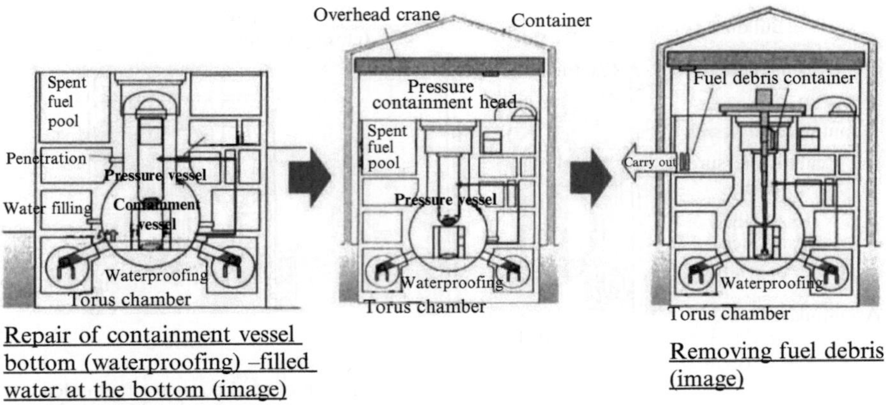

Fig. 9.9 Work steps for removing fuel debris

foreseeing the generation and progress of structural material corrosion and taking corrosion prevention measures are important to maintain and manage the RPVs and PCVs for an extended period.

According to the fuel debris removal work schedule, PCV leakage locations will be investigated and repaired, and after sealing, the PVP and PCVs will be flooded (Fig. 9.9) [16]. This flooding for debris removal will increase the reactor total weight and change reactor gravity center, which will apparently affect the RPV and PCV seismic resistance. In addition, the damage of the RPV lower head that contacted the molten core after the accident may have decreased the structural strength. Assessments of the seismic resistance of each unit, including the decreased structural strength, need to be performed.

For seismic resistant assessment, the effects of high temperature caused by the accident and corrosion wastage (thinning) on the PRV support skirt, on bolts fixing it to the RPV pedestal, on RPV pedestal reinforced concrete, on RPV/PCV stabilizers to prevent horizontal oscillation during an Earthquake, and on PCV walls need to be considered (Fig. 9.10) [17, 18]. In particular, estimation of corrosion damage to equipment materials exposed to water containing seawater elements and radiation is necessary. At the same time, corrosion inhibition measures effective even under such environment are urgently required. As a corrosion inhibition measure, hydrazine injection to the spent fuel pool cooling water circulation system to reduce its dissolved oxygen concentration has been performed since May 2011. Also regarding the PCVs, hydrazine injection to the reactor water injection system started in August 2013. At the PCVs, nitrogen injection to the containers and nitrogen bubbling to injected water are being performed to prevent hydrogen explosions. These can reduce the dissolved oxygen concentration in the water in the reactors and are expected to restrict corrosion.

Currently, in accordance with the R&D plan for decommissioning the Fukushima Daiichi NSP Units 1 to 4, R&D projects for the integrity assessment,

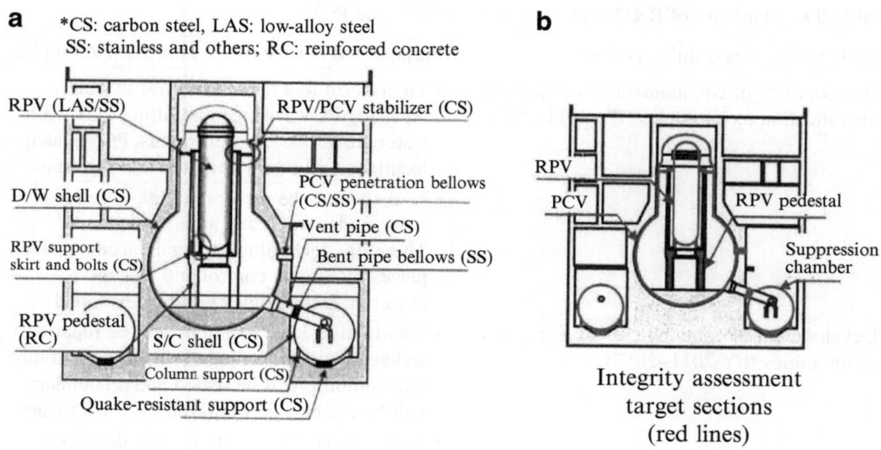

Fig. 9.10 Reactor structure and materials (**a**) and integrity assessment target sections (**b**). (**a**) Fukaya [17] (**b**) Council for the Decommissioning of TEPCO's Fukushima Daiichi NPS [18]

internal research and repair of each system and equipment are being implemented. Projects for maintaining and managing the PCVs and RPVs include the development of pressure vessel and containment vessel integrity assessment technologies, containment vessel leakage identification technologies, containment vessel repair technologies and containment vessel internal research technologies. Among these, in the project for RPV/PCV integrity assessment, data necessary for assessing and foreseeing the corrosion deterioration progress at both containers and their support members have been obtained through corrosion tests using diluted artificial seawater, and data on the corrosion of concrete structure reinforcing steel and on concrete deterioration have been obtained, which data are being used to RPV and PCV integrity assessment (this project is partially performed as utility joint research). Moreover, the applicability of corrosion inhibition measures is being studied along with radiation effect assessment through gamma ray irradiation corrosion test [19]. Table 9.6 summarizes the plan for implementing each R&D project and Table 9.7 summarizes long-term storage issues and related R&D projects.

As shown in Table 9.6, measures generally required for the long-term maintenance and management of major equipment and facilities (PCVs, RPVs and reactor buildings) are being taken via the R&D projects based on the Mid-and-Long Term Roadmap for Decommissioning the Tokyo Electric Power Co.'s Fukushima Daiichi Nuclear Power Station, Units 1–4. However, to fulfill the projects, the following points need to be further improved:

- The results of detailed investigations into the reactors should be publicized immediately and in detail to reflect them in the activities and programs of each project constantly, to utilize them to foresee new tasks and determine problems as early as possible.

Table 9.6 Summary of R&D projects related to RPV and PCV

R&D projects (scheduled period)	Purpose and outline of the implementation plan
Development of containment vessel leakage identification technologies (FY 2011–2014	• To perform fuel debris removal in water, repairing PCV leakage and filling PCV with water are necessary. Before this, PCV leakage locations will be investigated and identified
	• Leakage may be present at high-dose locations, under water and at narrow places. Therefore, technologies to gain access to such places by remote control and leakage-detection technologies will be developed
Development of containment vessel repair technologies (FY 2011–2017)	• Identified leakage locations will be repaired, leakage between the reactor building and turbine building will be sealed and a boundary will be established to fill the PCV with water
	• Leakage may be present at high-dose locations, under water and at narrow places. Therefore, technologies and techniques to gain access to such places by remote control and carrying out repair will be developed
Development of containment internal research technologies (FY 2011–2016)	• Distribution of fuel debris is yet unknown. To remove it, the location and status of fuel debris in the PCVs need to be investigated in advance and the status of pedestals supporting the pressure vessels also need to be checked
	• During the development of PCV internal research technologies, technologies applicable in terms of environment (narrow place, high dose, etc.) will be investigated and then inspection and investigation devices will be designed and manufactured. Concurrently, to ensure investigation work safety, measures for preventing radioactive materials from scattering will be studied
Development of pressure vessel internal research technologies (FY 2013–2019)	• To remove fuel debris, the status inside the RPVs (status of e.g. fuel debris, in-reactor damage, and contaminated equipment) must be determined
	• To investigate the status in the RPVs such as fuel debris status, technologies applicable to anticipated environment (high radiation dose, high temperature, high humidity, etc.) will be researched. Based on the investigation into the PCVs, devices for checking inside the pressure vessels will be designed and manufactured

(continued)

Table 9.6 (continued)

R&D projects (scheduled period)	Purpose and outline of the implementation plan
Development of pressure vessel and containment vessel integrity assessment technologies (FY 2011–2016)	• RPVs and PCVs into which seawater was injected are anticipated to be exposed to diluted seawater environment for an extended period also in future. Until the fuel debris is removed, equipment integrity must be ensured and stable cooling must be continued. Reinforced concrete structures that support RPV, namely RPV pedestals, need to be checked for the effects of temperature and seawater immersion
	• Corrosion data necessary to evaluate and foresee RPV and PCV corrosion degradation progress will be obtained. In addition, data on the corrosion of RPV pedestal reinforcing steel and concrete degradation will be obtained to assess their structural integrity. Corrosion and deterioration inhibition measures will be applied and their effectiveness will be checked

- In seismic resistant performance assessments, it should be taken into account that the gravity center of each pressure vessel has changed due to core meltdown and especially the seismic safety of each lower head is no longer ensured due to damage or interaction between the lower head and molten core. The effects of the molten core drop onto the RPV pedestal, which is a phenomenon never experienced even in the TMI-2 accident, on the concrete structures and their seismic resistance should be investigated from various perspectives.
- Corrosion damage has many problems common to the long-term storage of RPV and PCV as well as fuel assemblies and waste, meaning a cross-sectional study is necessary and effective. Methods to solve these problems should be examined immediately, the interrelationship and coordination among the projects should be enhanced further and effective measures should be taken.
- Current measures have been taken mainly by plant manufacturers, the Japan Atomic Energy Agency and TEPCO. To enable multifaceted studies by experts from broad-ranging fields, relationships with universities and academic society as well as the cultivation of human resources for allowing long-term measures should be strengthened further.

9.5 Long-Term Healthcare of Residents and Workers

Section 5.3.3.2 discussed measures taken and the situation of the radiation exposure and healthsurvey of the residents and workers immediately after the accident. Section 6.7.2 discussed the results of epidemiological research performed to date,

Table 9.7 Tasks and measures for the long-term stable storage of the RPVs and PCVs

Target equipment and task	Necessary measures	Relevant R&D project
Reactor pressure vessel (RPV) damage prevention	• Pressure vessel internal investigation	Development of pressure vessel internal investigation technologies
	• Assessment of the severe accident effects on pressure vessel steel	Development of pressure vessel and containment vessel integrity assessment technologies
	• Assessment of and measures against pressure vessel corrosion	
	• Assessment of seismic resistance and remaining life	
RPV pedestal damage prevention	• Assessment of concrete strength degradation	Development of pressure vessel and containment vessel integrity assessment technologies
	• Assessment of and measures against reinforcing steel corrosion	
Primary Containment vessel (PCV) damage prevention	• Investigation into containment vessels and of their leaking locations	Development of containment vessel internal investigation technologies
		Development of technologies for identifying containment vessel leaking sections
	• Repair of containment vessels and development of water-stopping method	Development of technologies for repairing containment vessels
	• Assessment of and measures against containment vessel corrosion	Development of pressure vessel and containment vessel integrity assessment technologies
	• Assessment of seismic resistance and containment vessel remaining life	
Prevention of the damage of water injection and cooling piping	• Measures to restrict piping corrosion	Development of pressure vessel and containment vessel integrity assessment technologies
Basic study for long-term maintenance and management of major equipment	• Assessment of the effect of high-temperature seawater injection on the steel materials	Development of pressure vessel and containment vessel integrity assessment technologies
	• Assessment of the effects of radiation from fission product and fuel debris on corrosion	
	• Study on long-term material deterioration prediction methods such as accelerated test	
	• Development of technologies to monitor damage during long-term equipment management	

internationally accepted radiation doses such as by the ICRP and UNSEAR, the concept of radiation-induced human health effects, and issues in reconstructing the dose estimation and initial exposure doses of the residents and workers.

The Government, local governments and other related parties have performed decontamination, developed food-related standards, reviewed disaster preparedness guidelines, and enhanced the monitoring of the environment, food and individuals so that the public and residents can live safely without worry. They also have taken measures such as the dose assessment and health survey of residents and workers and the establishment of a related database.

The Atomic Energy Society of Japan has made various proposals through the Nuclear Safety Investigation Committee and the Fukushima Projects so that that the dose assessment, exposure control and healthsurvey of the residents and workers are ensured. It has also organized meetings with the residents under cooperation from local governments to gain their understanding and support regarding decontamination and health issues.

Health survey must be continued for an extended period also in future, related measures appropriate for each situation must be taken, and new tasks that will emerge as the recovery progresses must be solved appropriately according to each situation.

With regard to individual exposure dose assessment, initial-stage radiation monitoring data, which should be already in place basically, is scarce hence the radiation doses of residents and workers after the accident have yet to be clarified. Dose estimation based on the results of limited individual monitoring and environmental radiation monitoring conducted to date still has significant uncertainties and the progress of studies that can clarify the substantial post-accident environmental effects is expected. Such studies will play an important role in finding effective emergency measures.

Because the full-scale return home of residents will be launched in the near future, the dose assessment and health survey of the residents after their return home from evacuation sites need to be adequately planned while determining the situation of areas where residents will have returned home ahead of others.

How the dose assessment and management of persons engaged in emergency readiness should be has not been adequately discussed yet. However, the dose assessment and health examination of such workers should be performed in the same level as those for plant workers. Their results should be managed in an integrated database. Also for an extended period in future, many workers will be engaged in decontamination and decommissioning work. These workers will be exposed to various types of high-dose radiation. Moreover, various exposure types (internal exposure, external exposure, non-homogeneous exposure, etc.) are anticipated. Dose mitigation measures should be taken to particularly reduce the doses of emergency workers who will be engaged in such work as far as possible. Also exposure dose management and healthcare should be ensured thoroughly.

To ensure that the doses and health management survey data on both residents and workers will be centralized, that the data will be appropriately provided to related parties, and that future health survey will be appropriate according to each situation, various proposals should be made to the Government. In addition, human resource development as well as biological experiments, epidemiological studies, and R&Ds related to dose assessment for clarifying the effects of low radiation doses and low radiation dose rates on health should be promoted. These are also the duties of the parties related to the Atomic Energy Society of Japan, who have experienced the accident.

References

1. Tokyo Electric Power Co., *Polynuclear Species Removal System*. http://www.meti.go.jp/Earthquake/nulclear/pdf/120328_02i.pdf
2. M.J. Fluss, N.D. Dubey, R.L. Malewicki, Phys. Rev. C **6**, 2252 (1972)
3. The Atomic Energy Society of Japan (ed.), *Reactor Water Chemistry Handbook* (Corona Publishing Co., Ltd., 2000)
4. Y.B. Katayama et al., *Status Report on LWR Spent Fuel IAEA Leach Tests* (PNL-3173, Pacific Northwest Laboratory, 1980)
5. Hachinohe Institute of Technology, *Final Report by the Working Group for Studying Nuclear Fuel Issues in Meltdown Accident*. Atomic Energy Society of Japan. 2013 Fall Meeting. (Hachinohe Institute of Technology, 2013)
6. A.G. Croff, ORNL-5621 (1980)
7. K. Nishihara, H. Iwamoto, K. Suyama, *JAEA-Data/Code 2012-018* (2012)
8. K. Suyama, T. Kiyosumi, H. Mochizuki, *JAERI-Data/Code 2000-027* (2000)
9. K. Okumura, K. Dobashi, K. Kaneko, *JAERI-Data/Code 96-015* (1996)
10. K. Okumura, in *2012 Nuclear Data Symposium*, Kyoto University's Research Reactor Institute (2012)
11. Atomic Energy Society of Japan, *Severe Accident Thermal Hydraulics Assessment* (Atomic Energy Society of Japan, 2001)
12. Annual Safety Research Report FY 2011, *JNES-RE-2012-0001-Rev-2* (2012), p. 404
13. Atomic Energy Society of Japan, in *Processing and Disposal of Radioactive Waste Generated from the Fukushima Daiichi Nuclear Power Station Accident Special Technical Committee, Processing and Disposal of Radioactive Waste Generated from the Fukushima Daiichi Nuclear Power Station Accident – Concept for Identifying R&D Issues and Their Solution* (2013)
14. Nuclear Information Archives NUCIA. http://www.nucia.jp/index.html
15. S. Tamura (ed.), *Chemical Substance and Plant Accident Example Handbook* (Maruzen, 2006)
16. Council for the Decommissioning of TEPCO's Fukushima Daiichi NPS, *Mid-and-Long-Term Roadmap towards the Decommissioning of TEPCO's Fukushima Daiichi Nuclear Power Station Units 1–4 (key points)* (Council for the Decommissioning of TEPCO's Fukushima Daiichi NPS, 2013)
17. Y. Fukaya, in *Proceedings of the 60th Japan Conference on Materials and Environments*, vol. A-101 (2013)
18. Council for the Decommissioning of TEPCO's Fukushima Daiichi NPS, *2012 Performance Assessment and Direction of 2013 Review Regarding the R&D Projects* (2013)
19. Japan Atomic Energy Agency, in *JAEA' Efforts Related to the Technologies for Decommissioning the TEPCO's Fukushima Daiichi Nuclear Power Station (2012 version)* (2012)

Afterwords

The Atomic Energy Society of Japan (AESJ) has made collective effort to study and research from broad, divergent and professional perspectives through the Investigation Committee. Although the details of the accident have not been clarified completely and should be revealed during the future decommission process of the accident reactors, the AESJ thinks that major facts have been clarified. Based on analyses of the direct and indirect factors of the accident, the Investigation Committee has revealed the underlying root causes of the accident. To cope with such causes, as well as utilities and regulatory organizations, AESJ members must also make continuous efforts according to each role. Experts must fulfill their duties while facing the facts squarely and while retaining a humble attitude toward technologies and their progress.

We hope that, based on the analytical results and recommendations of this report, efforts to improve nuclear safety will be promoted, the safety of nuclear facilities will be verified specifically, and the continuous improvement of nuclear safety, which is a common objective, will be achieved.

The Fukushima Daiichi NPS accident has revealed the various negative aspects of using nuclear energy. Faced with the nuclear disaster at the nuclear facilities, Japan is now urged to decide whether it should continue nuclear energy utilization or not. The lessons learned from the accident clarified in this report must be reflected in specific countermeasures steadily. At the same time, the decommissioning of accident reactors and restoration of the disaster areas, which are also important tasks, must be faced.

Conversely, specifically showing positive aspects of the utilization of nuclear energy is also a duty of the AESJ. To ensure nuclear power plant safety, comprehensive efforts must be made while taking into account the characteristics of a giant complex system. Developing human resources talented with comprehensive vision, developing cross-sectional research infrastructure, broadly disseminating their achievements, and promoting various dialogues are important. We strongly hope that, via such efforts, nuclear energy will contribute to the health and welfare of humankind, social safety and peace, and the sustainability of the global environment and restore public trust in it and contribute to the international society.

The Fukushima Daiichi Nuclear Accident: Final Report of the AESJ Investigation Committee, DOI 10.1007/978-4-431-55160-7

Appendix 1
List of the Members of AESJ Investigation Committee on the Nuclear Accident at the Fukushima Daiichi Nuclear Power Station

Chairperson

Satoru Tanaka (**)	Department of Nuclear Engineering and Management, the Graduate School of Engineering, the University of Tokyo

Board of Directors

Akihiko Kimura (**)	Institute of Advanced Energy, Kyoto University
Takanori Tanaka (*, **)	The Institute of Applied Energy
Tadashi Narabayashi (**)	Division of Energy and Environment Systems, Graduate School of Engineering, Hokkaido University
Kazuhiko Yamamoto (**)	Department of Nuclear Research and Technology, Japan Science and Technology Agency

Committee Members Designated by the Chairperson

Seiichi Koshizuka (**)	Department of Systems Innovation, the Graduate School of Engineering, the University of Tokyo
Masashi Hirano (**)	Japan Nuclear Energy Safety Organization

Committee for Investigation of Nuclear Safety—Technical Analysis Subcommittee

Koji Okamoto (**)	Nuclear Professional School, Graduate School of Engineering, the University of Tokyo
Akio Yamamoto (**)	Materials, Physics and Energy Engineering, Graduate School of Engineering, Nagoya University

The Fukushima Daiichi Nuclear Accident: Final Report of the AESJ Investigation Committee, DOI 10.1007/978-4-431-55160-7

Committee for Investigation of Nuclear Safety—Task Group on Radiological Aspects of the Fukushima Daiichi Nuclear Power Plant Accident and the Division of Health Physics and Environmental Science

Sumi Yokoyama — Faculty of Radiological Technology, School of Health Sciences, Fujita Health University

Committee for Investigation of Nuclear Safety—the Cleanup Subcommittee and the Reprocessing and Recycle Technology Division

Tadashi Inoue (**) — Central Research Institute of Electric Power Industry
Miki Umeda — Nuclear Science Research Institute, Japan Atomic Energy Agency

Public Information Committee

Junko Ogawa — Nuclear Safety Engineering Department, Tokyo City University

Ethics Committee

Kyoko Oba (**) — Academy for Global Nuclear Safety and Security Agent, Tokyo Institute of Technology

Standards Committee

Hiroshi Miyano (*, **) — Graduate School of Engineering and Design, Hosei University
Akira Yamaguchi (**) — Division of Sustainable Energy and Environmental Engineering, Graduate School of Engineering, Osaka University
Tadahiko Kawai — Nuclear Safety Division, Japan Nuclear Safety Institute

Reactor Physics Division

Ken Nakajima — Research Reactor Institute, Kyoto University

Fusion Engineering Division

Satoshi Konishi — Institute of Advanced Energy, Kyoto University

Division of Nuclear Fuel

Shinsuke Yamanaka	Department of Sustainable Energy and Environmental Engineering, Graduate School of Engineering, Osaka University

Division of Nuclear Fuel Cycle and Environment

Hiroshi Rindo	Nuclear Cycle Backend Directorate, Japan Atomic Energy Agency
Yuichi Niihori	Department of Quantum Science and Energy Engineering, Graduate School of Engineering, Tohoku University

Thermal Hydraulics Division

Isao Kataoka	Department of Mechanical Engineering, Osaka University

Division of Radiation Science and Technology

Hiroyuki Takahashi	Department of Nuclear Engineering and Management, the Graduate School of Engineering, the University of Tokyo

Subcommittee of Human-Machine Systems Research

Akio Gofuku	Graduate School of Natural Science and Technology, Okayama University
Kunihide Sasou	Nuclear Technology Research Laboratory, Central Research Institute of Electric Power Industry

Subcommittee on Particle Accelerator and Beams Science

Mitsuru Uesaka	Nuclear Professional School, Graduate School of Engineering, the University of Tokyo

Social and Environmental Division

Tsutomu Sata	Public Relations Department, Japan Atomic Energy Agency
Muneo Morokuzu (*, **)	Graduate School of Public Policy, the University of Tokyo

Division of Health Physics and Environment Science

Takumaro Momose	Nuclear Fuel Cycle Engineering Laboratories, Japan Atomic Energy Agency
Takeshi Iimoto	Division for Environment, Health and Safety, the University of Tokyo

Nuclear Data Division

Satoshi Chiba	Research Laboratory for Nuclear Reactors, Tokyo Institute of Technology
Kenya Suyama	Nuclear Science and Engineering Directorate, Japan Atomic Energy Agency

Materials Science and Technology Division

Hiroaki Abe	Institute for Materials Research, Tohoku University

Operation and Power Division

Norihiro Manago	Nuclear Power Department, The Federation of Electric Power Companies

Computational Science and Engineering Division

Norihiro Nakajima	Center for Computational Science & e-systems, Japan Atomic Energy Agency

Division of Water Chemistry

Shunsuke Uchida	Nuclear Science and Engineering Directorate, Japan Atomic Energy Agency
Takashi Tsukada	Nuclear Science Research Institute, Japan Atomic Energy Agency

Nuclear Safety Division

Naoto Sekimura (*, **)	Department of Nuclear Engineering and Management, the Graduate School of Engineering, the University of Tokyo
Toshimitsu Homma (**)	Nuclear Safety Research Center, Japan Atomic Energy Agency
Takashi Nitta	The Japan Atomic Power Company

Advanced Reactor Division

Hidemasa Yamano	Advanced Nuclear System Research and Development Directorate, Japan Atomic Energy Agency

Committee on Nuclear Non-Proliferation, Safeguards and Security

Kazunori Fujimaki	Former director of Nuclear Security and Safeguards Division, Reprocessing Business Department, Japan Nuclear Fuel Limited
Yusuke Kuno	Department of Nuclear Engineering and Management, School of Engineering, the University of Tokyo / Department of Science and Technology for Nuclear Material Management, Japan Atomic Energy Agency

Observers (Participation on an As-Needed Basis)

President of Atomic Energy Society of Japan

Hiroshi Horiike	Department of Sustainable Energy and Environmental Engineering, Graduate School of Engineering, Osaka University

Vice President of Atomic Energy Society of Japan

Ichiro Ikemoto	Central Research Institute of Electric Power Industry
Reiko Fujita	Toshiba Corporation Power Systems Company
Hiroshi Uetsuka	Japan Atomic Energy Agency

Director and Secretary General

Takashi Sawada	Atomic Energy Society of Japan

Secretariat

Person in charge of AESJ investigation committee

Shigeki Arai	Atomic Energy Society of Japan

This committee member list was prepared on October 1, 2013. The names of members who retired before this date are as follows (name of organization: organization where the person belonged during the person's tenure):

Yu Aoki	Nuclear Power Department, The Federation of Electric Power Companies
Toyoshi Fuketa	Nuclear Science and Engineering Directorate, Japan Atomic Energy Agency
Kotaro Matsuoka	Nuclear Power Department, The Federation of Electric Power Companies

Note: *and ** marks show Committee Director and a member of core group respectively.

Appendix 2
Past Records of the Activities of the Investigation Committee

Date	Major activities
2012	
June 22	**General meeting and Board of directors in Atomic Energy Society of Japan (AESJ)**
	Decision on establishing investigation committee
August 13	1st core group meeting
	Discussion regarding the policies of the investigation committee's activities
August 21	**1st investigation committee meeting**
	Discussion regarding the purpose and management policy of the investigation committee
August 22	2nd core group meeting
	Preparation for the 2nd investigation committee meeting and discussion regarding investigation committee management
August 30	3rd core group meeting
	Preparation for the 2nd investigation committee meeting and discussion of study items
September 4	**2nd investigation committee meeting**
	Hearing from the Government investigation committee
September 14	4th core group meeting
	Discussion regarding information requests to related organizations
September 20	**3rd investigation committee meeting**
	Introduction regarding the contents of the Nuclear Safety Division seminar, discussion of study items and others
September 28	5th core group meeting
	Discussion regarding important items
October 22	6th core group meeting
	Discussion regarding the preparation for and measures against emergency situations
October 24	**4th investigation committee meeting**

(continued)

The Fukushima Daiichi Nuclear Accident: Final Report of the AESJ Investigation Committee, DOI 10.1007/978-4-431-55160-7

(continued)

Date	Major activities
	Discussion of important items, reports on the review progress of each division, and others
November 5	7th core group meeting
	Introduction of interim report, and discussion regarding each division's review
November 12	8th core group meeting
	Discussion regarding the 5th investigation committee meeting proceedings and defense in depth
November 19	**5th investigation committee meeting**
	Hearing the summary of the accident, etc. from TEPCO
December 4	9th core group meeting
	Preparation for interim report, reports on each division's review progress and discussion regarding questionnaires in the AESJ
December 12	10th core group meeting
	Analysis of the social aspects of the accident and discussion regarding questionnaires
December 17	11th core group meeting
	Discussion regarding source term and safety basic principles
December 21	**6th investigation committee meeting**
	Introduction regarding the interim report of the Nuclear Safety Division, discussion regarding the contents of questionnaire survey, and others
2013	
January 13	12th core group meeting
	Discussion regarding Chapters 2 to 6 of the report
January 17	13th core group meeting
	Final stage discussion of questionnaire and discussion regarding basic safety principles
January 25	**7th investigation committee meeting**
	Discussion regarding interim report, report on-site visit, and discussion regarding nuclear safety
February 10	14th core group meeting
	Discussion regarding the each item of Chapter 6
February 16	15th core group meeting
	Adjustment of the Table of Contents of the report
February 18	**8th investigation committee meeting**
	Adjustment of each item of interim report
March 2	16th core group meeting
	Discussion regarding presentation (interim report) during the annual AESJ meeting in spring
March 6	17th core group meeting
	Discussion regarding preparation for and measures against emergency situations, and regarding decontamination and environmental remediation
March 10	18th core group meeting

(continued)

(continued)

Date	Major activities
	Discussion regarding presentation (interim report) during the annual AESJ meeting in spring
March 18	19th core group meeting
	Discussion regarding presentation (interim report) during the annual AESJ meeting in spring
March 19	**9th investigation committee meeting**
	Discussion regarding interim report
March 27	**Open session on the Annual AESJ meeting in spring**
	Interim report
April 10	20th core group meeting
	Discussion regarding policy for final report
April 20	21th core group meeting
	Discussion regarding the final report Chapters 2 to 4
April 24	**10th investigation committee meeting**
	Discussion regarding final report preparation policy and each item
May 15	22th core group meeting
	Discussion regarding nuclear industry human resource problems and others
May 19	23th core group meeting
	Discussion regarding human factors and others
May 29	**11th investigation committee meeting**
	Discussion regarding final report items and measures considering questionnaire results
June 9	24th core group meeting
	Discussion regarding analysis simulation, isolation condenser and others
June 15	25th core group meeting
	Discussion regarding the final report chapter 5, nuclear security and others
June 19	**12th investigation committee meeting**
	Discussion regarding each item of the final report and others
June 25	26th core group meeting
	Discussion regarding the relations with international society and research subjects
July 3	27th core group meeting
	Discussion regarding nuclear industry human resource problem
July 6	28th core group meeting
	Discussion regarding accident management, source term, regulation system, nuclear security and others
July 11	**13th investigation committee meeting**
	Discussion regarding each item of the final report and others
July 18	29th core group meeting
	Discussion regarding decommissioning, contaminated water (containing tritium) cleanup and treatment, accident progress assessment based on simulation, and others
July 21	30th core group meeting

(continued)

(continued)

Date	Major activities
	Discussion regarding final report (draft) presentation materials
July 27	31th core group meeting
	Discussion regarding IC, information dissemination, RPV/PCV long-term stable storage, and final report (draft) presentation materials
July 29	32th core group meeting
	Discussion regarding the SPEEDI
July 31	**14th investigation committee meeting**
	Discussion regarding each item of the final report, final report (draft) presentation materials and others
August 7	33th core group meeting
	Discussion regarding how the AESJ should be in future, final report (draft) presentation materials, and others
August 17	34th core group meeting
	Discussion regarding final report (draft) presentation materials and others
August 21	**15th investigation committee meeting**
	Discussion regarding final report (draft) presentation materials and others
September 2	**Final report (draft) presentation**
September 4	**Open session on the Annual AESJ meeting in autumn**
	Final report (draft) presentation
September 26	35th core group meeting
	Discussion regarding external event countermeasures and the system of AESJ, and the scheduling of overseas review
October 2	36th core group meeting
	Discussion regarding fuel debris, plant design, cooling system diversification, and instrumentation system
October 7	**16th investigation committee meeting**
	Discussion of each item of the final report and final report publication schedule
October 20	37th core group meeting
	Discussion regarding nuclear safety system, root cause analyses, resident healthcare and others
October 29	38th core group meeting
	Discussion regarding radioactive material release, the chief engineer of reactors and others
November 5	**17th investigation committee meeting**
	Discussion regarding each item of the final report and overseas review results
November 24	39th core group meeting
	Discussion regarding recommendations and others
December 15	40th core group meeting
	Final adjustment of the final report and others

Appendix 3
List of Abbreviations

Abbreviation	Full word
ABWR	Advanced boiling water reactor
ADS	Automatic depressurization system
AEC	Atomic Energy Commission
AM	Accident management
AMG	Accident management guideline
AOP	Abnormal Operating Procedures
AOV	Air operated valve
APD	Alarm pocket dosimeter
APRM	Average power range monitor
APWR	Advanced pressurized water reactor
ARI	Alternative rods injection
ASME	American Society of Mechanical Engineers
ATWS	Anticipated transients without scram
BAF	Bottom of active fuel
BWR	Boiling water reactor
CAMS	Containment atmospheric monitoring system
C/B	Control building
CCS	Containment cooling spray system
CDF	Core damage frequency
CFF	Containment failure frequency
CR	Control rod
CRD	Control rod drive mechanism
CRM	Crew resource management
CST	Condensate storage tank
CS	Core spray system
DBA	Design basis accident
D/DFP	Diesel-driven fire pump
D/G	Diesel generator
DGSW	Diesel generator sea water system
DOE	United States Department of Energy

(continued)

The Fukushima Daiichi Nuclear Accident: Final Report of the AESJ Investigation Committee, DOI 10.1007/978-4-431-55160-7

(continued)

Abbreviation	Full word
DS pit	Dryer separator pit
D/W	Dry well
ECCS	Emergency core cooling system
EDG	Emergency diesel generator
EOC	Emergency Operation Center
EOP	Emergency Operating Procedures
EPZ	Emergency planning zone
ERC	Emergency Response Center
ERSS	Emergency Response Support System
FAO	Food and Agriculture Organization of the United Nations
FCS	Flammability control system
FP	Fission product
FPC	Fuel pool cooling system
FP	Fire protection system
FW	Feed water system
HF	Human factor
HPCI	High pressure coolant injection system
HPCSDG	High pressure core spray system diesel generator
HPCS	High pressure core spray system
HVAC	Heating and ventilating air conditioning and cooling system
IAEA	International Atomic Energy Agency
IA	Instrument air system
IC	Isolation condenser
ICRP	International Commission on Radiological Protection
INES	The International Nuclear and Radiological Event Scale
INPO	Institute of Nuclear Power Operations
IPE	Individual plant examination
IPEEE	Individual plant examination for external events
IRRT	International Regulatory Review Team
JAEA	Japan Atomic Energy Agency
JAEA/ NEAT	JAEA/Nuclear Emergency Assistance & Training Center
JAMSTEC	Japan Agency for Marine-Earth Science and Technology
JANSI	Japan Nuclear Safety Institute
JAXA	Japan Aerospace Exploration Agency
JNES	Japan Nuclear Energy Safety Organization
LOCA	Loss of coolant accident
LPCI	Low pressure coolant injection system
LPCS	Low pressure core spray system
LUHS	Loss of ultimate heat sink
M	Magnitude
MAAP	Modular Accident Analysis Program

(continued)

(continued)

Abbreviation	Full word
M/C	Metal-clad switch gear
MCC	Motor control center
MCCI	Molten core concrete interaction
M/DFP	Motor-driven fire pump
MOV	Motor operated valve
MS	Main steam system
MSIV	Main steam isolation valve
MUWC	Make-up water system (condensate)
NEA	Nuclear Energy Agency
NEI	Nuclear Energy Institute
NRC	Nuclear Regulatory Commission
NUPEC	Nuclear Power Engineering Corporation
NUREG	Nuclear Regulatory Commission Report
OECD NEA	OECD Nuclear Energy Agency
O.P.	Onahama Peil
O.P.	Onagawa Peil
PAZ	Precautionary action zone
P/C	Power center
PCV	Primary containment vessel
PLR	Primary loop re-circulation system
P/P	Physical protection
PRA	Probabilistic risk assessment
PSA	Probabilistic safety assessment
PSR	Periodic safety review
PWR	Pressurized water reactor
R/B	Reactor building
RCIC	Reactor core isolation cooling system
RHRC	Residual heat removal cooling water system
RHRS	Residual heat removal sea water system
RHR	Residual heat removal system
RPS	Reactor protection system
RPT	Recirculation pump trip
RPV	Reactor pressure vessel
RW/B	Radioactive waste disposal building
SA	Severe accident
SAM	Severe accident management
SAMPSON	Severe Accident Analysis Code with Mechanistic Parallelized Simulations Oriented towards Nuclear Field
SARRY	Simplified active water retrieve and recovery system
S/B	Service building
SBO	Station black out
S/C	Suppression chamber
SFP	Spent fuel pool

(continued)

(continued)

Abbreviation	Full word
SGTS	Standby gas treatment system
SHC	Shutdown cooling system
SLC	Standby liquid control system
SOP	Severe Accident Operating Procedures
S/P	Suppression pool
SPDS	Safety parameter display system
SPEEDI	System for Prediction of Environmental Emergency Dose Information
SRV	Safety relief valve
TAF	Top of active fuel
T/B	Turbine building
TMI	Three Mile Island nuclear plant
T.P.	Tokyo Peil
UNSCEAR	United Nations Scientific Committee on the Effects of Atomic Radiation
WANO	World Association of Nuclear Operators
WASSC	Waste Safety Standard Committee
WBC	Whole body counter
WHO	World Health Organization
WSPEEDI	Worldwide Version of System for Prediction of Environmental Emergency Dose Information
W/W	Wet well

Index

The Fukushima Daiichi Nuclear Accident: Final Report of the AESJ Investigation Committee, DOI 10.1007/978-4-431-55160-7

MIX
Papier aus verantwortungsvollen Quellen
Paper from responsible sources
FSC® C105338

If you have any concerns about our products,
you can contact us on
ProductSafety@springernature.com

In case Publisher is established outside the EU,
the EU authorized representative is:
Springer Nature Customer Service Center GmbH
Europaplatz 3, 69115 Heidelberg, Germany

Printed by Libri Plureos GmbH
in Hamburg, Germany